GENIUS IN
THE SHADOWS

A Biography of Leo Szilard,
the Man Behind the Bomb

GENIUS IN THE SHADOWS

A Biography of Leo Szilard,

the Man Behind the Bomb

WILLIAM LANOUETTE

WITH BELA SILARD

Foreword by JONAS SALK

SKYHORSE PUBLISHING

Skyhorse Publishing books may be purchased in bulk at special discounts for sales promotion, corporate gifts, fund-raising, or educational purposes. Special editions can also be created to specifications. For details, contact the Special Sales Department, Skyhorse Publishing, 307 West 36th Street, 11th Floor, New York, NY 10018 or info@skyhorsepublishing.com.

Skyhorse® and Skyhorse Publishing® are registered trademarks of Skyhorse Publishing, Inc.®, a Delaware corporation.

Visit our website at www.skyhorsepublishing.com.

10 9 8 7 6 5 4 3 2 1

Library of Congress Cataloging-in-Publication Data is available on file.

ISBN: 978-1-62636-023-5

Hardcover ISBN: 978-1-62636-378-6

Printed in the United States of America

To my parents

Between the idea
And the reality
Between the motion
And the act
Falls the Shadow
    T. S. ELIOT,
     "The Hollow Men"

Contents

Illustrations

Founders of the Emergency Committee of Atomic Scientists at the Institute for
Leo Szilard with Cyrus Eaton at the first Pugwash Conference on Science and
World Affairs in July 1957
Leo Szilard in the Rocky Mountain National Park in the 1950s
Aaron Novick and Leo Szilard
Leo Szilard as sketched by Eva Zeisel
Leo Szilard on the Atlantic City boardwalk, March 1948
Leo Szilard with Jonas Salk
Leo Szilard with Matthew Meselson and Leslie Orgel
Leo Szilard with Trude at Memorial Hospital, 1960
Leo Szilard at a Pugwash conference banquet in Moscow, November 1960
Szilard with Inge Feltrinelli in Italy
Jacques Monod and Leo Szilard
Leo Szilard at the Dupont Plaza Hotel in Washington, DC
Caricature of Leo Szilard in a bathtub by Robert Grossman
Leo Szilard reading to a young girl
Henry Kissinger, Leo Szilard, and Eugene Rabinowitch
Eleanor Roosevelt and Leo Szilard
Leo Szilard with Michael Straight
Leo Szilard with Francis Crick and Jonas Salk, spring 1964
Szilard speaking at a Salk Institute seminar, February 1964
Leo Szilard's ashes at Kerepsi Cemetery in Budapest
Leo Szilard's tombstone at Lake View Cemetery in Itaca, New York

Introduction
to the 2013 Edition

Nearly half a century after Leo Szilard's death, and two decades after this biography's first edition, Leo Szilard remains, by and large, an obscure figure for the public. Not so for historians and many scientists, however; his ideas and spirit still offer us helpful scientific, political, and moral perspectives—and a few surprises.

In Szilard's political satire *The Voice of the Dolphins*, he imitated Edward Bellamy's utopian science-fiction novel *Looking Backward* by predicting correctly in 1961 how the US-Soviet nuclear arms race would wind down in the late 1980s. Nuclear weapons still imperil humanity, but now in newly bizarre ways that defy the Cold War's deadly logic. Szilard would applaud one Cold War outcome as—all too slowly—nuclear arsenals are being put to better use. In debates during the 1960s about whether to test nuclear weapons, Szilard joked that you *should* test them! Test them all! Every last one! That never happened, but by 2012, the weapons-grade uranium equivalent of 18,000 Russian warheads had been recycled into fuel for US nuclear power plants by the Megatons to Megawatts Program. And he would still urge nuclear powers to test them all!

On the fiftieth anniversary of Hiroshima and Nagasaki in 1995, scientists who strived to control nuclear weapons were honored when the Nobel Peace Prize went to the Pugwash Conferences on Science and World Affairs and its founding leader Joseph Rotblat. Szilard had helped start Pugwash in 1957, then fostered its approach of candid, private intellectual discourse among scientists, insinuating, he said, the "sweet voice of reason" to the nuclear arms race. He was one of its most ardent members.

But in 1995 a new history revealed how Szilard's influence had spread unforeseen but decisive results—not by intellectual discourse but by humor. With tongue in cheek but still in moral earnest, Szilard had unknowingly inspired the Russian H-bomb designer Andrei Sakharov to become a champion for arms control and human rights. This tale begins in 1947, when Szilard wrote the political satire "My Trial as a War Criminal" to dramatize how scientists are responsible for their creations: in his case, the A-bomb. Historian Richard Rhodes wrote in *Dark Sun*, his 1995 history of the H-bomb, that when Szilard's satire was republished in 1961, Sakharov read it in translation and embraced its moral imperative, prompting the heroic political activism that earned him the Nobel Peace Prize in 1975.

Szilard's influence was explained by Sakharov's friend and colleague, Viktor Adamsky, who had translated the satire. "We were amazed by [Szilard's] paradox," Adamsky told Rhodes. "You can't get away from the fact that we were developing weapons of mass destruction. We thought it was necessary. Such was our inner conviction. But still the moral aspect of it would not let Andrei Dmitrievich [Sakharov] and some of us live in peace." In this way, Rhodes disclosed, Szilard's story "delivered a note in a bottle to a secret Soviet laboratory that contributed to Andrei Sakharov's courageous work of protest that helped bring the US-Soviet nuclear arms race to an end."

Adamsky so admired Szilard that when he read *Genius in the Shadows*, he decided to translate it and had it published in Russia. In his introduction, Adamsky praised Szilard's "deep feeling of responsibility for the consequences which could follow from the uncontrolled use of scientific research." Citing both nuclear weapons and nuclear power, Adamsky called Szilard "a man who at each stage of development foresaw and understood the technological and political aspects of the problem more quickly and more clearly than others. This is one of those rare cases when the work of someone with no official political role substantially affected world politics." (*Гений в тени*, 2003).

Also in 1995, a fiftieth-anniversary program at the National Archives in Washington brought together three Manhattan Project scientists who had signed Szilard's July 1945 petition to President Truman, raising moral questions about using the new weapon. (In all, 155 scientists signed it, and in a poll the Army took, 83 percent of them favored a demonstration before bombing cities.) This petition was also highlighted in 2005, when in John Adams' opera "Doctor Atomic," about J. Robert Oppenheimer and the A-bomb, scientists at the Los Alamos nuclear weapons lab distribute and debate Szilard's petition. Szilard is portrayed occasionally in documentaries and docudramas about the bomb in film, television, and documentaries. His effort to prevent the use of atomic weapons at the end of World War II, by confronting President Truman's mentor James F. Byrnes, is now dramatized in "Uranium + Peaches," a play by Peter Cook and myself.

Gently and gradually, the genius that was Leo Szilard has emerged from the shadows. Szilard's centenary was celebrated grandly in his hometown of Budapest in 1998, and also in the United States. A Leo Szilard Centenary International Seminar at Eötvös University included praise by several Nobel laureates, colleagues, family, friends, and scholars. In his honor, a countrywide "Leo Szilard Physics Competition" was established for high school students, requiring contestants to use computers and conduct laboratory tests to solve modern physics problems. At Columbus, Ohio, that year the American Physical Society celebrated Szilard's centenary with a seminar at its annual meeting.

In Budapest, a commemorative stamp appeared in 1998 to honor Szilard's birth, and a plaque was mounted by the entry to the apartment building in Pest where he was born, at 50 Bajza Utca. In the city's historic Kerepesi Cemetery, a formal service featured Hungary's Minister of Culture and Education as Szilard's ashes were interred in the section reserved for members of the Hungarian Academy of Sciences. "Today his countrymen see Leo Szilard as a latter-day Erasmus coming home at the end of decades of wandering," the minister said. "Just like Erasmus centuries before him, he had the courage to send letters to chief executives of Great Powers when at stake was how to do good for, or save the peace of the world." And like Erasmus, "Szilard considered peace to be the most valuable asset of humanity."

In retrospect, Szilard's humanity was more aggressive and abrupt than that of Erasmus. Although both men were humanists who dabbled in diplomacy, Szilard was not always diplomatic, and at times he could be downright rude. His faith in rigorous reason sometimes blinded him to life's random emotional forces and to the feelings of others. The brilliant logic he applied to the chess-like arms race between the USA and USSR now seems passé against militant terrorists seeking nuclear materials to blow up or contaminate cities. Too, Szilard's profound episodes of self-absorbing thought—he called it "botching"—alienated both family and colleagues. He suffered neither fools nor other geniuses gladly, and this attitude left him with many distant admirers but few close friends. Szilard's need since childhood to be different and disruptive, played out in science and in politics, kept him from gaining the support that institutions might have offered to reinforce his creative efforts and enhance his reputation. In an unfortunate way, Szilard's whole creative style led him to the forefront of science but left him by the wayside of society. Although not a "mad scientist," he could be maddening.

And yet, despite his personal quirks, Szilard's moral views have radiated through many media. He was bold enough to write a personal version of the Ten Commandments, and in 1992 physicist Freeman Dyson used them as epigraphs in *From Eros to Gaia,* his book on science and society. Szilard's second commandment describes his own life: "Let your acts be directed toward a worthy goal but do not ask if they will reach it; they are to be models and examples, not means to an end." Szilard's own exemplary ways prompted the editor of the National Academy of Sciences quarterly *Issues in Science and Technology* to write in 1997 about "the power of the individual" in public life. "The key to Szilard's effectiveness and influence was that his sense of responsibility for making the world a better place compelled him to work so hard to advance his ideas," wrote Kevin Finneran. "What Szilard did was to approach public policy with the same rigor, determination, and persistence with which good scientists approach science. What works in advancing science can also work in improving policy."

Szilard's creations survive around the world as well. From his ideas and insights to create the European Molecular Biology Organization, there is now a Szilard Library at EMBO's European Molecular Biology Laboratory in Heidelberg, Germany. Each year since 1974, the American Physical Society awards the Leo Szilard Lectureship Award "to recognize outstanding accomplishments by physicists in promoting the use of physics for the benefit of society in such areas as the environment, arms control, and science policy."

Szilard was also remembered and praised in London, in 2008, at a seventy-fifth-anniversary celebration and conference for the Council for Assisting Refugee Academics, which Szilard helped found and run in 1933 as the Academic Assistance Council. In the keynote address, medical scientist Sir Ralph Kohn noted how "Szilard somehow always turned up at the right place and at the right time in the 1930s—a true man of destiny . . ." Esther Simpson, Szilard's colleague at the Council in the 1930s, saw him as a "bird of passage" alighting to share his fears and foresight.

Throughout his life, Szilard created institutions for the public good. Politically, his most successful was the Council for a Livable World, America's first political action committee to raise campaign funds for Senate and House candidates who favor arms control and disarmament. The Council celebrated its fiftieth anniversary in 2012, with a record of helping elect 120 candidates to the Senate and 203 to the House of Representatives, and that year raised more than $2 million for candidates' campaigns.

For all his benefits to the commonweal, Szilard is still remembered best as a creative and ingenious scientist, first in physics and then in biology. In 1996 the National Inventors Hall of Fame in Akron announced: "Statistical mechanics, nuclear engineering, genetics, molecular biology and political science—Leo Szilard tackled them all. However, it is for his nuclear fission reactor that Szilard joins co-inventor Enrico Fermi in the National Inventors Hall of Fame." A 1997 *Scientific American* article and a 2009 *Discovery* television series on twentieth-century science featured the Einstein/Szilard refrigerator, which the two colleagues designed in the 1920s with an electromagnetic pump to circulate liquid-metal coolant. Too noisy for household use, their concept has long appealed to engineers: first, for cooling nuclear breeder reactors since the 1950s, and recently appearing as small-scale working models built by Malcolm McCulloch at Oxford University and Andy Delano at Georgia Tech.

Szilard's contribution to what became "information theory"—the study and application of how data and ideas can be codified and shared—is still applied by researchers today. Szilard's 1923 doctoral thesis took on the thermodynamic puzzle known as "Maxwell's Demon," and in a 1929 paper he described how entropy (disorder in a system) affects the use of information. He conceived what is called today "The Szilard Engine," a process for manipulating molecules, and its use continues to engage and challenge researchers. Physicist Mark Raizen and his colleagues at the University of Texas, Austin, use it to separate isotopes and control atoms in small groups or even individually. The continuing interest in Szilard's seminal concepts for information theory is credited and well explained in *Maxwell's Demon 2 Entropy, Classical and Quantum Information, Computing* (Institute of Physics Publishing, 2003).

In biology, Szilard worked as an "intellectual bumblebee" according to French molecular biologist Francois Jacob. With Jacques Monod and Andre Lwoff, Jacob shared a 1965 Nobel Prize for research on the human immune system based on ideas they credit to Szilard. "Szilard is extraordinary," said Jacob, "an incredible, surprising man. He had ten ideas a minute. He had ideas about everything."

At the time, Szilard received scant credit for one insight: in the 1950s he proposed a technique for cloning mammalian cells that was claimed, developed, and applied by others. That issue is now resolved in the scientific literature to Szilard's full credit. Citations to his scientific and political works continue to pop up in journals and books.

Szilard's speculations in biology were tested in 2009 by researchers in Sweden who re-examined his assumptions, the then available knowledge, and hypotheses for a 1959 paper on the causes of aging. They concluded:

> Szilard's model was the first serious hypothesis to posit accumulated genetic damage as an important cause of senescence, a view most researchers agree with today . . . and in spite of being oversimplified and on some aspects wrong, it also predicts population-based death rates in a quite accurate manner. In the literature, Szilard is often described as a visionary man with a remarkable ability to make accurate predictions on grounds not always clear to his fellow scientists. In a sense, his model for the aging process may be another example of this extraordinary gift. (Zetterberg, et al., "The Szilard Hypothesis on the Nature of Aging Revisited," *Genetics* 182: 3-9, May 2009)

Indeed, Jonas Salk admired Szilard's freewheeling intellect when he was one of the Salk Institute's first fellows. Salk said that Szilard "could effect chain reactions both in atoms and in human minds." James D. Watson, who with Francis Crick gained a Nobel prize for discovering the genetic code DNA, said recently that Szilard had influenced him more than anyone else: not for any particular scientific idea, but by his creative imagination. Watson said he loved being around him because "Leo got excited about something before it was true."

Scholars continue to discover that ingenuity in the Leo Szilard Papers at the University of California, San Diego, Mandeville Special Collections Library (libraries.ucsd.edu/speccoll/testing/html/mss0032a.html). Several internet sites provide information on Szilard's life and science. A respected website containing various online documents has been created by independent scholar Gene Dannen (http://www.dannen.com/szilard.html), and Nelson H. F. Beebe at the University of Utah has compiled a comprehensive *Selected Bibliography by and about Leo Szilard* that seems to expand by the week (http://www.math.utah.edu/~beebe/).

Other Szilard documents, still in private hands, have become attractive to collectors. In December 2012, an auction that included manuscripts by Thomas Jefferson, Vincent van Gogh, George Washington, Adam Smith, Karl Marx, Beethoven, and Tchaikovsky also offered for sale Leo Szilard's January 1939 letter to Lewis L. Strauss that first described nuclear fission as "a very sensational new development in nuclear physics" that could lead to electrical generation, medical isotopes, and "unfortunately also perhaps to atomic bombs." This Szilard letter sold for $240,000, only barely out-priced by Jefferson and Van Gogh.

At Kerepesi Cemetery in Budapest, only half of Szilard's ashes were buried in 1998. The other half went to Ithaca, New York, to join those of his wife Gertrud Weiss Szilard and her family. Speaking at Lake View Cemetery in May 1998, Szilard's friend and colleague Hans Bethe recalled how in England, in 1933, Leo seemed ubiquitous as he raced about working to settle the academic refugees expelled by Hitler. "We thought he was seen at two places at the same time," Bethe recalled, and now, with ashes in Budapest and Ithaca, that suspicion was demonstrably true.

Szilard's wish was that his ashes not be buried at all, but go off in balloons because, he said, "it is more pleasing for people to look up rather than to look down," and that way "at least it will delight all the children." This was achieved in Ithaca as grandnephews and grandnieces placed some of his ashes in an airmail envelope (on which they wrote the epitaph Szilard preferred, "He Did His Best") and launched them with colorful balloons that drifted over treetops toward the Atlantic. A month later, Szilard's wishes were granted again when relatives at San Diego let fly some ashes lofted by balloons that rose above the Pacific.

Although a bird of passage whose genius shaped science and history at critical times, Szilard and his legacy are also timeless. His mode of being informs us anew. "There are all too few like Leo Szilard," his friend and colleague Jonas Salk recalled, and his example provides "a role model for others of his kind for which the world is now in great need."

—William Lanouette
San Diego, CA

Foreword

by Jonas Salk

Leo Szilard's life spanned the first two-thirds of the twentieth century, the two great wars, and beyond. His experiences extended from physics to biology and from academia to the halls of world power. In his special ways, through a quest for knowledge and through the force of his pragmatic idealism, he sought to create a more peaceful world.

He was equally himself with colleagues, friends, or heads of state, and possessed the wit and the wisdom to be listened to and appreciated because of his sincerity. However, because of the intensity with which his intellectual vigor was delivered, his intentions were sometimes misunderstood. His scintillating mind was always turned on, and as if through the power of his mind—knowing how rarely it might otherwise have occurred—his inoperable cancer appeared to vanish through his will to live and to continue to give of his genius.

This biography of Leo Szilard reveals a multifaceted singularity, a person who could effect chain reactions both in atoms and in human minds. Few came into his presence who left unchanged. The essence and spirit of this fascinating man who played so vital a role at a critical time in the history of humankind is captured in these pages. This book makes up for the fact that for all the people he influenced Leo remains a curiously obscure figure.

He was a true scientist/humanist and a philosopher as well, a natural philosopher who caused things to happen. That he was also an artist of science is evident in his statements: "The creative scientist has much in common with the artist and the poet. Logical thinking and an analytical ability are necessary attributes to a scientist, but they are far from sufficient for creative work. Those insights in science that have led to a breakthrough were not logically derived from preexisting knowledge: The creative processes on which the progress of science is based operate on the level of the subconscious."

His desire to associate with people of like mind was reflected in an ambition that we both shared, to bring scientists and others possessed of a deep human concern together in a scientific institution such as was established in La Jolla, California, in which he found peace in the last days of his life. The Salk Institute was for evolvers rather than maintainers of the status quo, and Leo was one of the most evolved of evolvers. Leo's suggestions for colleagues at the institute quickly revealed his preference for iconoclasts, a reason for the kinship that we felt. I discovered that we had even more in common one day when we were discussing an idea that I had, and Leo said there were those who would never forgive me for being right. I knew at that moment he was speaking about his own experiences as well, another reason for the understanding we shared.

There are all too few like Leo Szilard. This poignant story of his life will inform and inspire, providing a role model for others of his kind for which the world is now in great need.

Preface

It was hot and humid in Manhattan on the morning of August 2, 1939, yet Leo Szilard, a round-faced Hungarian-born physicist working at Columbia University, appeared at the curb in front of his hotel near the campus wearing a suit and tie. Szilard always wore a suit and tie. He hopped into a blue Plymouth coupe, and the car sped through Harlem, across the new Triborough Bridge, and past the glittering World's Fair grounds. Szilard was headed for a small cottage on Long Island to visit a friend. His driver that morning was Edward Teller. The friend was Albert Einstein.

This was Szilard's second visit to Einstein's cottage in three weeks. Szilard had studied with Einstein in Berlin in the 1920s, and their research had produced more than thirty joint patents for a refrigerator pump with no moving parts. The two friends had met but rarely since Szilard moved from England to America more than a year before. During that first call at the cottage, the forty-one-year-old Szilard had stunned his sixty-year-old mentor by describing the concept of a nuclear "chain reaction." Szilard had conceived this in London, in 1933, but had tested the idea using uranium at Columbia only in the spring and summer of 1939. "I haven't thought of that at all," Einstein had replied, seeing at last a mechanism and a natural element that gave his 1905 equation for $E = mc^2$ a new and terrifying reality.

On that first visit, Szilard and Einstein agreed to warn the Belgian royal family—personal friends of Einstein's, whom he called "the Kings"—that Nazi Germany coveted the world's largest uranium supply in the Belgian Congo. In the days between those visits, however, Szilard had decided that US officials should also be warned about German intentions, and through friends he had met Alexander Sachs, a Wall Street banker who claimed easy access to the White House. So, in the white-shingled cottage near Peconic Bay that sultry August day,

Einstein, wearing an old robe and slippers, and Szilard, now down to shirtsleeves and tie, hunched over the dining-room table, sipped iced tea, and wondered how to tell an American president about something that might change the world forever.

"We did not know just how many words one could put in a letter which a president is supposed to read," Szilard recalled later. That afternoon, Einstein dictated in German a brief, preliminary draft, leaving Szilard to write both a long and a short letter to Franklin Roosevelt.

"Some recent work by E. Fermi and L. Szilard . . . ." Einstein's now famous letter began, "leads me to expect that the element uranium may be turned into a new and important source of energy in the immediate future." It went on to explain how physicists Enrico Fermi and Szilard had just demonstrated that a nuclear chain reaction could probably occur. "This new phenomenon would also lead to the construction of bombs, and it is conceivable—though much less certain—that extremely powerful bombs of a new type may thus be constructed. A single bomb of this type, carried by boat and exploded in a port, might very well destroy the whole port together with some of the surrounding territory."

Einstein's letter ultimately led to the federal government's top-secret Manhattan Project to design and build atomic reactors and atomic bombs. When that letter was written, Einstein and Fermi were both acclaimed Nobel laureates in physics, while "L. Szilard" was unknown outside a small circle of physicists. He remains a shadowy historical figure today, even though he did the most among scientists of his generation to help forestall, then create, and then control the atom bomb.

Indeed, Szilard personifies the nuclear era and its fateful consequences as no other individual can. He successfully forestalled a nuclear arms race with Hitler by having the British government make his 1934 chain-reaction patent a military secret; and beginning in 1935, he angered many scientists outside Germany (including Fermi) by urging them not to publish their nuclear research.

After German scientists experimentally split or "fissioned" the uranium atom in 1938, Szilard worked frantically to start the very arms race he had feared, using Einstein's letters, Fermi's connections, and his own feisty imagination to beat Hitler to the A-bomb. He drafted a second Einstein letter to Roosevelt in early 1940, when federal interest in atomic research was flagging, and he code-signed with Fermi the world's first nuclear reactor. (Their joint US patent was issued publicly in 1955 to "E. Fermi et al.")

Yet by the spring of 1945, with Germany all but defeated and its race for the A-bomb lost, Szilard began efforts to forestall another nuclear arms race—this time with Russia. He drafted a third Einstein letter to Roosevelt, which failed to reach the president before his death. Then, with Truman in the White House, Szilard led two other initiatives: the atomic scientists' unsuccessful petition to the president to prevent the atomic bombing of Japanese cities and their successful lobbying of Congress to shift postwar control of the atom from military to civilian hands.

Szilard's many efforts to create peaceful uses for the atom are also lost in the shadows of modern history. A few weeks after the first Fermi-Szilard nuclear reactor surged to life in a squash court at the University of Chicago, in December 1942, Szilard, then chief physicist of the Manhattan Project's secret "Metallurgical Laboratory," proposed, named, and designed the "breeder" reactor—an atomic power plant that makes more fuel than it consumes. The breeder's wartime mission was to make plutonium for bombs; in peacetime it would generate electricity and fresh fuel for other nuclear power plants. Yet the first commercial breeder, when built near Detroit in the 1960s, was named for Fermi.

After serving as Szilard's chauffeur in 1939, Edward Teller worked as his agent in Washington, tracking federal decisions once Roosevelt had acted on Einstein's letter. Teller also joined Szilard in early meetings of the government's Advisory Committee on Uranium and in research for the Manhattan Project. But by 1945, with the race against Germany won, the two split over the need for arms control; in later years, they disagreed bitterly about nuclear policy, yet still remained personal friends.

From the war's end until his death in 1964, Szilard worked tirelessly to control the A-bomb he had helped create. He proposed or helped to found publications and institutions that today still influence science and politics: one is the *Bulletin of the Atomic Scientists,* a monthly magazine first published at Chicago's Metallurgical Laboratory in 1945; another is the Council for a Livable World, the first political-action committee for arms control, which he founded in Washington in 1962. Today the council leads and coordinates a network of groups working for demilitarization and peace, and each election year it raises more than one million dollars for congressional candidates.

Szilard's efforts were personal, clever, at times outrageous, and usually made behind the scenes. Justice Department officials threatened to jail Szilard (by now a United States citizen) in 1947 for writing a letter to Stalin that called for radio broadcasts by US and Soviet leaders to each other's countries, an idea finally realized by presidents Reagan and Gorbachev in 1988. Szilard succeeded more promptly when, at a private meeting with Nikita Khrushchev in 1960, he gained the Soviet premier's personal assent to a Kremlin–White House "hot line" that opened three years later.

Szilard is remembered by friends and colleagues for his wry and puckish humor, a twinkle in his eye as he deflated pompous egos or described playful (and sometimes even practical) scientific or political inventions. They remember with glee Szilard's tongue-in-cheek explanation for Grand Central Terminal's pay toilets, published in a 1961 collection of his satires called *The Voice of the Dolphins,* or his proposal that the National Science Foundation pay second-rate scientists *not* to conduct research and *not* to publish articles. To microbiologist François

Jacob, Szilard was an intellectual "bumblebee," cross-pollinating ideas and institutions with his unconventional wisdom. One Szilard idea earned Jacob and his French colleagues a Nobel Prize. Another idea—a research center combining science and human affairs—became the Salk Institute for Biological Studies in La Jolla, California, where Szilard spent his last days. Yet for many reasons, among them his own maverick personality, Szilard's story could not be told until now.

Since childhood Szilard enjoyed being "different" and independent, unbound by common ideas and conventions. In later life, this independence led to Szilard's being overlooked by the Manhattan Project's official history, in part because of his many policy disputes with its headstrong director, Gen. Leslie Groves. Szilard was for years a forgotten man. Documents about his physics research and political activities remained military secrets for decades, and after his death, his personal papers and letters were sheltered from would-be biographers by his widow, although she worked to publish his scientific and professional papers.

It was not until 1983, two years after Gertrud (Trude) Weiss Szilard's death, that I began work on a biography and all papers became available through her brother, Egon Weiss. Still, the papers were in chaos. Szilard lived most of his life in hotels and stashed his files in suitcases, which he left with friends and relatives around the world. Until recently the Szilard papers were organized by the color of the bag in which each document was found, and a reorganization by subject was not completed until after I finished writing this book. Luckily, in 1987, Egon Weiss rediscovered more than 350 personal letters to Trude by Szilard, and these were translated for use in this volume.

Szilard's widow failed to tell historians or writers about another vital source: Leo Szilard's younger brother, Bela Silard. Fortunately, Bela and I met through his son John, a Washington lawyer, soon after I began work on this biography. Had Bela Silard's writings and recollections, and his collection of family photographs and records, not become part of this work, no complete biography of Leo Szilard could have been possible. For this reason I credit my authorship "with Bela Silard."

"Leo Szilard was a very complex personality," his friend and fellow physicist Nobel laureate Hans Bethe remembers. "He was one of the most intelligent people I have ever known. His mind worked quickly and profoundly, and he was able to come to ideas that most of us appreciated only after many hours of talk. This was his strength and, of course, also his weakness. He was always ahead of his time. His ideas often were expressed in paradoxes, and the paradoxes were not always understood."

Both Szilard's candor and his cockiness made him a ready target of federal agents, who shadowed his actions for years without results. (Szilard's FBI files are, in turns, both painful and hilarious to read.) Szilard irked his Manhattan Project superiors by both criticizing and ignoring some measures imposed to maintain secrecy and by complaining to congressional committees and the press after the war that the army's "compartmentalization" of the scientists' work had delayed America's first atomic bomb by more than a year. After World War II, Szilard promoted sharing US atomic secrets with Russia in order to bring about international arms control agreements—hardly a popular idea at the time or through the cold war. And he predicted accurately the time needed by Russia to catch up with US nuclear developments. Fearlessly he mocked anti-Communist paranoia on university campuses. And in his lifetime Szilard publicly criticized every US president since Truman for escalating the nuclear arms race.

Now, more than half a century after Szilard's meeting with Einstein that started America's nuclear age and more than a quarter century after Szilard's death, many of his insights and ideas have been adopted by his successors. So, too, have his efforts to lessen US-Soviet tensions and to share technology in the pursuit of peace. During the 1940s and 1950s, Szilard devised novel schemes to verify nuclear arms control agreements, including the kinds of on-site inspections begun only in 1988. Szilard's many attempts at "citizen diplomacy" began weeks after the bombing of Hiroshima and Nagasaki, pioneering a technique that enjoys enthusiastic acceptance today. Yet during and after his lifetime, Leo Szilard remained elusive and is now largely a forgotten man, a genius in the shadows of the world he helped create.

WILLIAM LANOUETTE
Washington, DC
October 1992

PART ONE

 1898–1933

CHAPTER 1

The Family

 1898–1912

With an umbrella swinging from his V-neck sweater—so as to keep a hand free for his younger brother, Bela—Leo Szilard strode out into a cloudy fall morning in 1910. For the excursion he had planned, Leo sported a brown felt hat—flat brimmed, flat topped, and uncommonly formal, yet stylish for a twelve-year-old. Most teenage boys in Budapest wore cloth caps, like the one that ten-year-old Bela tugged down against the breeze. But not Leo, whose headpiece made him feel both "dressed up" and "different." Yet while daring in his dress, Leo was still cautious in his demeanor. He had that umbrella, just in case.

This stroll was the two boys' first without a parent or governess, and Leo had a special destination in mind. At the wrought-iron front gate to the yard, the Szilard boys left behind their family's towering stucco-and-timber art nouveau "villa" and turned right onto the Fasor, a broad allée linking Budapest's café and shopping district with the lush and spacious City Park. Under six rows of horse chestnut trees, the Fasor was divided into five pathways: a roadway down the center for cars, trucks, and horse-drawn carriages; two shaded pedestrian lanes; and along the sidewalk, the tracks for small electric trolleys. (Budapest's traffic clanked and rattled along on the left then, as it would until the occupying German army ordered a change to right-hand driving in 1941.)[1]

On the long block from their villa to the park, the Szilard boys walked past decorous mansions owned by down-at-the-heel Austro-Hungarian aristocrats and wealthy Jewish merchants, mansions set in green plots that gave this neighborhood the name Garden District. In the park, they followed shaded footpaths that rambled among manicured flower beds. They passed a restaurant beside a lake and, in the water, saw the spires of a castle built in ancient Transylvanian style but actually finished in 1896 for Hungary's millennial celebration. Beyond the Gallery of Fine Arts, the boys crossed a footbridge over the city's new subway line, the first on the European continent, then strolled by another boating lake and the elegant, new Gundel's Restaurant toward the Budapest Zoo.

At the Turkish-style entrance gate—an arch hoisted on the backs of four head-high stone elephants, surmounted by a ring of crouching bears— Leo bought tickets, and once inside, the boys rushed to a mosquelike brick structure that was home for the elephants and the hippos. Several elephants stood in the open yard, begging across a moat for nuts from children. Nearby, but standing alone, was the object of the boys' journey: the largest and oldest elephant, which Leo adored for being both wise and playful.

Rain began to fall as the boys approached the moat, and while Leo fidgeted to undo his umbrella strap, a gust of wind caught his prized brown hat, skimming it over the metal railing and into the moat. Leo grabbed his umbrella by the tip, leaned against the railing, and in a few seconds had hooked his hat onto the curved handle. But just as quickly the old elephant thrust his trunk for the hat. Boy and beast tugged, and despite the uneven match, Leo persisted, his youthful sighs matching the old animal's low grunts. After a few seconds, the elephant stepped back slowly and nearly yanked Leo into the moat. His chest sore from scraping the railing and afraid he might lose both his umbrella and his arm, Leo stopped pulling and stared as the elephant, with seeming satisfaction, waved the hat around, stuffed it into his mouth, and began to chew.

Leo's brown eyes glared as, enraged, he strained for composure against Bela's laughs. Ignoring the laughter, Leo rubbed his forehead in thought as raindrops spattered his hand and face. Then he turned to Bela and announced: "I have figured out that the hat cannot come out at the other end of the beast before tomorrow. Let's go." Leo snapped open his umbrella and tramped homeward through the park in brooding silence.[2]

Bela Silard is fond of that episode because it reveals many of his brother's characteristics at once: a need to be different and independent; an impulse to solutions, undaunted by the odds for success; a resort to reason in a conflict; and a desire to conceal personal emotions. Leo would change in the years ahead, and by his ideas and actions he would help change the world as well—into a more perilous and, eventually, a more peaceful place.

But throughout his life, Leo Szilard would also remain as he was that rainy morning at the Budapest Zoo: impetuous, determined, clever, rational, and private. Indeed, these contrary qualities would empower Szilard to challenge—against opponents far more daunting than a hungry elephant—the scientific and political establishments of our century. At times, his feistiness would aid Szilard's success; at times, it would assure his failure.

Throughout his life, Szilard tried to ignore the past and concentrate on the future. For his childhood playmates he constantly invented new games and new rules for old ones. As a teenager, he became enthralled by physics, the new science of his day. As a scholar, he pushed for novel theories in thermodynamics and mathematics and enjoyed posing ideas in economics and politics that experts said couldn't be realized—but sometimes were. He delighted in devising inventions: some profound, many impractical. And, in later life, he resisted writing an autobiography, saying the past is unimportant, the future is all.

Yet in ways that he may never have realized or may have surmised and then suppressed, Leo Szilard's own past held the forces that shaped his richly inventive yet frustrating life. From his intense and exuberant child-hood in Budapest's elegant Garden District he gained the means to be a pioneer in science and in politics. He also inherited from his forebears mannerisms that would both charm and confuse associates and friends, quirks that sometimes helped to advance his proposals but also made his pioneering ideas toilsome to accept. For by the time he fought that elephant, Leo Szilard had another quality that would both limit and advance his life: a passion to be honest and frank.

Late in life, Szilard traced this consuming candor to his mother and his maternal grandfather:

Very often it is difficult to know where one's set of values comes from, but I have no difficulty in tracing mine to the children's tales which my mother used to tell me. My addiction to the truth is traceable to these tales. . . . My mother was fond of telling tales to her children and she always had some particular purpose in mind. Why she wanted to inculcate addiction to truth in her children is not clear to me.

I remember one story, which made a deep impression on me, about my grandfather. My grandfather was a high school student at the time of the Hungarian Revolution in 1848. In high school, when the children were waiting for the teacher to turn up, it was customary in Hungary for one child to keep watch. It was his task to keep a list of those children who were disorderly, and when the teacher came to class he was supposed to submit the list of these disorderly children to the teacher for punishment. In the particular case of the story my mother told me, the Hungarian Revolution of 1848 was on. A troop of soldiers was marching by the school [in Debrecen, Hungary's second-largest city] and a number of children violated orders by leaving the class and lining the street and cheering the soldiers. My grandfather, who was supposed to keep watch on disorderly children, joined those who left the school building and cheered the soldiers. When the teacher turned up for class, all the children were back in the classroom and my grandfather rendered his report. He gave the teacher the list of those children who had violated orders and went out to the street, and this list included his own name. The teacher was so much taken aback by this frankness that nobody was punished.[3]

Exemplary behavior by Sigmund Vidor became a compulsion in his daughter Tekla Vidor Szilard and a lifelong bond between her and her son Leo. This honesty gave them a way to be close without being sentimental or emotional. Sigmund Vidor's example became more than an action to emulate. It was a standard to surpass. Szilard remembered his maternal grandfather fondly; they lived in the same spacious villa on the Fasor until Leo was twelve years old.

But his paternal grandfather, Samuel Spitz, who had died twelve years before Leo's birth, probably endowed him with just the stubborn self-assurance he needed to be so certain about everything, including his maternal grandfather's honesty and the courage to fight elephants. The maternal grandfather Szilard knew was gentle, intellectual, and socially admired, while the paternal grandfather he heard little about was a restless vagabond—querulous, strong-willed, daring, adventurous, and proud. The grandfather Szilard knew may have engendered a sense of duty and public responsibility, but the grandfather he never knew gave him a resolute will. And it was that will as much as Szilard's rigorous honesty that drove this restless genius to try to change our world.

The ancestors Leo Szilard never knew, his paternal forebears, came in the mid-eighteenth century to Slovakia, which lies between Hungary and Poland. They came "from the East," Szilard's father Louis noted tersely in his otherwise detailed memoirs, but from just where was never certain.[4] Probably from Galicia, the Austrian sector of divided Poland that is now part of Ukraine. Before that westward move, they, like many eastern Jews, did not use a last name and probably took Spitz as their family name only after settling in Slovakia.

Leo Szilard's paternal great-grandfather, who was born before 1800, was a shepherd in Nagyfalu (now Vel'ka Ves nad Ipl'om), a small village in the Slovak county of Arva (Orava), just below the high ridge of the Carpathian Mountains, near today's Slovakia-Polish border. He was said to have lived more than eighty years and to have driven his sheep to the mountaintops for grazing until the last summer of his life. Along with some wonder about his nomadic and restless ways this old shepherd's sturdiness continued in family legend.

Samuel Spitz, that old shepherd's only son and Leo Szilard's paternal grandfather, was born about 1810 and grew up in Arva. There he married a Jewish girl named Leontine Klopstock, who came from Matina, a hamlet in the Tatra Mountains that divide Poland and Slovakia. After their marriage, Samuel and Leontine Spitz moved to Turdossin (Tvarožná), a small village a few miles down the valley of the river Vag (Vah) in western Slovakia. In Turdossin the Spitz family grew rapidly, eventually filling a large farmhouse and barnyard that Samuel had built to contain both his family and a dairy herd.

To earn money, Samuel turned to farming and leased from a Hungarian noble an estate that included a half-ruined medieval castle called Végles (Viglas), near Detva, Slovakia. Perched on the side of a mountain, Végles included a drawbridge and two huge courtyards. It was large enough for the Spitz family, but the estate had poor soil—only good enough for growing oats, rye, clover, and hemp as cash crops. Samuel also planted large areas with potatoes for use as stock in a distillery on the property. In this castle Samuel Spitz raised ten daughters and four sons. After about seven years, however, Samuel's fortunes completely reversed: his distillery leveled in a boiler explosion, his cattle killed in a barn fire, his sheep lost to drought. Nearly bankrupt, Samuel moved his family into Detva, and soon he moved again, to Körmöczbánya (Kremnica), near Zólyom (Zvolen), today in central Slovakia but then part of the Austro-Hungarian Empire.

This restless and determined man was an entrepreneur, albeit on a small scale, for Samuel had little luck, and more often than not his failed schemes only left him ever deeper in debt and desperate to concoct new enterprises. He held contracts for timbering on large forest tracts. He operated a village tavern, supplied by the still he ran on the Végles estate. And in years when lumber was in demand, he tried his luck as a lumber merchant, often botching that enterprise. At times, he distributed liquor or imported coal. But with repeated failures, Samuel Spitz in later life grew stern and domineering, bitter about his financial perils and brutish with gout. Sometimes he was so impatient that his wife and children avoided him. When Samuel died an unhappy man, in Kremnica in about 1890, his widow moved to Budapest to live with her eldest daughter.

Grandfather Samuel's strong-willed tendency to dominate his family and surroundings continued in his progeny along with the greatgrandfather's physical sturdiness: not only in Samuel's son Louis, Leo's father, but also in his grandsons, Leo and Bela. More urbane Budapest relatives would later say that the Szilard tenacity "ran in the family," in contrast to *their* more artistic and pliant natures.

Leo's father, Louis Spitz, was born in Turdossin, in the Tatra Mountains, on December 23, 1860. Louis remained his familiar name, although he used Lajos, the Hungarian form, in business. Louis was his parents' third son and as a child became his mother's favorite. The Spitz family spoke German, and as with many Jews in Slovakia, their cultural leanings were toward Vienna rather than Budapest or Prague. The family read many German literary classics, and the children never learned more than a few words of Slovak, picked up from farm hands and yard workers. Samuel Spitz's four sons attended Jewish elementary schools, where they were taught in German. None of the girls attended school, but when finances allowed, they studied at home with German-speaking tutors. The two older boys, Arnold and Max, were later sent to commercial schools, one in Germany, the other in Austria. When they turned ten, the two younger boys, Louis and Adolf, were enrolled in high school in Kremnica. There, for the first time, they had to struggle to learn Hungarian, a language unlike any in Europe whose origins lie in central Asia. Louis Spitz was a poor student who satisfied a sweet tooth by selling decals to his classmates.

Even less is known about the maternal grandparents in Leo's father's family. His grandmother, Leontine Klopstock Spitz, was the eldest of six children born in the Tatra Mountains to Joseph Klopstock and his first wife. Leontine had also three stepsisters and a stepbrother, children of Klopstock's second wife.[5]

In 1880, when he was twenty, Leo's father, Louis Spitz, graduated from high school in Kremnica and left Slovakia for Pest. The city of Budapest, which had united only seven years earlier, consisted of Buda, a hilly area west of the Danube, and Pest, the level area to the east. Louis enrolled at the Institute of Technology (later called Technical University) to study engineering, and despite his poor command of Hungarian, a language that he would never speak easily, he was diligent, ambitious, and persistent, soon becoming one of the best students.

But because his father, Samuel Spitz, was usually in debt, Louis had to earn a living in his spare time. He supported himself by tutoring high school students from Budapest's middle-class families, teaching them the German language and its classical literature. Louis's effort and sacrifices paid off, for shortly after he graduated from the institute as a civil engineer, he could choose from among good jobs with several of Hungary's leading building contractors. But eventually he grew restless working in large organizations, decided to establish his own business, and in this way became a successful and prosperous contractor. His specialty was to plan and construct bridges and embankments for railway lines.

Louis Spitz first met Tekla Vidor in Budapest on Christmas Day, 1896. He had just turned thirty-six, and she was twenty-five. Meant to appear a coincidence, their introduction had been arranged by Tekla's cousin Willie. Tekla was an intensely introspective and moralistic girl who thought little about romance or "husband hunting." To many friends and relatives she appeared an earnest adolescent, perhaps too serious for her years. The couple met at other gatherings that holiday season, and by the time Tekla discovered Willie's designs, she so admired Louis as a "self-made man" and so enjoyed his conversations about a range of topics—but never his personal feelings—that she quickly forgave her cousin's matchmaking.

Tekla wrote later that she liked Louis's "big brown eyes and the tan that made him appear so different from the young men of the day who frequented tea parties and coffeehouses. . . ."[6] Soon she "began to think about the possibility of his proposing, and it was a disturbing thought that it would be up to me to make a decision. . . . I was shaken by doubt, restlessness, agitation."

Invited one day by Tekla's father to call at the Vidor house, Louis accepted and quickly became a regular afternoon visitor who often stayed for dinner. Only when Louis remarked that a true marriage must include children did Tekla see "the first shaft of light piercing the armor of his rather closed-in personality."[7] Then one night, when they were left alone—she seated under a lamp knitting, he passing her the yarn—Louis rose and began speaking in an earnest but oblique way. Louis struggled for words, and to ease his anxiety, Tekla set down her knitting and offered her hands, "thus giving him my answer without waiting for him to finish his discourse." Yes, they would marry. Then came the bouquets from Louis and "the valuable presents that earned him a scolding instead of thanks." Just a month after they met, the couple were formally engaged. Receptions for callers followed, then big dinners in their honor thrown by proud family members, and later a search for an apartment near Tekla's home in the Garden District.

On the day they married, Louis and Tekla first came to a "civil wedding" before a county registrar in the Office of the Registrar of Vital Statistics. It was a Sunday morning, April 25, 1897. This was the first year such a ceremony was required by the state, and the registrar decided to process four couples at once. Louis remembered the occasion as informal and "Very festive," but to Tekla the drab walls and the registrar's officious manner were "cold and uninspiring." She even feared throughout the brief bureaucratic ceremony that somehow her name would "be coupled with that of a total stranger in my marriage certificate." Then, at three that afternoon, the couple handed their marriage certificate to a royal notary, who read a marriage contract that he had drafted using information provided by the two families. For the first time, Louis learned that his wife had "a considerable fortune." The sum had been deposited in Louis's account at the Kommerzialbank, the receipt handed to him by the notary.[8]

Next, Tekla returned home, dressed in a flowing bridal gown, and rode in a carriage with her parents through the bustling city to a synagogue. There Louis Spitz married Tekla Vidor in a Jewish ceremony, arranged by the bride's father, Dr. Sigmund Vidor, a member of the governing body of the Jews of Budapest. With Louis's family in mind, the ceremony was performed in German by the German chief rabbi Kaiserling. After the ceremony, the couple took a coach ride through City Park, attended a party given by the bride's mother at her home, and finally, about midnight, moved into their apartment nearby at 28 (now 50) Bajza Utca, greeted by their new cook and maid.

Louis and Tekla Spitz took no honeymoon because he considered the custom frivolous. Tekla didn't mind, she later wrote, and found her first months of marriage delightfully "harmonious," perhaps "because we were not allowed to spend enough time together to turn mushy with sentiment." Resolutely and with a renewed desire to be "useful," Tekla focused her nervous energies on home and husband. Early each morning she rose to prepare Louis's breakfast. All day she puttered around the apartment, sewing and reading. Often she prepared lunch for Louis whenever he worked in the neighborhood. Most evenings she walked arm in arm with him along the Garden District's tree-lined boulevards. And each evening Tekla visited her parents at their apartment nearby.

All the while, Tekla Spitz was uneasy about her sudden change from "a girl innocent in body and mind" to a married woman. She was "repelled," in conversations with her married friends, by "frivolous remarks of double or unmistakable meaning"; she "felt that to make a joke of what was private and sacred was a brutish thing."[9]

The couple's first apartment was in a large corner building, an Italian Renaissance-styled stone structure set back from the street by a small lawn. That spring, Louis was able to spend more time there, for he moved his contracting office to a small room in their apartment. This he shared with Tekla's older brother, Emil Vidor, who was just beginning a career as an architect.

The Garden District in which Louis and Tekla Spitz lived was by the end of the nineteenth century both showy and shabby, an eclectic mixture of new "Villas" built by the upper middle class and older city mansions belonging to Hungary's ostentatious but often impoverished nobility. The neighborhood was a meeting place for two clearly different cultures: ambitious and successful merchants, many of them Jewish, whose wealth and lives centered on the city; and Magyar aristocrats, who depended on vast country estates for support but whose social and political pretensions also drew them to the bustling capital. The Garden District fostered a symbiotic relationship of another kind: The daughters of wealthy Jewish merchants married the titled sons and grandsons of counts, often at the time when their parents loaned money to the titled families. In turn, the Magyar court bestowed titles (usually Baron) on the more successful merchants, making them members of the Parliament's House of Lords or including them in the aristocratic civil service. By 1904, Jews owned more than one-third of the country's arable land.[10] Anti-Semitism would arise later, in part as a reaction to the access Jews gained to commerce and society. But for the moment, "assimilation" for Louis and Tekla Spitz was an easy step. Louis belonged to a Masonic lodge in Pest and over time rose to hold several offices. The couple attended evening card parties and dances there, and Louis repaired to the lodge once a week for a stag evening.

Budapest itself was a tense crossroads of monarchy and modernity. Buda and its dramatic hills featured palaces and citadels, a proud outpost for the co-capital (with Vienna) of the Austro-Hungarian Empire. Pest, flat and modern with its art nouveau apartment blocks lining broad Parisian-style boulevards, was a cultural and commercial center in its own right. Few other cities could boast a modern subway, the banks and stock exchange were influential throughout Europe, and the artistic and political life flourished in Budapest's many concert halls and elegant coffeehouses.

In early July 1897, Louis and Tekla Spitz left Budapest for a vacation in Alt-Aussee, a resort in the Austrian Alps, and while there, the couple delighted to discover that Tekla was pregnant. By the fall, Tekla Spitz suffered morning sickness and sometimes became ill after meals. But striving to prepare for her first child, "she bought linen for diapers, things for crocheting, and material for bibs and panties; and she started working on all this indefatigably," Louis wrote later. He helped "by cutting out patterns from paper and by doing the basting for her to make the hems." Tekla's father, an eye specialist, was medical director of Stefánia Children's Hospital in City Park, and on his way home at noontime he dropped in almost daily to visit his daughter, who regularly brewed for him a cup of strong beef broth.[11]

By the beginning of February, Tekla's nausea had stopped, and she was eager for the birth, although she later admitted to knowing little about how a birth might actually occur. Following her mother's advice and assisted by the midwife who had delivered all the Vidor children, Tekla strained to concentrate on avoiding what was "ugly" about childbirth and focused on "what was beautiful and good." Once his wife's labor began, Louis spent an anxious and excited night in an adjacent room, and the next day, February 11, 1898, Tekla was "glowing with happiness" as she proudly presented him with their first son, Leo.[12]

Giving birth may have been a startling ordeal, but handling infants was nothing new to Tekla, as she had tended her own younger siblings and her sister Margaret's children. She nursed Leo easily and followed the advice of Dr. Karman, the Dr. Spock of his day, from a book he had presented to her. From the beginning, she had definite ideas about Leo's upbringing. Tekla would "rebel" against anything she thought was "over-done," and she abhorred the fuss made by bourgeois families over the firstborn. As parades of relatives called with gifts, she broke tradition, refusing to dress Leo in white clothes or to display him in a new white carriage. Her strong views about rearing children also created early (and lasting) disagreements with Louis: Whereas she was "more decisive" because of her experience handling other babies, he was "the more tender" and approached them "gingerly"; she wanted to "harden" and discipline her children, while he wanted to "pamper" them; she adopted a "puritan" approach to living, with modest and simple dress, while he was more bourgeois, spending money freely on rich foods, fancy outfits, and expensive vacations.[13] The city girl was rejecting her family's luxurious traditions. The country boy was making up for lost time.

That summer, Louis had to travel to Újvidék, in southern Hungary (now Novi Sad in Serbia), where he was working on a new railway line. Louis rented a two-room apartment, hired a nursemaid, and moved Tekla and Leo there along with her youngest sister, Juli. Louis also rented a horse-drawn carriage to drive the few blocks to the large restaurant where they took most of their meals.

The excursion to Újvidék was elegant but short-lived. Juli, who slept in their second room, wanted to retire one night but found the landlord's daughter asleep in her bed. The girl refused to leave, forcing Juli to sleep on a hard sofa. Louis was outraged, concluding that his Serbian landlord was behind this contretemps. He swore so obscenely that the landlord lodged a complaint in court. Upset by the whole incident, Tekla left by train for Budapest the next day, taking Leo and Juli with her. For Louis, the Újvidék job ended in September—his first success as a contractor, his first failure as a husband. After this, Louis's quick temper and stern temperament caused Tekla to become more docile and introspective.

The young family had another unsuccessful summer in 1900, when they moved across the Danube to Szabadsag-hegy, one of the many hills in Buda. Seeking a restful retreat for Tekla, who was then pregnant for a second time, Louis rented a small villa just below the upper terminal of a cog railway. But the villa was shady and damp, and Tekla's right arm soon became swollen and inflamed. After she developed a high fever, Louis returned her to their apartment in Pest and confined her to bed.

To compound their troubles, Leo, who was two and one-half, came down with whooping cough and had to stay with Grandmother Vidor in the elegant and imposing Fonciere Building. The doctor advised that Leo spend several hours a day outdoors, so each morning, Louis picked him up and rode the ferry to the large park on Margaret Island in the middle of the Danube, between Buda and Pest. During their hours together Louis Spitz noticed his son's unusual "thirst for information," for Leo questioned not only his father but also his fellow passengers on the boat.

Each day's outing was a joy, Louis recalled later, but back at the Vidor apartment, Leo "wanted to stay with me in the worst way." There was no elevator at the Fonciere, and Leo refused to climb the stairs. This conflict ended each day when Leo's aunt Juli took him in her arms and carried him upstairs—the only way to quiet his screams, which echoed through the marble halls. Concerned over Leo's selfish behavior, his grandparents tried to teach him to say "please." Leo refused. "It was beneath my dignity," he recalled later, and the old couple had many arguments with him over this.[14] Leo remained away from home until after the birth of his brother, Bela, on September 6, 1900.

Less than a month after that, on October 4, Louis Spitz changed his surname to Szilard, yielding to growing government pressure for the Magyarization of foreign-sounding names. In Hungarian, Szilard (pronounced 'Sil·;ard) means "solid." Two of Louis's brothers changed their names at this time—to Arnold Salgo and Adolf Szego—but the fourth remained Max Spitz.[15]

Summers were remembered by the Szilards for misfortunes, and in 1901 another mishap occurred. As in the previous two years, the trouble came about despite the most pleasant intentions. The family—Louis and Tekla with three-year-old Leo and nine-month-old Bela—moved by train to Velden, on the west end of the Wörther See, a large lake in Carinthia, near Villach in southern Austria. The Vidors and their daughter Juli came along, as did Louis's niece, Aranka Spitz. Work prevented Louis from staying at the resort for the whole summer, but as Tekla was then pregnant with her third child, the Vidors remained nearby. They lodged at the Grand Hotel, while the Szilards rented a small house with a lawn and garden. There Leo could play while Bela was watched in his pram by the family nursemaid, Zsuzsa, a peasant woman nicknamed Auntie Susie.

One sunny day, a leather manufacturer's wife and her two sons— students at Budapest's Technical University—invited the Szilards for a boat ride to visit a coffeehouse across the lake. Tekla preferred her parents' company, but Louis fancied the offer and persuaded her to come along. With the two students at the oars and their mother tending the rudder, Tekla and Louis (with Leo on his lap) sat in the middle seats with Aranka. Tekla, who had never learned to swim, was nervous throughout the crossing and sighed with relief when she climbed from the boat at the far shore.

After a pleasant two-hour stay, the party returned to the boat. First Tekla climbed in, cautiously. Then Louis followed, with Leo cradled in his arms. But Aranka leapt in from the dock, tilted and rocked the boat, and spilled Tekla over the gunwales and into deep water. Leo began screaming as his father gripped him with his left hand. With his right hand, Louis grabbed the horsehair bustle at the back of Tekla's dress and held her above the cold water until the frightened students fetched a rope and hauled her to shore. Wrapped in borrowed blankets, Tekla was carefully set in the boat again, and the party crossed the lake with no further scares. At Velden, the Szilards' cottage was only fifty yards from the landing, and Tekla was quickly put to bed and warmed with hot bricks wrapped in linen. By the next morning she was cheerful and healthy and never mentioned the accident again. Nor did Leo, although he remained frightened of water for the rest of his life.[16]

CHAPTER 2

View from the Villa

 1901–1912

In the fall of 1901, Tekla Szilard's older brother, Emil Vidor, received his first commission as an architect—a contract from his parents to design and build an elegant residence. In it the entire family would live: the elder Vidors, their three daughters with their growing families, and Emil. The Vidors encouraged Emil to design the most beautiful house he could, with no regard to the cost, and committed part of Mrs. Vidor's inheritance to the task. The Vidor Villa, as it came to be known, was intended to display Emil's talents, with his office on the ground floor so that prospective clients might admire his work. "Give your artistic ambition free rein," Dr. Vidor instructed his son, and he did.[1]

There was a growing need for larger quarters, especially after the Szilards' third child, Rozsi (Rose), was born, on December 15, 1901. Rose's arrival sharpened the debate that Louis and Tekla had about child rearing. "He was anxious to give them all the comforts and advantages gained by his hard work, and that was his delight," Tekla wrote years afterward. "Spoiling the children was not my idea of being good to them. I did my best to give them happy experiences of a different nature." She wanted Leo, Bela, and Rose to be "plain, unassuming people" and to "avoid giving them the feeling that they were special and having them become conceited about their abilities and accomplishments." The couple's disputes rarely erupted into angry arguments, but over time battle lines were drawn. Because Louis traveled frequently to construction sites throughout Hungary, he often saw his children only on weekends. Then he lavished sweets and rich foods on them, often during excursions to the city's many parks. But come Monday, Tekla "took firm hold of the reins" to guide her children's development throughout the week. In this way, the three Szilard children were alternately spoiled and scolded, swayed by both their father's bourgeois tastes and their mother's sterner spirituality.

Their new home was as bourgeois as could be. True, Emil's Vidor Villa had to serve the practical needs of several families. But it did so exuberantly. Set on a large plot of 600 klafters (about half an acre) along the tree-lined Fasor, Emil's creation soared five stories in a mock-rustic, art nouveau style. Outside, curved stucco walls were fitted with pointed and rounded stained-glass windows; wooden turrets and porches seemed to pop out at unexpected places, braced by rough-hewn beams and hand-carved tree branches. Inside, the three floors of apartments were linked by an airy, curved stairway that was decorated with an intricate wrought-iron rail. The house was an architect&rsque;s fantasy, a child's delight.

But to Louis Szilard, a practical and very outspoken civil engineer, Emil's designs raised doubts from the start. When Louis eyed the plans and criticized using wooden beams rather than steel to support a ceiling, Emil took offense and refused to change the design. This "disagreeable discussion," Louis regretted years later, created "mutual ill feeling" between the two men that "never disappeared entirely." Even without Louis's criticisms, Emil's innovations and constant changes slowed construction. These delays forced the Szilards, who had broken their apartment lease on Bajza Street, to live for three months in borrowed rooms, so Louis was grateful when his family did occupy the villa in December 1902, because in addition to gaining a large apartment, he had a new office for himself near Emil's. (The villa, at 33 Fasor, now houses the Bela Bartok Collegium of Music, offering spacious quarters for living, practice, and recitals. Emil's name can still be seen, carved in a ground-floor archway that once led to his office, and Leo Szilard is commemorated on a plaque.)[2]

For Leo and Bela, the move to the Vidor Villa meant a new world to explore and conquer. From the third-floor bedroom they shared, which overlooked the rear garden, the two boys could wander along back stairs or charge the spiraling front hall. They could hide in the angular attic, romp through the narrow servants' corridors, or hang from the ornate porches and turrets. Dangers lurked throughout the villa, and Leo, like his father, was quick to find most of them. But he also found delights and surprises in the huge house and a sense of security he would never know again.

When the Vidors' extended family moved into the new home, Dr. Vidor, Leo's maternal grandfather, was soon the center of attention. A well-known ophthalmologist of about sixty-five, he resigned his post at the Stefánia Children's Hospital and donated his large medical library to the Medical Association, which he continued to serve as president. Life for the old man was a succession of delights: In the morning he visited his daughters and played with his grandchildren; in the afternoon he played at the Chess Club, of which he was president. Most evenings, he gathered his daughters, sons-in-law, and older grandchildren in his apartment and told them stories about his life. Leo remembered his grandfather best for the classroom episode in Debrecen, when he scrupulously reported his own truancy. But Dr. Vidor also regaled his grandchildren with tales of Hungary's 1848 revolt, when as a young man he joined the partisans led by the fiery patriot Louis Kossuth.[3]

For Leo, who was about to turn four, the move to the villa suddenly surrounded him with new playmates: his cousins the Scheibers. Quickly the children divided by age, with Imre, Otto, and Leo the "Big Ones" and Bela, Gabor, Karoly, and Rose the "Little Ones." They competed for years, usually in a friendly way, over control of the villa's rear-garden play yards. One battleground was a large, circular lawn ringed by linden trees; the other was a small hill against the back fence, with footpaths running up each side. Leo, the source of all legal arrangements, assigned the hill to the Big Ones and the playground below to the Little Ones. He further decreed that neither group could step on the turf of the other. But one day the Big Ones invaded the playground with strips of roofing slate lashed to their shoes, and when the Little Ones complained, Leo waved away their cries smugly. "We never stepped on your turf," he announced, "just on our own property—the tiles."

In another test of his legal agility, Leo directed a game in which the children sat in a circle and traded individual collections. A stickler for precision, Leo complicated the bargaining with conditional offers. When Bela wanted Leo's small pocket knife and offered in exchange his most valuable possessions—a dozen beautiful marbles—Leo replied: "If you do not give me the marbles, I will definitely not give you the knife; however, if you give me the marbles, I may give you the knife." Leo's cleverness seemed irrepressible even at play. When the Big Ones and Little Ones played *Pöce*(pronounced "Potze"), a game that involved rolling different-sized cardboard disks, they made each child's disk proportional to his or her age. But when it was Leo's turn to make the disks, he scaled them not only for diameter but also for thickness.

Leo turned his mind to more serious worries as well, often acting as the "conscience" for the Vidor Villa's children. Soon after moving in, when

Leo was almost five years old, the Szilards' nanny, Auntie Susie, took Bela and Rose to City Park in a baby carriage built for two. This large, oval wicker basket on wheels had seats at both ends for the children to face each other, which prompted Bela and Rose to take turns standing up to make menacing faces at one another. Walking along with his hand in Auntie Susie's, Leo became alarmed and warned them to stop their dangerous play. When Bela leapt up while Auntie Susie was distracted, she lost grip of the carriage, and it flipped suddenly, spilling the two young-sters to the ground. Bela and Rose crawled out from under it, laughing wildly. But Leo was by now so upset that he ran home to report the mishap to their mother.

Bela remembers that from the time he could walk Leo took him by the arm whenever they left home, even in the company of a parent or governess. Leo continued to grip Bela's sleeve, and Rose's, long after they thought it unnecessary. Leo's siblings and cousins often chided him for worrying too much, but he seemed unable to stop thinking about every possibility for any situation, especially the potential dangers. When one of the children heard that Australia was directly opposite Hungary on the other side of the earth and they all decided to dig a shaft to reach it, Leo had to ponder every contingency. At first, the experiment excited him. He began by picking the site to dig on a dividing line between the Big Ones' and Little Ones' territories so that fame and honor might befall both groups. But when the hole was deep enough to cover Rose, the smallest playmate, Leo began to fret. "If you dig much deeper," he warned, "the wall will cave in on you." Leo wanted to halt the project at once, but his playmates ignored him and only abandoned the enterprise when a thun-derstorm washed dirt into the hole.

Leo's aversion to danger—real and imagined—was matched by a dislike for manual work, and from an early age he considered it his role to think and instruct while others toiled or played by his designs. When Bela, Rose, and the Scheiber cousins enjoyed collecting discarded horseshoe nails from the pavement around a nearby stand for carriages, Leo insisted on having only the new, shiny nails that the coachmen dropped accidentally. He even expected his playmates to wash them before they entered his own "special collection." The Szilard and Scheiber children also enjoyed collecting horse chestnuts that dropped ripe from the trees lining the Fasor. Leo instructed his playmates to make "chains" of the fallen nuts but never shared in the tedious work of piercing and threading them himself. He merely took credit for the idea.

Leo's bossiness hid a serious lack of manual dexterity. He had early trouble writing legibly, and in play with his cousins and siblings he could never match their physical skills. As a draftsman, Leo sketched only two things, over and over in a clumsy style: the hand of a man smoking a cigar and a meat cleaver. A favorite pastime for all the children was building dollhouses and rooms, but whenever Leo tried to make furniture from matchsticks and bits of cloth, he dribbled the Arabic gum and crushed his lopsided creations. Instead, he laid out plans and prescribed the furniture's size. When he ventured to paint some of the inside walls with patterns he designed and claimed they represented wallpaper, his playmates repainted the walls once he had left.

Leo also refused to join his villa playmates in the wild running games they enjoyed on their daily walks to City Park. Instead, he sat on a bench with their governess and talked endlessly in German. Bela and Rose, who couldn't understand Leo's serious behavior, suspected he just wanted to be the first at the bread, butter, and the chocolate milk that the governess brought along for the children's second breakfast. His appetite aside, it seemed that Leo simply preferred sitting and talking to exercise. The only sport Leo seemed to enjoy was ice-skating, which he and Bela learned with their father at a rink in City Park. Their father decided to learn the sport at the same time and probably gave Leo the necessary example of patience and practice. Father Szilard cut an impressive figure on the ice, moving slowly but gracefully in his long fur coat and tall sealskin cap. As Leo learned to skate on his own, he glided ahead while Bela held their father's hand. The Szilard children took several tumbles on the ice, but, in their memories at least, their father always glided steadily.

During his early years in the Vidor Villa, Leo learned little about his mother's family beyond the lives of her parents. Her father, Sigmund Lustig, whom Leo knew as Dr. Vidor, was an only son born to Jewish parents in Debrecen before 1840. When he was about nineteen, Lustig moved to Budapest to enter medical school. After graduating in about 1865 he married Jeanette Davidsohn, the daughter of a wealthy Jewish couple in Budapest, and on that occasion he changed his last name to Vidor (the Hungarian equivalent of the German *lustig*, meaning "joyous" or "merry"). In time, through skill and dedication, Sigmund Vidor became one of the best-known ophthalmologists in the capital, a highly respected citizen who directed charitable organizations but devoted most of his free time to the Chess Club. This became such an addiction that on his deathbed Dr. Vidor lamented to his children that he had neglected both them and his profession because of chess and warned his daughters to keep their children from any acquaintance with the game.

Growing up in the villa, the Szilard children had almost constant contact with their grandparents. Grandpa Vidor was a mild man, respected and revered by both Leo and his mother. Leo also liked the old man's refined manners and the elegant appearance of his long, neatly combed white beard. Even so, the Szilard and Scheiber children were much closer to their maternal grandmother. They visited her apartment almost daily and many evenings sat at her feet as she darned stockings under a gas lamp pulled down to aid her vision. As she moved her needles, she told the children fabulous stories in her clear and dramatic German. (She, too, spoke only basic Hungarian.)

In later years, Grandma Vidor tried to teach Leo the piano, an instrument that intrigued him when his cousin Otto began lessons. After months of effort, Leo still failed to appreciate the importance of volume marks as he pounded out all pieces in the same loud and constant style. She stressed that "f" for *forte* meant loud, "like the voice of a lion," while "p" for *piano* meant soft, "like the voice of a mouse," Leo never learned the difference and before long dropped his lessons.

Grandma Vidor was better at teaching the children family history; she told many stories about the large circle of Davidsohns living in Budapest, and her sisters and their children were frequent visitors to the busy and friendly villa. The Szilard children were shaken when Grandpa Vidor died suddenly from a brief illness in 1907. After that, the Szilard family moved from the villa's third floor to the first, sharing a larger apartment with Grandma Vidor. But she became ill in the fall of 1909 with anemia and a heart condition and died early in 1910.

Leo's affinity for the Vidors came mainly from a close bond with his mother, nurtured by her continuing ties with her parents and her love of family traditions. Born in Budapest in 1870, the third of five children who survived infancy, Tekla Vidor in childhood and adolescence had been oversensitive, self-doubting, and frequently unhappy despite the affection of her loving parents; in later life she was extremely conscientious about family relationships and often guilt ridden about her affections— or anger—toward relatives. To the end of her life she brooded about once being called a "good-for-nothing" or "useless" child, when, in Hungarian, the same word might only mean "naughty." Indeed, she surmised that this childhood slur had driven her lifelong mission to be useful.[4] She was devoted to her three children and worried about their education. The nanny she employed, Auntie Susie, was a peasant woman with artistic talent and common sense. From her the children learned Hungarian; from governesses they learned German, French, and English.

Even keener than Tekla's concern over education was the worry that her children be honest, truthful, and considerate—qualities she admired in her father. She was, however, uneasy preaching the many virtues she hoped they would adopt and resorted, instead, to delivering moral precepts in stories whose protagonists led only sterling lives. These tales formed a continuous string of informal lessons, most often recited as she washed her three children at bedtime. Years later, the children forgot most of their mother's stories but remembered the washings vividly, as she always rubbed their naked bodies with cold water. This washing and storytelling was her special effort to be close to her children, since most upper-middle-class families in Budapest left all daily child care to household employees.

Neither the Szilard nor the Vidor families were actively religious, despite respectful attendance at Jewish weddings and funerals. Turn-of-the-century Budapest was a center of cultural ferment, a crossroads of old and new, East and West. Most members of the Jewish middle class relaxed in this setting and abandoned their religious traditions. For the Szilards, this "assimilation" was rapid and total.

Leo's paternal grandfather had been a weak believer in the Jewish faith, but he and his wife still had observed all the traditions with their sons. Around the villa, Louis and Tekla Szilard maintained an acquaintance with the social, if not the spiritual, aspects of their Judaism by sharing the conventions of her parents. As they left home, however, the three Szilard children quickly lost these social traditions, but even before that, Louis Szilard had become emphatically irreligious and resented that his sons were required to attend the synagogue once every two weeks, as part of the public high school curriculum. Leo was unaffected by this compulsory education, as his religious teacher, a young rabbi, discovered when he asked, "Do you possibly have not even a mezuzah at your door?"

"No, sir," Leo answered. "My parents do not like to show off." Leo never bothered to learn the Hebrew letters, let alone the required readings of simple Hebrew texts. To him, organized religion was just another convention from which he should be "different" and "independent."

Leo's mother and her family were also nonbelievers, although her father was a director of several Jewish organizations in Budapest. Tekla Szilard never attended Jewish services, but instead practiced what she called her "natural religion." These ideas she based loosely on the teachings of Jesus and explained to her children many times, but they seemed unable to understand just what she meant. As a result, all three Szilard children grew up with no personal religious beliefs or traditions, yet from their mother they did acquire strongly held ethical values. Leo nurtured these from childhood training, evolving an instinctive "sense of proportion" that guided his life's decisions.

Although Tekla spent much of her time with Rose, whom she nicknamed "Rozsika," it was clear to the rest of the family that she was mostly concerned with Leo, whom she called "Lajcsi." Like his mother, Leo was sensitive and was considered "different"—a child often preoccupied and fervently involved with abstract ideals, such as honesty and loyalty. He was also independent minded and defied social conventions. Yet he looked to her as a model of self-reliance. Both Louis and Tekla were active in the Masonic Reform Lodge, and after she had put the three children to bed, they enjoyed meetings and parties there.[5] The Szilards went often to the theater. And they frequently attended dinner parties among a large circle of friends. Once a week Louis Szilard spent an evening at the National Association of Engineers, a prestigious group that he had cofounded in his bachelor years. But whenever the Szilards left home at night, Leo acted unhappy, became moody, and sometimes could not fall asleep until his parents returned.

More squeamish than his brother, sister, or cousins, Leo feared not just the sight of blood but the mere threat of violence. When Rose tried to imitate the Big Ones by leaping off the high step of a fake well in the villa's front yard and instead flopped into the bushes and cut the corner of her mouth, Leo ran to help her, carried her indoors, and then fainted on the floor. Leo complained when his cousins killed earthworms in the garden and was just as squeamish about mistreating inanimate things.

"Stop! Stop!" Leo screamed as Bela set about to convert one of Rose's blond dolls into Hercules. Leo was queasy when Bela blackened the doll's hair with India ink and painted on sideburns and a menacing mustache. Leo cursed angrily when Bela and Rose poked out the doll's shiny blue eyes. But Leo's sharpest discomfort came as Bela and Rose replaced the doll's eyes with shiny black ones painted on paper. To do this they had to cut open its head, glue in the new eyes, and continue to open and close its skull for adjustments by lifting the wig.

Because physical violence so frightened and upset him, Leo spent much of his time with his siblings and cousins berating them for their enjoyable but destructive pursuits. He complained as they dug an underground hideaway in the villa's garden, covered the entrance with a trap door, and camouflaged the door with sod, fearing that an unsuspecting adult might fall in. He fumed when they toppled a tall outdoor jungle gym to snatch a birds' nest from its top rungs. He complained when they pried off the boards of the garden fence and roamed through the "wilderness" of adjacent, overgrown lots. He ordered them to stop their experiments with a cat, shouting orders they ignored as they dropped it from higher and higher steps of the villa's staircase to see whether it would still land on its paws. And while Leo joined his playmates in slapping down stag beetles as they flew around the garden, he cursed the children for trapping them under flowerpots; this, he insisted, was "unnecessary torment."

In his recollections, Szilard considered himself a "Very sensitive child and somewhat high-strung." He also wrote in later life: "I couldn't say that I had a happy childhood, but my childhood was not unhappy either." In fact, as he thought about it, Szilard concluded that his childhood continued to be a part of his adult life, especially as it impelled his scientific imagination. "As far as I can see," he wrote in 1960, "I was born a scientist. I believe that many children are born with an inquisitive mind, the mind of a scientist, and I assume that I became a scientist because in some ways I remained a child."[6]

If Szilard's childish social development troubled his parents, they were still proud of his early intellectual adventures. He spoke and read German and French by the time he began formal education at age six and studied English a few years later. His early instruction, by private tutor, gave him both the challenge and the freedom to pursue personal discoveries and questions. And, sensing his own success as a student, he spent his spare time tinkering with scientific gadgets. And reading.

One book Szilard read as a child amazed him and later in life shaped his mental and moral speculations. From it he also gained something that neither religious nor family influence had provided: a personal sense that his own life had some transcendent, global purpose. In 1860 the Hungarian poet Imre Madach completed *The Tragedy of Man,* a long and dramatic epic in iambic pentameter. The work is often compared to Goethe's *Faust,* for it concerns the Devil's temptation of Man and his redemption by faith in God. A theological spectacle in fifteen scenes, Madách's play involves a bizarre cast that includes a Choir of Angels, the Lord, the four Archangels, Lucifer, Adam, Eve, a Slave, Eros, two Demagogues, the Roman poet Catullus, a Monk, Witches, a Skeleton, the

Emperor Rudolph, Robespierre, a Showman, a Trollop, two Hucksters, a Gypsy Woman, a Mountebank, a Scientist, the Voice of the Spirit of the Earth, Luther, Plato, Michelangelo, and an Eskimo. It is the Eskimo that Szilard remembered best.

The Tragedy of Man, with its stilted poetry and its compressed array of philosophical and ethical concepts, is hard reading for adults. Yet while Szilard was only ten when he first discovered the poem, he found some of the scenes riveting and in later life considered it "apart from my mother's tales the most serious influence" on his life. "I read it much too prematurely and it had a great influence on me, perhaps just because I read it prematurely," he later wrote. "Because I read it, I grasped early in life that 'it is not necessary to succeed in order to persevere.'"

Szilard remembered vividly scenes in which Lucifer shows Adam the future of mankind. He recalled, in particular, a scene in the frozen north, with the sun slowly losing its heat and the earth gradually cooling. In the end, only Eskimos are able to survive on the frozen planet, and even they must confront the problem of too few seals to feed their population. Yet in their behavior and despite a prophecy of doom, the Eskimos show that mankind can maintain a sense of hope. It is this "narrow margin of hope" that Szilard would refer to, quote, and depend on during his own struggles to create a more livable world.[7]

While Szilard as a boy pondered the fate of mankind, he seemed less able to care about the individuals around him, and his mother worried about his becoming lonely and aloof. This aloofness was tempered somewhat by his cousins, the Scheiber boys. He especially admired Imre, the eldest, a gifted painter and musician who was given early to philosophical thoughts. But Leo's relationship with Otto Scheiber, whom he nicknamed "Molci," was more intense. The second-oldest Scheiber boy, Otto became Leo's first real and loving buddy and perhaps the most intimate and trusted friend he would ever have. The two cousins spent hours together, talking and reading to each other, but after little more than a year, Leo became restless with Otto and one day ended their steady companionship abruptly. This tie set a pattern for Szilard as an adolescent and as an adult: intense friendships that later faded quickly as he lost interest.

Louis and Tekla Szilard gave constant attention to their children, and when several childhood diseases spread through the capital city, they vowed to protect them. At the time, diseases such as whooping cough, measles, diphtheria, and scarlet fever were widespread, and since it was

25

many years before the introduction of antibiotics and specific vaccinations, a child contracting an illness faced a high risk of death. The Szilards hired a private tutor to come to the villa daily. First Leo, then Bela and Rose, received instruction at home covering the four years of elementary school (for ages six to nine). Then they took the year-end "private students examinations" at public school. This opportunity allowed Leo to study at a pace that suited his quick mind and, more importantly, gave him the freedom to pursue ideas as long and as thoroughly as he wished.

The tutor hired for the Szilard children was Mr. Abranyi, an earnest, soft-spoken retired teacher with brown hair and a beard. His three pupils quickly grew to love him. Eager to satisfy him with their achievements, Leo, Bela, and Rose studied hard and learned their lessons well. Each weekday, Abranyi would meet with the children in the dining room, assign readings and tasks, quiz them orally, then bid them good-bye. In theory, all three children expected Abranyi's attention, but Leo, acknowledged by siblings and cousins alike as "the brainiest among us," always seemed eager to jump ahead of the assignment, his mind never focusing on a topic for longer than it took him to understand it. The Scheiber cousins called Leo "number headed," and his intense cleverness at times disrupted even the patient Mr. Abranyi's formal presentations.

"A mother of four goes to buy shoes for all her children, for two korona a pair," Leo was told when studying for the year-end examination at age nine. "When it comes time to pay, she finds that she has no smaller denomination than a ten-korona bill, which she gives to the storekeeper. What happens?" Ignoring the computation of change expected of him, Leo answered: "The woman needs one more child!" Later, Szilard was handed a closed, complicated, hollow glass contraption in which a purple liquid was sealed. When the bulb was placed in his palm, the liquid boiled rapidly. The exam question: If water boils at 100 degrees centigrade, but your body temperature is about 37 degrees centigrade, why is the liquid boiling? "Either you have invented a liquid which boils at my body temperature," Szilard answered, "or you may have stolen the air out of this device, thereby reducing the air pressure and letting that purple water boil at a reduced temperature."

Szilard was just as unconventional when he did not know an answer. By the age of eight he had adopted a standard reply for teachers and adults who asked him to solve a problem above his capabilities. *"Nem tudom kikaparni a hasamból,"* he declared. "I can't scrape an answer out of my belly."[8]

While Szilard excelled at math, he also enjoyed literature. By age six he read children's stories in German, and within another two years he was fluent in the language, which all the adults in the villa spoke well. When nine, Szilard began to speak and read French, taught him by a governess; and he read French stories by his tenth birthday. But in literature Szilard discovered frustration as well as fantasy. He enjoyed hearing and reading fairy tales but became annoyed by certain details. He complained repeatedly that Hans Christian Andersen's story "The Snow Queen" was "illogical." In the tale, the Snow Queen gives Kay as a present "half the world and a pair of skates." This, Leo said, was absurd. "Since half the world must already contain at least a million pair of skates, it is ridiculous to add one more pair." Leo annoyed the other children by making this point every time he heard the story.

On his own, Szilard read eagerly from age four. After the fairy tales, he soon switched to adventures. An early delight was the work of Karl May, a German writer who created utterly fantastic tales—unrealistic accounts of the Far East, the American West, and other distant places—without ever leaving his German hometown. Then Szilard relished dramatic accounts of the troubles German engineers had in constructing large projects. (As a civil engineer, his father expanded Leo's readings with exciting tales of his own.) Leo's early developed technical interests led him soon to read avidly *Der Gute Kamerad,* a monthly from Germany written for adolescents. He usually ignored the mildly patriotic stories and turned right to the instructions and drawings for do-it-yourself household gadgets. With his clumsy hands, Leo seldom actually built these items; that was Bela's duty. More often Leo treated the plans as topics for intriguing daydreams. Like Karl May, Leo let one thought run to another, creating landscapes of gadgets in his busy mind.

Tinkering became an important part of Szilard's education, and as he and Bela matured, they devised and actually built some fanciful contraptions. From an uncle who traveled regularly to London, the Szilard brothers collected hundreds of pieces for a Meccano Set, the English forerunner of the Erector Set. Leo began by proposing a grand scheme, which he expected Bela to build, and when they had too few pieces for the design, Leo posed alternatives but all the while refused to raise a finger to help with the construction.

Leo supervised work on a camera that Bela made with the lens from their mother's opera glasses. They found an old oilcloth table cover and snipped it into the shape of a bellows, which allowed the lens to focus

27

by moving back and forth in front of a glass negative plate. This camera had no shutter but operated when the boys uncovered the lens for a few seconds. Their favorite subject was the children's English teacher, Miss Cochrane, who posed several times while demonstrating to Rose how to form her lips to pronounce "what," "which," and "while." Leo also tinkered with an electric teakettle that he and Bela designed for their mother's aunt Ilka. After days of wiring, cutting, and soldering, the boys produced a handsome silver pot atop a heating element. Their contraption resembled a skinny samovar and worked for more than twenty years.[9]

In 1912, at age fifty-one, Louis Szilard retired early from his business as a building contractor in order to devote all his time to his family and particularly to his sons' education. He expected Leo and Bela to become engineers but in this conviction seldom pressed the point. He checked their daily homework and practiced algebra and other mathematics with all three children, trying to give Rose as much help as he gave to the boys. This was made easier because Leo seldom needed his work corrected.

Louis and Tekla Szilard seldom disagreed, but when they did, it was usually over how to rear their children. In fact, the children so dominated their conversations that Tekla became impatient to read books. But Louis thought this would distract her from Leo, Bela, and Rose. Only years later did she complain, in her writings, that Louis "once he got home was totally absorbed in the worship of his children. He never talked to me about anything else except their care and feeding,"[10]

Especially their feeding. Louis Szilard savored rich foods, and Tekla, despite her plainer palate, easily acquiesced to serve lavish dinners at the villa for their friends from the Masonic lodge. With Tekla helping their cook, the Szilards served meals that often began with soup in elegant china cups. Favorite appetizers included jellied fatted goose liver, jellied herring fillets, or hot beef tongue. Three or four main courses usually followed, and the guests frequently sampled them all. Main dishes included a freshwater fish; a fish, such as carp or sturgeon, with tartar sauce; a roast ham, sliced dramatically from one piece by Louis and served with gobs of fresh horseradish sauce; and sometimes even a small piglet roasted on a spit in the family's huge coal- and wood-fired range. Vegetables included cauliflower polonais with butter and bread crumbs; peas; white asparagus tips, also served polonais; potatoes, either home fried in goose fat or boiled and served with butter on the side; and *pommes frites* fried in goose fat.

Desserts included a choice of rich home-baked cake; *palacsinta,* a very thin crepe fried in pork fat and filled with chocolate sauce, lemon and granulated sugar, nuts, or various preserves; assorted cheeses and breads; small, hard rolls; and a special pumpernickel just for cheeses. The meal ended with café au lait and strong Turkish coffee.

Louis Szilard lavished his attention on these feasts, and Tekla remembered, with some regret, how he applied the same "supreme importance" to his children's diets. "Our children were blessed with moderate appetites, and what with the emphasis their father put on the offering of food and the constant persuasion to eat, eat, the children began to conceive of eating as a big task," she wrote. "The poorer the appetite of the children, the more their daddy insisted that I give them delicacies to eat and the more he tried to arouse my interest in such earthly delights." This tension took a dramatic turn once, when Louis said:

"It's a lovely evening, Tekla. Why don't you walk over to Andrássy Avenue," a broad shopping boulevard. "I saw some splendid geese there at Brief's. Have one cut open, and if it has a really fine liver, you may pay more than the usual price." As Tekla and the cook stood before Brief's window, Tekla thought the force-fed geese were too expensive but never stepped inside because just next door she spotted—in a used-book shop's window—luxury editions of works by Goethe, Schiller, Heine, and Boerne "at a ridiculously low price, the price of a fatted goose." When Tekla returned home, Louis was amazed at the size of the package with the foie gras, his own favorite delicacy. But when Tekla untied the string and he saw the books, he had to laugh at his grand expectations. From then on, however, Louis did all the shopping for fowl.[11]

Leo grew to love both Goethe *and* goose liver, but of the two, he clearly preferred the goose liver. He also loved chocolate and all other confections, developing as he matured an unmanageable sweet tooth. His love of sweets once led to a scene when the Szilards and Scheibers joined relatives for supper at an aunt's summer cottage on Margaret Island. On a long table on the front porch were spread ham, sliced salami, black bread, butter, and a lettuce salad. Leo and Bela sat at the foot of the table, where they could snitch unlimited amounts of their favorite delicacies unnoticed. On that evening, Leo ignored the rich meats and reached instead for a large lettuce leaf. Stealthily, he maneuvered a tiny sugar bowl to his elbow, spooned the white powder into the leaf, and carefully folded the lettuce around it, rolling the handful like a fat green cigar. Expectantly, he bit the crunchy delight, then coughed, choked, and sputtered, alerting all the relatives to his secret doings. That little bowl wasn't full of sugar, he discovered, but salt.

Besides sweets, Leo craved most other rich foods and delicacies. He liked the large, fat livers of Hungarian geese that were force-fed with sweet corn kernels. He enjoyed heavily smoked "English bacon," which he ate cold in thick slices as a breakfast treat. Some Sundays, in pleasant weather, Leo's father passed up breakfast at home and went, instead, to Weingruber's (later Gundel's), a fancy restaurant in City Park with a terrace. There Louis Szilard always feasted on coffee, croissants with butter, and two boiled eggs. Whenever Leo came along, his father took care to open the egg and let him dip some of the croissant into the yolk.

Leo also enjoyed eating "home bread," which was prepared by the family's cook at the villa but taken to a nearby bakery to be baked overnight. Usually the maid carried the two large, unbaked loaves in baskets. Leo was the only youngster in the villa permitted to go along, because while not the oldest he was the most responsible and the maid had no hand free for a smaller child. He went not to be helpful with the dough but simply to savor the marvelous bakery smells. And to nibble. Even before the unbaked bread was placed in the baskets, Leo twisted off a piece of the dough from the bottom of the loaf, where it wouldn't show. He flattened the dough into a pancake, then had the cook salt and fry it in goose fat. This delight, called *langos*, Leo ate sizzling hot as it came from the pan.

The rich appetite that Szilard developed as a child was at times laughable, at times perverse, for it marked him as a person who could be fiercely rational in most situations but also ruled by impulsive cravings. Stronger still, and ultimately more dominant in his life, was Szilard's voracious appetite for fresh ideas, and he became impulsive and impatient for these as well—both his own ideas and those of others. He found ideas every bit as tempting and irresistible as sweets. They were intoxicating. Satisfying. And, unlike the food he so often craved, ideas always seemed delicious. To enjoy ideas any time of the day or night, Szilard had discovered by an early age, all he had to do was sit and think.

CHAPTER 3

Schoolboy, Soldier, and Socialist

 1913–1919

Even before his adolescence, Leo Szilard began leading a life of the mind: conversing more with adults than with other children, reading deeply and widely, and, on his own, thinking about questions both philosophical and practical. By his early teens, the playful tinkering and inventing he enjoyed with brother Bela had evolved from curiosity to purpose, from creating things for fun to creating things that worked to make life easier. Electricity was just replacing gas for lighting in Budapest's middle-class homes, and the novelty of this power source intrigued the Szilard boys. Leo bought a thick gray handbook on *Theory and Practical Applications of Electricity*[1] and read it eagerly, comprehending every point immediately. But he only showed Bela the simpler diagrams and explanations and insisted on directing all their experiments.

Their first project was an electric bell powered by a Leclanché cell battery they built themselves.[2] When the Vidor Villa's electrical supply was changed from direct to alternating current, the boys wired four empty Vin Bravais medicine bottles together to make a rectifier that restored the DC power source they needed for their experiments. Then they wired electrodes in a glass jar to make their own hydrogen and oxygen production plant. But once hydrogen and oxygen were bubbling up through the water, the boys realized they had no idea how to capture the gases separately and dropped the project. "That's just as well," Bela recalled, "because Leo's next step was to explode the gases with a match to enjoy a 'big bang.'" Leo also enjoyed watching Bela make long sparks by cranking a static-electricity generator called a Wimshurst's machine. But what Leo really liked about this, Bela concluded, was inhaling the pungent ozone smell that the sparks produced.

A more peaceful project was the Szilard boys' two-way crystal radio telegraph, intended for Morse code messages between the ends of the family's huge apartment. Leo proposed the device to help Rose when she was sick in bed, but once it was built and tested, the radio telegraph sat unused because Leo refused to learn the Morse code. His own code had one buzz for "yes" and two for "no," but this required a person to shout a question through the rooms, or deliver it by messenger, before the answer could be sent. More practical was the boys' wind-up alarm clock powered by an electromagnet. At six each morning and evening, the villa's calm was broken as sparks leapt from the clock's hands and bits of metal clanked onto the magnet.

In 1908, the fall after he turned ten years old, Leo ended his home tutoring and entered an eight-year high school. In Hungary at the time, high schools offered two choices: Both taught basic skills in language and mathematics, but a Gymnasium also covered classical subjects, while a "real" (practical) school taught science and technology. Under pressure from his father, Leo planned to become an engineer. The Szilard boys walked the five blocks from their villa to the Sixth District Realschool at Rippl-Rónai and Szondi streets by crossing busy Andrássy Avenue. The school building was new when the Szilard boys first attended and was noted for having the best scientific equipment in Hungary.

From the start, Leo was exalted in class for his studious ways. The year he entered school he scored perfectly in his report card, earning a "1" in every subject. This success led Leo to act aloof, and he never made close friends, although as time passed, he helped classmates with their math and science homework, and for that and his gentle manner he was widely admired.

At times, his teachers were unsure about his intellectual skills, and one teacher, when pressed by Leo's parents to describe their son's abilities in school, could only say that he was "unique." Although Leo usually earned a 1 in every subject, there were three years—at ages twelve, sixteen, and seventeen—when his perfect scores slipped. He earned a 2 in calligraphy, freehand drawing, gymnastics, and religion; and, once, a 3 in calligraphy.[3] Leo's freehand drawings were so sloppy that the instructor nicknamed him

"Salami" instead of Szilard. Leo's writing and sketches were impulsive, and his mechanical drawing assignments often required redoing by Bela to pass. Despite his clumsiness, Leo's gym teachers gave him high grades in order to boost his average.

One of the few sports Leo enjoyed—at least for a while—was fencing, a discipline that tied his generation to the fading traditions of Hungarian nobility and was considered part of any middle-class education. Leo took lessons at a studio across Andrássy Avenue from Budapest's opera house. Every time Leo arrived, the fencing master teased him about not bringing a gold watch as a present, and during classes he poked Leo in the stomach with a foil just for fun. Leo mastered the sport's movements quickly and seemed to enjoy outscoring Bela in matches with the teacher. But one day, after the master urged his pupils to be much more aggressive, Leo became upset, then stubborn. Driven by his intense fear of physical violence, Leo listened, thought a moment, and reacted abruptly. He threw his foil and mask into a corner, bowed stiffly to the teacher, and walked from the room without saying a word.

Detesting violence, Szilard even avoided strenuous exercise whenever he could. He never learned to ride a bicycle or to swim, and in school he shunned most sports. He agreed to sprint in gym class but refused to join the long-distance run, the high jump, the long jump, and exercises on the bar and horse. Instead, he sat by the sidelines and read a book. As he would later discover, however, there was more to sport than exercise. "One of the favorite sports of the class . . . was playing soccer," Szilard recalled later.

> I was not a good soccer player, but because I was liked there was always a rivalry between the two teams: On whose side would I be? I was sort of a mascot. They discovered early that I was, from an objective point of view, no asset to the team, and it didn't take them long to discover that I could do least damage by being the goalkeeper. So up to the age of 15, when I finally refused, I played every soccer game of the class on one or the other side, very often on the losing side. In thinking back, I have a feeling of gratitude for the affection which went so far that my classmates did not mind occasionally losing a game for the sake of having me on their team.[4]

Off the field, Szilard was more remote from his classmates, seldom paid attention to their adolescent musings or revealed his own emotions, and only talked easily when discussing their studies.[5] Some social graces were expected of middle-class boys in the Szilard family's circle, so when Leo was about eleven years old, he and his friends were sent to dancing school. Leo's movements on the dance floor were almost graceful, Bela recalled, but appeared awkward; he seemed propelled by duty rather than enthusiasm. Leo was polite but aloof to the young girls, yet revealed no enjoyment when holding them in his arms.

Leo seemed fond of girls his own age, or at least he liked the idea of being around them. They found him handsome and lively but also shy and inhibited, sensing that his mind and his energies were focused elsewhere. As emotions of friendship and romance stirred in his heart, Leo masked them with a forced but genteel formality. His humor was sly and curt, inspired by discomfort over what to talk about and often edged with self-defensive mockery.[6]

A favorite place for the Szilard and Scheiber cousins to meet and to play croquet and other lawn games was the Eppinger family's garden just across the Fasor.[7] Alice Eppinger was a close friend of Rose Szilard's, and this tie brought Leo and Alice together. Gradually, Leo befriended Alice, enjoyed her company, and began to relax when around her. More gradually, he found a way to be affectionate. Alice was also a brilliant student, especially at math, and as the months passed, she became Leo's first female friend. His first love. And, for years, a loving friend.

As his sister's closest companion, however, Alice evolved a complex relationship with Leo, one that took years to play out. When Leo and Alice were together, she sometimes eyed other young women jealously. Leo only laughed at this, but her fears seemed to come true once he began to notice Alice's friend, Leona (Lola) Steiner. "You should marry Lola," Alice insisted, trying to provoke Leo into a choice. But he only laughed and changed the subject. When pressed to show his true feelings, Leo hesitated, then told Alice that his emotions were ruled by his mind, in a controlled state of "active passivity."[8]

Moving in the social circles that included his large family and his parents' many associates, Leo had occasion to meet several young women as he grew to manhood. But aside from Alice, whom he viewed in a fondly familiar way, and Lola, who intrigued him for many years, Leo suffered his most intense—but distant—passion for a third young woman: Mizzi Gebhardt Freund. An engaging beauty with sparkling eyes, she put his "active passivity" to its ultimate test. She was clever and coquettish. She was humorous. And she was married—to Emil Vidor's brother-in-law, the prosperous brewer Emil Freund. Mizzi Freund was adept at social conversation but was also probing with her questions, and Leo found her utterly fascinating. Although he rarely saw her, when they did meet, it was usually in just the social situations that made Leo uneasy: large family dinners or parties with relatives.

His brooding romanticism for Mizzi amused Bela and Rose, who staged a humorous photograph to mock Leo's formal self-control. Arranged when Leo was about seventeen, the picture was staged and taken by Bela in a warehouse at the Freund brewery. Its theme was "Cooling the Passions," with a bucket and watering can on hand to quench both Leo's love and Mizzi's flirtatiousness. "We wanted to dress Leo in sackcloth, but he refused," Bela recalled. "He was not amused by this at all. He was furious."

In the photograph, Mizzi, the focus of attention, sits in the center, under an A-framed ladder. Her hands are folded demurely on her lap, but her legs cross in a jaunty pose. She smiles with radiant delight, no doubt amused by this odd grouping and perhaps aware of Bela's scheme. To the right, leaning forward on a cane and sporting a bowler hat, stands Mizzi's husband, Emil Freund, looking stern and prosperous.[9] To the left stands Leo, his feet shifting forward, his left arm almost touching Mizzi's elbow. Leo wears a heavy double-breasted topcoat, a stiff-collared shirt, and a formal suit. His bowler hat is perched atop the ladder, and by his right foot stands a watering can. But his expression—as he seems to stare into the camera, and through it—reveals a soul straining to remain detached from a scene while knowing how painfully he is stuck within it.

Behind this personal triangle hangs a painted theatrical backdrop showing flower-filled urns and distant ruins. From beneath the ladder peers Leo's older cousin Otto Scheiber, winking through a monocle and sporting the Austro-Hungarian army uniform he had just received as a conscript in the world war. On the left side of the ladder, near the top, stands Rose Szilard, almost grinning as she balances a bucket above Leo's shoulder. On the right side stands Emil Vidor's wife, Rezsin (Regine), Emil Freund's sister, in a mock-serious pose.

Leo was trapped in this humiliating situation and tortured by his stifled love for Mizzi, but from most other emotional ordeals he was able to escape into the manifold mysteries of science, which he pursued with fresh intensity as he matured. Once a high school course piqued his interest in physics, Leo's father bought him scientific equipment for birthday presents. And although Leo understood most theories behind class demonstrations well before the instructor could explain them, he never lost his fascination with the subject, remained spellbound by the experiments, and often tried to repeat them at home. A laboratory "demonstrator" for the physics and chemistry teachers, Leo escaped the usual resentments against students who earn top grades because, he surmised, his academic success came without really trying.

Humor afforded Leo another escape from emotional pressures, and he embraced and nurtured an ironic wit for the rest of his life. He saw comical forces in everything, which meant that his own pointed and sardonic jokes were frequently misunderstood. For enjoyment he read and reread *Max und Moritz* and other works by Wilhelm Busch, a droll and caustic turn-of-the-century German humorist and cartoonist.

For more serious entertainment, Leo read German novels and the recollections of Van Eyck, a celebrated engineer. One favorite, *Hinter Pflug und Schraubstock* (Behind the Plow and Vise), described celebrated engineering problems and misfortunes, including hopeless efforts to substitute tilling by plow and oxen for hand-hoe cultivation in the Egyptian cotton fields and a huge bridge's collapse at the Firth of Tay in Scotland. Still later, Leo devoured utopian books about new kinds of worlds and sat for hours devising better social and political organizations of his own. H. G. Wells excited him with both science fiction and political inventiveness, especially in *The Time Machine*. Leo also enjoyed Edward Bellamy's futuristic speculation in *Looking Backward*, a technique he would employ years later in *The Voice of the Dolphins*. He liked reading histories of the French Revolution, Plato's *Republic*, books by economist Henry George, Arthur Conan Doyle's mysteries, and George Bernard Shaw's plays and political writings. And for most of his life Leo was intrigued by the Peloponnesian War and mentioned often its lessons for modern times.

Beginning in adolescence, Leo Szilard was also preoccupied with what he boldly called "saving the world." As a boy of ten or twelve, the world to be saved contained only his family, his younger siblings, and his cousins as they lived and played together in the Vidor Villa and its gardens. But by high school, a growing sense of cynicism about politics, especially as played out by the empires and monarchies that clashed in the world war, led Szilard to worry about the fate of kingdoms and nations and the pivotal alliances they forged. In June 1914, Archduke Franz Ferdinand, heir to the Austrian throne, was assassinated at Sarajevo, Bosnia. Tensions and misperceptions compounded the crime, and by July 28, Austria-Hungary had declared war on Serbia, beginning World War I.

"When war was declared we were on vacation in Velden [on the Wörther See in Austria]," Louis Szilard remembered.

> Leo brought us the news the very moment it became known, pale with excitement, running to us from the post office, where he had heard about it. Everybody wanted to get home in a hurry. We managed to get on the train to Vienna even though it was already overcrowded. In Vienna, the demand for room on the trains was so great that we could get to Budapest only by boat. We arrived there on the morning of the second day, and back in our apartment, we were happy to have the trip behind us. I hastened to my bank in order to have all my securities sold on that very day and so to avoid even greater losses; 25 percent of my assets were already lost by then.[10]

Leo remembered that wartime journey home to Budapest for another reason. "I was certainly remarkably free of emotional involvement during the First World War," he recalled in 1960. On the ride to Vienna,

> more and more troop trains pulled alongside the train or passed us. Most of the times, the soldiers in all these trains were drunk. Some of the fellow passengers, looking out of a window and seeing the troop trains pass by, made a remark to my parents that it was heartening to see all this enthusiasm; and I remember my comment, which was that I could not see much enthusiasm but I could see much drunkenness. I was immediately advised by my parents that this was a tactless remark, which I am afraid had only the effect that I made up my mind then and there that if I had to choose between being tactless and being untruthful, I would prefer to be tactless. Thus my addiction to the truth was victorious over whatever inclination I might have had to be tactful.[11]

Back in Budapest, Leo was considered "almost a traitor" when he chided his older relatives with frank opinions about the inferior military strength of the allied Austro-Hungarian and German Central Powers. And at school that fall, while Leo was as confused as his classmates about what the war meant, he still knew with certainty "how the war ought to end." The Central Powers should be defeated, he declared, but so should their enemy, Russia.

> In retrospect I find it difficult to understand how at the age of sixteen, and without any direct knowledge of countries other than Hungary, I was able to make this statement. Somehow I felt that Germany and the Austro-Hungarian

Empire were weaker political structures than both France and England. At the same time I felt that Russia was a weaker political structure than was the German Empire.

He also relied on intuition about right and wrong and later wrote that it seemed unlikely that "the two nations located in the center of Europe [Austria-Hungary and Germany] should be invariably right and practically all the other nations should be invariably wrong," in ascribing blame for the war's beginning.[12]

The war became less theoretical for Leo on January 22, 1916, when an express telegram arrived, assigning him to the Fifth Fortress Regiment.[13] Szilard may have feared military service but outwardly struck a cavalier pose. "Austria-Hungary will lose this war soon," he told Bela and his friends, "so there was no danger of my being in uniform."

A few of Leo's classmates resented his rational, unpatriotic candor, but most of them admired him for both his knowledge and his direct, unconventional thinking. At a farewell party at the Vidor Villa after high school graduation, those closest to Leo emptied several bottles of champagne and fell into good-natured kidding. They hailed him with a ditty:

> The trouble with you dear Leo, is this:
> Your ears are long and hairy like lynx's
> You dont ever ask questions but give us the answers.
> And those are always right. Damn you and thank you.

Physics, chemistry, and mathematics had become Szilard's favorite subjects in the last years of high school; but when he graduated with highest honors, in June 1916, his choices seemed dismal.[14] "There was no career in physics in Hungary," he wrote later.

If you studied physics, all that you could become was a high school teacher of physics—not a career that had any attraction for me. Therefore I considered seriously doing the next best thing and studying chemistry. I thought that if I studied chemistry I would learn something that was useful in physics and I would have enough time to pick up whatever physics I needed as I went along. This I believe in retrospect was a wise choice. But I didn't follow it, for all those whom I consulted impressed upon me the difficulty of making a living even in chemistry, and they urged me to study engineering. I succumbed to that advice, and I cannot say that I regret it, because whatever I learned while I was studying engineering stood me in good stead later after the discovery of the fission of uranium.[15]

The Palatine Joseph Technical University (Kir. József-Muegyetem), now the Budapest Technical University, stands on the Danube embankment in Buda, just below the gray cliffs of Gellért Hill. Then, as today, the grounds resembled a Victorian college campus. Szilard rode the streetcar there each day, beginning in September 1916: from his neighborhood in the Garden District, around the broad Museum Körút, over the ornate wrought-iron Franz Josef (now Szabadság) Bridge, to the busy plaza by the neoromantic Gellért Hotel. At the university, Szilard enrolled in civil engineering, and within a month he entered a national competition sponsored by the Hungarian Association for Mathematics and Physics. Without effort he won second prize. At the time, physics was a subject he had been inclined toward even as he studied other topics.[16] Ironically, this was a good time to study physics in Hungary, Szilard noted later, because the courses were so bad that a student was forced to develop independent thinking and originality.[17]

In the Faculty of Civil Engineering for the academic year 1916–17, Szilard had no choice of courses and studied chemistry and general mechanics.[18] During his second semester, early in 1917, he studied analysis and geometry, projective geometry, drawing, and strength of materials experiments. But he was becoming impatient with some subjects and bored with the practical studies. As in high school, he enlisted Bela's help and on many Saturdays took him by streetcar to the university. There, in the drafting room for first-year engineering students, Beta executed Leo's drawings while Leo sat nearby chatting about politics with his colleagues. A favorite topic was resistance to the Austrian monarchy in Vienna.

For no apparent reason, Szilard decided not to take his exams at the end of his first semester, in January 1917, but instead transferred to the Faculty of Mechanical Engineering, where he studied harder and attended new lectures and workshops. Here he was a good but not brilliant student, earning the grade of 6, the highest, in analysis and geometry, and in chemistry. He earned 5 for both theory and practice of projective geometry, and 6 for theory and 4 for practice in drawing. Szilard also conducted strength of materials experiments in the spring of 1917 but received no grade.

Academic and military activities began to conflict for Szilard during his second school year, in the 1917–18 semesters. Although his report card shows that he studied analysis and geometry, machine elements, workshop practice, electrical measurements, and steel structures, he took no exams and received no grades. (The grades he earned in four other courses came only by taking exams after the end of the world war.) In the 1917–18 school year, Szilard again did well but not brilliantly, earning a 6 in physics with little effort. In mechanical drawing he earned 6 for theory and 5 for practice; in mechanics, 5 for theory and 6 for practice. But for practical courses Szilard's performance was poor: 4s in chemical technology and in casting and forming of metals; a 3 in electrotechnics.

The war that Germany's Kaiser Wilhelm II predicted would end by Christmas, 1914, had persisted more than three years before Szilard was drawn into it. Although notified in January 1916 that he would serve, he was allowed to continue engineering studies until the summer of 1917. On September 27, 1917, Szilard entered the Austro-Hungarian army as a one-year volunteer[19] in the Fourth Mountain-Artillery Regiment, Unit 18. Immediately he was assigned to Reserve Officer School in Budapest and so continued to visit his family; he sometimes even sat in on classes at the Technical University.[20]

Szilard was hardly a model officer candidate, for he still assumed that the conflict was about to end, yet he managed to impress his instructors at the Reserve Officer School. They liked the way he grasped scientific and technical problems, and they remembered him as the only student who could explain how the telephone worked. By his wits Szilard finished third in his class.[21] Later, Szilard would make light of this success, but Bela and his parents were surprised at how easily he took to barracks life and to military studies. Every few Sundays, Szilard arranged to come home to the Vidor Villa for dinner, and around the table he entertained the family with tales about the camaraderie with classmates and the hilarious stupidity of some senior officers. He quickly made friends with other trainees and even joined in their singing and drinking. He also became visibly stronger, impressing Alice Eppinger and her friends with his handsome appearance in uniform.[22]

In November 1917 the Bolsheviks seized power in Russia, and a month later, an armistice was declared on the eastern front between Russia and the Central Powers. War in the east was over, and in the west defeat seemed more likely when the United States declared war against Austria that December. Szilard's unit remained in Budapest all that winter and the following spring, allowing him to mix military studies with visits to the Technical University. Barracks life was informal. A photograph taken there reveals fourteen cadets grouped casually around an artillery piece. Some stand erect as they stare at the camera, but one man lies back, his head pillowed on the barrel. Another wraps his leather boot around his saber. Szilard stands to the side, leaning his elbow on the cannon's wheel and his head in his hand, acting nonchalantly bored by the occasion. His belt buckle is tugged to the side, and his boots need a shine.

Military action finally came in May 1918, when Szilard boarded a troop train in Budapest and found himself at a camp of the Fourth Artillery, near Kramsach, high in the Austrian Tyrol, near Innsbruck.[23] Assigned to study explosives as a *Feuerwerkenkadetsaspirant* (ordnance cadet) and wearing the rank of sergeant, Szilard managed to live a tranquil life under a captain who had few fixed ideas about instruction. "One of these ideas was that it was not becoming to an officer to walk along the street with his gloves in his hands," Szilard recalled later.

> He should either wear no gloves or he should put the gloves on. He was also concerned about our being properly dressed, and he was indignant when he learned that we didn't bring with us our dress uniforms. From then on it became customary to ask for leave of absence to go home in order to pick up one's dress uniform. If the war had lasted long enough we would all have ended up with our dress uniforms, ready for festive occasions. Those who went home to Hungary to pick up their dress uniforms were also expected to bring some flour. There was a shortage of foodstuffs in Austria, while Hungary still had plentiful food. These leaves of absence usually amounted to about one-fifth to one-third of the school being absent on trips home.[24]

Judging from his postcards home, Szilard's life in the mountain camp was uneventful. In one card he thanked his father for sending a food package but reported that it had been opened en route. He also reported that he had thought of going into Innsbruck, "but it was raining."[25] After three months at Kramsach, Szilard was transferred about twenty miles to Kufstein, there to study the art of saber charging. As the war raged elsewhere, these trainees had little more to do than take daily hikes in the nearby Kaiser Mountains.

Trained in artillery and the saber, Szilard knew almost nothing about small firearms. But once, when serving as officer of the day, he was expected to wear his Luger in a holster. During Szilard's watch an Italian war prisoner who was working in the kitchen tried to escape. The prisoner dashed toward the woods, and Szilard, the only soldier in charge, was expected to give chase. As he ran, Szilard pulled the Luger from his holster but found it too large and heavy to hold level and take aim. He had never fired a pistol and solved the problem by deciding to run "at my own speed" so that the prisoner could slip into the woods. As officer of the day on another occasion, Szilard had to chase a deserter from his own regiment. He ran through the winding streets of Kufstein, this time with his Luger drawn and cocked, as the soldier ducked in and out of buildings and finally disappeared. But in this chase Szilard slowed down for another reason: He didn't want to overtake the man because he knew the Luger was not loaded.

One morning in late September, Szilard awoke with a painful headache and fever. Sickness always frightened Szilard, but this time he also worried about recovery: If he had to enter a hospital, he would prefer to be near home, not in Kufstein. He applied for a short home leave, hoping that if he were still sick in Budapest, he would be treated there. Slumping with pain as he lined up in a corridor at his captain's residence, Szilard had to wait an hour for him to turn up. By then Szilard was about to keel over, but somehow he stayed upright until the officer walked and talked his way down the line to him.

"Sir," Szilard said, "I request a one-week leave of absence to be at home when my brother undergoes serious surgery. I want to be there to give him, and my family, moral support."

"No objection, Cadet," the captain replied. "But right now there are just too many men on leave. Wait until a few people come back."

"My brother's operation cannot be postponed," Szilard said quickly. "Instead, I request a two-day leave to be with him the day of his operation." This shocked the captain, who seemed confused, then agreed to a leave of absence. As Szilard later noted, ". . . while it was perfectly all right to lie [about the operation], it was not customary to insist if the request was refused. However, just because he was taken aback and didn't know quite what to think, I got my leave of absence."[26]

The next train for Vienna did not leave Kufstein until about midnight, but by late afternoon Szilard could hardly stand. He enlisted a few comrades to hold him erect as they walked through town, and they propped him up until they could push him onto the train. Slumped in a seat in the corner of a compartment, Szilard slept most of the way to Vienna.

"Do you feel better now?" an officer said as dawn brightened their compartment. "You were pretty drunk last night."

"I was not drunk. I was ill," Szilard insisted, but the officer seemed unconvinced.

As the train glided through the Vienna woods outside the city, Szilard took his temperature again. To his horror, it was below normal. He spent the night in Vienna and found a doctor to check his condition. "It's not pneumonia," said the doctor, "and you're not in bad shape." Later the next day, Szilard's train arrived in Budapest. Dazed and exhausted, he made his way home. His temperature was still dropping, but now he had a persistent cough. Using family connections, Szilard arranged to be admitted to an army hospital, in Köbánya, a suburb about three miles southeast of his home.[27] Doctors there decided that he had the Spanish flu. At that time, treatment was the same for pneumonia or any other respiratory illness, and Szilard was wheeled into a ward that resembled a laundry, draped with wet sheets that hung between the beds. There he rested while the room's humidity supposedly cured him. Generous servings of his mother's home-cooked food, delivered almost daily by Bela, also aided his recovery.

Lying among the wet sheets, Szilard wrote to his captain in Kufstein, sending regrets that he could not "return to the cause." Szilard's commanding officer replied on October 10 with an affectionate "Dear Comrade" letter, commending him for his exemplary service and wishing him well in his military career. A week later, another letter arrived, this one reporting that Szilard's class had been dissolved and his unit sent to the Italian front. Late in October, the captain wrote again, advising Szilard that a Spanish flu epidemic had practically shut the school. The last letter he received came in early November, a few days before the armistice that ended the war. Szilard's regiment had come under heavy attack along the Isonzo River near Trieste. There all his comrades had disappeared.[28]

Leo was still sick on November 6 when military authorities in Budapest granted him a three-month medical furlough. The war ended five days later, on November 11, 1918, and a week after that, Szilard was demobilized from the Austro-Hungarian army.[29] At that time, he was "medically examined and found to be in good health," and within a few weeks he was back in the lecture halls at the Technical University.[30]

After living most of his life in the comfortable Vidor Villa, Szilard gained a new sense of impermanence and danger while moving about in the army. "Perhaps because such an important part of my life evolved during the First World War," he wrote years later, "I had the tendency to limit my possessions to what could be held in two suitcases. I think I would have preferred to have roots, but [when] I couldn't have roots, I wanted to have wings, and to be able to move at a moment's notice became important to me."[31]

Even Szilard's home was not safe at war's end; soldiers returning to Budapest from the front were tired, angry, and dejected and sought revenge on anyone they could identify as an officer. The only clothes that fit Szilard were his uniforms, and to avoid attacks by the marauding soldiers, he had his mother cut all the insignia from his military jacket. Next she altered the collar and sleeves to resemble the uniforms of ordinary soldiers. Then Szilard threw away his high-topped officers' boots. Still frightened, he had his mother cut down one of his father's jackets to his size. For a while Leo thought he was being followed and organized a night watch to protect the Vidor Villa, enlisting Bela and the Scheiber cousins to take turns. Their arsenal had but two pieces: Szilard's army Luger and an uncle's small, silver-plated six-shooter. The boys spent the nights in different turrets and bay windows of the huge house, fortified by a pitcher of hot coffee and a sandwich—Leo also required a large chocolate bar. At first, the night sentries read to stay awake, but they frequently nodded off despite the gunfire that echoed among the Garden District's looming mansions and through the leafless trees.

※　　※　　※

The returning soldiers' fury marked an upheaval in Hungary—and throughout Europe—that the prolonged war had only deferred. Even before the armistice, revolution was sparked in Russia, a democratic government declared its independence in what was then Czechoslovakia, a republic was proclaimed in Poland, and in Germany, Kaiser Wilhelm II abdicated, and the majority socialists founded a republic. Karl, who was emperor of Austria and king of Hungary, the last of the Hapsburg monarchs, abdicated on the day the armistice was signed, and his empire of half a century collapsed. The next day, German Austria proclaimed itself a republic.

For the Hungarians an end to the Dual Monarchy with Austria completed a liberal quest for self-determination that had been waged since the mid-nineteenth century. But Hungary was still the most aristocratic country in Europe, its rural peasant class dominated by a tiny minority of court-appointed nobles and influential capitalists. After Istvan Tisza, the country's pro-war prime minister, was murdered, the new emperor appointed Count Mihály Károlyi as premier. Károlyi declared Hungary an independent republic and vowed to reconcile his country's polarized conservative and liberal political factions.

The years of war had also fractured Hungary's economy. Inflation soared in 1918, the korona had lost much of its prewar value, and government bonds became worthless, destroying most of Louis Szilard's remaining investments. With almost no income, his family tried to maintain their only asset, the posh apartment in the Vidor Villa.

Tekla Szilard recalled years later:

> There was no anthracite [coal] for the central heating. Our teeth chattered in the oversized rooms; only in one of them was there a fire, in an ugly iron stove. Daddy no longer wielded his scissors cutting coupons; his securities had become worthless. I, too, had stopped punching holes in fine linen to make lace. I had my hands full mending old linen, darning socks and stockings. Old clothes were taken out of chests. They had to do me for years; there was hardly any money for shopping. If I decided to get some new hosiery, I did not buy it by the dozen as I used to do. I got one pair at a time from a place that advertised "Slightly defective, good buy." . . . Like a jack-of-all-trades, I made everything myself, beginning with slippers. When the toe of a small statue broke off, I replaced it with one made of bread dough. . . . What good was it to keep up painfully the old framework when the happy old life was missing from it?[32]

But Louis Szilard was too proud to take in boarders, at the time the family's only potential source of income. The Szilards tried to keep up appearances, hoarding the last of their savings to pay for the children's education. Rose, a talented painter, studied at the Academy of Fine Arts; the boys still aspired to be engineers.

Out of the army hospital by late November 1918, Leo attended a few lectures at the Technical University, all the while following political developments in the daily papers. At night, he and Bela attended the Galilei Circle, a group founded before the war by reformer Oszkár Jászi as an "association of freethinking, socially and progressively minded students" to discuss and propose social and political reforms.[33] Szilard had little to contribute but joined in the lively discussions as a critic—by constantly seeking to clarify the conversations. Once, when Julius Pickler, director of the Hungarian Statistical Office and a leader in the

Radical party, called for economic reforms, Szilard listened intently and at the end of the talk rose to speak.

"Say, sir," he asked, "what is the exact meaning of 'radical' in your party's name?"

"A radical person wants very badly that which he wants," Pickler answered.

"But what is it that the party wants?" Leo persisted.

"That is an unnecessary question," said Pickler. "Everybody knows that!"

At the university that fall Szilard attended classes in steel structures; machine elements; cutting metals and wood; hydraulic machines, compressors, and steam turbines; workshop practice; mechanical technology; and electric batteries. But repeatedly he found himself daydreaming about a new subject: economics. For several years he had read and respected the works of Henry George, the American economist noted for advocating a single-tax system. Now Leo sought out economists at the Galilei Circle, at school, and at the New York Café—a favorite meeting place of artists, academics, and politicians. Standing among students in the university's corridors or seated around the small, marble-topped tables in the rococo splendor of the New York Café, Szilard assaulted his colleagues with basic questions about economics and political reform.

Szilard bemoaned the "inexactitude of the most inexact science of all: economics." He complained about its "lack of logic," pointing out that the most illogical premise of all was the belief most economists held, namely, that by observing the past and present they would find clues to the future. "The most important facet of the future," he said repeatedly, "is that it is inherently *different* from past and present." Furthermore, "it is impossible to know *in what way* the future will be different." And, in his hours of discussions, Szilard grew angriest when someone began his projections by saying, "Everything else remaining the same . . ." "Nothing *ever* remains the same," Szilard declared. "Nothing!"

Hoping to quell student unrest, Count Károlyi's government suspended classes at the universities in November and December 1918, but this merely gave the political activists—Szilard among them—more time to debate and plan for the upheavals they saw coming. Freed from engineering studies, Szilard found in economics a new way to "save the world"—in this case, his homeland, then devastated by postwar depression and inflation. He devised tax schemes and monetary reforms, his freewheeling imagination seeking basic and novel systems. These schemes he posed to colleagues in the cafés, where he spent more and more time. Leo's academic effort in the fall semester of 1918 involved only attending a few lectures and writing an exam for a chemical technology course he had taken in 1917.

When the universities reopened in January 1919, the Szilard brothers both enrolled at the Technical University: Leo continuing the education interrupted by the war and Károlyi's decree, Bela a first-year mechanical engineering student. Leo enrolled in several courses, including: steel structures; machine elements; hydraulic machines, compressors, and steam turbines; elements of geodesy; spinning and weaving fibrous materials; cranes and elevators; workshop practice; and mechanical technology. But he apparently attended few lectures and earned only one grade that semester, a 5 for a course in the encyclopedia of construction. Clearly, he was not preoccupied by science and engineering but by Hungary's political and economic crisis.

One spring day Leo brought to Bela a handwritten manuscript and asked him to type it. As usual, the scrawl was nearly illegible, and Bela found the ideas just as difficult to follow. Leo had devised a clever tax scheme that seemed at once both logical and bizarre. After more changes and retypings, the brothers found a printer in Pest who was sympathetic to Leo's efforts. He, too, lost the arguments as he proofread the manuscript but willingly set the tract into type. He even offered some leftover paper, free.

"Do you mind if the back side of your flier is orange colored?" the printer asked Leo when he visited the small shop.

"Yes, I mind," Leo replied. "Print the text on the orange side and leave the back side white. This proposal comes rather close to a socialist solution; therefore, it is only appropriate that it appear on paper whose color is close to red."

Once the fliers were printed, the Szilards dragged the heavy bundles from the shop to the streetcar stop and rode home. Leo decided that mailing the thousands of fliers was impossibly expensive, and Bela's suggestion—handing them out at the two entrances to the university—would take too long. Leo thought a while longer, then proposed that *others,* as members of an organization, distribute the fliers. An intellectual elitist by his tutoring and social milieu, Szilard looked to socialism as the cure for Hungary's economic and political problems. But he shunned the Socialist party for its ties to the new Soviet state in Russia. Instead, he decided to found his own group, at first with only the two Szilards as members, just for the purpose of distributing the tract on taxation. With India ink and a brush from his engineering drawing kit, Bela lettered notices on four-by-six-foot posters, which the brothers tacked to the university's two main bulletin boards. They announced an organization meeting of the "Hungarian Association of Socialist Students." Leo and Bela also stood at the two main entrances, handing out leaflets about the meeting. These leaflets, also printed on orange stock, urged students to withhold their support from all political parties until the issues became clearer.

Dozens of curious students came to the meeting at the university, and the lecture hall rang with loud arguments and intense discussions. Through the din, a few radical students were elected to be officers. The Szilards stayed in the background, except when Leo handed around samples of his orange tract, then formally asked the association officers to take charge of distributing it in and outside the school. Volunteers came forward, although few who read the tract understood it, and Leo's ideas on taxation were never debated by the association. Indeed, the association never met again: Its first general meeting—scheduled for the huge glass-domed inside courtyard of the university—was canceled following Hungary's political upheavals.

Outside Hungary the politics of dissent and disillusionment took new turns. Early in January 1919, German independents and other radicals took to the streets in the "Spartacist revolt." By the nineteenth, German voters passed a measure that sent some four hundred delegates to frame a new constitution for the Reich, in the city of Weimar. In Russia, Red and White armies clashed in civil war. Szilard tried to focus his mind on his studies, but the turmoil around him in Budapest—and alarming news from other capitals—prevented his doing so. He managed to write examinations in physics in January 1919, earning a 6, and in February in electrotechnics, earning only a 3; both exams for courses taken while he was in the army.

Count Károlyi's postwar Hungarian government was faltering despite its promise of sweeping reforms. A liberal politician from an ancient noble family, Károlyi had disbanded his estate and distributed forty thousand acres to the peasants in order to set an example for the country. But no other nobles followed his bold move, and class inequity continued. Hungary's minorities also refused to cooperate with Károlyi, and he was demoted to provisional president. Then Hungary's victors in the war assigned two thirds of the country's territory to its neighbors. Faced with an invasion from Rumania and forced to choose between the conservatives and the

Communists, Károlyi reluctantly invited the Communists under Béla Kun to form a government for national defense with the Social Democrats.

Kun had first adopted Bolshevism as a prisoner of war in Russia, and after the 1917 Revolution he was sent home to Budapest as a propagandist. At thirty-two, Kun projected the image of youthful energy as he strode around the halls of the neo-Gothic Parliament building. Once his "dictatorship of the proletariat," called the Hungarian Soviet Republic, took power in March 1919, he nationalized banks, large businesses, and the nobles' estates. His Red Army ruthlessly suppressed political opponents, then invaded Slovakia to reclaim territory lost in the postwar settlement.

In Budapest, Kun applied agitprop techniques he had seen in Russia: opening propaganda centers, decorating trucks with slogans, forming people's theaters and arts workshops. Parades and demonstrations wound through the streets. Statues and buildings at intersections were draped with red banners. His administration's high point came on May Day, when a massive people's procession flowed through workers' and commercial districts for hours. Day-care centers were created for workers' children, who were taken, free of charge for the first time, to play in the vast park on Margaret Island. Kun delivered fiery speeches for hours, engaging Hungary's liberal intellectuals in his great cause.

Socialists and liberal students demonstrated for his programs before the Parliament building and marched through the city's streets waving red banners and shouting slogans. It was a springtime of anticipation—and anxiety. Even Bela and Leo were so caught up in the excitement that they borrowed a truck, draped it with banners for their Hungarian Association of Socialist Students, and drove around Budapest.

Although Szilard feared the emotional power that the marches and demonstrations generated and found most socialist rhetoric tedious, he was intrigued by the idea that intellectuals could rule his homeland, and through the Galilei Circle he was acquainted with a few of the incoming ministers. Moreover, Kun and many leaders in his government were Jewish, and Szilard could appreciate the intellectual tradition that they brought to everyday politics. Still, if infatuated by the Kun government, Leo apparently played no active role beyond supporting it in debates with other students, for he recoiled at the brutality Kun's Red Terror imposed in the name of "the people," and he distrusted the simplicity of their economic reforms. Life was much more complicated, and uncertain, Szilard believed; and economics should be, too.

Living where he did, among nobles and Jews alike, Szilard also sensed quickly the conservative and anti-Semitic backlash that Kun's ideas and methods inspired. And, as Szilard followed politics closely, he saw the folly of Kun's military conquest. Once the Red Army invaded Slovakia, it sparked a counterrevolution. Czech and Rumanian foes threw their support behind Miklós Horthy de Nagybánya, an admiral and former commander of the Austro-Hungarian fleet. A government formed around Horthy in the southern city of Szeged, and sympathetic Rumanian troops were soon marching toward Budapest.

In the capital that summer of 1919, Szilard sensed Kun's imminent defeat and, fearing the repression that might follow, decided to leave Hungary for engineering studies in Germany. During the first week in July, Szilard took more examinations at the Technical University, seeking to complete his degree quickly. His mechanics and mechanical drawing exams went surprisingly well; in each he earned a 5 in theory and a 6 in practice. At the same time, he applied through family friends in Berlin to the Technische Hochschule (Technical Institute), and through acquaintances in the Kun government he sought to obtain a passport and a visa.

On July 24, as Rumanian troops began their offensive against Budapest, Szilard went to the Reform church in his neighborhood and applied to change his religion from "Israelite" to "Calvinist."[34] Because he had no emotional attachment to Judaism, Szilard could easily take an action then popular among Jews, who feared anti-Semitic repression once Kun's government fell. Szilard's mother found this conversion painful despite her own lack of interest in practicing Judaism and her infatuation with the life and teachings of Jesus.

Kun's Red Army and thousands of civilian volunteers streamed out of Budapest on streetcars, trucks, and horse-drawn wagons to defend the city. Ill equipped and poorly trained, they quickly fell to the invading Rumanian army. As the Rumanians marched into the capital on August 1, Kun fled to Vienna, and his Soviet Republic collapsed.

In September 1919, Leo and Bela Szilard returned to register at their engineering school, now renamed the Budapest Technical University, but on the main building's front steps they were stopped by more than a dozen students.

"You can't study here," one shouted. "You're Jews."

Leo protested, trying to reason with the group. "We're Calvinists, not Jews, and have the papers to prove it," he said. But that only angered the students more, and they rushed at the two brothers, kicking them as they crawled and tumbled down the broad marble steps.

At home in the villa, Tekla Szilard was pained by her sons' blood and bruises for being Jews but also by their religious conversion. Yet reluctantly she, too, converted. "Only later did I come to understand that my loyalty to my ancestors had been overcome by my concern for the well-being of my children."[35]

When Szilard reapplied in September for an exit visa to study abroad, the Horthy regime refused, citing his activities with socialist students during the Kun regime. Szilard did not know it at the time, but Horthy's detectives had already investigated student activists, and an August report listed eighty-four "Communist students at the Technical University," the "most aggressive and dangerous ones" marked by blue crosses. The five names marked were: György HALASZ, Vilmos ROSENBERG, György SZAM-UELY, Bela SZILARD, and Leo SZILARD. On August 22, Budapest's public prosecutor issued an order to begin an investigation of the students on the list.[36] Later that fall, Leo suspected he was being followed, and Bela was warned that a detective had called on his employer. Leo was understandably anxious and again organized an armed watch at the villa.

In November, Szilard returned to the Technical University and from a sympathetic professor obtained an exit certificate that listed the courses he had taken and the grades he had earned. Then, pleading to trusted friends with connections in the Horthy government and offering bribes, Szilard reapplied for a passport. His answer from the passport office came five weeks later, on December 18. He could travel to Berlin. But he had to act swiftly: The permit to leave Hungary was valid only between December 25 and January 5. And he had to travel indirectly via Passau, a German city on the Austrian border, to avoid a Czechoslovakian boycott against Horthy's government.

To complete his travel pass, Szilard sat for a picture that captured a quizzical, apprehensive young man, his face half in shadow, eyes peering over his left shoulder. Looking at the passport and its scrawled entries, Szilard must have been both elated and sad; elated that he could escape the persecution he had suffered for months and could study engineering in Berlin but sad that he must now leave his family for the second time in less than two years. Szilard's mother assured him—and herself—that study in Berlin was the safest path open to him. Knowing that the state police had her sons under surveillance, she urged Bela to follow in Leo's footsteps, and he, too, applied to the Technical Institute in Berlin.

Still fearing the police and not knowing how "official" his travel papers actually were, Szilard counted the days until Christmas, the first date he could leave. Into a large suitcase he stuffed more than a dozen books, some clothes, and a few papers. His father handed him a wad of 100-pound sterling notes, then most of the family's savings, which Leo hid under the innersole of a thick-soled pair of shoes that he tucked among the books. At the Vidor Villa he bade his parents and sister farewell and with Bela's help hauled the bulky suitcase to the streetcar stop at the corner. The state police might be watching Budapest's long-distance train stations, Szilard thought, so he rode the streetcar through Pest to the Danube embankment. At a wooden kiosk by the shore, Szilard bought a one-way ticket for a daily excursion cruiser to Vienna. He hugged Bela, dragged his suitcase over the metal rungs of the gangplank, and was soon standing by the railing, waving good-bye.

The excursion boat was not crowded that Christmas Day, but on shore the holiday season was in full swing. As the steamer chugged into the center of the Danube, Szilard could see Christmas lights strung from the lampposts along the embankment. Against the strong current the steamer pushed under the ornate Chain Bridge and slid past Margaret Island—in Szilard's youth, a green and generous playground, now a dark and leafless shore. On up the river, Christmas greens and colored lights hung about the landings. Although Szilard and his family were now nominal Christians, the holiday itself meant nothing to them, and they never considered exchanging gifts.

The country Szilard left behind that chilly day further depressed him, he later recalled. He thought of the sorry history he had just survived— war, revolution, counterrevolution, and economic depression. And as Szilard sat in the steamboat's warm cabin, brooding about his past and future fears, he caught the eye of an old farmer seated on the opposite bench. The farmer seemed dressed for the fields nearby but was actually a Hungarian émigré on a return visit after living in Canada for forty years. The old man smiled and, seeing that Szilard seemed sad, asked him why.

"I am leaving my country, perhaps for good," Szilard answered.

The old man grinned. "Be glad!" the farmer said in a jovial voice. "As long as you live you'll remember this as the happiest day of your life!"[37]

CHAPTER 4

Scholar and Scientist

 1920–1922

Pushing into the Danube's surging current, the excursion steamer from Budapest passed below the hillside city of Bratislava, cleared customs at the Czechoslovak-Austrian frontier, and chugged by Vienna's low wharves and looming warehouses. On board and happily free of Hungary, Leo Szilard still faced problems during his escape, but arrest was not one of them. He had to drag his book-filled suitcase to a streetcar, ride to the center of the city, find an affordable room, and make visa arrangements for the rest of his journey to Berlin.

Szilard's personal sense of relief and exhilaration drew on the city's flippant spirit. "I was greatly impressed by the attitude of the Viennese," he remembered later, "who in spite of starvation and misery were able to maintain their poise and were as courteous as they have always been, to each other as well as to strangers."[1] Szilard was also impressed by Vienna's political climate, now a socialist city in contrast to the reactionary regime he had just fled. Happily he ran his errands, first securing a student visa to enter Germany, then buying a ticket for the circuitous rail route spelled out on his Hungarian passport. On New Year's Eve, with visa and ticket in hand, Szilard lugged his case to the train station to continue his journey.

Szilard may have enjoyed the train ride along the Danube, through snow-bright fields, but after he crossed the German border at Passau, on New Year's Day, the bitter-cold weather and a miners' strike soon turned his trip into a gruesome ordeal. Coal shortages halted his train unexpectedly, for hours or days, each time quenching the steam heat. A trip that should have taken all day stretched to nearly a week as Szilard huddled in his frigid passenger compartment, living on whatever snacks he could buy when his train stopped in a town.

Cold and exhausted, Szilard finally arrived in Berlin on January 6, 1920, to find a city and its people drained by the war and appearing as desperate as he felt.[2] Hunger marked the faces in the crowds that shuffled through the snow. Szilard himself found food expensive, restaurants crowded, and wondered how long his 100-pound notes might last. But in the bustle Szilard also discovered a new spirit, energy, and speed that Budapest never had. Berlin's traffic seemed to roar and swoosh around the circles and along the broad boulevards. The city had more bright lights, more noise and music, more beggars, more sleek buses, more style and schmaltz to assault his attentive mind. If Budapest was sedate, bending under its aristocratic past, then, to Szilard, Berlin seemed sassy, thrusting itself on his senses and sensitivities. Here was a new world where past and future collided.

Germany's loss of the war had also destroyed its imperial way of life. Unemployment spread rapidly after the collapse of the kaiser's armies. Revolts in 1919 had shaken the gloom of defeat as the socialist Spartacists had marched on the capital. Hundreds died in the streets, combatants and civilians both caught in the crossfire. By 1920, Berlin was tense with social and political strife.

The first lodging Szilard found was a room in the Bohemian neighborhood around the busy Berlin Zoo Railway Station and the cafés of Kurfürstendamm. His address at 5 Savigny Platz put him on a lively and pleasant tree-ringed square of small shops, restaurants, and cafés. Once his bulky suitcase was stowed, Szilard strode north along the western edge of Berlin's Tiergarten, a huge park of birds and lakes and strollers. At the "Knee," the busy intersection of Berlinerstrasse and Hardenbergstrasse, by the park's western tip, he climbed the steps into the imposing Technische Hochschule (Technical Institute) of Berlin to register. "The number of foreign students who were admitted was limited," and "the attitude towards foreign students was not friendly. . . ." Only by "having to bring to bear all the pressure I could through such private connections as I was able to muster in the city" was I admitted, he recalled.[3]

Szilard also needed a permit to live in Berlin, and at the police station he produced documents from his Hungarian Association of Socialist Students to argue that he had been exiled by the reactionary Horthy regime, whose anti-Communist White Terror was then purging anyone involved in the previous Kun government. During his first busy days in

Berlin, Szilard also called on business contacts of his father and on family friends. He even tracked down his third cousin, Paul Heller from Budapest, and through him was invited to a fancy-dress ball, given each year by the son and daughter of Dr. Felix Frankel, a prominent physician and public health official. When Szilard called at the Frankels' large home at 82 Kurfürstenstrasse, near the zoo, he was ushered upstairs to a large ballroom. As a newcomer, Szilard was expected to meet the adults but then mingle with the dozens of young people as they chatted and frolicked. Instead, he took a seat in a corner of the huge room, struck a pose of aloof boredom, and sat there most of the night.[4]

By the end of January, Szilard began the first of a dozen moves when he left Savigny Platz and sublet a room at 11 Lohmeyerstrasse, just off busy Berlinerstrasse, by the stately Charlottenburg Castle.[5] Back in Budapest, where Bela was trying to secure his own exit visa for study in Berlin, detectives in the Budapest State Police filed a report on the "Affair of Leo Szilard and his accomplices" that concluded:

> Leo Szilard, born in Budapest, Calvinist, unmarried, twenty-three years old, student of the Technical University, did not have any assignment at the university during the Commune [Kun's government]. He seemed to agree with the spirit of the Commune, however he did not make any propaganda or behave extraordinarily in any other way. No evidence proving his being a Communist could be obtained.

The detectives wrote the same verdict for Bela, and five days later, Budapest's public prosecutor declared the investigation closed, "as no evidence of their guilt could be found."[6] This, and a few bribes from Louis Szilard, cleared Bela's exit visa to join Leo at the Technical Institute.

Conspiracy of another sort was just beginning elsewhere that month: a plot that would eventually send Szilard fleeing Europe and lead to the horrors of World War II. In Munich, Germany, on February 24, Adolf Hitler proclaimed to a meeting of his followers twenty-five points of the National Socialist program. At the time, these "Nazis" and their manifesto were generally ridiculed, if noticed at all.

Anticipating Bela's arrival, Leo moved to a larger apartment in Charlottenburg on March 1, just two blocks from the Technical Institute. To reach their new apartment, the Szilards entered a door at a custodian's booth on the street at 2 Herderstrasse, then walked through a courtyard and archway to a back garden and climbed four flights of stairs. The four rooms were home to Mrs. Else Dresel—a living room, a salon and kitchen, her bedroom, and, at the back, a small bedroom for Leo and Bela. Their window overlooked a barren weed plot and the back of an apartment on Liebnizstrasse to the west. It was much grimmer than their back-garden view in Budapest, but with housing scarce and rents rising, the Szilards were grateful for a room of their own with a window.

The widow of a banker with two sons and a daughter, Mrs. Dresel provided the Szilards room and board. Their room was small and comfortable, but the Szilard brothers spent most of their time in an adjacent salon, a large, formal room with worn but elegant silk-upholstered furniture where they worked on their school assignments each night. The board was minimal, leading Szilard to wonder if the lodgers or their landlady was poorer.

Lunch at the Mensa, the institute's cafeteria, never seemed filling enough, so the Szilards walked home with big appetites. Weekdays they fixed dinner in their room: usually *Schrippen,* the Berlin hard rolls, with butter and cold cuts; occasionally, tinned sardines. On weekends, Mrs. Dresel cooked dinner, but all she could offer was fried potatoes, occasionally served with onions or a vegetable. This was their hardiest fare of the week. The bleak diet was relieved only rarely, when food packages arrived from Mrs. Dresel's brothers in America or on a few Sundays when the Szilards bought liverwurst and bratwurst for everyone—occasions they called their "festive meals." Mrs. Dresel was the same age as Mrs. Szilard and easily fell into mothering Leo and Bela. Leo, especially, seemed to need her comforting attention, but he alternately respected her kindly efforts and then became impatient with her concern.

Leo soon became impatient with the maid as well for removing a bedside chair on which he routinely dropped his clothes. He continued to drop his clothes where the chair had stood, insisting to Bela that this was logical behavior. "As it is not my job to place the chair where it *ought* to be," Leo announced one night, "it couldn't be my duty either to check whether that job has been done. So I just do what I have always done and let my clothes drop on what is supposed to be there."

This kind of persistent "logic" often drove conversations around the Dresel household, for Leo seemed to ponder everything that was said, no matter how mundane.

"Close the window; it's cold outside," Bela complained one evening.

"I will close the window," Leo answered immediately, "but that will not make it less cold outside."

Being busy and curious, the Szilards soon made friends with some fellow students. They also dispensed advice and aided anyone who asked. Bela helped with drafting; Leo, with math problems or counsel about careers. As their advice became more popular, however, the Szilards had to post a sign in the salon: Friendly Invitations to Stay a Little Longer Should Not Be Taken at Face Value.

In Berlin, Leo became friends with Budapest acquaintances he had first met during high school science competitions. One was Albert Kornfeld, an institute student for whom Szilard found a room on the floor below Mrs. Dresel's. But because Kornfeld's eccentric landlady alternately pestered and pampered him, he studied most nights with the Szilards. Many nights the three sat around the salon for hours, devising solutions to "problems" that others might not even imagine. To simplify and speed haircuts, for example, they agreed that barbers' chairs should be charged with a slight electric current that would stand the customer's hair on end for a simple mowing operation.

One problem that did exist was a drastic shortage of apartments, and Szilard had met several young men in Berlin who had married after returning from the war but could still only afford to live with their parents. His solution was a "Bride-and-Groom Exchange," a scheme for young married couples to contract with others to swap quarters. It is only "logical," Szilard reasoned, that living with another person's parents would create less strife than living with your own in-laws. Szilard also devised a way for lovers to communicate secretly. His "Bride-o-Mat" was a coin-operated mail drop for people living with their families who could not risk receiving love letters at home.

During his musings in the salon at Herderstrasse, Szilard even thought of a way to hold up women's stockings. He noticed that women in Berlin seemed preoccupied with hiking up their stockings but was mystified by how they were suspended. With no lady friend to ask about such matters, he turned to his own imagination. Szilard's system had flexible iron threads woven into the stocking tops and two powerful magnets in the woman's lower jacket pockets.

With such schemes Szilard probed the thin line between seriousness and humor. Rarely seen laughing or smiling as a youth, Szilard nonetheless forced humor from situations he saw as absurd or surprising, his jokes often delivered with tongue in cheek. Indeed, for most of his life Szilard admitted having trouble knowing when he was serious and when he was joking, a trait that confused both his companions and himself. His hyperactive imagination found release when Szilard asked his friends simple questions, then offered clever answers.

"Question: What is an optimist?" he often asked new acquaintances. "Answer: one who thinks that the future is uncertain." In that sense, Szilard was not an optimist, for he usually predicted that things would get worse before they might improve.

Little had changed for the Szilard brothers since the time in 1919 when they first studied together at the Technical University in Budapest. Bela was industrious and conscientious, while Leo spent most of his time just sitting and thinking, seldom reading course books, and rarely attempting the practical exercises that were essential to engineering studies. In fact, Leo seemed not to study at all except when he struggled with quick pencil sketches for his engineering-drawing assignments. Then he became so exasperated he dropped them on Bela's lap, expecting him to finish and copy them in India ink.

While sitting and thinking one night, Leo came up with a scheme that the Szilard brothers found very helpful. Unfamiliar with German prices, they had trouble deciding whether something that cost, say, two marks and fifty pfennig was expensive or cheap. And ceaseless postwar inflation made all values relative. To cope with this everyday confusion, Leo decreed that all prices be converted to the price of a *Schrippe,* the familiar hard roll sold in Berlin.

Szilard soon found himself impatient with the practical subjects taught at the institute and began daydreaming about physics. "I really lost interest in engineering," he recalled years later, finding it too much "routine application of already established knowledge. . . ." Physics was not taught at the Technical Institute but at the Friedrich-Wilhelm University (now the University of Berlin), across from the opera house on Unter den Linden in the heart of the city, and at the Institute of Physics, behind the Reichstag (Parliament) building near the Brandenburg Gate. Late one afternoon, Szilard wandered through the Tiergarten to the institute and sat in on a physics colloquium—a meeting where the latest articles from scientific journals were summarized and discussed. The topics and talk were fascinating, and soon Szilard made a weekly excursion to the colloquium. When the spring semester ended at the Technical Institute, he picked up a university catalog, eyed the pages of listings for courses and seminars, and applied. His application was delayed for two months, however, while Szilard ordered more transcripts and certificates from home, and this kept him from classes until two weeks after the winter semester began, in October 1920. But the wait was soon rewarded.

Berlin in the early 1920s was the capital of modern physics, attracting to the university and nearby research institutes the pioneers of the generation. At the university were Max Planck, Max von Laue, Walther Nernst, Fritz Haber, and James Franck. Planck, who had won the Nobel Prize in physics in 1918, was considered the founder of quantum theory. He had speculated that atoms, then considered the basic building blocks of nature, did not absorb and emit energy continuously but only in discrete bundles, or "quanta," an idea that revolutionized modern physics. With the work of Planck, a divide opened between classical (Newtonian) and modern (quantum) physics. Classes and discussions at the university constantly probed and defined that division.

Von Laue, a student of Planck's, had received a 1914 Nobel Prize in physics for discovering X-ray diffraction by crystals. Nernst had established a "third law" of thermodynamics to describe how matter behaves at temperatures near absolute zero, for which he would soon receive the 1920 Nobel Prize in chemistry. Haber, who had directed Germany's chemical-warfare research in World War I, received the Nobel Prize in chemistry in 1918 for his process to synthesize ammonia from its elements. And Franck, a physicist who was then visiting Berlin from his teaching post in Göttingen, specialized in atomic structure and photosynthesis. Franck would share with Gustav Hertz the 1925 Nobel Prize in physics for their studies of electron collisions within the atom. At the time, Albert Einstein studied on his own at the Prussian Academy of Science and directed theoretical physics at the Kaiser Wilhelm Institute, a prestigious research center in Berlin's western suburb of Dahlem. But Einstein gave a weekly seminar at the university—a meeting always crowded by faculty and the brightest invited students—and he attended other colloquia. He would receive the 1921 Nobel Prize in physics.

Szilard was excited about joining this lively academic community and valued the intellectual energy and talent around him, although he was not overawed. "I only want to know the facts of physics," Szilard told Planck in November 1920, when he called on the professor to apply for his course. "I will make up the theories myself." When Planck related Szilard's remark to Franck, the two men laughed. Whether he was serious or not, they could not help liking Szilard's cocky exuberance.[7]

During Szilard's first semester at the university, from October 1920 to March 1921, he registered for 'Theory of Temperature,' Planck's "Exercises in Mathematics and Physics," "General Theory of Relativity and Non-Euclidian Geometry," von Laue's "Proseminar in Physics," and a "Seminar in Applied Mathematics" with Richard von Mises, a pioneer in developing a philosophical approach to probability theory. He also attended "Exercises in Integral Calculus," a "Colloquium for Advanced Students," "Electrotechnical Experiments" and "Electrical Engineering," and "Practical Exercises for Beginners." Szilard enrolled in the university's philosophy faculty, not that of natural sciences, and in his first semester studied the "General History of Philosophy." In fact, Szilard continued taking philosophy and ethics courses throughout his work for a Ph.D.[8]

Despite this heavy course load, Szilard seemed to have almost nothing to study when he returned each night to his rooms at Mrs. Dresel's. In the salon after supper, while Bela and his colleagues from the Technical Institute pored over engineering assignments, Leo routinely sat in an armchair and thought. Sometimes he read a book; sometimes he joined in their debates. But most often he just stared ahead, as if looking through all objects in his gaze. At the university, he attended lectures and followed his professors' remarks with intense concentration. But his true excitement with physics came in the seminars and colloquia, the often noisy meetings where scientists discussed and debated their latest ideas. When the physics colloquium convened each Thursday afternoon at the Institute of Physics, Szilard was usually there. In the front row sat the heroes of his new profession—von Laue, Nernst, and Einstein—who chided one another with pointed comments and corrections. After the formal presentations, everyone met for coffee, cakes, and more discussion.[9]

Szilard came to enjoy his lectures and conversations with von Laue, a formal but wry gentleman whose stiff appearance was belied by a surprising twinkle in his eyes. From von Laue, Szilard learned Einstein's general theory of relativity and became fascinated with thermodynamics, especially for its statistical complexities and subtleties. And through von Laue, Szilard confronted "Maxwell's demon," a creature that would come to challenge his own impishness.

At the time, a student could take any courses he fancied and faced an examination only after as many as four years' study. For a doctorate, a thesis that involved original scientific work was also required, on a topic posed by the student or suggested by a professor. At the beginning of the 1921 winter semester, Szilard asked von Laue to become his thesis adviser, and he suggested a topic concerned with the theory of relativity. First posed by Einstein in 1905 and expanded in 1915, the relativity theory asserted that space and time are not absolute but are relative—both in themselves and to each other. In 1911, von Laue had been the first person to write a monograph on the theory of relativity, and now he enjoyed teaching a course on the subject and leading students to understand, explain, and quantify it. But the thesis topic that von Laue posed troubled Szilard, perhaps because the basic premise had already been proven and he preferred to disprove scientific tenets.

Szilard's mind was also engaged by statistical mechanics, which he had boldly asked Einstein to teach to a few friends that semester. When Einstein had agreed, Szilard invited the brightest people he knew, among them three Hungarian friends who would later revolutionize science: Eugene Wigner, a 1963 Nobel laureate in physics; John von Neumann, creator of game theory and developer of the modern computer; and Dennis Gabor, inventor of holography and a 1971 Nobel physics laureate. Wigner, who became a close friend of Szilard's for much of his life, thinks that in Einstein's seminar Szilard confronted a fundamental problem with his own scientific career. Szilard was "pleased to have helped inspire the course," Wigner concluded years later, and he "made strong proposals in the seminar and began a long friendship with Einstein," often visiting him at home. Yet the seminar also troubled Szilard because "it seemed to convince him he wasn't good enough at mathematics to change theoretical physics." Szilard might have been lazy, which he could be as often as he was stubbornly creative. Or he might have preferred to pursue other topics. "There is no need to study mathematics," Szilard told Gabor at this time. "One can always ask the mathematician!"[10]

Whatever Szilard learned from Einstein in the fall of 1921 convinced him that von Laue's thesis topic was a waste of time, for despite weeks of hard thinking, Szilard recalled, "I couldn't make any headway with it. As a matter of fact, I was not even convinced that this was a problem that could be solved." For almost three months, Szilard forced himself to work on the problem, but the more he thought about it and the more he scribbled his calculations and fiddled with his slide rule, the more he seemed to lose his way to a solution. When the university holidays began in December 1921, Szilard decided to stop work entirely. Back in Budapest, the Horthy regime was more repressive and anti-Semitic than ever, so a visit home seemed too risky. "Christmastime is not a time to work," Szilard thought, "it is a time to loaf." He decided to "just think whatever comes to my mind."[11]

Through December's damp and blustery winds Szilard walked and wondered, pacing the Tiergarten's broad promenades and weaving through narrow streets in the nearby residential districts of Charlottenburg and Wilmersdorf. Szilard in motion was a peculiar sight: his gait more a stride than a shuffle; his head cocked high, chin out, chest breathing in air and exhaling confidence. There was nothing pensive about these walks; they were times to expound on the world and its wonders, to expand on pet theories and peeves. The more he walked, Szilard discovered, the further his mind wandered from von Laue's relativity topic. Soon Szilard was musing on the state of science itself, on the gaps between the "old" physics of Newton and the "new" physics of Planck. How, he wondered, might they be reconciled?

Szilard later recalled:

> I went for long walks and I saw something in the middle of the walk. When I came home I wrote it down; next morning I woke up with a new idea and I went for another walk; this crystallized in my mind, and in the evening I wrote it down. There was an onrush of ideas, all more or less connected, which just kept on going until I had the whole theory fully developed. It was a very creative period, in a sense the most creative period in my life, where there was a sustained production of ideas.

Returning to the rooms on Herderstrasse each evening, Szilard sat in the salon scribbling his thoughts. Soon he saw a pattern, then a thread to unite his calculations, and finally a thesis of his own. He would try to use statistics rather than commonly accepted experimental evidence to confirm the second law of thermodynamics. The law posits that some heat energy is always lost when converted into mechanical work, and the proof he derived statistically seemed to apply not only to temperature fluctuations but to all calculated variables. "I had produced a manuscript of something which was really quite original," Szilard wrote later. "But I didn't dare take it to von Laue, because it was not what he had asked me to do."[12]

Instead, Szilard turned for advice to a man of comparable authority but a gentler disposition, a man he was destined to share a close relationship with as both a scientist and a friend. He called on Einstein. At first, Szilard asked for the chance to describe his recent work. The courtly professor nodded and listened intently to the calculations and their results. When Szilard had finished, he looked to Einstein for a response.

"That's impossible," Einstein said. "This is something that cannot be done."

"Well, yes," Szilard replied, "but I did it."

"How did you do it?" Einstein wondered. "It didn't take him five or ten minutes to see, and he liked this very much," Szilard recalled.

This reaction from Einstein gave Szilard the courage to face von Laue once the Christmas break ended, in January 1922. Although he rarely chose to, Szilard could be deferential when necessary. Now it was necessary. He waited until after class, approached von Laue, and confessed that he had not written the paper assigned to him. But he *had* written a paper. Something original in thermodynamics, which was von Laue's field. Might this new paper be considered as his dissertation for a doctorate? Von Laue looked at Szilard quizzically but took the manuscript in hand, agreed to read it, and walked out.

Early the next morning, Mrs. Dresel's telephone rang. It was Professor von Laue, calling for Szilard. "Your manuscript has been accepted as your thesis for the Ph.D. degree," von Laue said.[13]

Both Einstein and von Laue were surprised and pleased with Szilard's thesis, for it refuted a scientific concept that had puzzled physicists for half a century. He had "exorcised" Maxwell's demon and done it in a way that united both classical and modern physics.

Thermodynamics is the study of heat and its conversion to mechanical, chemical, and electrical energy. Certain principles, or laws, of thermodynamics describe how heat behaves, and of these the second law asserts that it is impossible by any continuous self-sustaining process for heat to be transferred spontaneously from a colder to a hotter body. The energy lost, or unavailable, was called *entropy* by the German mathematical physicist Rudolf Clausius in 1850, when he formulated the second law. In all natural systems, entropy (disorder) increases over time. Clausius even assumed that as entropy increased the universe would eventually "die" by cooling, a notion that Szilard had first encountered at age ten in the epic Hungarian poem *The Tragedy of Man* by Madách, years before he studied physics. For the rest of Szilard's life, he remembered Madách's scene in which starving Eskimos are the last survivors of the human race as the earth steadily cools. The entropy concept had momentous personal meaning for Szilard, quite apart from its significance in science, and this concern may explain why he gave the subject so much thought.

In 1871, fifty years before Szilard wrote his thesis, the Scottish physicist James Clerk Maxwell published *Theory of Heat*. In it he posed a way to defy the second law of thermodynamics with an impish creation. Maxwell conceded that the second law "is undoubtedly true as long as we can deal with bodies only in mass, and have no power of perceiving or handling the separate molecules of which they are made up." But he postulated that a "being" small enough to manipulate molecules—his demon—should be able to defy the second law and use all available heat energy without expending any in the process, in effect creating a perpetual-motion machine.[14] To prove his point, Maxwell described a vessel with two chambers, A and B, which were connected by a tiny hole. Then he proposed that his demon "opens and closes the hole, so as to allow only the swifter [hotter] molecules to pass from A to B, and only the slower [cooler] ones to pass from B to A. He will thus, without expenditure of work, raise the temperature of B and lower that of A, in contradiction to the second law of thermodynamics."[15]

At the time Szilard encountered Maxwell's demon, physicists explained the second law in two ways: One was phenomenological; the other, statistical, or atomistic. The phenomenological theory started from the principle that heat is energy and always flows spontaneously from hot to cold regions. Laboratory experiments could demonstrate this, and using various ingenious arguments, many diverse phenomena could be explained by this principle. Clausius had formulated an abstract concept of entropy this way when he wrote that "the energy of an isolated system remains constant, and its entropy always increases."[16] With this, a theory that first explained the nature of heat was extended to include all forms of energy.

This phenomenological argument was simple and elegant, but necessarily abstract, and by the turn of the twentieth century, some physicists sought a more concrete formulation of the same idea. In doing so, they turned to the frontier of physics at the time, to the study of atoms, and within this field they described various physical events in terms of the random motion that atoms were known to perform. Atoms could not be observed in a laboratory, however, so physicists had to resort to models of atomic behavior by using statistics and probability. It was (and still is) impossible to describe the nature and position of each atom at a given time; the best physicists could do was describe the probability that certain events and conditions are occurring on the atomic level, making their method statistical. Around the turn of the century, Austrian physicist Ludwig Boltzmann was the first to quantify heat and entropy as statistical probabilities that he related to the randomness found in atomic movement. [17]

With this, two different explanations were available to describe thermodynamic equilibrium: In a phenomenological explanation equilibrium is static, whereas in a statistical explanation it is dynamic. Szilard had considered both approaches to entropy and in his thesis demonstrated that the atomic fluctuations that are part of a detailed statistical equilibrium can be included within the elegant but abstract phenomenological theory as well. And he did this without making any reference to probability or other mathematical models of atomic behavior. As he explained it later, he "showed that the second law of thermodynamics covers not only the mean values, as was up to then believed, but also determines the general form of the law that governs the fluctuations of the values."[18]

In his calculations, Szilard also demonstrated that entropy must increase for Maxwell's demon, using proofs from both the "old" and the "new" physics and forging a link between classical and quantum mechanics. The second law, he showed, does not become inexact when demonstrated statistically but "evolves in some higher harmony" to explain not only thermodynamics but also its fluctuations—and other changes as well. "On the Extension of Phenomenological Thermodynamics to Fluctuation Phenomena" was accepted as his doctoral thesis in 1922 and in 1925 appeared as his first published paper, in *Zeitschrift für Physik,* Europe's leading physics journal. The thesis was noted as "eximia," the highest honor.[19]

Historians of science have found other achievements in Szilard's thesis as well. To give a mathematical expression of "thermal equilibrium," Szilard coinvented the concept of *sufficiency* (a statistic that summarizes all the information in a sample) at about the same time as the English mathematician Ronald Aylmer Fisher did, but entirely independently.[20]

While Szilard struggled with these abstract mathematical concepts, he was forced each day to confront a more practical numerical problem: inflation. By 1922 it had risen to undermine even the lives of the wealthy, and the cost of meeting League of Nations reparation payments for the world war further eroded the German mark's value. To survive, both Leo and Bela collected food coupons at their schools and redeemed them for sugar, flour, and other commodities, which they gave to Mrs. Dresel. The economy in Hungary was almost as weak, and the Szilards could only rarely send food packages to their sons in Berlin. "I remember," Szilard told a friend years later, "walking around Berlin, looking at the food in store windows, looking at the people in restaurants, and not being able to afford a thing. To satisfy my appetite, I didn't eat. I just looked."[21]

It is a mystery to Bela how Leo survived as a student once his roll of 100-pound notes was spent. Bela earned money tutoring fellow students in engineering drawing and as a part-time draftsman for engineering companies. Except for the occasional help he gave to students needing tutoring with their math, Leo appeared to have no income at all. Yet somehow both Bela and Leo managed to send small sums of money to their parents in Budapest. From their first days together in Berlin, Bela did most of the letter writing home; Leo would only sign "*Gruss,* Leo" (Greetings, Leo) at the bottom of the last page. After a while, as Leo became involved in his lectures and thesis, even this became too demanding, and he scrawled only "Gr. Leo."

Within months of submitting his doctoral thesis, in the summer of 1922, Szilard wrote a second paper on thermodynamic equilibrium: "On the Decrease of Entropy in a Thermodynamic System by the Intervention of Intelligent Beings."[22] This work extended the calculations in his thesis from physical phenomena, such as heat and gases, to "information" by paying more attention to the activities that Maxwell's demon might perform. This work is now considered to be the earliest known paper in what became the field of "information theory" in the 1950s and 1960s. "In eliciting any physical effect by action of the sensory as well as the motor nervous systems a degradation of energy is always involved, quite apart from the fact that the very existence of a nervous system is dependent on continual dissipation of energy," Szilard wrote.[23] Or as Szilard's friend in Berlin physicist Carl Eckart later summarized: 'Thinking generates entropy.'

Here Szilard addressed Maxwell's challenge directly, seeking the essential interaction that might allow the demon to decrease his system's entropy. Szilard found that the interaction is "a kind of memory" inherent in the "measurement" that the demon must perform when he decides to let one molecule through the hole but to block another. To measure, he argued, is to establish and memorize. Szilard assigned a value of y for the demon's memory/measurement, then identified this activity according to the time (t) it occurs. "If we are not willing to admit that the second law [of thermodynamics] is violated," he wrote, "we have to concede that the action which couples y and t—i.e., establishes the 'memory'—is indissolubly connected with production of entropy."[24] Szilard then postulated that any decrease of entropy in other parts of the system would be compensated by an increase in that of the memory/measurement process itself. Therefore, Maxwell's demon could *not* decrease entropy in the system and thus could not violate the second law of thermodynamics.

"In this paper," Eckart noted, "Szilard had eradicated the ancient dichotomy of mind and matter, just as Einstein had already eradicated the less ancient dichotomy of energy and matter."[25] When von Laue and Planck reviewed Szilard's work on entropy and information, they found it an example of independent thought and recommended "warmly" to the faculty that he be admitted. The two reviewers did note, however, that Szilard "obviously chose some of the experiments not very carefully . . . but the way he did it is correct."

While Szilard saw the key elements of information theory some three decades before it became popular, there is no evidence that the theory's early developers were ever aware of his pioneering work. For although he first wrote about his insights in 1922, he did not present them publicly until he gave a "habilitation" lecture in 1926 to qualify for a teaching post at the university—and this Szilard did not publish in *Zeitschrift für Physik* until three years later.[26] If any link did exist, it may have come through the extraordinary mind of John von Neumann, the mathematical genius whom Szilard knew and taught seminars with in Berlin during the late 1920s. He may have recalled Szilard's bridge between statistical definitions of motion and the interchange of ideas as von Neumann designed and built the world's first computers in the 1940s. In 1951 the information theorist Leon Brillouin reviewed Szilard's 1929 paper; he apparently heard about it from Warren Weaver at the Rockefeller Foundation. Brillouin then met and chatted with Szilard around Columbia University and after these conversations restated the idea: An intelligent being, whatever its size, has to cause an increase of entropy before it can effect a reduction by a smaller amount. Claude E. Shannon, who spelled out detailed relationships between information and entropy in the 1950s, also later acknowledged that Szilard's paper had proposed the basis for his new field of study.[27]

The way that Szilard set aside his insightful paper on information and entropy shows the impatience—even audacity—of his restless mind. He rarely followed an idea from beginning to end, taking his intellectual discovery to its practical application; had he done so, Szilard might have tried to design a computer in the 1920s—a Bride-o-Mat for thoughts rather than emotions. Thinking that some of his ideas had practical value, he did apply for patents now and then, hoping revenue from his inventions would finance free time to continue brainstorming. But most of Szilard's thoughts were never even scribbled in his hasty hand, and he rarely kept a notebook or journal, as many scientists do. Instead, he freed his ideas in conversation and wondered on.

In addition, Szilard seems to have sought something even broader than what we know as information theory. Influenced by Einstein's quest for unifying principles, Szilard was after a more general concept for understanding life itself—and on the grandest possible scale. Szilard was fascinated with kinetics, the science of objects in motion, and for him the "equilibrium" that any activity reaches—whether molecules in water, people in a railway station, or stars in a galaxy—tended in some way toward order and away from chaos. He seemed to crave universal laws that would explain equilibrium and define order and no doubt enjoyed discovering that the second law "evolves in some higher harmony" beyond classical nonstatistical principles. In this pursuit, the dynamics of information were just one of many activities that gave mystery to the world.

Szilard may have even seen entropy as a concept to explain and unify other scientific and personal understandings. What we call disorder or chaos might actually be order, if order is seen as a random distribution and not as a static, idealized condition. Often he spoke to colleagues in Berlin about *freie Weglänge,* the free path of molecules as they move between collisions, suggesting that at this moment of their escape a kind of universal order occurs.[28]

Colleagues such as Eckart have even suggested that Szilard could be so brilliantly original because on his own free path of thought he was able to personify abstract theoretical concepts. In a sense, his analysis of the problems facing Maxwell's demon became an exercise in psychophysics, just as years later he would use a personalized approach to try to trick mammalian cells into dividing with a kind of psychobiology.[29]

He received his doctorate in physics on August 14, 1922, and wondered what to do next. At first, he "thought it would be interesting" to earn a second doctorate, this time in economics. But when he applied at the university, he was referred to one official after another, each doubting that it could be done. As Szilard persisted, he eventually was shown into the office of the rector.

"We would like to oblige you," said the rector, "but I don't see how we can do it. When we gave you the degree of doctor of philosophy, we certified that you are a man who is able to acquire any kind of knowledge that he desires, is capable of independent judgment, and has the maturity of a scientist and scholar. I don't quite see how we can certify the same thing twice."[30]

Still, Szilard enjoyed the atmosphere around the university and relished talking and arguing with his colleagues. Although shy in most social situations, he could be glib and entertaining among intellectual peers, posing impossibly simple questions or improbably complex answers in the chatter of academic discourse. One colleague with an esoteric sense of humor called Szilard the university's *Katholisator,* a medieval title bestowed on the academic official charged with making sure that students and visiting scholars met all the right people. And with informal relations already begun among Planck, Einstein, and other great men of science, Szilard seemed at ease in this role.[31] He would play it for the rest of his life.

CHAPTER 5

Just Friends

 1920–1932

Women bored and befuddled Leo Szilard. They were, for him, both weird and wondrous creatures; weird because they didn't think logically, as men did; wondrous because they lived in a seemingly fanciful and sensuous realm that he could only begin to fathom. Szilard became infatuated with several women in his life, especially when he could match wits and banter—his closest approach to flirting. But for most of his life the intimate honesty he first shared with his mother was never captured in another relationship—never, it seems, even seriously pursued—until years after he came to know the woman who would be his wife.

This meant that Szilard's friendships with both men and women were practical, candid affairs based on certain shared interests. When those interests waned, so did the friendships as Szilard's mind chased on to new curiosities. "Leo would pick up people, suck them dry of ideas, and like an empty orange peel, toss them aside," his brother, Bela, recalled of their years together in Berlin in the early 1920s. For Leo, friendships were mostly extensions of his mind, not of his heart.

Shy and introverted except in situations he could clearly control, Szilard behaved formally with most of his friends. As a teenager in Budapest and as a young man in Berlin, he seemed awkward at the few social gatherings he bothered to attend, by turns withdrawn, aloof, gregarious, or distracted. His giddy and brooding adolescent infatuation with Mizzi Freund, a coquettish woman who was older than he and married, left Szilard reluctant to open his heart. His teenage companion from across the Fasor, Alice Eppinger, was more a fixture in his family life than a personal sweetheart—first his sister Rose's best friend and only rarely the object of his own attentions. At the time of Leo's first acquaintance with Alice, Bela and Rose Szilard, their cousins the Scheiber boys, and Alice took hikes in the Buda hills every week. But Leo never joined them, and his own excursions with Alice seemed strained, almost furtive.

Alice admired Leo in high school, and she thought him especially handsome in uniform during the war. Sometimes directly but more often through Bela, Alice wrote to Leo in Berlin. She seldom received a direct reply. "I missed him terribly," she recalled years later, "and decided to do something bold."[1] When her high school studies ended with a baccalaureate degree in mathematics in the summer of 1921, Alice boarded a train in Budapest and headed for Berlin, seeking a way—somehow—to be closer to Leo. A gifted math student, Alice was accepted at the University of Berlin for the winter semester of 1921–22, the same time that Leo was struggling unsuccessfully with von Laue's assigned thesis topic for a doctorate in physics. Her father's business contacts brought him to Berlin so often that he owned a house there and knew the city well. He arranged a room for her in an elegant dormitory for university women on the Kurfürstendamm, the lively and broad boulevard near the zoo and the Tiergarten—Berlin's central park. Mr. Eppinger liked Leo and thought, as Alice did, that if she studied in Berlin, Leo might see more of her and perhaps consider marriage. Mrs. Eppinger had no such expectations but, to satisfy her husband and daughter, made the trip to Berlin, inspected the dormitory, and grudgingly approved Alice's move.

"You only studied math because you're in love with Leo," Leona (Lola) Steiner, her friend from the Fasor, teased Alice.

"That's true . . . probably," she admitted. But in Berlin, Leo seemed more distant to Alice than if she had stayed in Budapest. He always seemed to be rushing somewhere, to a class or seminar, walking about and talking quickly or thinking in a pensive trance. She was surprised at how Leo seemed to take on the pace and the energy of the city itself—its brusque and edgy style, its frantic tempo, and its halting, cloudy moods. His mind, Alice surmised, raced so frantically that his body was merely keeping up. On a few evenings, Alice did manage to meet Leo and Bela in the salon of their apartment, and once or twice Alice and Leo walked in the Tiergarten. They also strolled Berlin's wide and busy boulevards in the Charlottenburg and Wilmersdorf neighborhoods around where they lived but rarely stopped for a meal in the bustling restaurants, rarely took a moment to sit and sip tea or cocoa in a sidewalk café. Leo seemed so restless, so reserved, that when they did chat, the encounter was more instructional than intimate.

"You must read *Buddenbrooks!*" Leo insisted during one walk. "That's how you can understand the Germans. And you must read *Babbitt.* That's how you can understand the Americans." Alice nodded, and Leo talked on as arm in arm they wove through the strollers on the sidewalk.

"The greatest book ever written is *War and Peace*," Leo declared. "You must read it."

"But I like Dostoyevski better," Alice insisted. "*The Idiot* is so exciting, so passionate."

"When you are older, you will like Tolstoy better," Leo declared, ending the conversation.

Leo once teased Alice about her interest in math as a way to test her feelings about him. He teased her for studying something she clearly couldn't put to use and seemed to confirm by this query that she liked him. But while this discovery made him uncomfortable, Alice's company also gave Leo a chance to seem important. One spring day in 1922, as they stood chatting in the main entry hall at the University of Berlin, the huge door to Unter den Linden swung open, and in walked Einstein.

"Would you like to meet Professor Einstein?" Szilard asked Alice, and not waiting for her reply, took her arm and led her across the large lobby. "Say, 'Good day,' and shake hands," Leo whispered.

"Professor, I would like to introduce Miss Eppinger from Budapest."

"Good day," Einstein said.

"Good day," Alice replied, and shook hands. Nothing more was said, but from this encounter Leo grinned with pride.[2]

Alice Eppinger also saw Leo through their friend John von Neumann, a student in one of her math classes. She knew von Neumann from Budapest and found him jolly company, if a bit absentminded, as when he rushed up to Alice at the university one afternoon to ask her for "five pfennig for the streetcar."

"I don't have five pfennig, Jancsi," she said. "But here's a mark."

Von Neumann studied the coin a moment and frowned. "No. No," he said. "I only need five pfennig," and walked away, unable to see that he could change the coin.[3]

When Alice, von Neumann, Szilard, and his new Hungarian friend Eugene Wigner met around the university and conversation drifted to their lives and families in Budapest, Szilard fell silent. He claimed no interest in the past, his or theirs, and dismissed their conversation as sentimental small talk. Life for Szilard was too serious, too puzzling, too grim—even among friends. His jokes had a serious point or ironic twist as well, and the only times he laughed out loud were at the movies, where he enjoyed Charlie Chaplin's slapstick comedies.

After a year of studying mathematics in Berlin, Alice Eppinger returned to Budapest, feeling more distant from Szilard than ever. Still, despite his few hasty replies, she wrote him often. In one letter, Alice compared their love to a delicate thread, and to her delight Leo answered that this thread could be woven into a net that could not be unbound. "I have made a braid of the string that will never break," she wrote back. But nothing more direct than this was ever said of their friendship.

Fearing the Fascist regime of Admiral Horthy that had scared him from the country in 1919, Szilard was reluctant to return to Budapest and stayed away until 1923. On his first return home, he called on Alice and her family at their villa across the way from his own, and he invited her for a walk in the handsome park on Margaret Island. On this walk they paused during their brisk hiking, sat on the lawn, held hands briefly, and even kissed and cuddled. But back on the Fasor that evening, when Szilard's sister, Rose, and his mother asked about the outing with Alice, Leo seemed evasive and troubled.

Szilard's second return to Budapest, in August 1924, was no easier for the young couple, and after calling on Alice, his relatives noticed that he was uneasy and distracted.[4] Dutifully, Szilard called on Alice whenever he came home, and each time he was well received by her parents. Both families expected that the two would marry eventually, but nothing was ever said openly. When Alice mentioned that she wanted to have children, Leo ignored her remark. The two took long walks in City Park and along bustling Andrássy Avenue, near their families' homes, and on that broad boulevard one sunny afternoon Leo finally spoke about their affair.

"Listen," he announced as they strode along. "I want to tell you a story. . . . About bees."

"About the 'birds and the bees?" Alice asked, half-teasing and half-intrigued.

"No, no! Just about bees," Leo said, now speaking and walking rather stiffly. "Bees. There is a book by Maeterlinck, all about bees. It tells how a family of bees lives." Alice was delighted to hear the word "family."

"In each family there are three kinds of bees," Leo continued. "A queen, workers, and drones. Imagine this is a family of bees and I am a worker."

"A father?" Alice asked impatiently.

"No, a worker. I am a worker! Don't you understand?"

"Well," she said, mystified, "I suppose."

A few blocks up the busy boulevard, he finally halted and turned to her. "Listen, Alice, I am not the marrying kind. I do not want to have children. I am a worker, not a drone."[5]

Alice was confused by this declaration and heartbroken, for she sensed—correctly—that Leo had little interest in her. His distance—his personal defenses and rational domineering—had pained her for most of their relationship, and their separateness, even when "together" in Berlin, only confirmed her fear that their love could never be mutual. But in his intense honesty she found some solace, eventually recognizing that Szilard could not defy his inner convictions and could not hide his own fears and foibles. For his part, Szilard spoke about their break with no one; he soon left Budapest for the Austrian Alps, then traveled on to Italy and France. He wandered next to spend two weeks in Vienna, a city he enjoyed for its lively atmosphere and its liberal political style, and once back in Berlin, in October 1924, he kept moving, vacating his room on Leibnizstrasse for another at 11 Geisbergstrasse.[6]

Alice returned to Berlin in the spring of 1926 when her father tried to separate her from a poor cousin with whom she had fallen in love, but she failed to gain Leo's attention. Leo and Alice were together that summer, when he joined a hiking excursion to the Italian Alps organized by Bela and his girlfriend (later wife) Elizabeth Fejer. Also along on the trip, as chaperon, was Mrs. Eppinger and Alice's friend Lola Steiner. Leo enjoyed the mountains around Lake Como, the fresh air, the cool temperatures, the dramatic change in scenery from Berlin's cloudy grime. At twenty-eight years of age, Leo appeared fit and athletic, his build solid in hiking boots, knee socks, and knickers. Yet he seemed to shun the group's easy sociability, preferring to hike and think as if alone. At the town of Lecco on Lake Como, one photograph from this trip caught Leo sitting with his back to the others, nibbling a piece of fruit. In another picture, he is lying on his back, his left arm shading his eyes as the others around him chat and pose for the camera. Alice seems pensive, her elbow touching Leo's but her stare remote. At Leo's right arm, Lola leans forward, laughing with the other young women.

The summer of 1927 found Leo and Lola together again, this time when he took the initiative to invite her on an alpine excursion and called on her parents to ask permission. From Budapest the couple rode a train west toward the Ötztal Alps on the Austrian-Italian border. Lola later recalled:

> At a small station we were supposed to change trains. But we missed the connection and had to stay there overnight. Two separate rooms would have cost much more [than one], and we conferred about that. I had to be very parsimonious at that time because my father's business prospects were poor. . . . I felt that I must decide for thriftiness and chose the common room. . . . [But] aside from some smooches, nothing happened, and this strange situation continued for the rest of our summer.[7]

By then Alice realized that Leo had forgotten their relationship entirely and suggested to Bela that Leo should marry Lola. Alice soon married Theodor Danos, a Budapest businessman.

For his part, Leo had no interest in marrying anyone. He was courteous to the women he encountered around the university, although sometimes brusque with those he met socially. The only exception was the tall, dark-eyed, dark-haired Gerda Philipsborn, a good-natured woman with a lively sense of humor who remained a mystery to all Szilard's friends, even Bela. When the three met for lunch in a restaurant on Kurfürstendamm, Gerda told Bela she planned to move to India, to see how she might help the poor. Leo had met her in the fall of 1924, when he sublet a room in her mother's apartment on Geisbergstrasse.[8]

There are no other records of their relationship until February 1929, when Gerda wrote on Leo's behalf to Albert Einstein, from a Brixton Hill hotel in south London, about English patent negotiations for the electromagnetic pump that he and Szilard had designed.[9] She and Leo apparently returned to the city a year later (or perhaps Miss Philipsborn had moved to London in 1929 and Leo had visited her). In a letter to Einstein that Szilard wrote from Gerda's Queens Gardens address in the Bayswater section of London in March 1930, he described efforts to promote to English academics his idea for a "Bund" of young intellectuals—a group entirely separate and different from the pro-Nazi Bund.[10] Szilard wrote again from that address in April, and back in Berlin, when the Philipsborns moved to Prinz-Regentenstrasse, still in the Wilmersdorf neighborhood, Szilard moved with them.[11]

Acting as Szilard's secretary, Gerda Philipsborn wrote to Einstein from Berlin in October 1931. She reported that Szilard wanted to turn his US visitor's visa into one for permanent immigration, and she asked for a recommendation note to the American consul. This Einstein provided, and when Szilard landed in New York in December 1931, he gave Gerda Philipsborn's name and address as his European residence. She, in turn, reported his move to the Berlin police for him.[12] By October 1932, Szilard was living with the Philipsborns at a new apartment on Motzstrasse.[13] But then, Miss Philipsborn left for Palestine, by December had landed in Bombay, and the next month began work at Jamia Millia Islamia (National Islamic University) near Delhi, where she taught until her death from cancer in 1943. Szilard moved to the Harnack House, the faculty club for the Kaiser Wilhelm Institute in the western suburb of Dahlem. A few months later, Szilard's friend Eugene Wigner would note "how easily he parted from Miss Philipsborn. . . ."[14] From London, in 1934, Szilard sent her a ten-pound money order; at about that time, he was also considering teaching physics in India but soon dropped the idea.

The woman Szilard would eventually marry, in a private ceremony in 1951, he first met more than two decades earlier. It was 1929 when he encountered a shy and charming Viennese woman named Gertrud (Trude) Weiss. Their relationship, once begun, continued to swing like the repelling and attracting poles of two dangling magnets, not only during their long courtship but also well after their marriage.

Born in 1909, Trude grew up as a daughter of a successful and widely respected physician. A talented math and physics student, she enrolled in these subjects at the University of Vienna in 1928 but a few weeks into the first semester dropped her studies. A family friend, psychoanalyst Hanns Sachs, asked Trude if she would consider working as a governess in Switzerland. His offer intrigued her, and she moved to Vevey, a town on the north shore of Lake Leman, between Lausanne and Montreux. Trude's employers, Kenneth and Annie Winifred Macpherson, were caring for the eight-year-old daughter of their friend the American poet Hilda Doolittle, who wrote under the name "H.D." Trude enjoyed teaching the young girl, named Perdita, and in off hours took language courses at the University of Lausanne.[15]

Mrs. Macpherson, an English author who used the pen name "Bryher," decided in the summer of 1929 to move to Berlin and found a magazine on modern film. Trude agreed to come along as her part-time secretary and translator and in Berlin helped produce *Close Up*, a distinguished and authoritative English-language quarterly. Trude and Bryher also began work on a book of German lessons for English speakers, *The Lighthearted*

Student.[16] ("Hundreds of people have learned to understand German talkies with its engaging help!" one ad boasted.)

In Berlin that fall, Trude met Leo Szilard through their mutual friends Karl and Michael Polanyi and decided to enroll in university science courses. She also called on her mother's brother, Paul Schrecker, a renowned Leibniz scholar whose wife, Claire Bauroff, was an avant-garde dancer. Trude agreed to translate a German manuscript into English for Szilard, perhaps his proposal for the Bund, and when she had finished, she asked Aunt Claire for advice.

"How much should I charge Dr. Szilard for this translation?" Trude asked. "A hundred marks? Fifty?"

"What does it matter?" her aunt answered. "He'll pay you, but what will you do next? Buy a dress. Wear it once. Hang it in your closet and forget it." She paused, staring at Trude knowingly. "But if you charge him nothing, he will be in your debt forever." Trude charged him nothing.[17]

In November 1929, Trude enrolled in the university to study physics and biology for the winter semester and completed her courses the following March. Szilard was teaching "New Questions of Theoretical Physics" that fall, and Trude, who had enrolled for six other courses, sat in on his presentation one Friday afternoon. She attended for a few weeks and apparently asked enough questions to engage his interest. Two friends from that time, physicist Victor Weisskopf and chemist Hermann Mark, recall that Szilard brought Trude to their parties, sometimes called her *kedves,* Hungarian for "dear," and engaged her in deep conversations about film— then the most radical of the arts.

Szilard often became interested in the lives of people he met, if only as a way to dispense helpful advice, and during one conversation he asked Trude what she planned to do when her courses ended.

"I can't decide between physics and medicine," she admitted.

"You are too dumb to go into physics," Szilard said. "Go home to Vienna and study medicine."[18] This she did, and during her stay at the University of Vienna, Trude twice made—and broke—engagements with young men. When she graduated, in 1936, she was still infatuated with Szilard.

Szilard's impulsive personal advice, a "service" he provided to family and friends throughout his life, was not always followed. But it was usually appreciated and recognized by many as the most intimate response he could make to those he met. Like many shy and private people, Szilard sheltered his own emotions by focusing on details in the lives of others.

He was uneasy with intimacy, yet by the gift of his advice Szilard was able to be both clever and personal without actually becoming close. Typically, he addressed the women he knew by their first names and the men by their last names, striking a familiar, avuncular tone with his female friends, a formal tone with the men.

Szilard's closest friend for much of his adult life was Eugene Wigner, the son of a Budapest tannery manager. As a teenager, Wigner had studied at the Lutheran Gymnasium, a few doors down the Fasor from Szilard's family villa, but the two did not meet until both were at the Technische Hochschule (Technical Institute) in Berlin, where Wigner enrolled in 1921. Wigner's father wanted his son to learn chemical engineering, to prepare for work at the tannery, but once in Berlin, Wigner's studies led him to the "borderlines of knowledge" and such disciplines as physics, statistics, and mathematics.[19]

Although both young men studied at the institute, and Wigner remembers that they first met in a physical chemistry class there "during a brief flirtation of Szilard with this subject," the two really became acquainted during Einstein's seminar on statistical mechanics.[20] The two also attended Max von Laue's Thursday physics colloquium. At first, Wigner "did not understand a word, but somehow it had a fascination for me, and soon enough I understood." (In 1963 he won the Nobel Prize in physics.) From the start, Wigner and Szilard found affinity in their love of physics and their reserved manners. Both enjoyed walking and talking in the Tiergarten, from the Technical Institute at the west corner to the Brandenburg Gate and university district at the east end, through broad allées and winding paths, around lakes and streams; Szilard at a quick shuffle, Wigner at a lope. Almost all their thoughts spun around theoretical physics, and if the conversation touched on their common Hungarian roots and homes, Szilard was quick to change the subject.[21] "He seldom spoke about his parents," recalled Wigner, "but when he did, his affection for his mother rarely failed to come through."[22]

On their many walks Szilard and Wigner also talked about German politics and the fate of the depressed economy. Szilard read the Berlin newspapers every day and enjoyed following the many parties' intrigues. He worried about how the new constitution, devised at Weimar in 1919, would survive economic and political pressures, especially the humiliation and drain of postwar reparations to the Allies. How, he asked Wigner, could this new Weimar Republic make its constitution work? It all seemed too complicated—even for Szilard's inventive mind—with its strong president responsible to all the people, a chancellor responsible to the Reichstag (Parliament), a proportional-representation scheme to protect minority parties in elections, and only limited autonomy for the once-mighty German states. How would it work? What would keep it all together?

Although Wigner acknowledged Szilard's "great influence," he also considered Szilard "a very queer person" for the political and personal twists he gave to science. "He was a very able person," Wigner recalled, "but his interests were concerned very strongly [for] himself and for influence. . . ."[23] The two friends also disagreed about human nature, Szilard arguing that much could be achieved through applied reason, Wigner questioning this faith in the powers of the mind.

"You are a pessimist, Wigner," Szilard declared during one walk.

"But am I not usually right?" Wigner replied.

"It is hard to be right and be a pessimist," Szilard concluded.[24]

Another Lutheran Gymnasium graduate whom Szilard met in Berlin, John von Neumann, was one grade behind Wigner but two grades ahead in math. In Berlin, too, von Neumann's love of math inspired many of their conversations. All three shared the background of middle-class, nominally Jewish families in Hungary that had enrolled in Christian religions to deflect the swelling anti-Semitism after World War I: The Szilards became Calvinists; the Wigners, Lutherans; the von Neumanns, Roman Catholics. But von Neumann's family also had more money and loftier social pretensions; his father, a successful mortgage banker, adopted the title Neumann von Margitta, which John continued to use in the late 1920s when he taught courses with Szilard at the University of Berlin. The three were usually formal, in the style of German academia, addressing one another by last names. Szilard had no nickname in Berlin; but von Neumann was sometimes called "Jancsi," a Hungarian nickname for Johnny; and Wigner was "Jeno," Hungarian for Eugene, and later "Wigwam."[25] Behind his back, Wigner's friends also called him "Pineapple Head" for the shape of his pointed crown.

In Berlin, "Johnny led a life very different from ours," Wigner recalled. "He was sort of a bon vivant and went to cabarets and all that. Szilard was also different from me. Very different."[26] Yet the three Hungarians shared an intense intellectual vigor and often met on Saturdays—at Wigner's rented room or at Mrs. Dresel's, where Leo and Bela lived—to talk over studies and to joke about politics. They argued about anything, as long as the replies could be clever and the thinking quick. Indeed, the three became so nervously astute together that a thought could scarcely be uttered before it was understood and refuted or refined by another voice—a routine they shared into middle age.

From what they learned at Einstein's seminar on statistical mechanics, von Neumann developed his pioneering ideas on quantum mechanics, and Szilard tried concepts that he used in his doctoral thesis.[27] A few years later, this Hungarian circle around the University of Berlin expanded to include the chemist Michael Polanyi, an acquaintance of Szilard's from Budapest, and the Russian-born economist Jacob Marschak. The group was eager to learn economics as a way to understand Communist social developments in the Soviet Union, and in their freewheeling discussions some ideas arose that would prove important years later. At one gathering, Marschak remembered, when he was talking about the "classical" Marshallian concepts of "demand and supply equations," von Neumann stood up and ran around the table, saying: "You can't mean this; you must mean inequalities, not equations. For a given amount of a good, the buyer will offer *at most* such-and-such price, the seller will ask *at least* such-and-such price!" That focus on inequalities rather than equations was later developed as "mathematical programming," Marschak reflected in the 1960s, a technique also called "feasibility sets."[28]

Szilard had first met Michael Polanyi just after the war, when he and his brother Karl led political discussions in the Galilei Circle. Michael Polanyi had trained as a doctor and served as an army medical officer, but at war's end he returned home to complete a doctorate in chemistry, then left for Berlin. There he began research and teaching at the Kaiser Wilhelm Institute for Fiber Chemistry in Dahlem. Wigner worked as a researcher for Polanyi there and asked him to be his thesis adviser. Szilard was good friends with Polanyi and sometimes dined with him and his wife, Magda, at their apartment. Szilard also became friends with Hermann Mark, a chemist with whom he conducted X-ray experiments at the Kaiser Wilhelm Institute.

Szilard had no office at Dahlem or anywhere else, but wandered about the institute chatting with researchers and suggesting experiments they should try. This practice earned him the title *Generaldirektor,* a name for the heads of large companies. Szilard savored this nickname. He also liked spending evenings in Dahlem, when he stayed to dine at his colleagues' homes. Mark's wife, Mimi, sometimes saw Szilard twice a day, first when she organized sandwiches, coffee, and tea at the laboratory's lunch room and later when he dined at their apartment, four blocks from the institute on Werderstrasse. Both high-spirited Viennese, the Marks enjoyed Szilard's jokes and wry comments.

Like many Jewish men of his generation in the Austro-Hungarian culture, Szilard relished clever jokes about rabbis, and in Mark's opinion Szilard was a champion humorist. Mark's favorite Szilard story involved two Jewish boys who called on their rabbi to settle a dispute. The rabbi was dining with his wife but consented to hear them, and the first boy explained his side of the argument.

"You're right," the rabbi said. Then the second boy stated the opposite position.

"You're right," the rabbi said. At this, the rabbi's wife protested.

"You can't do that," she complained. "You can't tell one he's right, then tell the other that he's right."

"You know," the rabbi told his wife after a thoughtful pause, "you're also right."[29]

Jewish colleagues, such as the Viennese-born physicist Victor Weisskopf, turned this storytelling into a friendly sport. And they usually lost. At a party in Weisskopf's Berlin apartment one night, Szilard and his host matched each other joke for joke for more than an hour. When Weisskopf could recall no more, he retreated to the bathroom, where he had hidden a Jewish joke book, and by leaving the room a few more times, he kept up with Szilard. But eventually Weisskopf realized that he was whipped: Szilard was making up his jokes as he went along.

For his part, Szilard delighted in his colleagues' company and their cuisine. With no kitchen of his own in his rented rooms, he tried to reciprocate for the hospitality by inviting friends to cafés and restaurants. Szilard also enjoyed a night at the movies, especially to see the comedies of Charlie Chaplin, Buster Keaton, and Harold Lloyd. *Those Three from the Gas Station,* a German comedy heavy on slapstick, left Szilard shaking with belly laughs. He also liked to read and talk about current authors, among them the Hungarian dramatist Ferenc Molnár and Polish-born novelist Sholem Asch.[30]

During the 1920s in Berlin, Szilard was thin and vigorous, in contrast to his later roly-poly appearance. He was a light and erratic eater, at first from penury, later because he wasted little time at meals and seemed almost constantly in motion. Most of his nervous energy he spent moving around the university and Kaiser Wilhelm Institute. For a while he even applied his verve to tennis games with Mark and Polanyi. On the court, Szilard's busy legs helped to compensate for poor coordination and style. After work at the institute labs in Dahlem, Mark and Szilard often walked the few blocks south to a sports club in Lichterfelde and on some afternoons made a party of their game. For these occasions Mark invited his wife, and Szilard brought a date: usually Gerda Philipsborn and once or twice Sylvia Jacobson, an attractive blond woman who worked in a downtown department store.[31]

Szilard befriended Eva Striker, a niece of Michael Polanyi's who also knew Szilard's sister, Rose, and Victor Weisskopf. An artist and ceramic designer, Eva was passing through Berlin in November 1927, on her way to a business meeting in Hamburg, when she called on her uncle and through him met "a young man in a trench coat who spilled over with advice." Szilard told her: "Rose is in Hamburg. She says the city is unsafe. But there is one *pension* which is both safe and cheap. You *must* stay there."[32] Eva Striker worked at pottery factories in Hamburg and in the Black Forest and returned to Berlin in 1930. There she rented a large fifth-floor studio apartment on Tauenzienstrasse, five doors from the Romanisches Café, an artists' and intellectuals' haven surveying the Kaiser Wilhelm Memorial Church in Kurfürstendamm. Szilard, who lived a short walk away, often dropped by the café, and Eva made a point of inviting him to her lively parties.

Eva's studio was bright by day, with a tall window and two terraces that overlooked a courtyard. By night the huge, dark, high-ceilinged room resembled the set of a modernist German film; gooseneck lamps lit the spare white walls, and an ornate cast-iron stove warmed one corner. Guests streamed in and out: artists and politicians, playwrights and professors, scientists and pottery workers. Weisskopf, who sublet a room in the studio for a while, remembered Eva as "magnetically attractive, both physically and intellectually, a beauty not in the sense of the movies but in the sense that her intellectual interest and openness shone through. She was always present, always completely attentive."[33]

Expressionist Emil Nolde had a studio in the same building and often dropped in for a cup of tea or a glass of wine. Also on hand was the brooding Hungarian playwright Arthur Koestler, for a while Eva's lover, in 1930 an editor at *Berliner Zeitung am Mittag* and later well known for writing *Darkness at Noon,* a novel about Stalinist oppression. Michael Polanyi came by, along with Russian physicist Lev Landau. Both men enjoyed chatting with Szilard. Physicist Erwin Schrödinger, then teaching a course with Szilard at the university, joined some of these talks to give science a philosophical twist. At Eva's parties Szilard also met Manes

Sperber, a politically active Jewish author from Paris, and the Austrian physicist Alex Weissberg, whom Eva would marry a few years later when they moved to the Soviet Union.[34] Eva met Szilard's friend, lawyer Hans Zeisel, during one evening gathering and would later marry him, in 1938, after Weissberg disappeared in a Soviet prison camp.

Politics sparked comments and queries from almost everyone in 1930 as the Weimar Republic tottered into the worldwide Great Depression. With political factions multiplying, almost everyone felt obliged to take sides, and arguments about the many parties flared among the academics, artists, and politicians who crowded Eva's noisy room. But through this uproar Szilard remained strictly independent, refusing to back any party. Although he relished the give-and-take of political discourse, by this time Szilard espoused his own solution to Europe's impending chaos: the Bund, an apolitical institution to train and employ intellectuals as government advisers and leaders.

Szilard's intellectual powers often ruled his social behavior, too, sometimes yanking his attention away from friends and conversations in pursuit of some fresh idea. At Eva's parties, Szilard argued—and listened—intently; yet he could quickly fall silent, staring in a trance of concentrated thought. If the studio's roiling conversations suddenly seemed a bore, Szilard retreated in mid-sentence to a corner, sat down, and began reading a book or paper through the smoky light. When even reading seemed banal, Szilard omitted formal farewells and simply walked out the door, seeking new friends in the multitude of his own thoughts.

CHAPTER 6

Einstein

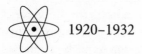 1920–1932

In 1920 Max von Laue's colloquium at the University of Berlin's Institute of Physics attracted the undisputed masters of the field, including Planck, von Laue, and Einstein. Seated in the front row of the classroom, these three defined and refined quantum mechanics and nuclear physics week by week. When Leo Szilard first sat in that spring, he had joined other students in the back row of the hall. But by the time he enrolled for a doctorate at the university that fall, he had eased to the middle and then to the front row. As one session ended, Szilard approached Einstein with a question, and within a few weeks of their first encounter the two men—both intensely shy but also both childishly open with their ideas—became nodding acquaintances. This introduction began a dialogue, then a friendship, and at times a mutual dependence that would flourish for years.[1]

Despite Germany's academic formalities and the differences in their status and ages, Einstein, then forty-one, and Szilard, twenty-two, soon fell into informal and friendly discourse. Szilard enjoyed Einstein's skeptical attitude toward science: "Oh, no," he often said to speakers at the colloquium. "Things are not so simple."[2] Within a year of meeting Einstein, Szilard was comfortable enough to ask him to teach a seminar in statistical mechanics; and Einstein was pleased to accept. This same openness prompted Szilard to abandon the thesis topic on relativity that von Laue had assigned him and to turn to Einstein for assurance that a completely different formulation was better.

Szilard's relationship with von Laue was formal; he never cracked jokes in his presence and only rarely contradicted him. Von Laue respected Szilard's keen intellect and invited him to dinner a few times but was himself so reserved that he never engaged Szilard's freewheeling mind in rambling conversation.[3]

To foster a friendship, Szilard often walked Einstein home after the colloquium: from the Institute of Physics building behind the Reichstag, around the Brandenburg Gate, and along the edge of the Tiergarten to his apartment in the nearby Schöneberg neighborhood. Einstein taught no formal courses but enjoyed meeting researchers at home, and every Wednesday he and his wife served tea and pastries for students. Chemists Hermann Mark and Michael Polanyi attended from time to time, but Szilard appeared every week. Szilard enjoyed the conversation and on his paltry budget no doubt appreciated the tasty food, but he also gained something more from Einstein's company. For in rejecting engineering, his father's profession, and adopting physics, Szilard saw Einstein as a father figure he could revere.

It helped that both men were intellectual and social misfits. They enjoyed challenging the tenets of modern physics, they shared an ironic sense of humor, and gradually they also discovered a deeper bond—the comfort that links one outsider to another. For despite his international reputation, Einstein then worked outside the mainstream of atomic physics and of German science, and Szilard, who strained to be independent in his own thoughts and actions, admired this detached quality in Einstein.[4] At the same time, Einstein enjoyed Szilard's startling and playful suggestions, which he attributed to lateral thinking, to seeing connections and similarities in apparently disparate events. Einstein played the straight man to Szilard's impetuous proposals and comments, and as it evolved, theirs became a relationship that both men seemed to savor. For a while, the two met almost every day.[5]

Besides ideas, Einstein and Szilard shared the bond of shyness. "I'm not much with people," Einstein admitted at about this time.[6] And Szilard, despite his clever verbal turns in classrooms and cafés, was for much of his life a distracted loner. Their mutual friend Eugene Wigner concluded that "Einstein was a naturally solitary person who didn't want his weaknesses to show and didn't want to be helped even when they did show."[7] He could have said the same for Szilard. Yet when Szilard slumped in an armchair in Einstein's book-lined study, fresh thoughts and rare sympathies abounded.

Mental agility and intensity came naturally in this cozy, cluttered room as both men forgot the strictures and courtesies imposed by German academic life and played with their own quirky ideas.

When alone, this shared directness was a source of joy, but outside Einstein's study it sometimes shocked their colleagues. In a laboratory at the Kaiser Wilhelm Institute in Dahlem, where Einstein was a consultant and Szilard a visiting researcher, Einstein one day walked to the board to sketch out and explain his ideas for an X-ray experiment.

"But, Herr Professor." Szilard interrupted from the front row, "what you have now said is just nonsense." Einstein paused, thought for a second, and smiled but said nothing. Szilard was right but rude. During another seminar at the institute, Einstein and Szilard discussed the Compton effect—an increase in the wavelengths of X-rays when they collide with electrons in matter.[8] Each man in turn made his point by striding to the blackboard, scribbling an equation, and sitting down again to listen. But when Szilard responded to one of Einstein's points and he failed to reply, Szilard paced over to his chair and glared down.

"Herr Professor?" Szilard asked. "Are you opposed to this statement, or don't you understand it?" Einstein looked cowed.

"Yes, yes," he muttered, peering up to Szilard. "Really, you are quite right."[9]

Einstein realized that Szilard's impertinence made him a maverick in the formal world of German academic life and once suggested a novel alternative. Soon after Szilard earned his Ph.D., Einstein asked: "Why don't you take a job in the patent office? That would be best for you; it is not a good thing for a scientist to be dependent on laying golden eggs. When I worked in the patent office, that was my best time of all. "[10] While working as a clerk in the Swiss patent office, Einstein had developed his theory of relativity.

Although both Einstein and Szilard sought ways to unify scientific knowledge with elegantly abstract theories, they could be intensely practical as well. After Szilard read in a newspaper that a Berlin family had perished when a toxic refrigerator coolant leaked in their apartment, he and Einstein wondered how such an accident might be prevented. A failed circulating pump had leaked, so, theoretically, the simplest solution must be to build a pump that cannot leak. To prevent leaks, they reasoned, their pump must have no moving parts and thus no gaskets and sealants that might fail. The more they thought about it, the more intrigued the two became. They considered the human heart, which works by muscular contractions. What if these spasms could be duplicated, not by mechanical surfaces that move but by another force? Electromagnetic waves can be spasmodic. Might these waves propel a liquid coolant through the system's pipes? They knew that a magnetic field would not move water. But several metals have a liquid form. Might a liquid metal, driven by magnets, circulate as the refrigerator's coolant?

After brainstorming with Einstein a few more times, Szilard called in a draftsman—an engineer he had first met in Budapest named Lazislas Bihaly—and as the three men sat together in Einstein's study, their ideas for an electromagnetic pump took shape on paper. Bihaly sketched a rough plan; Einstein and Szilard stared in thought, inserted details, questioned the lines and the scale. Then they sent Bihaly off to draft a final sketch, and Szilard, punching clumsily on his landlady's typewriter, wrote out a patent application.[11]

The heart of the Einstein-Szilard refrigerator design was a pump that circulated a liquid-metal coolant by creating electromagnetic pulses in an array of coils around a long tube. Just the kind of elegant simplicity both men esteemed. But would it work? Once Szilard had filed a German patent, he called on Siemens, the international electronics firm, and when they turned him down, he approached General Electric of Germany (AEG). They bought the idea, hired Szilard as a part-time consultant, and invited him to their laboratories. There Szilard built a prototype, and after months of work, a test model. The pump was set on a shop bench and switched on. The liquid metal, a sodium-potassium alloy, began to gurgle through the tube. Szilard grinned with delight for a few seconds. The machine was working beautifully, but as the metallic stream accelerated, the men around the bench heard a low hum that grew to a whine and finally swelled to a shriek. "It howled like a banshee," Szilard reported to Bela. Hardly the right sound for a family kitchen.

In all, Einstein and Szilard filed eight joint patents for their electromagnetic pump and its related parts. But the domestic-refrigerator market was soon dominated by General Electric's Monitortop design, with motor and cooling coils in a turbanlike array above the cabinet. Not until the late 1940s would their pump prove its worth to circulate liquid-metal coolants in an advanced nuclear power plant called a "breeder," a reactor that Szilard himself would both conceive and name during World War II.

Besides refrigerators, Szilard and Einstein shared a sense of wonder about religion and cosmology. "As long as you pray to God and ask him for *something*, you are not a religious man," Einstein said during one of their many talks. Szilard agreed.[12] Instead, they sought to understand the universe by discovering its essential nature. To a student in Berlin,

Einstein had declared: "I want to know how God created this world. I am not interested in this or that phenomenon, in the spectrum of this or that element. I want to know His thoughts; the rest are details." And to philosopher Martin Buber, Einstein had once said that what physicists strive for "is just to draw His lines after Him."

Einstein's God was "subtle" but "not malicious," meaning that He was consistent in His rules. So, by his often quoted thought that "God does not play dice with the universe," Einstein affirmed that His world had rules that were knowable and then predictable—if only you could ask the right questions. When a New York rabbi once asked Einstein about his faith, he answered that he believed "in Spinoza's God who reveals himself in the harmony of all that exists, not in a God who concerns himself with the fate and actions of men."[13].

As biographer Ronald W. Clark put it, "Einstein's God thus stood for an orderly system obeying rules which could be discovered by those who had the courage, the imagination, and the persistence to go on searching for them."[14] Baruch Spinoza, the seventeenth-century Dutch philosopher and lens maker who combined the ancient Hebrew Scriptures with science in a quest that was profoundly spiritual, became a model for Einstein and Szilard as they mused about God and nature. They read *Ethics,* Spinoza's principal work, and they read his letters. Notes about Spinoza creep into their own letters and postcards, for to them he was another independent mind in search of all that was knowable.[15]

The most sustained contact between Szilard and Einstein began in 1926 as they worked to design their electromagnetic pump, and this friendship of occasional visits and letters continued until after the last related patent was filed in 1930.[16] This collaboration overlapped with Einstein's own pioneering work on a unified field theory, which he published to widespread press acclaim in January 1930. Szilard's letters during their early friendship were prosaic; Einstein's, often ironic and sometimes even whimsical: In 1928, Einstein addressed Szilard as "Dear Dulder," slang for sufferer or toiler, and chided him for being too logical. He "tends to overestimate the role of rational thought in human life," Einstein once wrote of Szilard and his political reform scheme.[17]

Szilard and Einstein continued to write to one another about their refrigerator patents, and in 1931, Szilard enlisted his mentor to write a letter to the American consul in Berlin to support a visa application to visit the United States. Szilard's draft letter for Einstein had the famous scientist declare, "In that the purpose of this projected trip is the advancement of our commutual work, I have a direct personal interest in the granting of an entrance visa."[18] Einstein's letter worked, and Szilard sailed for New York. On this first American trip Szilard visited his friend Eugene Wigner, then a math and physics teacher at Princeton University. Szilard traveled to Washington on patent business. And he dropped in on the physics department at New York University.

Restless and unsure what to do and unwilling to seek work in America, Szilard returned to Berlin in the spring of 1932, where he resumed writing and calling on Einstein. The two kibitzed about physics and their colleagues. But while Szilard wondered aloud about his own future, he would never ask Einstein for help directly.[19] Szilard resumed teaching courses at the university but only dabbled in research, and when the semester ended in July, he traveled about Europe aimlessly. From Zurich that summer Szilard wrote to Einstein about their refrigerator designs and sought his signature on a letter to secure partial payment for patent rights. An AEG contract, Szilard wrote, would allow him "to tour America for about a year" or otherwise "possibly spend some time on an intermezzo in England."[20]

By the fall of 1932, Szilard seemed to have more doubts than ever about his career, but he still hesitated to ask Einstein for help. Instead, Szilard wrote to Wigner, who also worried about his friend's seeming inability to seek—or even decide on—a job. From Princeton, Wigner wrote to their mutual friend Michael Polanyi in Berlin, and Polanyi raised the matter with Einstein.

13 October 1932

Dear Professor,

I am writing to you in the interest of my friend Szilard, who finds himself in a rather difficult situation. Now that the matter of the refrigerator is coming to an end without having furnished him with the hoped-for financial independence, he has to find a way for his further livelihood that suits his peculiar character and gives him the opportunity of earning money. I believe that you, Professor, have particular goodwill for Szilard, and thus I permit myself to submit to you the following:

To me it appears possible that Szilard could obtain a position at the Flexner Institute [the Institute for Advanced Study] in Princeton if you support him in this quest. I also believe that this would be a suitable solution for him, since he could do theoretical work there and he could at the same time be a guest at the experimental institutes of the university. Life in America would probably satisfy also his social aims.

I have discussed this plan with Wigner, who thought of it even earlier. However, we did not want to do anything before your connection, dear Professor, with the Flexner Institute becomes publicly known. [Einstein would join the Institute the following year.]

Now that Wigner is gone to America, it occurred to me that I should turn to you. (Szilard refused to ask you a favor where his own interest is involved.)

Thus, may I ask you how you feel about this matter? Do you want to talk to Szilard about it? Or can I do anything in this matter, perhaps by inquiring about the local situation in Princeton, by asking Wigner or [Rudolph] Ladenburg? I believe that your decision will be of great importance for Szilard's fate.

Respectfully yours,
Michael Polanyi[21]

Replying within a week, Einstein declined to recommend Szilard to the institute. Despite an admiration for Szilard's talents, Einstein thought him more gifted in technical sciences and experimental physics than in mathematical physics. But he agreed to recommend Szilard if an opening occurred at Princeton for an experimental physicist.[22]

Szilard was so restless and impatient at this time that he left his rented room in Schöneberg and, as Polanyi's guest, moved into Harnack House, the Kaiser Wilhelm Institute's faculty club in Dahlem.[23] But within a few days Szilard left his Harnack House room vacant and traveled to London, where he wrote again to Einstein for help with a visa application to return to the United States.[24] By now Szilard felt comfortable asking Einstein for advice and letters of support on almost any intellectual or legal matter, from promoting the Utopian political reform program he called the Bund to renewing a visa. Yet Szilard still hesitated to ask his mentor for help finding a job—not, it seems, because he doubted that Einstein would comply but rather because Szilard himself had no clear idea what he wanted to do.

CHAPTER 7

Restless Research and the Bund

 1922–1932

In September 1922, a few weeks after receiving his Ph.D. in physics from the University of Berlin, Leo Szilard moved out of the rooms he shared with Bela at 2 Herderstrasse, in the city's residential Charlottenburg neighborhood, and sublet a tiny furnished room from an elderly couple at 61 Zimmerstrasse.[1] This put him in the center of the city and within a dozen blocks of the university. His new room was much cheaper, an advantage at a time when Szilard had no predictable income. But the savings came at a price: To reach his own room, and the bathroom, he had to walk through the landlord's bedroom, a passage that offended Szilard's acute sense of modesty.

The move was one of a dozen Szilard—stirred by a restlessness that upset both his personal and professional life—made during his twelve years in Berlin. With a doctorate, Szilard briefly considered teaching physics and in October asked his friend Michael Polanyi to write to a professor he knew at the University of Frankfurt. "I would like to recommend in the highest terms Dr. Szilard," wrote Polanyi. "He has a thesis worked out entirely on his own, a study in thermodynamics for his doctorate, which is of a fundamental nature, and on his own initiative. Prof. [Otto] Stern . . . calls him a fabulously wise man. . . ."[2]

There is no record that Szilard pursued a job in Frankfurt or anywhere else, for he preferred his life in Berlin, passing the days in lectures and seminars and the nights brainstorming with colleagues in cafés. Szilard later enjoyed sitting in the front row of a quantum mechanics seminar given by his friend John von Neumann, asking questions and posing his own clever answers. At Szilard's side for many of these sessions sat Erwin Schrödinger, a theoretical physicist from Vienna. (Schrödinger would soon develop a theory of "wave mechanics" as a new way to view the behavior of atomic matter, and for this work would share with Paul Dirac the 1933 Nobel Prize in physics.) With conjectures and criticisms flying among von Neumann, Szilard, and Schrödinger, students must have felt bewildered, for this trio was defining new aspects of quantum theory as they went along.

The three would also teach a course together a few years later and pair off to teach other topics.[3] Schrödinger enjoyed Szilard's participation and later recalled that he ". . . took an important part in all the discussions of our colloquia and seminaries. He was one of those to whom one always listens with greatest interest, for what he had to say was always of a profound and original kind. He very often points to an important point or to a view of the subject matter that would not occur to anybody else."[4]

Two days after Christmas in 1922, Szilard received a permit from the Berlin police that allowed him—as a long-term resident alien—to rent a furnished room in the city. Once assured that he could return to Berlin, Szilard boarded a train for his first trip home in more than three years. During two weeks in Budapest, he visited high school friends and chatted by the hour in the rococo splendor of the New York Café. At the family villa, he found his parents gloomy with financial problems; reluctantly, they had taken in a boarder to help make ends meet, and Louis Szilard, who had retired at age fifty-one in 1912, was back at work as an engineering consultant. The only consolation that Leo could give his parents was that life was far worse in Germany.

While Szilard was home, French and Belgian troops marched into the Ruhr Valley in northwest Germany to enforce reparations payments to the victorious Allies of the world war.[5] It was a bad beginning to a year that would see more economic and political unrest in the country. Inflation that began after the war now made life chaotic: When Szilard returned to Berlin in January 1923, the exchange rate was 7,000 marks to the US dollar and falling. Communists prompted uprisings in the northern state of Saxony. French agents backed separatist conspiracies in the Rhineland. And from a beer hall in Munich, a disgruntled veteran of the war, Adolf Hitler, led a putsch (revolt) to force local officials to swear allegiance to his National Socialist (Nazi) "revolution." Army troops broke up the confrontation, and Hitler spent nine months in prison, where in angry confinement he wrote *Mein Kampf* a book that brought national publicity to this young reactionary and his grandiose ideas.

Through the spring and summer of 1923, both Leo and Bela Szilard struggled along with most Berliners to simply survive. With the mark at 1 million to the US dollar and still dropping, people were paid daily and rushed out to spend their money, knowing its value would be even less by morning. Wallets were useless; shoppers used suitcases and wheelbarrows.

Yet amid this turmoil Szilard seemed unperturbed. One spring day in 1923 he met Michael Polanyi and rode the U-Bahn (subway) out to the Kaiser Wilhelm Institute (KWI) for Physical Chemistry, where he was conducting research. Set in suburban Dahlem, southwest of the city center near the forest and park called Grünewald, the KWI had been created in 1910 as a privately funded center for research that would aid German industry's scientific and technological development. In contrast to research at the state-run universities, work at the institute was to be practical and applied. But many scientists shuttled back and forth between the university downtown and the campus in Dahlem, in their brief commute closing the traditional gap between theoretical and applied research.

"This is Dr. Leo Szilard, who has done his thesis in physics under Professor von Laue on a theoretical subject," Polanyi told chemist Hermann Mark. "He would like to see what experimental methods are used today in modern physics and chemistry." Mark, a bright and energetic Viennese who was investigating the structure of fibers, walked Szilard through the laboratories, explaining the elaborate X-ray diffraction equipment, then the most powerful and precise in the world. Szilard looked and listened intently and, in a room filled with X-ray scanners, eyed the gleaming machines carefully.

"What are you doing with those instruments?" Szilard asked.

"Analyzing cellulose and silk fibers," Mark boasted, "and the behavior of metallic materials—zinc, copper, tin—when they are pressed and deformed." Szilard looked disappointed.

"You are wasting your time," he declared. "With such an excellent apparatus, you should not work on the practical application of X-rays but on their own fundamental nature and basic properties. You should study the X-rays themselves!"

At first, Mark resented Szilard's arrogance, but a few moments later he realized its value. During the next several months Mark persuaded some of his colleagues to listen to Szilard, who soon became a regular visitor to the KWI. Some researchers continued to think Szilard haughty and intrusive as he wandered through the laboratories, asking questions and proposing new experiments for them to do. But once it became clear that his comments were often right and sometimes practical, Szilard was accepted, although with the title *Generaldirektor.*[6]

Szilard also annoyed some fellow scientists when he filed for patents on his ideas. He did so, Szilard later wrote, to secure the money needed for independent research. But to many colleagues this practice seemed selfish and unscientific. Szilard filed his first patent, on October 9, 1923, for an X-ray sensor element he conceived after viewing Mark's equipment.[7]

If Szilard's motives seemed suspect, he could hardly be blamed for financial anxiety at the time. With no steady income, the occasional consulting and research fees Szilard earned were scarcely enough for both food and rent. In April 1923, Szilard left his awkward room on Zimmerstrasse and rejoined Bela at a new sublet at 211 Kaiser-Allée (now Bundes-Allée), a broad boulevard running south from the zoo railway station between the residential neighborhoods of Wilmersdorf and Schöneberg. Their landlady, a Mrs. Salomon, was a widow who could not maintain her large and elegant apartment against the storm of inflation and to meet expenses had dismissed the maid and rented out her room.

Theirs was a large and airy room with high ceilings, and on a small table by a window most mornings the brothers shared a simple breakfast. Leo used student tickets to buy lunch at the university, and he attacked whatever was set out at teas and parties. Bela could eat a hearty lunch at his part-time job in an electronics company, and when together at night, the Szilards fixed a simple cold meal of cured meats, bread and butter, sometimes with tinned fish or fruits. Bela's hand slipped one evening as he opened a sardine tin; the lid slit the knuckle of his left little finger, and blood spurted onto the tablecloth. When the bleeding continued, Leo became alarmed, took Bela by the arm, and led him a few blocks to a city-run first-aid station. Leo sat in a chair as Bela's hand was doused with iodine and bandaged, and when the treatment was complete, Bela turned to find Leo in a dead faint, his head against the wall, his feet sprawled out across the tile floor.

Although Szilard needed to find work, he was loath to seek a full-time job, fearing that it might imperil his cherished independence. In May 1923, he registered at the Technical Institute, where he had begun engineering studies three years before, perhaps seeking a place to conduct experiments,[8] and began to monitor theoretical physics research at the

KWI. At the same time, he sought collaborations with researchers at the KWI for Fiber Chemistry, for which Mark arranged to pay Szilard a modest fee as a scientific consultant. This arrangement suited Szilard well, giving him the freedom to be in two places at the same time—or in none.

Szilard's personal ties to the KWI grew in several ways. Von Laue followed Mark's experiments on X-ray diffraction carefully because they related to his own work on the structure of crystals. Einstein visited the labs frequently, following Mark's study of the Compton effect. Unexpectedly, as people dropped in, discussions began about different problems in contemporary physics. Szilard's friend Eugene Wigner, a doctoral candidate at the Technical Institute, performed his thesis experiments in Dahlem and was on hand for informal chats. Satyendranath N. Bose, an Indian physicist working closely with Einstein, met and talked often with Szilard about relativity and quantum physics. But in this brilliant company Szilard often became so enchanted with the implications of the dialogue that he shut his eyes in thought—a habit some mistook for boredom or fatigue.

Cadging meals and crashing conversations, Szilard lived as an intellectual vagabond. He owned almost nothing: one or two suits, shoes that he wore until they frayed and fell apart, a few dozen favorite books, and an untidy array of papers and periodicals strewn around his rented room. In this way, Szilard was free to be himself: a mind and spirit constantly on the move. In April 1924, after living with Bela for a year, Szilard left 211 Kaiser-Allée for a room at 104 Leibnizstrasse, near Kurfürstendamm. Six months later, he moved to 11 Geisbergstrasse, a modern five-story apartment building a few blocks south of Wittenberg-Platz and the busy cafés and stores of Tauenzienstrasse. There he rented a room from the Philipsborn family and began a sensitive and abiding friendship with their daughter, Gerda. After four months with the Philipsborns, Szilard moved on to 32 Barbarossa-Platz, a modern apartment over a corner store and within a few blocks of Einstein's apartment, and by the following summer Szilard had moved again, this time to 58 Pariserstrasse.[9] Then, in October 1927, Szilard moved to 95 Prinz-Regentenstrasse, to a room in the Philipsborns' new apartment. This stylish art nouveau building, near Prager Platz, finally became a home, of sorts, for Szilard; he lived with the family for the next five years and moved with them around the corner, in 1932, to an elegant, balconied building at 58 Motzstrasse, by the picturesque Victoria Louise Park.

Szilard's appointment in 1925 as von Laue's assistant was his first paid academic position. Although part-time, this post was prestigious because von Laue, a Nobel laureate in 1914, picked only one assistant at a time. With Mark, Szilard published two papers on the anomalous scattering of X-rays in crystals and on the polarization of X-rays by reflection on crystals.[10] But beyond this, the patent applications for an electromagnetic pump that he invented with Einstein, and expansions of his thesis, Szilard wrote very little throughout the 1920s.

A world of rented rooms and borrowed laboratories gave Szilard the sense of independence he had craved since childhood and with it his childish sense of freedom. Now in his late twenties, he could act impulsively for no apparent reason but to assert his will. When his sister, Rose, visited Berlin and telephoned from her hotel, she offered to walk the few blocks to his apartment, but Leo insisted they meet at her room.

"Leo, just tell me which street you will take and we can meet in between," she said.

"No," he answered. "I don't know which street I will choose until I choose it! I will meet you at your room."[11]

Szilard was footloose in Berlin; he scurried around the city, taking in lectures and scientific meetings as if they were light entertainments. If any money accumulated, he quickly spent it on train tickets, sometimes leaving Berlin in mid-semester for jaunts to Paris, Vienna, Zurich, or Danzig (now Gdánsk) in search of fresh ideas. He once took Wigner to Hamburg to hear physicist Werner Heisenberg speak and after the lecture quizzed him about his work in quantum mechanics.[12] And when Mark became assistant research director at an I. G. Farben chemical plant at Ludwigshafen, near Heidelberg, Szilard sought him out and brainstormed around the laboratory for two days.

"He was full of interesting results on the new quantum mechanics of Heisenberg and the wave mechanics of Schrödinger," Mark recalled later. "Our laboratory was part of a large industrial combine, and Leo was interested in low-boiling, nontoxic liquids that could be used in his refrigerators. He was equally sharp and knowledgeable in talking with the engineers in the plant as he was in talking with the scientists in the laboratory." From his conversations, Mark concluded, Szilard was "the early founder of nonequilibrium thermodynamics" through practical application of the statistics he had devised in his doctoral thesis.

Szilard thought constantly about the science that abounded in his narrow world, and with no quarters in which to entertain, he spent hours in cafés, drinking tea or coffee, eating sweet pastries, and talking. A favorite haunt, the Romanisches Café, was a meeting place for artists and intellectuals. The location itself was exciting: The café's awning faced the back towers of the neo-Romanesque Kaiser Wilhelm Memorial Church, looming on an island in the traffic where Kurfürstendamm and Tauenzienstrasse collide. Seated in the open or under the broad awning, Szilard and his friends could look up at the church spires and back to the café's own Romanesque towers; they could glance through the whirling traffic to the brightly decorated pavilions at the zoo; and they could peer along the crowded sidewalks of the two bustling thoroughfares.

This was hardly the cozy, sedate world of Budapest street life, where horse-drawn trams and wagons had gone clopping by the shaded coffeehouses. This was Berlin, a harsh playground for commerce and industry, a hectic intersection of shattered lives and swelling ambitions. Inside the Romanisches Café the small, round tables were clustered under brightly decorated Romanesque arches that echoed conversations and the clatter of cups and spoons. People hopped and shouted among the tables, came and went at all hours, sat by day and slumped by night in a swirl of talk and expectation. At the time, this bustling café was called "the place where everybody is somebody."[13] And for Szilard, whose rented rooms were just a few blocks away, it was the hub for his own frantic and fragmented life.

With his friend Dennis Gabor, a Hungarian physicist then studying electron beams at the Technical Institute, Szilard spent hours sitting in cafés and walking the streets, wondering aloud about new discoveries and experiments. Many of their daydreams are today's technology, and Gabor remembered one chat when Szilard suggested making a musical instrument that resembles today's devices for reproducing visible speech.[14]

"You know," Szilard told Gabor over a café table in 1927, now that it is possible to make electron lenses, "why do you not make a microscope with electrons?" At smaller and smaller wavelengths, you would achieve much more detailed resolution than is possible with microscopes using light. Gabor and Szilard pondered his idea for a few minutes, then agreed it would serve no useful purpose. After all, you could not put living matter into the kind of vacuum tube needed to control electron beams. Besides, they concluded, so much power would be focused in the electron beam that it would incinerate any sample.

But as Gabor later realized, with that idle suggestion Szilard had grasped the possibility of an electron microscope at least a year before anyone else. And of the incinerated sample, Gabor later wrote, "Who would have dared to believe that the cinder would preserve not only the structure of microscopic bodies but even the shapes of organic molecules?" Gabor would be remembered years later as the inventor of holography, for which he received the 1971 Nobel Prize in physics.[15]

Szilard himself filed a patent application for a simple variation on the electron microscope in July 1931, the same year that Ernst Ruska operated his first crude device. But, typically, Szilard failed to follow through this idea beyond the thought and the patent sketch, demonstrating his need to stay free from the drudgery of scientific invention—the calculations and drafting that must succeed creative impulses if ideas are to be converted into practical things.[16] Perhaps in Szilard's case this is just as well, considering some of the practical things he thought of making: He once proposed to Gabor connecting the bloodstreams of an old dog with young ones to produce a giant breed that would live forever.[17]

Solving personal problems also appealed to Szilard, provided they were someone else's. The Budapest artist Roland Detre, a close friend of Rose Szilard's, moved to Berlin in 1926 and often sought Leo's advice. But affairs became touchy when Detre and Rose, who was still living in Budapest, decided to marry. The Szilards despaired that their daughter should marry a starving artist and wrote to Leo for help. If at all possible, they said, try to dissuade him from marrying Rose. Szilard invited Detre to dinner and began their conversation by mentioning the letter from his parents. But poverty should be no reason to avoid marriage, Szilard said. Matrimony should be avoided on principle.

"I have decided never to marry," Szilard declared, "because I am convinced that marriage becomes a liability for someone who is dedicated to the creative life." But Detre protested that he loved Rose and wanted her as his wife. The two men argued into the night, until, exasperated, Szilard finally stared at Detre and said, "If you are so determined to live with Rose, why don't you bring her out here? You could live with her without getting married."

"But don't you think that this would make your parents feel even worse?" Detre asked.

"What do *you* care about *my* parents?" Szilard said, trying to conclude the discussion. Detre did care about the Szilards and soon returned to Budapest determined to marry Rose. The wedding was a cheerful celebration, attended by their friends. The Szilards refused an invitation but appeared at the last minute, reluctantly accepting that their daughter would marry a poor artist.[18]

When Roland and Rose Detre moved to Berlin and rented a garret room near the Romanisches Café, they met Leo often for coffee or a snack. All three loved to talk—and to argue. Rose considered herself "very religious" but without a formal religion. "Oh, yes," she would say, "I believe in everything." Detre shared Szilard's fascination with Spinoza and used his writings to devise a new religion of his own. In a Chinese restaurant one night, Szilard and Detre discussed his new theology throughout the meal and later talked on over cups and cups of tea. Szilard strained to find a rational way to refute Detre's system. Detre was delighted when Leo, acting miffed, admitted, "It would be possible."[19]

✳ ✳ ✳

In his hours of thinking and talking—questioning and quibbling at studio parties, arguing in cafés and classrooms—the spare and ad hoc life that Szilard led enabled his mind to range wildly across the scientific disciplines. Political and social forces were driving and changing the Berlin society he enjoyed, and while studying the newspapers and listening to the cafés' clientele, Szilard almost daily pondered Germany's fate. By the mid-1920s Szilard believed that the Weimar Republic, Germany's postwar government, was doomed to fail. Lacking procedures that might nurture and elevate new leaders, the overly complicated Weimar constitution could only grind itself down in a friction of disorder; in effect, an entropy of governance.

But Szilard still believed that democracy in Germany might survive for "one or two generations," and he devised a rational scheme to help postpone the republic's collapse and, perhaps, to prepare for transition to some healthier form of government.[20] Drawing on the example of the Youth Movement (*Jungendbewegung*) that had flourished in Germany before the world war, Szilard called his organization the Bund, to his mind a closely bound alliance of like-minded young people. When Szilard brainstormed with Polanyi about the Bund, he praised *The Open Conspiracy: Blueprints for a World Revolution* by H. G. Wells and thought that the first twenty pages of this book, which was published in 1928, posed succinctly the problems that the world faced.[21]

Polanyi arranged for Szilard to meet Otto Mandl, a wealthy timber merchant from Vienna then living in London, and during the Easter break in 1929, Szilard made his first journey to England. Mandl had discovered and enjoyed the writings of Wells, and moved by their visionary genius, he had arranged to publish them in Germany. Also, Mandl had married pianist Lili Kraus, who was born in Budapest and studied in Vienna, so Szilard found much to chat about when he called at the couple's home. During his London sojourn, Szilard ate with the Mandls almost every day, and a highlight of the visit was a dinner in late March attended by Wells. Writing Polanyi on April 1, Szilard reported that Mrs. Mandl and the children "are very nice each one for himself and compose a family altogether pleasant to look at." Szilard's English is stilted and affected, his spelling has German lapses, but through it all burst new enthusiasms for the ideas that might drive his own concepts for political reform. Mandl liked Szilard's notion of the Bund, but two other candidates Szilard tried to meet, biologist Julian Huxley and biochemist John B. Haldane, were abroad during the Easter holiday.[22]

Back in Berlin that spring, Szilard taught but one course, "New Ideas in Theoretical Physics," leaving him free to think about the Bund, and the idea sustained his attention much longer than the inventions proposed in physics. "What we want are boys and girls who have the scientific mind and a religious spirit," he wrote in detailed plans for the Bund.[23] He thought about the Bund throughout the summer of 1929 and continued to scribble notes and pose ideas that fall, when he also helped teach a course on "Problems of Atomic Physics and Chemistry." This course brought him to the KWI at least once a week, and at the Physics Institute downtown he joined his friend John von Neumann in teaching "New Questions of Theoretical Physics."[24]

Whenever talk turned to politics around the café tables, Szilard expounded his own analysis of postwar economic developments, seeking a balance between laissez-faire capitalism and the socialist elitism that he thought necessary to reform the world. The life-styles predominating in "civilized countries" were closely associated with the ruling economic system, Szilard argued, and were largely determined by underlying economic principles. He complained that there was no sense of community purpose to bind the needs and aspirations of the German nation. And he concluded that some form of parliamentary democracy must be maintained to support laissez-faire capitalism.

From all this brainstorming, Szilard easily became gloomy about the survival of German democracy; he had noted a serious danger sign in February 1929 when he read about a conference in Paris on the rescheduling of Germany's postwar reparations payments. Representing Germany was the flamboyant Hjalmar Horace Greeley Schacht, president of the

German Reichsbank, who declared that Germany would not resume payments until she recovered her former colonies. Szilard concluded that "if Hjalmar Schacht believed he could get away with this, things must be rather bad." The practical step Szilard took after reading about that was to transfer his money to a bank in Geneva.[25]

That move would save his personal finances, at least temporarily. But how, Szilard wondered, might he save Germany? Europe? The world? The Bund remained his answer. Szilard envisioned as the ultimate result of the Society of Friends of the Bund a utopia brought about in a society guided by an intellectual elite. "If we possessed a magical spell with which to recognize the 'best' individuals of the rising generation at an early age . . ." he wrote, "then we would be able to train them to independent thinking, and through education in close association we could create a spiritual leadership class with inner cohesion which would renew itself on its own."[26] His Bund would engage its members in a "life of service," but unlike a religious order, "the ruling opinion" would not "be retained for a very long time." Nor would this institution resemble "scientific academies," because they usually replenish themselves by "inside elections" that "preserve the ruling opinion" by selecting candidates for their "political" views.

What Szilard sought instead was a way for the Bund to constantly generate fresh ideas, then assume "a more direct influence on public affairs as part of the political system" either with or in place of government and Parliament.[27] The trick in all this Platonic speculation, he warned, is that the Bund's members "must not of course be entitled to a higher standard of living nor to personal glory."

Szilard's three-step proposal began with "the best" boys and girls being recruited from the top form of secondary schools at ages eighteen or nineteen to become "junior" Bund members. To spot the "best" candidate, he suggested looking for students who are considered "personalities" by their peers.[28] (In their independence and personality, these ideal candidates were remarkably like Szilard himself.) Juniors and seniors could interact in "club rooms" at centers for discussing "public affairs," with reading and lecture rooms also on site. But the juniors should also keep in touch with friends from the general population. Szilard hoped that in the club rooms young people "will learn to think for themselves and do so in areas where for most people passions and emotions prevent clear thinking."

In a group version of Szilard's own "active passivity," the Bund's juniors should meet the "finest representatives" of contemporary political movements but "should not be allowed to join . . . any political party or philo-sophical movement before they have reached the age of 30 years."[29]

As a second step, juniors would study at only a few universities in order to keep in close contact through seminars and workshops. Step three, an "Order of the Bund," Szilard hoped would allow dedicated members to indulge in "a life of sacrifice and service." The order's members must hand over all money they earn "above the base minimum necessary for their existence." And for about one-third of them "celibacy may perhaps be prescribed."[30]

Eventually, "cells" of thirty to forty members would hold elections to designate the "best" among themselves. Through the cohesiveness of the cells, the Bund would gradually overcome the tendency that individuals in capitalist societies have to "think of themselves in the middle of a struggle for everyone against everyone else." In this way, Szilard's desire for a "community purpose"—a decisive ingredient that he found missing in modern societies—might finally be defined and achieved. The way Szilard described this result, writing about 1930, sounds almost mystical:

> Thus it seems quite feasible that by the weight of the personality of the indi-vidual [Bund] members and by the cohesiveness of this group, the Order might represent some form of structure in public life, which would leave an imprint on the whole spiritual life of the community. It is probable that subsequently other patterns will arise spontaneously, which will safeguard the transmission of public opinion—forever renewed within the Order in difficult strife—to the general population.[31]

The Weimar Republic's own strife had led Szilard to devise the Bund, but he soon broadened the concept to include France and England. With members in "key positions in education and other fields of communica-tion," Szilard hoped, "the country as a whole would be responsive to the leadership of the Bund. . . ."[32]

Szilard's scheme for the Bund envisioned a transition of one or two generations. In fact, the time remaining for meaningful reform in Germany was but one or two years. The Weimar Republic's economy suffered new strains after the collapse of the world's stock markets in October 1929, and within a year unemployment soared. In England, too, unemployment rose, and when Szilard returned to London in March 1930, the Bund had an urgency there that it had lacked just a year before. H. N. Brailsford, a British socialist leader acquainted with Einstein, was cordial when Szilard called to explain the Bund but noncommittal after hearing about it in what must have been numbing detail. After this meeting, Szilard sent Einstein a twelve-page, single-spaced summary to illustrate his ideas about the Bund.[33]

Confused by what Szilard had said, Brailsford also wrote to Einstein. Szilard's "scheme interests me," Brailsford said, "but my difficulty in making up my mind about it is aggravated by the fact that I know nothing of the personalities connected with it." If there were "a very devoted and very able group of young men in Germany," Brailsford might be more disposed to provide help in England. "I have seen enough of Dr. Szilard to realize that he is an attractive personality, and he evidently has what he calls the religious spirit," Brailsford wrote. But he wondered if Szilard had the "capacity for this very difficult piece of work." And he asked pointedly what Einstein thought of the whole idea.[34]

Einstein replied that Szilard had "a circle of excellent young people" interested in the Bund but "as yet no organization of any kind." In this letter to Brailsford, Einstein was both friendly and candid when he assessed Szilard and his ideas.

First of all, I am of the opinion that [he] is a genuinely intelligent man, not generally inclined to fall for illusions. Perhaps, like many such people, he tends to overestimate the role of rational thought in human life.

As for the [Bund] itself, I, too, very strongly feel the incoherence of those who are magnanimous enough to wish for a more rational and more orderly power to determine the fate of mankind in the long run. But I do not really trust my judgment in practical matters of this kind; and I have no clear picture about the future prospects of the idea advocated by Szilard. Above all, I do not see a strong binding force that would give uniformity of action to a mass chosen in this way. On the other hand it would seem inappropriate to remain inactive in matters of such importance to mankind, simply to plant one's cabbages and watch the power hungry and the obsessed turn the face of this planet into ever-increasing ugliness.

This does not answer your questions, but no fool gives more than he has.[35]

For all its inflated faith in reason and its intricate rules, Szilard's Bund was remarkable in its time for two features: It included "boys and girls" on equal footing when European academic and government circles were almost exclusively male, and it required an internationalism by the participants that would not be echoed for almost two decades—and then mainly by Szilard's own colleagues in their attempts to curb the spread of the nuclear weapons they had helped create. In his world, the international scientific community, contacts among scholars in different countries were commonplace, peer review was rigorous, and rational progress was something that many assumed was achievable. Despite repeated rejections of his Bund, Szilard strained for the rest of his life to devise a self-selecting intellectual elite that might provide both national and international leadership.

In 1930, by virtue of being a *Privatdozent* (university lecturer), Szilard received German citizenship. Yet his new country was by then lurching further from the democratic goals he was straining to perfect. Hitler's Nazi party made surprising gains in the Reichstag, increasing its seats from 12 to 107 and becoming Germany's second-largest party.

In Berlin that winter, Szilard resumed his casual academic career by joining the Austrian-born radiochemist Lise Meitner to teach "Questions of Atomic Physics and Atomic Chemistry" each Thursday at the Kaiser Wilhelm Institute for Physical Chemistry in Dahlem. And, on Fridays, he led "Discussion of New Work in Theoretical Physics" with von Neumann and Schrödinger.[36] In the spring of 1931, Szilard joined with Dr. Emil Rupp at the Laboratories of the German General Electric Company (AEG) to study the behavior of polarized cathode rays in magnetic fields.[37] During the summer semester that began in April 1931, Szilard's teaching involved a single course on Friday afternoons when he led a "Discussion of New Work in Theoretical Physics" with Schrodinger.[38] In June, with Einstein's help, Szilard began efforts to secure an immigration visa to the United States. And, in July, he filed a patent for an electron microscope. No other achievements are recorded for this time in his curriculum vitae or in his papers. It seems likely that Szilard spent most of his energy thinking—worrying—about the shaky state of the world. An astute newspaper reader and amateur economist, Szilard watched in alarm as the Austrian and German banking systems collapsed in 1931. He watched as German unemployment soared, from fewer than a million in 1929 to more than 5 million in 1931. By that summer, US President Herbert Hoover declared a moratorium on Germany's war-reparation payments, but this came too late to stay the country's economic collapse.

During the winter of 1931, Szilard left Berlin for London, came back to Berlin, returned to London, then sailed to New York, where he filed immigration papers. In May 1932 he was back in Berlin with a fresh infatuation for nuclear physics. And fresh dread about the country's fate: Two months later, his political anxieties proved warranted as the Nazis doubled their strength in the Reichstag and became, for the first time, the largest party.

Szilard's acquaintance with Lise Meitner—around the KWI and at international conferences—had grown to a respectful friendship after they had taught together. He admired her research on the nature of beta and gamma radiation. He appreciated that she was just the patient experimentalist he could never be. And he fancied collaborating with her on some scientific experiments. But by 1932, Szilard's interest in physics seemed more science fiction than science. After reading and talking to H. G. Wells, Szilard believed that human survival depended on colonizing other solar systems, and the atom seemed the only power source capable of sustaining space travel.[39]

Yet after a dozen restless years as a physicist, Szilard had little to show for his efforts—a few papers, a few patents, and a fruitless political scheme. Although he savored brainstorming with colleagues, Szilard did almost nothing to develop his ideas from such playful banter. Szilard had thought up and patented a linear accelerator in 1928, about the time that Rolf Wideröe first envisioned his successful invention.[40] And Szilard had patented a cyclotron in 1929, a year before Ernest O. Lawrence built his first machine at Berkeley.[41] But for all this thoughtful turmoil, most of Szilard's ideas seemed to evaporate as quickly as they were spoken.

Ironically, if Szilard and Meitner had collaborated in 1932, as he had wished, the world might have had the atomic bomb years before it did, but with more dreadful results. After Hitler took office in 1933, Szilard left Berlin for London and there conceived the nuclear "chain reaction." He worked to keep this concept a military secret from Germany while searching for a natural element that would create a chain reaction, and in 1934 he enlisted Meitner (still in Berlin) to experiment with the release of neutrons from beryllium.[42] In 1936, Szilard was intrigued by Meitner's discovery of radioactive anomalies in uranium, which he saw as akin to his own research on indium—then the element he thought might fuel a chain reaction.[43] But it was 1939 before Meitner first recognized that uranium split (fissioned) to give off extra neutrons, and this was just the element that ultimately yielded the first nuclear chain reaction in 1942, in a device codesigned by Szilard and Enrico Fermi. Had Szilard and Meitner worked together, they might have made their fateful discoveries about uranium while still in Hitler's Berlin—a scientific possibility that we can be glad never occurred.

CHAPTER 8

A New World, a New Field, a New Fear

 1931–1932

In the spring of 1931, Leo Szilard was more restless than usual. Every Friday afternoon he and physicist Erwin Schrödinger led a weekly "Discussion of New Work in Theoretical Physics" at the University of Berlin, and at other times during the week Szilard visited the laboratories of the German General Electric Company (AEG) to study the behavior of cathode rays in magnetic fields. But neither activity fully engaged his quick and quixotic mind, which wandered more than ever—from physics to politics to the appealing thought of visiting America. In June, Szilard had his colleague and friend Albert Einstein send a letter asserting that "the advancement of our commutual work" required Szilard to obtain an entry visa to the United States.[1]

That summer, Szilard poked around Austria, spent the last week of August in Salzburg, and when he returned to Berlin, received an invitation to spend a year at Princeton with a group that was developing "mathematical physics."[2] For this Szilard received a US quota immigration visa on October 4, packed frantically, and caught a train early the next morning for London. There Szilard rented a room at the King's Court Hotel on Leinster Terrace, in the transient neighborhood of Bayswater, near Paddington Station. He went to the Bow Street Alien Registration Office in Covent Garden to file papers allowing him to stay in England, just in case that might be necessary.

While Szilard waited in London, he became convinced that he should immigrate to the United States, not just visit. He wrote Einstein seeking help to change his visa status and returned to Berlin. This time Szilard had Einstein write the American consul, saying: "My dear colleague Dr. Leo Szilard is planning to continue work, which we have started together, in America. As the carrying out of this work in which I myself have an interest will probably require some lengthy time, I am taking the liberty to support herewith his application for a nonquota immigration visa."[3] Once again, Einstein's letter worked.

For the winter semester that began in November 1931, Szilard was scheduled to give the "Seminar on Questions in Atomic Physics and Atomic Chemistry" that he had taught before with Lise Meitner. He was also scheduled to again lead the "Discussion of New Work in Theoretical Physics" with Schrodinger.[4] But at most Szilard appeared at only a few classes, for early on the morning of December 1, he boarded a train for London, where he bought a tourist-class ticket for the SS *Leviathan*. On December 19, 1931, Szilard sailed for America.[5]

Most immigrants who have steamed into New York harbor by ship retain vivid memories of their first impressions, and Szilard's memories of that moment were also clear, though oddly different from those of his fellow passengers on deck that morning. Years later, he recalled the moment as gravely prophetic. "As the boat approached the harbor, I stood on deck watching the skyline of New York," Szilard said in a 1954 interview. "It seemed unreal, and I asked myself, 'Is this here to stay? Is it likely that it will still be here a hundred years from now?'

"Somehow I had a strong conviction that it wouldn't be there. 'What could possibly make it disappear?' I asked myself . . . and found no answer. And yet the feeling persisted that it was not here to stay."[6] It was Christmas Day, 1931.

Once ashore, Szilard found Manhattan cold and dazzling. He made his way to the cavernous, neo-Roman Pennsylvania Station and from there rode by train to Princeton, where he checked into a hotel and called on Eugene Wigner, the Hungarian friend from Berlin who had suggested Szilard's name to the mathematics department. Szilard met the department chairman and dean of faculty, Prof. L. P. Eisenhart, and joined in his colleagues' seminars. But within a few days Szilard felt restless again. The young men working in mathematical physics were mostly interested in abstract mathematics, while he craved problems that were more concrete. And Szilard could not forget his brief time in New York; he was simply fascinated by the steely spirit of the place. Before long, he stopped his daily visits to the mathematics department and instead walked to Princeton's cozy railroad station and rode into the city. On one trip Szilard paid six dollars to rent a box at the Chase Safe Deposit Company on Fourth Avenue and Twenty-third Street.[7] On another visit he rode the subway to the stately campus of New York University, perched on a plateau over the Harlem River in the Bronx, and there dropped in on the physics department and its chairman, Prof. Richard T. Cox, a lively and friendly physicist who was about Szilard's age. Cox was studying polarization and electron scattering, and when he heard about Szilard's recent work at AEG, he wanted to duplicate it. Szilard, in turn, was beguiled by Cox's informal interest—electric eels. Cox had devised electrical systems to measure their voltage and enjoyed wiring his specimens to power neon lights. Szilard and Cox got along well, and on the first day of February, Szilard packed his bags in Princeton and moved to New York, checking into the Kenmore Hall Hotel on East Twenty-third Street.[8]

The Battle of Shanghai began in February 1932, when Japan attacked the Chinese port city. The raid broke a Chinese embargo against Japanese goods, but at a price Szilard thought intolerable. Upset by the brutality and by the breach of international order that he already saw as cracking in Europe, Szilard protested Japan's siege to several young scientists at NYU. He urged them to organize a protest. And he drafted a public statement to be signed by scientists, but with a twist: The signers committed themselves to boycott Japan only if a certain percentage of those approached actually signed the document. In this way, Szilard reasoned, most would be likely to sign, but no individual had to take the initiative.[9] There is no record that his boycott amounted to anything, but Szilard was quick to organize fellow scientists for many other political causes until near the end of his life.

When summer semester at Berlin University began in April 1932, Szilard was scheduled to teach the "Discussion of New Work in Theoretical Physics" with Schrodinger. But he was still puttering around New York, running errands, and paying calls on the NYU physics department. Szilard picked up a German passport at the consulate and traveled at least once to Washington, probably to inquire about his immigration status and patents.[10] It was May 4 before he boarded the SS *Bremen* to sail for Europe, his foray into mathematical physics now all but forgotten.

Back in Berlin, Szilard was eager to study nuclear physics and discussed a few possible experiments with Lise Meitner. And he resumed the "Discussion of New Work in Theoretical Physics" with Schrodinger, although his thoughts were elsewhere.[11] In the summer of 1932, no one Szilard approached seemed interested in his proposal to save Europe from political catastrophe with the Bund—not even the intellectuals he hoped would make it a reality. But if saving Europe was impractical, what about saving the human race?

The notion took shape in Szilard's mind that fall when Otto Mandl, the man who had introduced him to H. G. Wells in 1929, moved from London to Berlin. An engaging person who was, by turns, both practical and philosophical, Mandl met with Szilard, and during one "memorable conversation," Szilard later recalled, Mandl

said that now he really thought he knew what it would take to save mankind from a series of ever-recurring wars that could destroy it. He said that Man has a heroic streak in himself. Man is not satisfied with a happy idyllic life: he has the need to fight and to encounter danger. And he concluded that what mankind must do to save itself was to launch an enterprise aimed at leaving the earth. On this task he thought the energies of mankind could be concentrated and the need for heroism could be satisfied.

Szilard's own reaction to all this he remembered clearly:

I told him that this was somewhat new to me, and that I really didn't know whether I would agree with him. The only thing I could say was this: that if I came to the conclusion that this was what mankind needed, if I wanted to contribute something to save mankind, then I would probably go into nuclear physics, because only through the liberation of atomic energy could we obtain the means which would enable man not only to leave the earth but to leave the solar system.[12]

By the fall of 1932, Szilard's thoughts were so freewheeling, his professional options so uncertain, that he considered taking up a new field: biology. His first contact with biology had come through reading. He had enjoyed *The Microbe Hunters,* Paul Henry de Kruif's 1926 saga about the discovery and early use of the microscope in bacteriology, and he probably read *The Science of Life,* H. G. Wells's 1929 visionary story about genetic engineering. Besides meeting and liking Wells, Szilard had tried to contact Wells's coauthor, biologist Julian Huxley, when promoting the idea of the Bund in London. Undoubtedly, Szilard also knew British biologist John B. Haldane's inspiring 1924 essay *Daedalus, or Science and the Future,* about the human consequences of biological progress, since he had tried to interest Haldane in the Bund during the same London visit.

In 1932, Szilard's impetus to study biology may have come from the essay "Light and Life," published that year by Danish physicist Niels Bohr. This bold paper urged applying the "complementarity" principle from quantum mechanics to biology. Just as physics had abandoned conventional physical concepts and created new techniques from quantum mechanics to better explain the atom, Bohr reasoned, so might biology be understood in the same seemingly irrational but ultimately useful way.[13] A very remote but possible link between Bohr and Szilard may have come through physicist Max Delbrück, who returned to Berlin in the fall of 1932 from a year of study with Bohr in Copenhagen.[14]

Unlike most of his prominent colleagues in physics—among them Werner Heisenberg, Erwin Schrodinger, Wolfgang Pauli, Paul Dirac, and Victor Weisskopf—Szilard had never visited or studied at Bohr's Institute for Theoretical Physics in Copenhagen during the 1920s; but he greatly admired Bohr's work and his courage when confronting new ideas.[15] Bohr's ideas about biology gave Szilard a chance to be inquisitive, perhaps even heretically different. Whatever Szilard's route to biology, once he thought about it, he became intrigued by an analogy with modern physics. For just as physics had broken from analysis to theory in the last few decades, Szilard suspected biology might be ripe to make the same break, and he wanted to share this adventure.

Space travel was one of the many intellectual fancies that Szilard would set aside, and biology was one that he would eventually rediscover. But the idea that came to haunt Szilard to the end of his life also dawned on him in 1932: his fear of nuclear war. Like the idea of space travel, this danger came to him through Mandl from H. G. Wells. In 1914, Wells published *The World Set Free,* and when Szilard read the novel, in 1932, he saw science and politics in a new and frightful alliance. Mankind's fate may not necessarily be improved by research, he realized, but in science fiction, at least, it could worsen in catastrophic ways.[16] Wells's novel predicted—correctly—that artificial radioactivity would be discovered in 1933. In the novel, Szilard recalled later, Wells then proceeds to describe the liberation of atomic energy on a large scale for industrial purposes, and the development of atomic bombs, and a world war which was apparently fought by an alliance of England, France, and perhaps including America, against Germany and Austria. . . . He places this war in the year 1956, and in this war the major cities of the world are all destroyed by atomic bombs. . . .

Although Szilard regarded the book as fiction at the time, it jarred his thinking about war and peace and science, then and for years to come.[17] Indeed, while Szilard was speculating about biology, teaching theoretical physics with Schrodinger, and talking about nuclear research with Meitner, he was suffering such profound doubts about his own future that he questioned even continuing as a scientist. From his sublet room on Motzstrasse, Szilard wrote one of his most conflicted and self-revealing letters to his friend Eugene Wigner.

October 9, 1932

Care of Philipsborn
Berlin W 30
Motzstrasse 58
Tel: Barbarossa 0225

Dear Wigwam,

Here is the promised letter:

1. When the knowledge that right now we have more noble causes than to do science, when this knowledge has entered our blood, then I am afraid this knowledge cannot be distilled out of it. It is bad fortune when knowledge enters blood so easily.

Thus, once one is devoted to some work that so far has not yet been done and for which, therefore, there are no instituions yet, there is no justification for a complaint that such institutions do not yet exist.

So what can one do? Well, if one does not succeed in becoming financially independent, thereby getting into a situation that makes one a free man, then one must try to get a job that leaves one enough time and permits a sufficient amount of attention for the things that one considers more important. A professorship in India for experimental physics would be a good solution, because there it would not be necessary to prepare oneself for the scientific topics before the beginning of every semester and one would have thus practically one's energy freely available.

Whether one can get such a job, or a similar one with respect to peace and leisure, in a capital or on the East Coast of the US between Washington and Boston is known only to the rather imperfect gods. I have written to India.

2. Up to the time when such a "position" would offer itself, I could not, without having a bad conscience, devote myself to science.

This is the way I relate to scientific interests. Of course, physics interests me still one full magnitude more than refrigerators. It seems clear that anything having a structure that can capture one's interest could do so; and that if one can do something well, one does it gladly. But is this a passion that would move mountains?

More precisely, if the zero point is represented by [their colleague Rudolph] Ladenburg's interest for physics, and if the boiling point [of 100 degrees centigrade] is represented by my own interest in physics at the time I was eighteen years old, then my temperature is now thirty degrees.

If I decide now for physics, so then I will try to go after nuclear physics in some experimental institute with occasional excursions into the theories. I have never been for pure theoretical physics, probably because of lack of confidence in my abilities in mathematics, or should I better say disabilities? Also, it is an intolerable situation, as Einstein says, to have to rely on laying golden eggs.

I will write you soon, whenever I have a concrete problem. Please give my regards to Mr. and Mrs. Ladenburg as well as to Mr. and Mrs. [John von] Neumann.[18]

Yours
[Szilard]

Here Szilard struggled with a problem he did not resolve until the final days of his life: finding—or creating—an "institution" as free and noble as his own imagination. He admits failure with the hoped-for financial success of the electromagnetic pump, and he all but dismisses a conventional teaching or research "position," although life anywhere between Washington and Boston held some appeal. Reluctantly, he favors experimenting in nuclear physics but seems to fear the possibility of failure.

When Wigner received this confused and soul-searching letter, he was so pained that he wrote their mutual friend Michael Polanyi in Berlin. Independently, Polanyi had asked Einstein to recommend Szilard for a job in Princeton.[19] To Polanyi, Wigner wrote:

This moment I received from Szilard a rather depressed letter. I am under the impression that it would be good to do something soon. True, one cannot come to conclusions on the basis of one letter. I also believe—and you will understand me—that he is an ass in many respects. I answered him as best I could. I have just written to him just as well as I am able. Now I do think that it would be best if he could again settle down. I thought that Einstein would bring him here [to Princeton]. Of course, that would be wonderful.

Please let me know how you feel about this plan and whether you would find it correct for me to write this to Einstein. Or else do you find a "different" way better? Please, write to me about this. The whole thing worries me very much.

Your wife [Magda] should also talk to Szilard in a "wise" way; I feel that would be helpful. He seems to be rather frustrated. Of course, it is possible that nobody could help him, but it is as possible that one could do that. . . .

There is much talk here now about [John von] Neumann and I really should stay here for good, but I am rather undecided. . . .

Many friendly greetings from E. Wigner.

P.S. . . . Please do not show this letter to Szilard.[20]

Something besides his professional anxieties must have agitated Szilard during the fall of 1932. Perhaps it was a breakup with Gerda Philipsborn or her departure for India. Perhaps it was the ominous political events that Szilard followed so attentively in the daily newspapers: The Nazis were then consolidating their July victory as the largest party in the German Reichstag and by January would control the government. Or perhaps it was the realization that any academic appointment or industrial job, no matter how appealing it may at first seem, would soon lose its challenge.

In late October, just before the university's winter semester began, Szilard tucked his few belongings into two suitcases and moved from his sublet room with the Philipsborns in the Berlin neighborhood of Wilmersdorf to the Harnack House, the faculty club for the Kaiser Wilhelm Institute in suburban Dahlem. Registering as Polanyi's guest, he gave Budapest, not Wilmersdorf, as his home address and settled into the Muller Zimmer, a room for visiting scholars. He was restless and edgy: fascinated by the energies and enigmas he had discovered in America, excited by the theoretical leaps he might make in biology, intrigued by the experiments he might devise in nuclear physics. But he was also fearful that the Weimar Republic's political and social order would soon collapse. And he suffered profound doubts about his own life as a scientist. From this time on, Szilard was ready to move, but still unable to decide just when, or where, or why.

CHAPTER 9

Refuge

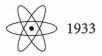 1933

Leo Szilard had little to celebrate with the New Year of 1933. Anxious, lonely, and uncertain about his faltering career in physics, his only employment was to lead a weekly seminar at Berlin University and to poke around the research labs at the Kaiser Wilhelm Institute in suburban Dahlem. Free to wander about the gray and cloudy city and among his friends, Szilard had plenty of time to worry. Michael Polanyi sat in his small apartment when Szilard called during the first week of the New Year. Think about leaving Germany, Szilard insisted. Things will get worse under Hitler. Much worse. But Polanyi demurred. After all, Hitler headed a minority party in the Reichstag, Polanyi reasoned. How much trouble could one man create? Like many others in Berlin at that time, Szilard recalled later, Polanyi thought that the "civilized Germans would not stand for anything really rough happening." Szilard took the opposite view, "based on observations of rather small and insignificant things." During his twelve years in the capital, Szilard had "noticed that the Germans always took a utilitarian point of view. They asked, 'Well, suppose I would oppose this, what good would I do? I wouldn't do very much good, I would just lose my influence. Then why should I oppose it?'"[1]

Szilard urged Polanyi to accept the professorship just offered him at Manchester University in England. But Polanyi hesitated, complaining that if he moved to a new lab now, he could not be productive for another year.

"Well," Szilard replied sarcastically, "how long do you think you will remain productive if you stay in Berlin?"

The two old friends disagreed, but Szilard persisted, urging that if Polanyi said no to Manchester, he should at least say it was because his wife opposed the move; she, Szilard argued, could always be said to change her mind. Still Polanyi hesitated, and his stubbornness only fueled Szilard's fears. In fact, Polanyi did follow Szilard's advice and declined the appointment because of "rheumatism," expecting "difficulties" with Manchester's "humid climate."[2]

Berlin's political climate changed on January 30 when the aging and senile president, Paul von Hindenburg, then weary of the squabbling among minority political parties, yielded to pressure by the National Socialist (Nazi) party and named its leader, Adolf Hitler, chancellor of Germany. When he learned about Hitler's appointment, Szilard returned to his cozy third-floor room at Harnack House, stuffed his belongings into two suitcases, and set them by the door so that "all I had to do was turn the key and leave when things got too bad."[3]

If his friends wouldn't listen, Szilard thought, maybe his family would, and off he went by train to Budapest, arriving at his parents' villa on February 3. "Hitler and his Nazis are going to take over Europe," Szilard warned his parents, his brother, Bela, and his cousins. "Get out now. Leave Europe—before it's too late!"[4]

But, like Szilard's Berlin friends, his family also hesitated. They could not understand Szilard's constant travel, his need to board a train or ship so impulsively, his agitated fright and flight. They had homes, families, and more important, cultures. They were "Hungarian," with roots in Buda and Pest and the remote mountain towns of their ancestors. Szilard, by contrast, seemed rootless and adrift, stirred by events of the moment or by visions of a dark future they could not see. While their sensitivities bent toward music and family and friendships, Leo's were to news and its nuances in science and politics. He was a man of the world, and this winter meeting in Budapest convinced them that his fast and frightful world had little to do with their own.

While at home, Szilard telephoned his longtime friend Alice Eppinger, who had married Theodor Danos in 1927 and now mothered three lively children. Szilard seemed nervous and preoccupied when he called at her spacious apartment.

"Hitler is a crazy, stupid man," Szilard warned her. "Everybody must flee! Under Hitler a terrible time will come!"

She heard his warnings but could give no coherent response. To Alice, the politics of Germany seemed remote. Hungary had economic troubles of its own. Changing the subject, Alice introduced her children, but Szilard, always unable to make small talk, just stared at the youngsters a long while.

"They are not very well developed," he finally said, breaking the silence. "What do they read now?" When Alice mentioned a few children's book titles, Szilard frowned.

"When I was six, I read *Faust,* in Hungarian," he told the children.

"Nobody else has told me they are not well developed," Alice said quietly.

As Szilard waited for Alice's husband to return, he said little and nervously paced around the parlor. When Theodor Danos arrived, the two men shook hands, then stood in silence. Szilard could only repeat his warnings about Hitler, and gaining no response, he excused himself. This was to be the last time Alice and Leo were to meet, although they did exchange letters after World War II and Szilard sent her money after learning that the Nazis had killed Theodor and one of their daughters.[5]

During his three weeks in Budapest, Szilard tried to revive his friendly and familiar routine as a student—the afternoons at the New York Café arguing politics and economics, the evenings by the fireplace in the parlor at the Vidor Villa, chatting with uncles and cousins. But he was too impatient and gloomy, his friends and family seemed inattentive, and by the last week of February he boarded a train that carried him through Prague and down the Elbe River valley to Dresden and Berlin.

The letter awaiting him brought more gloom: Siemens, the international electronics company, had written rejecting his proposal for research on an absorption refrigerator.[6] Harnack House seemed colder than usual to Szilard when he returned late in February. In contrast to the warm and cozy atmosphere of his family's home, the stucco-covered exterior reflected the gray winter sky and made the huge dormitory look inhospitable and Teutonic. Inside, the dark halls to the small scholars' bedrooms were dim in the thin winter light. In Szilard's room, under the eaves on the third floor, frost curtained the metal casement windows, obscuring his view of Dahlem's arching tree branches, now bare of leaves as they scratched the dark sky.[7]

Berlin itself must have seemed as tense and uneasy as he felt. Its population had swelled after the war, overcrowding both housing and public places. With unemployment severe, the prosperous and the poor collided on Berlin's noisy sidewalks, and the political and class tensions made the capital seem increasingly hostile. Economic distress had quickened tensions that winter in Germany and around the world.

In the United States, the stock market crash of October 1929 had plunged the nation's economy into a deep depression. The Democrats won a landslide election in November 1932, voting to the White House New York governor Franklin D. Roosevelt and returning both houses of Congress to their party for the first time since 1915. In January and February 1933, Roosevelt was assembling his new cabinet—he would be inaugurated on March 4, beginning his "Hundred Days" of reform legislation destined to expand the federal government's role in American life.

In England, economic distress fueled social unrest and new interest among intellectuals in both communism and fascism. This was the first worldwide depression, but nowhere was it more destructive than in Germany, a country still straining from the military and economic defeats of the First World War.

Szilard had been in Berlin when Hitler took power, and by the time he returned there from Budapest three weeks later, he sensed a new chill more potent than the winter's damp and biting air. He sensed, as never before, a pervasive anti-Semitism and anti-intellectualism in public speeches and newspaper editorials. To strengthen his minority government, Hitler had scheduled a Reichstag election for March 5, and for it, red, white, and black Nazi slogans and banners accented the gray city-scape. In the halls of Harnack House, Szilard heard about colleagues who were losing their teaching posts—forced out by timid academic councils that feared a rising wave of anti-Semitic policies.

Just four weeks after Hitler's appointment, on the night of February 27, the massive neoclassical Reichstag building burst into flames. Within minutes Hitler was on the scene to declare this "crime" the work of Communists, and the next day he persuaded Hindenburg to sign a decree "for the protection of the people and the state," suspending sections of the constitution that guaranteed individual and civil liberties.[8] Again Szilard called on Polanyi to warn about trouble ahead.

"Do you really mean to say that you think that the secretary of the interior had anything to do with this?" Polanyi challenged.

"Yes," Szilard insisted, "this is precisely what I mean."[9]

The Reichstag fire allowed Hitler to feature his anti-Communist theme in the March election campaign. With his inflated promise of economic recovery, his scapegoating of other political leaders, and his party workers' intimidation of voters, Hitler gained the Reichstag majority he craved. But their 44 percent of the vote still denied the Nazis the two-thirds majority needed to impose sweeping economic and legal reforms. So Hitler turned to other means, ordering the Communists subject to arrest and persuading Hindenburg to appoint Joseph Goebbels to a new cabinet post, Reich minister for people's enlightenment and propaganda. And, unnoticed by all but the nearby residents in the Munich suburb of Dachau, the Nazis opened their first concentration camp on March 20.

The next day, Goebbels and Hitler staged a "Day of the National Rising," important as both the first day of spring and the day on which Bismarck had opened the German Reichstag in 1871. Hitler began the first session of his own Third Reich in a solemn ceremony staged in Potsdam, the Berlin suburb once the residence of the Prussian monarchs. Hindenburg, in his spike-topped helmet, symbolized the old order; Hither, the new. That night, long torchlight parades flowed through the streets of Berlin, and on March 23, Hitler's public fawning complete, he proposed a "Law for Removing the Distress of People and the Reich." It passed 441 to 84, in a stroke shifting the Reichstag's powers to the new cabinet for the next four years. "Give me four years' time and you will not recognize Germany!" he promised.[10]

Four days later, Goebbels proclaimed a boycott against Jewish businesses in retaliation for "the anti-German atrocity propaganda which interested Jews have started in England and the United States against the new Nationalist regime." To silence protest from intellectuals, Goebbels announced, Jews will now be admitted "to universities and to the professions of attorney and physician" only in proportion to their numbers in the German population: then less than 1 percent. On the next afternoon, in Antwerp, Belgium, Albert Einstein arrived from the United States aboard the SS *Belgenland* and announced that he would not return to his Berlin home as long as anti-Jewish threats continued. For the next few months, Einstein said, he would stay in Belgium, a decision made more appealing by his friendship with the royal family.[11]

At Harnack House on the afternoon of Thursday, March 30, Szilard finally decided that things had gotten "too bad." He locked his two packed bags and lugged them down the broad flagstone stairs to the high-ceilinged lobby and out to the porte cochere. He hailed a taxi, which first meandered through Dahlem's suburban lanes, then charged into Berlin's speeding traffic. At the massive red-brick Anhalter Station, Szilard bought a one-way ticket on the night train to Vienna, booking a berth in the comfortable first-class wagon-lit. As a porter heaved the two bags onto the overhead rack, Szilard sank into the upholstered seat of his empty compartment, no doubt relieved he was leaving but still anxious about the nightlong trip ahead.

The express eased from the noisy terminal, glided through Berlin's suburbs, crossed rolling farmlands, paused briefly in Dresden, then snaked along the Elbe River under soaring sandstone cliffs. A police officer roused Szilard at the German-Czech border early Friday morning, read in the brown German passport that he was an assistant professor, asked a few routine questions, saluted, clicked his heels, and said, "Good night, sir!" Szilard sighed, recalling later that he had booked the more expensive wagon-lit for two reasons: He liked to travel in style, but more importantly, he hoped that officials would be less likely to interrogate first-class passengers carefully. This mattered to Szilard, for he had tucked into his bags small bundles of banknotes. Only when the train crossed into Austria and, later that morning, when it thundered across the Danube River bridge and swung into Vienna's Nord-Bahnhof could Szilard feel safe.

At the German-Czech border after midnight the next evening—as the Nazis' anti-Jewish boycott began—that same train from Berlin was overcrowded. Troopers questioned every passenger, holding back those deemed "non-Aryan" or seizing their most valuable possessions.[12] This close call so frightened Szilard that anxiety about his personal safety endured for the rest of his life. From then on, he always kept two bags packed. But to mask his fears, Szilard made light of his escape years later, only using the incident to boast: "This just goes to show that if you want to succeed in this world you don't have to be much cleverer than other people; you just have to be one day earlier. . . ."[13]

In Vienna, a city Szilard enjoyed visiting, he checked into the Regina, on Freiheitsplatz (now Rooseveltplatz), a small and elegant hotel overshadowed by the soaring spires and decorative tile roof of the Votivkirche. Built to resemble a French villa, the Regina had an imposing, high-arched dining room and an arcade facing the park in front. Immediately, Szilard cabled Bela in Budapest, calling him to Vienna for a "family conference." To the brothers, this meant that Leo planned to dispense some serious advice.

In the newspapers Szilard read about the Nazis' anti-Jewish boycott, which began on April 1, and from friends in Vienna he learned how academic dismissals in Germany were spreading. Szilard walked a block from his hotel to the University of Vienna's Chemical Institute, and there called on his friend Hermann Mark, whose research had moved from X-rays to the study of synthetic plastic polymers. The two agreed to collaborate again, not on research this time but to help academic refugees find places to work and study outside Germany. Mark wrote letters on Szilard's behalf and gave him money to help the cause.

From there Szilard called on the embassies of countries likely to accept academics: France, England, Switzerland, and the United States. At each stop he warned the political officer about Germany's anti-Semitism and the exodus that had just begun. The embassy officials were polite but noncommittal.[14]

Szilard also called on Michael Polanyi's brother Karl, an economist he had first met at Galilei Circle discussions in Budapest. Polanyi suggested that Szilard deal directly with the International Student Service (ISS) to create a system for relocating both faculty and students. He added that a personal friend, a charming Englishwoman named Esther Simpson, recently moved to Geneva and might help with an introduction to the ISS.

Szilard also took time from his rounds to call on Gertrud (Trude) Weiss, the woman he had first met in Berlin four years before. Following Szilard's advice to become a doctor, Trude was now a medical student. She lived near the Regina in her parents' large, comfortable apartment on Alserstrasse. She was delighted to see him and solicitous about the news of his close escape. The two had politely exchanged formal letters over the years and had begun to spark a friendship. Szilard called her T.W. or *Kind* (German for "child," which he later abbreviated "Ch."), but Trude found him captivating despite this condescension. In the years since she had left Berlin, Trude had remained infatuated with Leo. For his part, Leo was genial but seemed preoccupied with the academic refugees. "Now how can we get the others out?" Szilard asked Trude impatiently. "There're so many people stuck, and how do we find jobs for them once we get them out?"[15]

When Bela arrived, the Szilard brothers talked over a long and sumptuous dinner in the Regina's handsome dining room. Leo urged that they secure their savings in a Swiss bank, and the next day they boarded the elegant Orient Express for a ride along the Danube, through Salzburg, Innsbruck, and the Alps to Zurich. Bela alighted there, opened a bank account, and rented a deposit box. Leo rode on to Geneva, checked on the account he had opened three years before, and deposited the cash he had slipped out of Germany.

In Geneva, Szilard contacted the ISS through Polanyi's friend Miss Simpson, who agreeably typed several letters for him; one, to Einstein, then staying in a cottage at Coq-sur-Mer on the Belgian coast, described Szilard's plans to visit Oxford, where Einstein was soon to lecture.[16]

Back in Vienna a day or two later, Szilard met by chance a Berlin acquaintance, economist Jacob Marschak from Heidelberg University. A Russian who shared Szilard's sensitivities to political events in Germany, Marschak agreed to help academic colleagues still in the country. But for this they needed some money, Marschak said, and took Szilard to meet economist Karl Schlesinger, a rich friend, at his luxurious apartment in the Liechtensteinpalais. With Schlesinger that day was Gottfried Kuhnwald, to Marschak a "shrewd and mysterious" person, who was an economic adviser to the Austrian government.[17] Kuhnwald, who called himself "the last Austrian" and sported Franz Joseph sideburns, predicted that once the refugee flood hit, the French would pray for the victims, the British would organize their rescue, and the Americans would pay for it.[18] Szilard tried to lighten the mood with a favorite joke: A rabbi says, "Old man with baby, cold, hungry, prays to God, who makes milk flow from the man's breast." "Why could not the Lord simply give money?" his student asks. Says the rabbi, "Why give money when you can make a miracle?"[19]

Through Schlesinger, Szilard met the economist Ignatz Jastrow, who in turn suggested approaching Sir William Beveridge, director of the London School of Economics (LSE). A tall and formal Englishman, Beveridge was visiting Jastrow to collaborate on the history of market prices and was staying at the Regina. When Szilard called on Beveridge there, he learned that word about academic dismissals had already reached London and that the LSE planned to hire one or two refugees. Moreover, Beveridge said he would plan a joint effort to receive academics displaced by the Nazis. Schlesinger, Marschak, and Szilard met Beveridge for tea, and with much prompting from Szilard, Beveridge vowed to form a refugee settlement committee in England and urged Szilard to come to London to "prod" him in this effort.[20] A few days later, Szilard appeared in London.

"I intend . . . to stay in a hotel," he wrote Bela shortly after arriving. "I do not intend to rent a house; I like mobility!"[21] Szilard found a hotel near the LSE that seemed uniquely suited to his mood: the Imperial. On Russell Square in the heart of London's Bohemian Bloomsbury neighborhood, this eccentric-looking hostelry had a decorated Edwardian Gothic facade that mimicked the cultural confusion of Vienna.[22] Yet the London that Szilard came to resembled, in other ways, the Berlin he had fled. Unemployment was high, strikes were frequent, and the pressure for jobs was intense. How, he wondered, might hundreds of refugee scholars fare in this harsh climate?

Still, he would try to help, and from the Imperial, Szilard wrote to Beveridge, saying he had come to London "in order to meet Prof. Niels Bohr, of Copenhagen, and to discuss with him the whole situation."[23] The next day, April 23, Szilard wrote to Einstein, reporting that plans had advanced for aiding refugees, with funds possible from the Rockefeller Foundation in New York. He described his own ambitious scheme for creating new boards and review committees to settle refugees and recounted meeting Beveridge and their plans to collaborate. Beveridge, in turn, would alert other peers and academics, Szilard wrote.[24]

"Your plan doesn't really set me on fire," Einstein replied, chiding his younger colleague for devising an overcomplicated scheme. "I believe, rather, that one ought to try to form a kind of refugee Jewish University which would be best placed in England."[25]

Full of promise for settling refugees, Szilard wasted little time creating and mobilizing a network of concerned academics. From the Imperial, Szilard walked down busy Southampton Row to the LSE, where, with no formal introduction, he called on Harold Laski, a professor of politics and a prominent member of the British Labour party. The two agreed to identify famous persons who might serve on the board of a new group to aid refugees.[26] Although later history gives him little credit, Szilard appears to have first proposed an Academic Assistance Council (AAC) to Beveridge in Vienna, then helped make it work by his energy and prodding in London. Szilard had also proposed, and helped organize, the council's office in London as a clearinghouse to match refugees with placement offers.

Before receiving Einstein's reply, Szilard wrote again, this time laying out his thoughts for founding a Palestine University for Jewish refugee scholars, a project then being promoted in England by Chaim Weizmann, a chemist and active leader of the World Zionist Organization.[27]

Szilard continued to "prod" Beveridge by calling on him at the LSE and by writing him letters. One letter outlined Weizmann's plan to raise money for the Palestine University and another to fund an emigrants' university somewhere in Europe. Szilard suggested that the refugee committees being founded around Europe needed to be coordinated, perhaps from England.[28] During his first weeks in London, Szilard met with other academics, at times seeming to stage-manage the whole refugee effort. Szilard had no formal position and little power, but working as he preferred, behind the scenes, he pushed others to use theirs.

With news that "seems to open a new era" Szilard wrote Beveridge that physicist James Franck had resigned from Göttingen University and Michael Polanyi had left Berlin. Polanyi's move had special poignancy for Szilard, who had failed in January to persuade his friend to leave. "It appears," Szilard concluded, "that Hitler cured his rheumatism."[29] To Beveridge, Szilard assessed how he might be most useful: "I think it is not for me to represent officially our project" because of a "lack of knowledge and experience of the English way of doing things. . . ." Instead, Szilard vowed to contact other scientists and wealthy benefactors.[30]

Szilard's prodding behind the scenes paid off, a network of support was patched together, and on May 24 a statement in the newspapers announced the creation of the AAC, endorsed "with forty-one signatures of men of distinction in every branch of science and the arts."[31] The Royal Society, Britain's preeminent science association, offered the AAC two small rooms on the top floor of Burlington House, its neoclassic headquarters on Piccadilly in central London.[32]

Meanwhile, Szilard urged Benjamin Liebowitz, a physicist from New York whom he had met at New York University in 1932, to consult Niels Bohr in Copenhagen and, on his return to New York, to call on anthropologist Franz Boas at Columbia University. Szilard wanted Liebowitz to introduce Bohr and Boas, hoping these respected academics might publicize the refugees' plight and rouse support in America.

On his own, Szilard kept busy by visiting as many academics as would see him: in London, mathematician Harald Bohr, Niels Bohr's brother; in the small town of Harpenden, an hour's train ride north of London, Sir John Russell, an agricultural chemist; at Trinity College in Cambridge, G. H. Hardy, a mathematician; and back in Bloomsbury, physiologist Archibald V. Hill and Frederick G. Donnan, a physical chemistry professor and department chairman at University College. Szilard coordinated the AAC's efforts with the Committee of the Jewish Board of Deputies and met often with Weizmann about awarding fellowships to Jewish scholars.

When Szilard schemed to create a board of some twenty scientists to administer fellowships, he concocted, as usual, elaborate procedures for others to follow, including steps to select board members, review applications, and make awards. His rules allowed scholars to live anywhere in the world and only report their results annually. In this way, Szilard reasoned, the world's best academic talent might spread to India, Egypt, and other developing countries, not just to England and North America. [33]

"We were convinced that Szilard could be in two places at the same time," recalled physicist Hans Bethe, who arrived in London from Germany in 1933. "There was much talk at that time about 'making particles' [which disintegrate then re-form inexplicably], and Szilard seemed to prove the point. He was a person who could be annihilated at one place and appear at another place, being re-created." [34]

On May 12, Szilard disappeared from London and reappeared in Belgium on the first of several hasty trips to the Continent that spring. There he met with the rector of Liège University [35] and, ignoring Einstein's complaint about overwork, also called on him. The two colleagues reviewed their efforts to resettle German academics and discussed ways to found a refugee university—an idea they soon forgot as other placement efforts succeeded. Szilard left the meeting disappointed, without Einstein's agreement to use his name in fund-raising for the AAC but still determined to "ask for his help in such a way as I think fit." [36]

Back in London for a few days, Szilard then appeared around the University of Paris, and "on the train to Geneva" he wrote to Liebowitz in New York, urging that a refugee effort might be started in America by "next week." [37] In Geneva, Szilard brought together Dutch and Austrian academics with executives from the ISS and the Intellectual Co-operation Section of the League of Nations, also opening contacts among the ISS, the AAC, and the Jewish Board. [38] His "brief report" to Beveridge from Geneva noted meetings with rectors of all four Belgian universities and with the president of the University of Brussels. At the ISS, Szilard invited Esther Simpson to join his refugee-settlement work in London, and she accepted. Although the AAC could only pay one-third her Geneva salary, she later recalled, "with the chance to help the sort of people I'd played chamber music with in Vienna," his offer was irresistible. [39]

In the spring and summer of 1933, Szilard crossed and recrossed the English Channel, moving his savings among Swiss, French, and English banks along the way and finally collecting money and valuables from Bela's safe deposit box in Zurich. On July 17, Esther Simpson began working with Szilard at the AAC, and within three weeks they were joined by Walter Adams, a young instructor from the LSE. But almost immediately Adams went to Germany to check on a list of scholars Szilard had compiled, leaving Szilard and Simpson to run the AAC. [40]

The news from Germany was grimmer by the day. In July the Nazis had dissolved all other political parties. When in Paris, Szilard had met with refugee professors who had organized as the Association of German Scientists Abroad,[41] and in London he tried to affiliate this group with the AAC, offered to be its representative, and asked the AAC to find part-time secretarial help for the Germans.[42] To place scholars outside England, Szilard met in Cambridge with the Russian physicist Pyotr Kapitza and urged him to inquire about academic openings during his forthcoming visit home.

To some people, Szilard's own movements seemed amazing; to others, conspiratorial. In Cambridge that July, Prof. L. W. Jones, a representative of the Rockefeller Foundation then traveling in Europe, checked into his hotel and found Szilard waiting for him in the lobby. Szilard named several academics Jones should contact and questioned him about matching funds for refugee settlement. In his log for the foundation, Jones noted that Szilard "seemed to know intimately things which are not currently known to others in England." Later the same day, Szilard followed Jones to London and approached him again. Szilard "appeared to be omnipresent without any official connection to any country," Jones wrote. "Szilard had been in Switzerland, in France and in England and seemed to know everything being undertaken in these countries."

Jones questioned Szilard's "official connection" to the refugee groups, found that he had none, and learned that English academics were also "suspicious" of his activities.[43] Working independently gave Szilard the freedom he enjoyed but also forced him to pay for all his hectic travel that spring and summer, prompting the fear he "cannot possibly go on with this for very long." Yet his accomplishments made the sacrifice worthwhile. "I can be so useful that I cannot afford to retire into private life," he boasted.[44]

Indeed, Szilard's work to create and run the AAC was useful beyond his dreams: It placed more than 2,500 refugee scholars by the outbreak of war in 1939. The AAC exists today as the Council for Assisting Refugee Academics and still aids displaced scholars. "This activity suited my temperament," Szilard later reflected, "for I always found it easier to solve the problems of others than to solve my own problems."[45]

Behind Szilard's compulsion to "be so useful" during the summer of 1933 was his belief that a war with Germany was inevitable. He mentioned "the next war" in conversations and letters, wondering when

England and the United States might become involved. "I think most of my friends feel the burden of the situation and react by plunging deeper into their work and sealing hermetically their ears," he complained in one letter. "I feel rather reluctant to follow their example, but I may have no choice left."[46]

In fact, Szilard's choices for an academic career were very few, in part because his "sense of proportion" kept him from asking for help from the refugee committees he assisted. Szilard's reluctance also masked deeper confusion about what to do with his life, confusion that had troubled him for years. With his savings from Switzerland now in a Russell Square bank, Szilard had no need to find an academic post by the fall term. He could survive, he reckoned, "in the style in which I was accustomed to live" for about a year.[47]

Yet at age thirty-five he did worry about his future and tried to secure a position, at least indirectly. First, Szilard considered teaching at Dacca University in India, where he knew physicist Satyendranath N. Bose, a friend he had met through Einstein in Berlin.[48] Then Professor Donnan offered Szilard a lectureship at the University of London. Szilard's reluctance to ask for help confused the many colleagues who were eager to aid him, including his friend Eugene Wigner. When Szilard wrote Wigner to say that "somebody should also worry about myself, since I obviously cannot do that, because at any rate it would be incompatible with my present activity" for the AAC,[49] Wigner turned for help to their mutual friend Michael Polanyi, now in Manchester.

. . . I thought of you, since your hands hold so many strings, that perhaps you could know of something that would be good for him and which takes into consideration his personal characteristics, too. . . . He has my undivided respect because of his straightforwardness and his truthfulness. His selflessness is almost without parallel among my acquaintances. He has an imagination that would be of unusual value for any institute for which he would work. I would not know whether a purely scientific occupation would be the best for him, although even that should be considered. At any rate, it would be a great shame even from this point of view if his abilities would remain unused. . . .

I feel that it would be better if he would continue to busy himself [with] politics only incidentally and as a side line.

127

The above should by no means be understood as a doubting that he would do well in his political efforts.

As a main occupation, I was thinking about two possibilities: either [for him] to be active as a consultant to a large company; that would gain for the company considerable value, even though rather irregularly at various time periods. The second possibility I thought of would be for him to be working for a publisher. . . . I do *not* think that Szilard would do sloppy work; to the contrary, I believe that he could that way have an occupation in which his capabilities would develop excellently.

Of course, we have to ask him, too.

Yours, E. Wigner.[50]

When Donnan asked Einstein and others for recommendations, physicist Max von Laue, Szilard's doctoral thesis adviser, and Erwin Schrödinger replied with an enthusiastic joint letter.[51] Another testimonial, by Prof. Max Volmer of Berlin's Technical Institute, told Donnan that Szilard

. . . is worthy of special consideration, since he is one of the most capable and many-sided people I have ever met. He unites in a rare fashion a complete understanding of the development of modern physics with a capacity for dealing with problems of all fields of classical physics and physical chemistry.

Dr. Szilard is unique in his independent, original, and inventive attitude toward all problems.[52]

Even as this praise for Szilard came to those in London who could help him, he remained ambivalent about a science career. "I am not against going to America," he wrote in August, "but I would very much prefer to live in England. I have not dismissed the idea of going to India; neither has this idea grown stronger."[53] Then something appears to have changed with Donnan's offer: a second opening, perhaps, or a time limit on the first one. To Wigner, Szilard wrote:

I am sorry that our correspondence is so full of misunderstandings. I know that I am guilty because my letters are so short, but I cannot help it, as I am working against time. . . .

To begin with, you need not worry about my last letter. There was not a shade of reproach in it. . . .

Secondly, I did not make any serious attempt until now to get a scientific position here. All I did was to ask several people for testimonials to be at hand should occasion arise later. Donnan asked me about my plans and said that he would like to do something for me, so I suggested that he should first get testimonials. That's all. I should very much like to see you, and I shall stay in London all the time so that if it is convenient to you, we can easily meet, if you still sail from Southampton. Why are you sailing on the *Bremen?*[54]

Szilard reported that he was working at the AAC offices while others were on holiday, and "I never finish before 10:00 p.m." The same day, Szilard betrayed more anxiety about his future when he wrote to Einstein, enclosing a copy of Donnan's request for a recommendation. "Since he did not get an answer from you," Szilard wrote, "he asked me what to do; it seems that momentarily there is a chance that may not last for very long." Szilard also reported that he was "very much involved" with both the AAC and the Hartog Committee (the Jewish Board), "so that I could not have them give me a stipend." He closed by noting that "the secretaries of the [AAC] are on vacation, and I am substituting for them. It is pretty exhausting and full of responsibility but also fun."[55]

Einstein's prompt endorsement to Donnan was both warm and generous, describing Szilard as "a versatile and able physicist" who "has ideas on the experimental but also technical level and keeps his focus on the theoretically substantive matters." Szilard "belongs to the group of people who, through their richness of ideas, create an intellectual environment for others. I have come to value highly his abilities through my collaboration with him on technical levels." Einstein noted that during the economic and political pressures of the last few years Szilard "has also gained credit in human terms, as he has helped to take care of younger colleagues. It seems only just if he himself is not left out."[56]

The Austrian physicist Paul Ehrenfest heard from friends that "people in England are now trying to do something for Szilard" and sent his own enthusiastic reference. Ehrenfest was one of the first physicists to explain quantum theories in relation to older studies and at the time held the theoretical physics chair at Leiden University in Holland. "Szilard is a *very rare* example of a man," Ehrenfest wrote, "because of his combination of great *purely scientific* acumen, his ability to immerse himself in and solve technical problems, his fascination and fantasy for organizing, and his great sensitivity and compassion for people in need. . . ." Ehrenfest also praised Szilard's "extremely original, versatile, and innovative intellect" and noted that

> what I find so particularly enviable in him is that he reacts to any difficulty that may arise with immediate action rather than depression or resignation. For even though this procedure is not always successful, an energetic reaction is still vastly more fruitful than a passive attitude. I feel deeply ashamed when I see how wonderfully energetically he immediately set about doing everything in his power to work for the Jewish-German scholars. And I know only too well how much he would long for an opportunity to sit back and *quietly contemplate* those questions that interest him most.

Ehrenfest "would be most particularly happy," he concluded, "to hear someday that Szilard is working specifically with your laboratory and in addition discusses problems of the frontiers of science with you. A discussion on physics with Szilard is always a highly interesting occupation, particularly if it starts off with a heated argument"[57]

And Schrödinger wrote Donnan a second letter, ranking Szilard "among my true personal friends" and adding that "what he had to say was always of a profound and original kind" and "would not occur to anybody else." Schrödinger noted the same for Szilard's publications, which are few because of

> a rather wide range of interests (which are by no means limited to the science of physics) and the absolute refusal of publishing on a subject that he has not really thoroughly studied in all its details and connections to others, even far-removed subject matters. My personal hope with regard to Dr. Szilard is that he will find occasion to specialize his work and his interests a little more.

Schrödinger considered Szilard "an absolutely trustful and truly altruistic person who has all the qualities that would make him a valuable and esteemed member of an institution colaborating [sic] toward some common aim."[58]

With this stream of high praise from the giants of modern science, a position at University College seemed assured for Szilard—if he wanted it. Donnan liked Szilard. And Szilard liked London. "In spite of being rather tired, I feel very happy in England," Szilard wrote that August.

This is partly due to the phenomenon that I always feel very happy for the first few months in a foreign country, but probably also due to the deeper sympathy I feel with the country and the people. I am not yet sure about the sympathy being mutual, but this is only a matter of practical importance.[59]

Szilard liked the English because of their reserve. A very private and often shy person himself, he enjoyed their childish formalities and academic humor. If Budapest had been his natural home, Berlin his intellectual home, and Vienna his cultural home, then London was quickly becoming Szilard's spiritual home. Quite simply, the place and the people made him feel at ease.

During that busy summer in London, Szilard also took special pleasure in the comforts of the Imperial Hotel. For all his running about, he often began the day very slowly: eating breakfast in the ornate restaurant, then returning to soak in the bathtub. This Archimedean routine, he would later say, offered the solitary and relaxing atmosphere he needed to just think—the task that Szilard had considered his most important ever since adolescence. Some mornings he soaked for two or three hours.[60] And each morning, Szilard also took the time to read London's many papers, his inexpensive way both to improve his English and to follow world affairs.

In his work for the AAC, Szilard called on Prof. Archibald V. Hill at the University of London, and during their talk he learned that Hill had begun as a physicist, then changed to physiology, winning the Nobel Prize in 1922. Hill encouraged Szilard's dawning interest in biology and offered him a part-time job as a demonstrator. "Twenty-four hours before you demonstrate you read up these things, and then you should have no difficulty in demonstrating them the next day," Hill proposed. "In this way, by teaching physiology, you would learn physiology, and it's a good place to begin."[61]

Biology then appealed to Szilard because enough was finally known about its physical laws to allow analytic scientists to become theoretical, to postulate general principles from their detailed observations. Besides, Szilard later rationalized,

if you live in an orderly society in peacetime, the social pressures are such that it is very difficult for a man to change his field, even within physics, and even more difficult to change his field from physics to biology. But these were not ordinary times.[62]

He would have probably joined London University in some capacity when the academic year began that September. But within two weeks of collecting his many avid recommendations for Donnan, another idea so gripped Szilard's imagination that he deferred studying biology for more than a decade. That idea was the nuclear chain reaction.

PART TWO

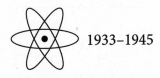 1933–1945

CHAPTER 10

"Moonshine"

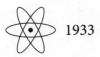 1933

Work for the Academic Assistance Council (AAC) kept Leo Szilard so busy during the spring and summer of 1933 that he had no time to read the physics journals, but when he heard from academics that Lord Ernest Rutherford was to speak at the annual meeting of the British Association for the Advancement of Science, he wanted to attend. Then director of the eminent Cavendish Laboratory at Cambridge, Rutherford was a nuclear physicist renowned for his early work in the structure of the atom. Szilard's feisty nature kept him from holding anyone in awe, but he deeply respected Rutherford as a nuclear pioneer.

At the beginning of this century, Rutherford had discovered and named two of the three kinds of radiation: "alpha," the positively charged radio-active particles, and "beta," which have a negative charge.[1] With chemist Frederick Soddy in 1906, Rutherford had postulated the radioactive "trans-mutation" of atoms—their disintegration and change when bombarded by radioactive particles. In 1911, Rutherford envisioned the atom as having a small, heavy core, or "nucleus," of positively charged "protons." Around the nucleus whirled lighter, negatively charged "electrons." Two years later, Niels Bohr combined Rutherford's view with new theories of quantum physics (explaining how energy is emitted and absorbed in discrete and impulsive amounts) to create a description of the atom still popular today. With Bohr's discovery, "atomic" physics became "nuclear" physics—the study of the atom's nucleus.

In 1919, the year he became director at the Cavendish Laboratory, Rutherford published the first evidence that the nuclei of atoms can be charged artificially and that they split off when they are bombarded with alpha radiation. But more had to be understood about the mechanism of atoms before their energy could be released, and two necessary steps in this direction both occurred at the Cavendish in 1932. The young physicists John D. Cockcroft and Ernest T. S. Walton managed to split atoms by a new method; not, as Rutherford had done, with naturally radioactive radium but by using high-voltage electricity to speed up streams of hydrogen protons that bombarded small samples of lithium, a light metal. This was the first nuclear transformation produced by purely artificial means when, as the two reported in *Nature,* "the lithium isotope of mass seven occasionally captures a proton and the resulting nucleus of mass eight breaks into two alpha particles, each of mass four. . . ."[2]

In the same year, physicist James Chadwick identified a third particle in the atom: the "neutron." His discovery gave a new, more complete description to the atom: a nucleus of *both* positive protons and neutral neutrons surrounded by orbiting, negative electrons. Having no charge, a neutron could, in theory, enter and leave an atom's nucleus more easily than a charged proton or electron. At first, Chadwick and others saw the neutron only as a research tool, although it would later provide the means of releasing the atom's energy. Szilard, too, only wondered briefly about the neutron's potential uses for research when he first read of it in *Nature*[3]

The morning of Rutherford's lecture, September 11, 1933, Szilard awoke with a bad cold and stayed in bed. But the next morning, curious about the talk he had missed, Szilard paged through *The Times* and spied this intriguing column of type:

THE BRITISH ASSOCIATION

— — — — — —

BREAKING DOWN
THE ATOM

— — — — — —

TRANSFORMATION OF
THE ELEMENTS

Farther down the column, Szilard saw:

THE NEUTRON
NOVEL TRANSFORMATIONS

He read on, about Rutherford's survey of "the discoveries of the last quarter century in atomic transmutation," to this summary:

HOPE OF TRANSFORMING ANY ATOM

What, Lord Rutherford asked in conclusion, were the prospects 20 or 30 years ahead?

High voltages of the order of millions of volts would probably be unnecessary as a means of accelerating the bombarding particles. Transformations might be effected with 30,000 or 70,000 volts . . . [and] we should be able to transform all the elements ultimately.

We might in these processes obtain very much more energy than the proton supplied, but on the average we could not expect to obtain energy in this way. It was a very poor and inefficient way of producing energy, and anyone who looked for a source of power in the transformation of the atoms was talking moonshine.

"Lord Rutherford was an expert in nuclear physics," Szilard thought, and "an expert is a man who knows what cannot be done." Szilard found that last paragraph "rather irritating because how can anyone know what someone else might invent?" Perhaps, thought Szilard, the famous Lord Rutherford is talking "moonshine."

In the days that followed, he pondered Rutherford's declaration in a routine favored for serious thought: long soaks in the bathtub and long walks in the park. There were no large parks near the Imperial Hotel, but its Bloomsbury neighborhood was a patchwork of Georgian terraces and tree-lined squares, so along and through this shaded cityscape Szilard walked and wondered that chilly September, seeking with each quick step a way to disprove the "expert" Rutherford.

"I was wondering about this while strolling through the streets of London," Szilard recalled. A week or two passed, and then it happened. Southampton Row, the main street in front of the Imperial, is a busy but narrow thoroughfare lined with banks and shops. "Walking along Southampton Row, I had to stop for a streetlight, and at the very moment when the light turned green, it occurred to me that Rutherford might be wrong. . . ."[4] Szilard's pause in his normally brisk movement would change his life, and ours, and ultimately the history and fate of the twentieth century.

137

He recalled:

> As I was waiting for the light to change and as the light changed to green and I crossed the street, it suddenly occurred to me that if we could find an element which is split by neutrons and which would emit *two* neutrons when it absorbed *one* neutron, such an element, if assembled in sufficiently large mass, could sustain a nuclear chain reaction. I didn't see at the moment just how one would go about finding such an element or what experiments would be needed, but the idea never left me.[5]

What he did see at that fateful intersection were two concepts needed to free the energy locked in the atom: the "nuclear chain reaction" and the "critical mass" needed to set off and sustain it.

Szilard quickly seized the implications: "In certain circumstances it might become possible to set up a nuclear chain reaction, liberate energy on an industrial scale, and construct atomic bombs." Suddenly the H. G. Wells novel he had read a year before had a grave new meaning. Atomic bombs were science fiction to Wells when he wrote *The World Set Free* in 1913, and they were frightful to contemplate when Szilard first read about them in 1932. But by the fall of 1933, Rutherford's challenge and Szilard's response were moving atomic bombs away from fiction to scientific fact. Atomic bombs, and the chain reaction that would power them, became Szilard's "obsession,"[6] pushing aside his plans for a new career in biology.

"The thought did not come entirely out of the clear sky," he said later, but seeing both the mechanism and its fateful implications when he did was Szilard's special insight. The chain-reaction concept was common in chemistry, studied by Szilard's friend Michael Polanyi and others, and an atomic bombardment process similar to Szilard's had appeared in the *Nature* account of Rutherford's speech on "transmutation":

> Beryllium, of mass 9 and charge 4, when bombarded, captures an alpha particle of mass 4 and charge 2, giving rise to a structure of mass 12 and charge 6 and emitting a *neutron* of mass 1 and charge zero.[7]

To enhance this process, Szilard substituted a neutron for the alpha particle that bombarded the beryllium. "I was wondering whether Rutherford was right when it occurred to me that neutrons, in contrast to alpha particles, do not ionize [or electrically charge] the substance through which they pass. Consequently, neutrons need not stop until they hit a nucleus with which they may react."[8] Szilard further assumed that when one neutron entered the nucleus, *two* might be expelled, creating a chain reaction that seemed awesome. Two calculations fired his imagination, and his fears.

First, if a neutron could strike an atom's nucleus with such force that it would emit two neutrons, then with each collision the freed neutrons might double in number. One neutron would release 2, which would each strike an atomic nucleus to release 4. These would each strike a nucleus, releasing 8, then 16, 32, 64, 128, and so on. In millionths of a second, billions of atoms would split, and as they tore apart, the energy that held them together would be released.

Second, the amount of energy released could be huge. If Einstein's calculations of 1905 were accurate, then his famous formula $E = mc^2$ assured that Energy equals mass multiplied by the speed of light (whose symbol is c) squared. The number for mass was minute, but the number for light speed in this equation is immense, and at least in theory, the amount of energy latent in matter was also immense.

What allowed Szilard to put together the stray clues about a nuclear chain reaction, clues that scientists working directly in atomic research had overlooked? No conclusive answer is possible, given the mysteries of Szilard's creative mind and the scant details he recalled. But this much is clear: While teaching discussion courses at the University of Berlin, Szilard followed developments in nuclear physics by reading scientific journals. He also questioned anyone who knew about a subject, often with the precision of a prosecuting attorney. Unlike his colleagues in nuclear physics, Szilard was no experimentalist. Instead, he was free to speculate haphazardly—intuitively—about the implications of other scholars' practical works, not bound to move each insight only as far as its next logical step and experiment.

In fact, Szilard's restless mind led him to consider studying biology in 1933 because he desired to be a theorist in a field still practical and analytic. Moreover, his early speculations about biology might have included the rapid multiplication of dividing cells. Szilard was both predisposed and eager to pick seemingly unrelated facts from diverse scientific sources, then blend them whimsically. And while living on his savings, Szilard had no other scientific or academic burdens and deadlines. No family. No close friends. No household chores. No pets. No hobbies. When he wanted to think about the chain reaction, he could. And did. For days and nights at a time.

Beyond his eclectic and intense work habits, Szilard's thoughts often sprang from his feisty spirit of defiance. Contradiction led him to pose an opposite view to whatever he heard. In novel ways he loved to play the constructive dissident. Rutherford must be wrong, Szilard assumed. But how? Perhaps by overlooking some basic mechanical process in nuclear transmutation, one even simpler than the structural changes that the experts at the Cavendish then studied.

Perhaps Szilard's idea for the nuclear chain reaction was even aided by the London traffic light itself—in a sequence that appears to move spheres of light from green to amber to red to amber to green. Viewed from across the street, the change of lights can resemble colliding billiard balls. Not a knocking, one against another, but something subtler; a blur of one into another, through another.

By whatever means Szilard conceived the nuclear chain reaction, as soon as he had, he turned his mind on himself and tried to prove the idea right or wrong. He retreated to his room at the Imperial. Thinking. Scribbling calculations. Sketching hasty schematic patterns. For a week or more he saw no one, broke his meditation only to eat meals sent up by room service, and each night fell exhausted into bed to sleep.[9] Szilard soaked for hours at a time in his bathtub, dozed and daydreamed on his bed, and forced his impulsive vision at the traffic light into twin hypotheses. Not only did he see a chain-reaction mechanism to release the atom's energy; he also realized why a critical mass of material was necessary: Only with many atoms close together could the neutrons reach other nuclei and not escape.

Having fled Nazi Germany that spring, Szilard also saw beyond his hypotheses to their political implications. He had feared for months that the Nazis were preparing for war and now worried that in the coming conflict Germany might be the first to build—and use—atomic bombs.

By mid-October, Szilard moved out of the Imperial Hotel, perhaps to find quieter or cheaper quarters, and rented a flat at 97 Cromwell Road, in a block of Victorian row houses a few doors west of Gloucester Road.[10] There he continued his calculations, but apparently found it difficult to concentrate. A glance at the morning papers would have been distraction enough. Germany had quit the foundering League of Nations, and rhetoric at the Nazi party's rallies in Nuremberg was increasingly anti-Semitic and bellicose. No longer just a political aberration, the Nazi party was now the German state.[11]

A few weeks later, Szilard hoisted his two suitcases and moved from Gloucester Road back into central London, to the Strand Palace Hotel, a white marble structure in the busy thoroughfare that links the newspaper and legal world of Fleet Street with the government offices around Trafalgar Square and Whitehall. The Strand Palace appeared more conventional than the eccentric Imperial, but its lower rates appealed to Szilard now that the academic year had begun and he was not working. Although he entertained friends in the hotel's elegant dining room, among them the AAC's secretary, Esther Simpson, to save money, Szilard rented a tiny room that had once been a maid's closet.[12] His new room had no private bathtub, but shared one down the hall, and it was there that Szilard continued his brainstorming, usually beginning each day with a soak around nine to "dream about the possibilities" of nuclear physics.[13]

"Are you all right, sir?" asked the maid, knocking on the bathroom door at about noon. Yes. Szilard was quite all right, thank you. He had just been thinking; in particular about beryllium, a steel-gray metal that he knew could give off neutrons. What if extra neutrons split off when beryllium was bombarded by neutrons? "The reason that I suspected beryllium of being a potential candidate for sustaining a chain reaction," Szilard later wrote, "was that the mass of beryllium was such that it could disintegrate into two other particles and a neutron."[14] This Szilard may have known before, but it, too, was suggested in the *Nature* account of Rutherford's speech, so he had good reason for his suspicion. The only difference was that the beryllium described in *Nature* was bombarded by alpha particles, not by neutrons. Yet elsewhere in the same article neutrons had been described as "a very powerful weapon of research."[15]

Other elements might also split and release extra neutrons, Szilard thought, and "this possibility intrigued me so much that I gave up the idea of shifting to biology. . . ." He credits this decision to three people: H. G. Wells, who showed Szilard "what the liberation of atomic energy on a large scale would mean"; and Frédéric and Irène Joliot-Curie, the French nuclear scientists who at about this time demonstrated that radioactivity could be created artificially and need not depend on nature's elemental design. Working in Paris, the Joliot-Curies had bombarded aluminum foil with alpha particles, then noticed that the foil continued to emit radiation after the stream of particles was shut off. "If elements could be made radioactive by bombarding them with alpha particles," Szilard reasoned, "then why shouldn't elements be made radioactive when they are bombarded by neutrons?" And if neutrons could make an element radioactive, then science had "a very simple tool" to study nuclear physics.

"In science it is not enough to think of an important problem on which to work," Szilard reflected years later. "It is also necessary to know the means which could be used to investigate this problem." In short, Rutherford's challenge to nuclear transmutation had driven Szilard to the chain-reaction concept, and the Joliot-Curies' discovery of artificial radiation had provided the means that would make the concept work.[16]

Cautiously, Szilard mentioned his "nuclear chain reaction" idea to a distinguished professor of physics he had met on his rounds for the AAC, George Paget Thomson at the Imperial College of Science in London. (Thomson would share the 1937 Nobel Prize in physics for discovering diffraction phenomena in electrons.) When Thomson showed little interest, Szilard called on Prof. Patrick M. S. Blackett in the physics department at the University of London. Blackett then specialized in the use of Wilson cloud chambers, instruments that detect ionizing radiation. (For this research, and for work on cosmic rays, he would receive the 1948 Nobel Prize in physics.) But from him, too, Szilard "couldn't evoke any enthusiasm."[17] Szilard then believed that while beryllium did not disintegrate spontaneously, it might yet split if "tickled" by a neutron.

"What one ought to do," Szilard told Blackett, "would be to get a large mass of beryllium, large enough to be able to notice whether it could sustain a chain reaction." But beryllium was so expensive that it was almost unobtainable, even in tiny quantities. "Look," Blackett finally complained to Szilard, "you will have no luck with such fantastic ideas in England. Yes, perhaps in Russia. If a Russian physicist went to the government and said, 'We must make a chain reaction,' they would give him all the money and facilities which he would need. But you won't get it in England."[18]

With this rebuff Szilard turned, in December 1933, to British industry. Szilard may have feared the chain reaction's use in a bomb, but he also fancied its commercial possibilities as a source of power. Through Michael Polanyi's contacts at General Electric (U.K.), Szilard approached the director of the company's research laboratories.[19] That same month, Szilard gave his friend Esther Simpson power of attorney to protect his patents and assets. At year's end those assets stood at 1,594 pounds, 8 shillings, and 9 pence.[20] "Who knows?" Szilard may have wondered as he dreamed about a new Industrial Revolution fueled by nuclear power. Someday it may be millions.

CHAPTER 11

Chain-Reaction "Obsession"

 1934–1938

"I feel somewhat depressed about Szilard," Eugene Wigner wrote to their mutual friend Michael Polanyi in January 1934, because "his last letter sounded a little more unstable than I am used to with him."[1] Wigner and others fretted about finding Szilard a steady job: at Princeton, where Wigner taught physics; at New York University, where Wigner hoped Szilard would "sooner or later feel quite happy."[2]

Szilard infuriated his friends, at first telling Polanyi he was "very sorry" when a Liverpool University appointment fell through but in the next sentence asking to visit him at the University of Manchester along with the Berlin physicist Fritz Lange, "more for the fun of it than for any special purpose. . . ."[3] Szilard had met Lange and his partner, Arno Brasch, in Berlin during the 1920s and brainstormed with them about accelerators, which use high voltages to speed up radioactive particles to bombard atoms. For maximum voltage, Brasch and Lange were trying to tap lightning at alpine research stations atop the Zugspitze on the German-Austrian border and on Monte Generoso near Lake Como—a project that both amused and frightened Szilard.[4]

At the time, accelerators seemed a promising way to manufacture radioactive isotopes for use in medicine, and during Lange's December visit to England, he and Szilard toured the General Electric (U.K.) research laboratory at Wembley, in west London.[5] Szilard considered working there, briefly, and he had fancied teaching physics in India, also briefly.[6] But despite fervent efforts by Wigner and other friends, Szilard had "decided to mark time" in his career.[7]

Even more than lightning-powered accelerators, however, it was his nuclear-chain-reaction concept that energized Szilard's thinking. His self-proclaimed "obsession" with chain reactions distracted Szilard from the refugee-settlement work for the Academic Assistance Council (AAC), and by March 1934 he had reduced his ideas to paper. In a fifteen-page patent application he named beryllium as the element most likely to be split by neutrons and, in turn, free other neutrons in a spontaneous frenzy of energy. In fact, beryllium would never split as Szilard expected; he was misled by incorrect published data about the element's atomic weight. But, intuitively, Szilard also named uranium and thorium—the only two natural elements that would eventually sustain chain reactions.[8] He filed the patent on Monday, March 12, 1934, handing in a typed manuscript sprinkled with penned corrections and x-ed-out words. Then he began to dream about the atom's new commercial uses, perhaps replacing coal and oil as the world's industrial fuel; about its social implications, perhaps bringing abundant energy to developing countries now starved for water and minerals; and—unavoidably—about its potential as a weapon of mass destruction, perhaps giving Adolf Hitler "atomic bombs" to terrorize the world.

His patent filed, Szilard wrote to Sir Hugo Hirst, founder of General Electric (U.K), then on holiday at the Carlton Hotel in Cannes, and in a playful letter urged him to read "a few pages" from *The World Set Free* by H. G. Wells. In the book Szilard cited an "interesting and amusing" section that had predicted in 1914 how artificial radioactivity would be produced in 1933, just as it had been by the Joliot-Curies.

> Of course, all this is moonshine, but I have reason to believe that in so far as the industrial applications of the present discoveries in physics are concerned, the forecast of the writers may prove to be more accurate than the forecast of the scientists. The physicists have conclusive arguments as to why we cannot create at present new sources of energy for industrial purposes; I am not so sure whether they do not miss the point. . . .[9]

But instead of proving this point, Szilard pursued another invention the same week and filed a patent for a "microbook" that would reduce whole libraries to minute images on a roll of film. He would later learn that Siemens had already patented the idea, as "microfilm."[10]

Also in mid-March, Szilard's mentor and friend Albert Einstein wrote to support a Rockefeller Foundation grant that would finance Szilard's

144

research at NYU, praising him as "an especially intelligent and many-sided scientist, extraordinarily rich in ideas." Einstein mentioned Szilard's "absolutely independent performance" in his Ph.D. thesis and his original paper on entropy and information, noting also that "he is the originator of an important feature of the modern quantum mechanics, namely, of the consideration of measurements in the formulation of the wave function." Szilard's "personality is estimated very highly by all his colleagues," Einstein wrote, "especially for his great unselfishness."[11] While praising Szilard's originality, Einstein would not learn about the truly original chain-reaction concept for another five years.

With no laboratory of his own to test his chain-reaction ideas, Szilard enlisted Fritz Lange and Lise Meitner in Berlin to arrange certain experiments into "the production of radioactive bodies." Szilard urged them to "take one after another all seventy elements and bombard them with cathode rays [X-rays] and see if there is any activity by using a Geiger counter or the Wilson [cloud] chamber."[12] In science at the time, such international collaboration was rare, although to Szilard it seemed the obvious thing to do.[13]

For all his involvement with nuclear research, Szilard could not ignore the pull of politics. It caught him when he picked up the *Manchester Guardian* on the morning of April 24 and read that Japan had rejected all interference, by the League of Nations or by any country, in its invasion of Manchuria. This invasion, and Japan's arrogance, upset Szilard's "sense of proportion," the moral and ethical balance that he sought in nature and in modern life. Szilard ripped the article from the page and with a letter mailed it to Lady Murray, the wife of classical scholar Gilbert Murray, then president of the International Committee of Intellectual Cooperation and chairman of the League of Nations Union. "We felt [at NYU in 1932] that a mere protest would not be of any value," Szilard wrote, "but that a definite pledge on the part of the leading scientists, though of rather limited value in itself, would serve to 'keep the faith' in the cause of justice."

In his current scheme, Szilard would try to politicize Nobel laureates for the first time. But this needed "the right person" to call for a scientists' boycott of Japan. Realizing how timid his scientific colleagues could be, Szilard added a twist to his appeal: The laureates' protest would only take effect if "eight-tenths" of all prize winners agreed to sign. Under his proposed boycott, scholars would refuse to send scientific and technical information and journals to Japan or to cooperate with Japanese students. [14]

Gilbert Murray liked Szilard's idea, "considering the extreme reluctance of governments to shoulder any burden or take any risk for the sake of world peace," and urged him to call on Dr. Maxwell Garnett, at the League of Nations Union, to consider approaching some famous scientist for the cause. But when Szilard met Garnett and realized that the plan appeared "too rigid and impractical" to him, he let his efforts lapse, although enlisting Nobel laureates in political causes would eventually become a popular technique.[15]

Wigner and Szilard met in London early in the spring of 1934, and the two talked for hours about the chain-reaction patent. From Wigner, Szilard obtained new calculations and—most important to him—encouragement from a brilliant colleague whom he respected. Szilard reduced the chain-reaction idea to a simple equation: $Be^9 + n = Be^8 + 2n$. Bombard beryllium *(Be)* with a neutron *(n)*. The beryllium changes atomic weight, but not its chemical identity, and in the process emits two neutrons. These neutrons penetrate other atomic nuclei, releasing four neutrons, then eight, then sixteen, and so on. Control this instantaneous reaction and you create immense quantities of heat; let it run wild and you transform matter to energy in a violent explosion.[16]

In May, Szilard warned Australian physicist Mark Oliphant about a possible nuclear explosion, but he seemed to make little of it.[17] During the last week of that month, Szilard and Wigrer visited the Cavendish Laboratory in Cambridge, where the director, Nobel laureate Ernest Rutherford, had agreed to meet Szilard "for a short time," although offering "little chance of . . . getting work" there.[18]

Their talk, at noon on Monday, June 4, 1934, might have changed the course of nuclear research and with it modern history. Szilard, the author of the nuclear-chain-reaction concept, met Rutherford, the authority who could have tested and developed that idea. But Szilard was overly cautious, perhaps nervous, and in the great man's presence apparently tried to appeal to Rutherford's scientific conservatism. Szilard described a chain-reaction method that was more conventional than the neutron theory that had flashed before him in Southampton Row. Szilard said that a chain reaction might be achieved in two stages: Alpha particles would bombard light elements to release their protons, as in the many transmutation experiments performed at the Cavendish Laboratory by Rutherford and Chadwick. These protons, in turn, would disintegrate lithium and release two alpha particles, as Cockcroft and Walton had done there when they first split atoms.

In this explanation, Szilard kept secret his original idea that two particles would escape with each bombardment. And instead of neutrons, he
used Rutherford's technique with alpha particles. It was a silly mistake.
Rutherford knew the atomic structures and energy efficiencies of the
systems Szilard described. Szilard did not. This apparent ignorance exasperated Rutherford, who became further upset when Szilard announced
that he had filed a patent on the chain-reaction concept. When Szilard
asked if he could work at the Cavendish Laboratory to test his ideas and
offered to provide recommendations, that did it. "I was thrown out of
Rutherford's office," Szilard later told his friend Edward Teller.[19]

Three days after this testy meeting, Rutherford's answer arrived by mail
at the flat of Szilard's friend Esther Simpson in north London, and fearing
it contained a rejection, Szilard sat down to write Rutherford, On one of
the few occasions in his life when he groveled, Szilard wrote that "to be
able to work in the Cavendish Laboratory for a year means so very much
to me that I feel extremely anxious that a decision about the possibility of
my entering there should not be based entirely on the few conversations I
have recently had in Cambridge." He promised to arrange recommendations from von Laue and Schrödinger. He promised, as Rutherford had
suggested, to meet with a Professor Fowler in Cambridge. But the "unusual
request" that followed must have enraged Rutherford even more than
their face-to-face meeting. "I have just heard that a letter from Cambridge
has arrived at my Halliwick Road address," Szilard wrote, "and I would
ask you, should that letter be from yourself containing your decision, to
allow me to return it to you unopened."[20] There is no record in the Szilard
or the Rutherford papers that Rutherford ever replied or that Szilard ever
pressed his request.

To financiers, Szilard appeared just as impertinent and impractical. He
asked investors to fund experiments into "a rather romantic enterprise on
which I embarked," one "which arose out of certain recent developments
in physics." These experiments, Szilard promised, "could in a short time
lead to a sort of industrial revolution," although "it is not possible to fore-
tell with certainty the outcome of the proposed experiments."[21]

Throughout the spring and summer of 1934, Szilard repeatedly
approached GE officials with offers they politely termed "somewhat vague."
These letters document Szilard's inability to communicate his mind's intellectual excitements. He could scarcely focus on the chain-reaction concept
himself, and during the six months he sought GE's help, he filed as many
"improvements" to his original patent: adding the names of elements

likely to release neutrons, giving the size of the beryllium block to be used in experiments, proposing to mix seventy elements together in order to isolate the radioactive ones systematically. While promising to tell the GE executives about the atom's new "industrial applications," Szilard would only describe his work on medical isotopes. The chain-reaction experiments, which Szilard called "the other more important issues," he kept secret, proposing instead to "give a detailed picture to some third person who is attached to one of the English universities and that you should get information from him about his views on the subject."[22]

Understandably, GE researchers could see nothing "new" or "practical" in the techniques to manufacture isotopes and asked, instead, for "concrete proposals" on Szilard's mysterious power source—provided his proposals are contained in patents.[23] This request prompted Szilard's "Memorandum of Possible Industrial Applications Arising Out of a New Branch of Physics," which predicted "liberating atomic energy" on "such a large scale and probably with so little cost that a sort of industrial revolution could be expected; it appears doubtful, for instance, whether coal mining or oil production could survive after a couple of years." Szilard mentioned holding patents in the field but would only cite those for isotope production, not for the chain reaction. No wonder GE and other potential investors were skeptical.

Had Szilard been more practical and disciplined, his chain-reaction concept might have been confirmed first in the Bronx, at NYU's physics department, which in the summer of 1934 had invited him to work for a year as a research associate. At the time, Szilard's "Suggested Experiments for the Detection of Nuclear Chain Reactions and the Liberation of Nuclear Energy" were in a memo that proposed both commercial and university work. But Szilard was unsure if NYU's laboratories were properly equipped, and he miffed physicists there by demanding the right "to resign at the beginning of the term if the equipment disappointed him."[24]

Szilard also annoyed GE executives, who told him that his accelerator work was not original and that his "larger issue" of power production—the chain reaction—"is so far outside the scope of a company's normal activities, that unless the proposition takes some much more definite shape, it would be impossible to participate."

"I am afraid I have to contradict almost every statement you make in your last letter," Szilard countered in his impolitic style. GE's research director eventually apologized for his "inadvertent error," but that ended the company's interest in Szilard's revolutionary new energy source.[25]

Seeking an academic laboratory for his chain-reaction experiments, Szilard called on University of London physicist George Paget Thomson and dazed him with "as many details as is possible in one interview."[26] Thomson, who would be a Nobel laureate in three years, appreciated Szilard's "quite interesting ideas" and offered him a place in the laboratory. The "commercial possibilities" of Szilard's ideas held "a fairly remote chance of a very important discovery," Thomson concluded, and he thought "that the chance of atomic disintegration becoming of commercial importance in the future is very real, and the type of experiment which Dr. Szilard proposes seems as likely to lead to it as any other."[27]

Despite this enthusiastic response, Szilard never pursued Thomson's offer, perhaps because the results would have to be published in open scientific journals and might alert the Germans to the chain reaction's possibilities. Instead, Szilard wrote to Francis Simon in Oxford, a German physicist he had known in Berlin, seeking through him a meeting with Frederick A. Lindemann, the physicist who directed Oxford's Clarendon Laboratory and who advised Imperial Chemical Industries (ICI) about its research fellowships. Lindemann had studied with Walther Nernst in Berlin and during the world war had invented a way to save an aircraft from an uncontrolled spin. He had also developed a formula for determining different materials' melting points.[28]

But before his university prospects in New York, London, or Oxford developed, Szilard had talked his way into another research laboratory. Interest in manufacturing isotopes for medical use led Szilard to St. Bartholomew's Hospital, a teaching institution near St. Paul's Cathedral. Szilard had met the director of the physics department, F. L. Hopwood, and in early August 1934 gained his permission to use radium samples for research during the summer holidays, provided Szilard team up with a hospital staff member.[29] Soon Szilard and Thomas A. Chalmers, a young research physicist, were studying beryllium—the element Szilard believed would produce a nuclear chain reaction.

At the time, Szilard had stopped working for the AAC but kept in touch with his colleague Esther Simpson. She had moved from north London to a flat on Brunswick Square in Bloomsbury. Through August and September Szilard divided his time between the hospital, where he worked longer and longer hours; his hotel, where he ate and slept and soaked in the bathtub; and Miss Simpson's, where he dropped by for tea and chitchat and also received mail.

To test whether gamma rays from radium would free neutrons from beryllium, Szilard and Chalmers surrounded an ounce of radium with a thin sheet of beryllium, then wrapped this with silver, iodine, or iridium.[30] They set a Geiger counter on the outer metal sheet to detect neutrons and during their collaboration made two important discoveries. First, they learned that beryllium did emit neutrons when exposed to the radium's gamma rays. Nothing like a chain reaction followed, but Szilard now saw for himself that neutrons could act to rearrange and transform (transmute) the structure of certain atoms.

In Rome a few months earlier, physicist Enrico Fermi and his colleagues had begun to bombard different elements with neutrons. They found that slow neutrons, with very low energy levels, were more easily absorbed by an atom's nucleus than were fast (higher-energy) neutrons. Once its nucleus absorbed a neutron, the atom became a different "isotope" of the element—having the same number of protons but differing numbers of neutrons and thus chemically related to its progenitor. This new arrangement within the atom usually made its structure unstable and resulted in a release of energy as radiation.

Sometimes Fermi's group noted that a neutron added to an atom's nucleus did not simply make it a heavier isotope of itself. Instead, it became an isotope of a different element on the periodic table. For example, when rhodium[103] absorbed a neutron it did not become rhodium[104]; it became palladium[104]. In this way indium[115] became tin[116]. This phenomenon of an atom absorbing a neutron and becoming a heavier element came to be called the Fermi effect.

Initially, Szilard aimed to bombard his elements with neutrons in order to create new compounds that would be medically useful. He hoped that their temporary radioactivity could help trace liquids and gases within the human body. But a few days' work convinced Szilard and Chalmers that neutrons made some new compounds so unstable that their constantly changing states were difficult to control or calibrate. This was not the Fermi effect, but something quite different. Faced with this unexpected result, many scientists would have abandoned their idea for some other project. But Szilard, in the words of his later colleague physicist Maurice Goldhaber, "turned this apparent defeat around, and it led him to a brilliantly simple method of isotope separation."[31] They had just devised a way to isolate radioactive and nonradioactive forms of the same element.

Some basic physics will help to explain their feat. The isotopes of an element have the same number of positively charged protons in their nuclei but differing numbers of uncharged neutrons. As a result, the isotopes' atomic weights can vary, yet they remain chemically indistinguishable. Beryllium offers a good example. All beryllium found in nature is Be^9, a stable element that is not itself radioactive. (In this case, the number 9 denotes a nucleus containing 4 protons and 5 neutrons.) Beryllium's unstable isotopes, having fewer or more neutrons in their nuclei, are Be^7, Be^8, and Be^{10}.

In the empty laboratory that summer, Szilard and Chalmers beamed neutrons from a radon-beryllium source onto iodine and measured their results. To Hopwood, then on holiday, Szilard wondered in a letter "whether we can separate the radioactive Iodine from the bulk of the bombarded Iodine," admitting that "this sounds 'blasphemous' to the chemist, but I hope it can be achieved. . . ."[32] Until Szilard and Chalmers began their work, different "isotopes" of the same element could only be separated through tedious and expensive methods. One was "gaseous diffusion" through a series of porous barriers; another was "electromagnetic" isolation involving costly apparatus. In 1934 it was deemed a major achievement when Mark Oliphant and others at the Cavendish Laboratory managed to separate less than one-ten-millionth part of a gram of a lithium isotope by using electromagnetic currents.

When Szilard and Chalmers tried to isolate iodine[128] by bombarding an iodine compound (ethyl iodide) with neutrons, they freed an irradiated isotope from the compound radioactively. Their results were published in the scientific journal *Nature* that September, giving researchers a simple method for separating isotopes. Known as the Szilard-Chalmers effect, this technique became widely used. With time they also recognized that slight amounts of the irradiated element sometimes do remain in the original compound, so their method had medical uses, after all—as "tracers" in the body.[33]

In September 1934, Szilard was praised for his discovery during an international conference on nuclear physics in London.[34] Although largely self-taught, "these experiments established me as a nuclear physicist," Szilard said wryly a quarter century later, "not in the eyes of Cambridge but in the eyes of Oxford." At Rutherford's Cavendish Laboratory, Szilard said, he was considered "just an upstart who might make all sorts of observations, but these observations could not be regarded as discoveries until they had been repeated at Cambridge and confirmed."[35]

At St. Bartholomew's Hospital, Szilard enhanced his "upstart" reputation by arguing with the staff about publishing "prematurely" the results of his experiments—a matter later resolved amiably. And he ignored certain hospital procedures and regulations. One required that the radium needles he and Chalmers used be returned to a safe between 6:00 P.M. and 9:00 A.M. But as they worked late many nights, Szilard found this routine impossible to follow.

"You must understand my point of view if I suggest to you that you are to pay more attention to the customs of the hospital," Hopwood scolded Szilard. "It is the point of view of a man who is very much aware of the fact that these walls you see from the window have been standing here for over five hundred years."

"I understand that very well," Szilard replied, "but please keep in mind that these walls may not be standing here ten years from now." Szilard remembered this tense conversation years later, after the area around St. Paul's was heavily bombed in World War II and the wall seen from Hopwood's window was destroyed.[36]

While still at St. Bartholomew's in the summer of 1934, Szilard asked his friend Michael Polanyi for permission to work as his guest in Manchester: another instance of Szilard's frantic indecision, for as soon as Polanyi obliged, Szilard changed his mind.[37] Instead, he searched for private investors to finance his chain-reaction research, and beginning that fall, Szilard made new overtures to Chaim Weizmann, the renowned chemist and Zionist leader. Weizmann was unable to raise the 2,000 pounds (then about $10,000) that Szilard needed to systematically bombard all seventy elements with neutrons. Michael Polanyi and the University of London's Frederick Donnan brainstormed one fall weekend to find Szilard "a very rich man, who would let you do just as you please, asking only for a dividend at the end," and on his own Polanyi sought out a possible "silent partner" for Szilard's research. But that fall Polanyi also reported a complaint against Szilard that would be heard for years to come: "Donnan told me that there is an opposition to you on account of taking patents," he wrote.[38]

By late November 1934, Szilard completed an experiment to test beryllium's potential for a chain reaction, drawing on research conducted in Berlin and London. This appeared in Nature on December 8, coauthored by Szilard, Arno Brasch, Fritz Lange, A. Waly, T. E. Banks, Chalmers, and Hopwood as "Liberation of Neutrons from Beryllium by X-rays: Radioactivity Induced by Means of Electron Tubes." Brasch, Lange, and Waly were in Berlin, assisted by Lise Meitner.[39] This work proved to be well ahead of its time by its use of X-rays, which Szilard knew about from work in the 1920s with Hermann Mark.

For this experiment, Szilard had his Berlin colleagues bombard beryllium (Be^9) with X-rays. This produced photoneutrons, which they beamed at a sample of bromoform, a colorless, heavy liquid used for chemical synthesis that is analogous to chloroform. Then Szilard had the bromoform flown to London, where he and Chalmers extracted the radioactive beryllium using their new isotope-separation method. This experiment was conducted using an incorrect measurement for the mass of the Be^9, which misled Szilard to speculate that beryllium might be "metastable" and release neutrons when it was bombarded.[40] Szilard was then alone among scientists in his belief that nuclear chain reactions might liberate the atom's energy; at the time, his friend and mentor Albert Einstein was touring the United States, where he told newspaper reporters that such efforts would be "fruitless."[41]

In late December, Szilard and Chalmers drafted an article for *Nature*.[42] Somehow rumors about their work reached Rome by the first week in January 1935, prompting Fermi and his colleagues to rush into similar experiments and to publish their results hastily. Although the Rome physicists looked upon Szilard with some amusement, they were also coming to respect his forays into the unknown.[43]

By then, Szilard had packed his two suitcases, checked out of the Strand Palace, and caught a train to Oxford, hoping to talk his way into Lindemann's Clarendon Laboratory and continue chain-reaction experiments. Szilard's first stop in town—as it was for many central European refugees—was the spacious cottage home of Francis Simon, a physicist who had studied and taught in Berlin and recently emigrated from Breslau (now Wroclaw, Poland). The Simons entertained students constantly in their two large sitting rooms and deep garden; there was always an extra place at dinner, and their Sunday teas were known for a predictable spread of goodies and an unpredictable array of guests. When Szilard decided to stay in Oxford for a few weeks, the Simons referred him to the English physicist James Tuck, who sublet a room with breakfast in his house on Banbury Road. At breakfast after the first night, Tuck said he hoped Szilard had had a comfortable night.

"Oh, not quite," Szilard said. "I didn't like the bed at all."

"Well, you'll get used to it," Tuck reassured him. "It takes a little time to get used to a bed."

At breakfast the next morning, Szilard reported, "It's still very bad." Again Tuck reassured him.

On the third morning, Szilard appeared pale and haggard. "There can't be anything wrong," Mrs. Tuck insisted, and went upstairs to check. There she discovered Szilard's trouble. The charwoman, when she had remade the bed for this new lodger, had forgotten to put back the mattress. Szilard had been sleeping on the bedsprings.[44]

Through Simon's introduction, Szilard met Lindemann at the Clarendon Laboratory, a neo-Gothic complex on Parks Road near the university cricket ground. Hoping for a fellowship, Szilard described his research project confidently, and when Lindemann could offer no support, Szilard decided to stay on in Oxford anyway. He drafted a fresh round of appeals to rich men, seeking support for a new organization that would combine social and scientific interests—in a way, a practical version of the Bund and a forerunner to his idea for the Salk Institute.[45]

Poking around the Oxford colleges, as he had at the research institutes in Dahlem, Szilard quickly sensed that his exuberance was not considered charming but quite odd. Even his friends had their doubts. "Terrible! Terrible!" Szilard announced as he walked in on Simon one day. "There's a case of cholera in London!"

"Okay," Simon said, nodding, not really believing him but already expecting the unusual from Szilard.

"I'm getting worried," Szilard said two days later when he stopped in again. "Fifty deaths from cholera! What should we do?"

"Oh, come, come," Simon replied. "You exaggerate."

But the next day Szilard returned to warn: "It's getting worse! More cases of cholera. More deaths."

"Szilard! Where do you learn these things?" Simon demanded. "I've not heard of this."

"It's in *The Times*," Szilard insisted. And it was. In the column of news from a hundred years ago.[46]

This mistake typifies Szilard's frantic life. Driven by the irrepressible energies and fears of his mind, he yearned to understand everything his senses touched, though seldom for very long. Many scientists are inquisitive as they try to comprehend and explain their universe and their own existence. But for Szilard, knowing and understanding were not enough. His thoughts about his world attained a reality all their own, and his life became an urgent struggle to animate these thoughts and perhaps control them. For many hours a day Szilard kept company with thoughts that drew him, logically and persistently, toward a future that often he alone could see.

Impatient that there were no openings at Oxford, Szilard packed his two bags again and sailed on the SS *Olympic,* landing in New York on February 21, 1935. There he hoped to explore the laboratory situation at NYU and tend to the US immigration papers he had filed in 1931. After a visit with Einstein at Princeton, Szilard wrote to Lindemann with an impertinent scheme for chain-reaction research at Oxford. Einstein had been offered a place at Christ Church College, Szilard wrote. Why not apply that money to Szilard's work at the Clarendon?

His wandering mind prevented Szilard from conducting any systematic research at NYU; the only work he finished was to build an ionization chamber to study the scattering of slow neutrons. As it turned out, slow neutrons would prove to be the key to a chain reaction, and Szilard may have already suspected this, but he was too impatient to take his suspicions the next logical step to a conclusive experiment. Within a few months, Szilard's hosts at NYU seemed impatient with him, too. "They emphasize that I could leave here at twenty-four hours' notice if required," he wrote Lindemann.[47] And by the end of March, when Lindemann offered a research fellowship at the Clarendon, Szilard promptly accepted.

Typically, Szilard had trouble focusing on any project for long, and in the months spent at NYU's labs he enjoyed discussing the chairman's experiments with electric eels and searched restlessly for something else to think about and to do. One research opening seemed possible at His Master's Voice, the electronics company that became part of RCA.[48] In New York, Szilard also visited the Research Corporation, a nonprofit group that reinvested profits from patents on scientific discoveries into new research—later a model for his own finance schemes.[49]

Before sailing for England, Szilard walked around the Columbia University campus, dropping in on physicist Isidor I. Rabi, whom he had met before in Berlin and New York, and on chemist Harold Urey, who had recently received the Nobel Prize for isolating the heavy hydrogen isotope deuterium, a discovery that fascinated Szilard.

"Szilard kept telling me which experiments I should conduct," Rabi recalled years later. "Finally, I said, 'Leo, if you really think this is so important, do it yourself.' Of course, he never did."[50]

Irresistibly, politics engaged Szilard again when Pyotr Kapitza, the Russian physicist who had worked at the Cavendish Laboratory for twelve years, was barred by Stalin in 1934 from returning to England after a visit to Moscow. Kapitza's detention was on Szilard's mind when he attended the first Theoretical Physics Conference in Washington during the spring of 1935 with Paul Dirac. A Cambridge physicist who had married Eugene Wigner's sister, Dirac had shared the 1933 Nobel Prize in physics with Schrödinger for their equations explaining quantum mechanics. He was friendly with Kapitza at Cambridge and a regular member of his weekly discussion "Club."

At the Washington meeting, Szilard and Dirac tried to alert their American colleagues to Kapitza's plight, proposing a scientific boycott of the Soviet Union. But their only ally was Nobel laureate Robert A. Millikan, a physicist (and president of the California Institute of Technology) who was noted for his anti-Soviet views.[51] When Szilard at NYU and Dirac at the Institute for Advanced Study in Princeton could muster scant support for Kapitza, Szilard devised a bolder scheme: smuggle Kapitza from Russia in a submarine.[52]

The "Washington meeting," held each spring since 1935 at the Department of Terrestrial Magnetism (DTM) of the Carnegie Institution of Washington, was, despite the awesome title of the locale, an informal and lively event in the physics community, a fair-weather gathering of the great minds in a small but expanding field. The conference was cozy, with no more than three dozen scientists attending, at the DTM's spacious grounds near the city's Rock Creek Park and at the National Academy of Sciences (NAS) by the National Mall. Here Szilard could talk with several nuclear innovators: Gregory Breit from Wisconsin, who had helped land his NYU appointment; Hans Bethe, then at Cornell; George Gamow, a Russian émigré to America whom he had met at Cambridge; and Rabi from Columbia. Szilard also met Edward U. Condon, a lively and humorous journalist-turned-physicist from Princeton, and Ernest O. Lawrence, the pioneer with cyclotrons from Berkeley. On the steps of the NAS building on April 27, Szilard urged Lawrence to keep secret his research on neutrons, at one point raising his index finger to his lips in a sign of silence.[53]

One sunny afternoon, Szilard and Bethe decided to take a stroll and ambled by the Lincoln Memorial through West Potomac Park. Stopping by the Tidal Basin, they watched as tourists in rented paddle boats glided by the famous cherry trees. And they talked physics. For the first time, Szilard tried to tell Bethe about the nuclear chain reaction but instead carried the idea a step further to describe how neutrons might escape by an entirely different process.

Until then, Szilard had proposed splitting heavy elements apart to release their neutrons, but that day he told Bethe how neutrons might also be released by forcing light elements together. Szilard suggested compressing Urey's deuterium by immensely high temperatures. At the Cavendish Laboratory a year earlier, Rutherford and Oliphant had studied deuterium, and they had bombarded it with other deuterium atoms.[54] But Bethe realized later that Szilard's suggestion was the first time he had heard about a thermonuclear reaction that might become self-sustaining. Bethe wondered about how such a process might occur, especially how it might be confined once the reaction began. Years later, the concept of magnetic confinement would make the idea of nuclear "fusion" a possibility. In 1935, by the banks of the Potomac on that sunny spring afternoon, it all seemed intriguing and romantically impractical to Bethe, like so many of Szilard's ideas.[55]

Back at NYU Szilard realized for the first time the significance of his work with Chalmers, and deciding to focus on studying the chain reaction in Oxford, he sailed from New York aboard the SS *Majestic* on May 23, 1935.[56] From London he wrote Lindemann an urgent letter about a matter of "great seriousness" involving nuclear physics. Indium—a soft, silvery metal extracted from zinc—had by now replaced beryllium as Szilard's candidate for creating a chain reaction, and he was eager to study it.[57]

This June 3 letter to Lindemann spelled out Szilard's strategies for conducting nuclear research while keeping the results secret among a small circle of atomic scientists "in England, America, and perhaps in one or two other countries. . . ." He proposed filing patents on discoveries but assigning control to "some suitable body" that would "ensure their proper use." Szilard said that "obviously it would be misplaced" to consider his own patents as "private property" or to "pursue them with a view to commercial exploitation for private purposes." Instead, he recommended that a £1,000 budget be raised from private donors to allow him to hire two laboratory assistants—all, ultimately, in the name of trying to "greatly accelerate the building up of nuclear physics in Oxford. . . ."[58]

A week later, Szilard admitted to science historian Charles Singer (an active member of the AAC) that he had abandoned efforts to finance nuclear research, yet urgently predicted that his work approached "the starting point of a new industrial revolution." He predicted a fifty-fifty chance that the proposed experiments might work out. But, Szilard wrote, "the disaster to which all this may lead is more imminent than the pleasant changes it may bring about, since applications for purposes of war are closer at hand than anything else and go beyond anything one is likely to conceive."[59]

Szilard invited Singer and his niece to lunch in London, visited his country house at Parr in Cornwall, and seemed to use him as a sounding board for ideas about social and political organization. To finance research, Szilard urged that it would be necessary "gradually to bring about something like a conspiracy of those scientists who work in this field," citing the Research Corporation as an exemplary institution for controlling dangerous information through patents. "I am looking for a Maecenas," he said. For his part, Singer was "somewhat mystified" by Szilard, although his wife, Dorothea, a classicist, found him "charming and convincing to my ignorance."[60]

Szilard was more personally revealing when he sent two letters to Singer, one about social engagements, the other about research, and admitted: "I wish it would be as easy as that to keep private and public life in watertight compartments."[61] Sometimes affecting an air of mystery about his affairs, Szilard enjoyed masking his personal reactions in social situations. When he unexpectedly met his AAC colleague Esther Simpson at a tea one afternoon, Szilard showed no surprise or reaction to her and afterward boasted that he was anxious to conceal his emotions. Miss Simpson assured Szilard that he had but failed to understand this behavior at all.[62]

Szilard's desire to keep his friendships in isolated compartments surprised many people who thought they knew him. He could be intensely personal, and playful, in one-to-one encounters but controlled carefully these unguarded moments. Throughout 1934, Szilard was infatuated by Jean dePeyer, a free spirit he had met through Trude Weiss. Occasionally, dePeyer came to the Strand Palace Hotel to type for Szilard, and one day she helped him with a scientific article.

"What is one word for looking glass?" Szilard asked.

"Mirror," she replied faintly, and then passed out in her chair. When Szilard tried to revive her, she began coughing violently, and he called a doctor he knew.

"Tonsillitis," the doctor announced, and Szilard promptly packed dePeyer into a taxi and off to her small flat in Holland Park. He called the next day, bearing a straw-covered bottle of Chianti and a bundle of fish-and-chips wrapped in newspaper. And every day thereafter he followed her recovery carefully.

With Jean dePeyer, Szilard could be giddy and "childlike" by sharing "a wonderful sense of the ridiculous," she recalled years later. He enjoyed telling the same silly stories over and over again. In one she remembered, he asked a young girl what she wanted to do with her life, and her mother interrupted with "Oh, she wants to go into politics." Szilard asked the girl how she meant to begin going "into politics," and again her mother replied: "Oh, she is taking elocution lessons!" Szilard laughed and laughed as he told and retold this story.

DePeyer felt close to Szilard and years later recalled one sunny spring day when they boarded an excursion steamer at Westminster Pier and cruised up the Thames to Richmond Lock. During a long and leisurely walk through Richmond Park, which she recalled as "full of laughter and sunshine," Szilard spied a hollow tree and quickly climbed inside— laughing all the while—then peered out impishly. Later, they noticed that several children had discovered the tree. "You see," said dePeyer, "*all* the children have to get into that tree!"[63]

As he did with people he liked, Szilard told dePeyer what to read: "Three Old Men," a story by Tolstoy, and *Grand Hotel*, by the Austrian novelist Vicki Baum. While humor enhanced their understanding, the two found a deeper bond in candor. DePeyer's boarding school had trained her to tell "only the strictest truth," a discipline that reminded Szilard of the rigorous honesty that bound him to his mother. And, like Szilard, dePeyer valued her privacy and felt comfortable around this "discreet and sensitive" man. Indeed, it was "because he distanced himself" that dePeyer "felt so easy and close to him."

"You only understand a person when you love them," Szilard announced one day.

"I think," dePeyer replied, "the more you love someone, the *less* you understand them." Szilard flashed a quick, sad look.

Still, Szilard's depth of understanding appealed to both Jean dePeyer and Trude Weiss. "He spoils you for other men," Trude once told her in London, meaning that he instantly understood so many things that most men never recognize. "Not only for other men," Jean later thought, "but for other people."[64]

Separated by World War II and other romances, Jean dePeyer forgot about Szilard until she saw him interviewed on television in 1960. She wrote. He replied, merrily recalling their escapades and remembering two friends whom Jean had forgotten. He reported trying to find her on his first postwar trip to Europe, two years before, and proposed a detour to see her if a planned trip to Moscow and Paris came about. "Why did you think that I may have forgotten about you?" Szilard added at his letter's end. "As a matter of fact, I have a very vivid memory of everything, including 'the color of your hair,' " a private joke.[65] Several people who knew Szilard say that he remembered some of their friends longer than they did, an indication that, though often brief, his concern and involvement with people were intense.[66]

Szilard's playful walks in London's parks and his earnest discussions about literature were escapes from his nagging concern for what his chain-reaction research might mean to the coming war. Ironically, he said later, he helped keep Germany from winning World War II. If he *had* raised the money and painstakingly tested all seventy elements, Szilard concluded, he could have discovered as early as 1935 or 1936 that uranium released neutrons—a fact not recognized until 1939. Such a discovery could not have been kept secret, and Germany, then planning for war, would likely be quick to apply this knowledge to building an A-bomb.[67] After the war, Szilard said jokingly that he, Fermi, and other physicists should receive the Nobel Peace Prize for *not* having conducted uranium experiments in the mid-1930s. Had they done so, Szilard said, Hitler might have conquered the world.[68]

Summoned to Paris in 1935 by news that his brother-in-law, Roland Detre, was ill with tuberculosis, Szilard boarded a plane at the Croydon aerodrome south of London, landed at LeBourget, and visited the Detres in their garret studio near Place de la Bastille. After giving them money for medical expenses, Szilard decided to walk into the Joliot-Curies' Radium Institute. Unannounced, he entered a laboratory and spotted Walter Elsasser, a German-refugee physicist whom Szilard had first met in Berlin in the late 1920s.

Szilard described his efforts to organize refugee scientists in England, then announced that he wanted to start something similar in France. "Form an association of some kind and the French government will be forced to treat you better, grant you better benefits, better jobs," urged Szilard. Elsasser listened carefully, for as a Sorbonne professor he held the senior position among the refugee scientists and was just the man Szilard should be speaking to. But as Szilard chattered, Elsasser felt "a sudden violent fright," perhaps the most ominous of his life. "This is the perfect setup for a fifth column," Elsasser thought.[69]

"This is a good idea, Szilard," he replied, "but I think it is very dangerous to confront the French with an organization. They may just get sour and throw all the refugees out."

"You must organize," Szilard insisted. "You must." This dialogue continued for more than an hour, with neither man softening his views. Then, just as suddenly as he appeared, Szilard stood up, walked out, and flew back to his English campaign.[70]

Unable to finance chain-reaction research, Szilard decided that his last chance to keep the concept secret lay in his 1934 patent. On September 16, 1935, he offered his patent to the War Office as a military secret and visited the army's research department at the huge Woolwich arsenal on the Thames east of London. There Szilard met with the director of radiological research and for several hours explained how nuclear chain reactions might be made explosive. When the army's Directory of Artillery wrote that "there appears to be no reason to keep the specification secret so far as the War Department is concerned," Szilard took his idea to the British Admiralty.[71]

Still the chain reaction haunted Szilard, and with no funds available to study it, Szilard continued haphazard research on his own. Back in Oxford, he rented two rooms in a Victorian row house at 8 Keble Road, around the corner from the Clarendon, and immersed himself in research to understand how neutrons interact with atomic nuclei. The only time in his career that Szilard conducted an experiment alone, he became so self-absorbed that almost everywhere he appeared he toted two black leather bags: a medium-sized valise for clothes and papers and a small, square case for his research equipment. In the small bag Szilard stuffed a Geiger counter, shaped-metal amplifiers, paraffin wax, and flat metal foils stacked in cigarette tins. Szilard was not a smoker, but whenever someone offered him a cigarette, he replied: "No, thank you, but I'll take the tin."[72]

For his experiment Szilard placed radon gas and beryllium together so that gamma rays from the radon released slow neutrons from the beryllium. He beamed these slow neutrons through a sixteen-inch tube lined with paraffin, which acted to slow them further before they were absorbed by thin sheets of cadmium or indium. Szilard measured the different absorption rates for the neutrons, noting that "residual neutrons," which are slowed but not stopped by the sheets, behaved very differently, depending on whether they passed through cadmium or indium. Completing his work on November 14, Szilard mailed a manuscript about it to *Nature* and, obviously pleased with his findings, sent another copy to Rutherford in Cambridge. In reply, Rutherford acknowledged that Szilard had advanced the understanding of slow-neutron absorption beyond the work of two Cavendish colleagues.[73]

Szilard's work with "residual" or "resonance" neutrons, which was performed independently by Fermi in Rome, soon contributed to a new understanding of the way neutrons are absorbed and how the atomic nucleus is structured—concepts expressed in 1936 by the Breit-Wigner formula for neutron absorption and by Bohr's compound model of the nucleus.[74] Szilard sent detailed notes on his work to Breit and Wigner, and Bohr replied a few weeks later, saying that when he visited Donnan in London, he hoped they could meet. "We have of course all been very interested in your beautiful recent researches on the nuclear problems," he wrote. Szilard's work, Bohr said, had contributed to his own recent search for "some simple views about the constitution of the nucleus," and this, rather than a more general subject, would be his topic at a forthcoming lecture in London.[75] Szilard's work may have contributed to Bohr's conclusion that neutrons might create "an explosion of the whole nucleus," although he would not foresee a weapon from this process until 1939.[76]

Szilard's paper on neutron absorption established him as a serious participant in the field of nuclear physics, a self-taught outsider finally respected as a serious researcher. John Cockcroft, one of the Cavendish physicists who had first split the atom, invited Szilard to address the Kapitza Club about his slow-neutron work.[77] Szilard was now advising Breit, Wigner, and Bohr—three respected pioneers in the field. Fermi mailed him a manuscript of his latest paper.[78]

In meetings around the Clarendon with Lindemann, Szilard repeatedly urged that nuclear scientists agree to prevent their work from being applied to "lethal purposes,"[79] Beginning in 1936, Szilard also pleaded with nuclear scientists outside Germany not to publish their neutron research. Such assertions, especially from an interloper in the field, were resented as being "unscientific." But Szilard kept pleading. He saw a war coming and feared that atomic weapons would determine the outcome. When Germany reoccupied the demilitarized Rhineland on March 7, 1936, Szilard heard that the British government would do nothing and concluded that Hitler could not be stopped. Szilard vowed that day to leave England.[80]

The same month, Szilard wrote from Oxford to Enrico Fermi, proposing to turn over his patents on atomic-energy discoveries to some new nonprofit research corporation. "I feel that I must not consider these patents as my private property and that if they are of any importance, they should be controlled with a view of public policy." He further spelled out for Fermi how research money should be spent: to investigate all the elements systematically, to rent radium sources for this work, and to pay travel costs for "any of us to move from one laboratory to another" to share necessary apparatus.[81]

But for all his pledges, Szilard's motives remained suspect. In Cambridge that March, Maurice Goldhaber reported to Szilard that someone had sent a copy of his neutron patents to the Cavendish Laboratory, where they were discussed at tea, "and, of course, your intentions were misunderstood to be financial or otherwise unscientific." Goldhaber defended Szilard by saying he intended to direct science "in a way which will be useful to science and not damaging to the public."[82]

In three important ways, Thursday, March 26, 1936, was a turning point in Szilard's life, a conjunction of professional and personal events that would shape his decisions for years to come. First, the director of navy contracts wrote to report that a certificate of secrecy for the chain-reaction patent had just been filed, along with the assurance that it would be reassigned to Szilard "if and when secrecy . . . is waived."[83]

Second, Szilard wrote to Bohr, alerting him to the possibility that the isotope uranium^{235} was "somewhat analogous to the case of indium," still Szilard's candidate for the element that might create a nuclear chain reaction. Szilard cited a paper by Otto Hahn and Lise Meitner that reported uranium has several "isomers," different forms all having the same atomic number. In this conclusion Hahn and Meitner were mistaken; they were actually observing the products of nuclear "fission," when the atom's nucleus breaks apart while releasing extra neutrons. But this process would not be understood or named until 1939. By pointing out that these "isomers" of uranium might have behaved in the way he predicted indium should, namely, through a chain reaction, Szilard edged very close to recognizing the fission process itself. And, by identifying uranium^{235}, he intuitively picked the one isotope in nature that would fission most easily.[84]

A third event on March 26 would change the course of Szilard's personal life: He wrote a long letter to Trude Weiss in Vienna, speculating about what she should do "after the war" and inviting her to join him in England. As a "push," he offered to pay her train fare. Trude had just graduated from the University of Vienna's medical school and had written Szilard for advice about her career. Szilard was, by turns, bossy and patient. He warned her that in two years she would not be able to work in Austria. He described life in England but recommended America instead. And he debated, then dismissed, her own idea about emigrating to Palestine. "You must *first* try that which you yourself prefer to do so that you know whether it works out or not," he insisted. "So, come immediately to England."[85] Within three weeks of writing this letter Szilard traveled by train across France and Switzerland to Vienna and sat Trude down and lectured her about a career as a doctor and about her survival as a European Jew. His warnings worked, and Trude moved to London, at first sharing Esther Simpson's flat on Brunswick Square.[86]

When Szilard returned to England two weeks later, he tried to enlist Fermi, Chadwick, Cockcroft, and Rutherford in an international scientists' research corporation, repeatedly assuring them that the chain-reaction patents were not his property. But he deepened their skepticism by remaining vague about the reasons for secrecy, refusing, for example, to describe his research before the Kapitza Club.[87]

"I am . . . in the uncomfortable position of a man who during a fire (either real or existent perhaps only in his imagination) tries to remove some jewelery [sic] which does not belong to him to some place of safety," Szilard admitted to Fermi's colleague Emilio Segrè. "Some passers-by who meet him in the street with the jewelery in his hands must inevitably take him for a thief, even if they are too well-bred to say so. While I am quite prepared to face this if necessary," Szilard concluded, "you will appreciate that I should like to get out of this situation as quickly as possible."[88]

But it was not until May 1936 that Szilard could bring himself to describe his formulations and fears. Then, writing to Rutherford, he enclosed a draft letter to *Nature* that speculated about the anomalous Fermi effect in indium. Should his research succeed, Szilard wrote,

. . . we would for the first time have to envisage the theoretical possibility of nuclear chain reactions. The prospect of bringing about nuclear transmutations on a large scale by means of such chain reactions is somewhat disconcerting. It is very unlikely that the misuse of chain reactions could be prevented if they could be brought about and became widely known in the next few years. I am quite aware that the view which I am taking on the subject may be very exaggerated. Nevertheless, the feeling that I must not publish anything which might spread information of this kind—however limited—indiscriminately has so far prevented me from publishing anything on this subject.[89]

CHAPTER 12

Travels with Trude

 1936–1938

One summer weekend in 1936, Leo Szilard distanced himself from his nightmarish fears about the nuclear chain reaction by inviting his friend Trude Weiss, who had just begun studying medicine in London, to visit him in Oxford, and together they explored the nearby countryside. By train they rode to the ancient town of Wendover and there visited Ye Olde Red Lion Hotel, a 400-year-old Tudor inn frequented by Cromwell. On a hill nearby, at the imposing South African Monument, Szilard posed for Trude's camera among the rusticated columns, peering into the lens inquisitively, staring in absent thought at the trench coat folded over his left arm, grinning self-consciously, gazing across the rolling green hills. In Oxford, Szilard introduced Trude to his friend Nicholas Kurti, a Hungarian-born physicist working with Francis Simon on low-temperature physics.

"Trude wants us to go on the river," Szilard told Kurti. "She has packed us a hamper with tea and cakes." Kurti owned a folding canvas canoe, which he offered to the couple. Trude was delighted, and off they walked across the meadows to the river Cherwell, a narrow channel lined with weeping willows and wildflowers that meanders through college playing fields and into the Thames. In an orchard by the riverbank, Szilard eyed the craft suspiciously. He reached down and wiggled it, cautiously placing one, then the other, foot inside, and sat down.

"What's the point of putting it in the water?" Szilard asked, grinning across the gunwales. "Let's have our picnic right here. We can sit in the boat and *see* the water; we don't have to be *in* it." And there they sat, munching on cakes, sipping tea, and looking about from their grass-bound canoe. At the time, Kurti thought Szilard overdid the logic of this situation, and only years later learned from Trude that it wasn't logic behind this escapade but fear. Szilard suffered from aquaphobia, which he masked with playfully intense rationality.[1]

In an avuncular way, Szilard was pleased to have Trude in England. He seemed to enjoy her company and introduced her to friends in Oxford and London. He apparently had no romantic need to be with her but was flirtatious in a nervous and adolescent way. His joy with Trude seemed to come from providing for her safety and well-being. He called her *Kind* and wanted to protect her from the coming war.

Nazi propaganda from the 1936 Olympic Games in Berlin that summer and Germany's treaty guaranteeing Austrian neutrality gave Szilard further evidence that Hitler was planning to go to war.[2] Trude had returned to Vienna in September, and when Mussolini and Hitler signed a treaty establishing the Rome-Berlin Axis a month later, Szilard feared for both her safety and that of his family. In November he returned home to Budapest and urged his parents and brother: "Come to America." But, again, they shook their heads, as they had to Leo's warnings when Hitler first took power. Leo worried too much, they thought. To him the future was one dark cloud. From Vienna, Szilard brought Trude back across the Continent to London, where she rented a room on Cromwell Road in Kensington and enrolled in medical classes at the Post-Graduate Medical College in nearby Hammersmith.

In Oxford, Szilard resumed his neutron experiments and, by late December, signed a formal agreement with Arno Brasch's wealthy uncle, financier Isbert Adam of Danzig, for isotope-separation patents. This contract paid Szilard more than $14,000 over the next year, providing his first financial security since leaving Germany more than three years before.[3] With that Szilard began "playing with the idea" of visiting America in the spring, ostensibly to attend the Third Conference on Theoretical Physics at the Department of Terrestrial Magnetism (DTM) in Washington, more likely to follow a restless urge he could scarcely comprehend or contain.[4] He did sail to America in the spring of 1937 but never attended the annual conference in Washington; for most of his time on this short visit he tried to find investors for his isotope-separation patents and to arrange for immigration papers for himself and Trude. His letters to Trude betray a fear that she might not follow him to America and, indeed, might be too distracted or confused to make a rational decision herself. "I will keep my fingers crossed tightly, hoping that you will surely come, too," Szilard wrote in his jumbled, free-association style just before his liner steamed from Southampton.

> Have a good look at England and make the most of it so that you can differentiate between the two when you come to America. . . . The third class of the Queen Mary is built *very* beautifully. I have discovered one sympathetic-looking man (not in the mirror), but I am hoping for more. So take care and be cheerful.[5]

The day before landing in New York, Szilard wrote to recount a so-so crossing.

> The weather was not good, but because of the combination of "depressing" action of the hyoscine [a drug Trude had recommended] and the "cheering up" action of your radiogram, nothing bad has happened so far. . . . Please do not fall to pieces again! Otherwise, I have just finished reading a book that you should buy right away and credit it to my account: *The Street of the Fishing Cat* [by Jolan Foldes]. . . .[6]

Once Szilard landed in New York he filed a declaration of intention to become a US citizen.[7] From International House, a comfortably formal hostel for overseas students above the Hudson River near Columbia University, he wrote Trude that "everybody here is very nice to me; because they feel that I do not want anything from them."[8] Szilard promised to check on New York medical-examination requirements and a week later wrote with detailed instructions about obtaining a visa and applying for a medical license. Szilard again urged Trude to visit New York and reported meeting a doctor friend of his sister's named Keresztury, who had offered Trude a place to stay.

> We have beautiful sunshine here, very bracing air; everybody is in good spirits and very helpful. The only doubt I have concerns a certain degree of anti-Semitism, especially in medical circles, which you would have to come to terms [with].[9]

While in New York, Szilard began to doubt his own plans:

> I am fairly reserved and low-key, which is probably not bad in terms of success, but the reason is (unfortunately) that I am not in high spirits in spite of the sun.

Today I dreamt the following: (As an illustration) I had a *carcinoma* under the skin on the left side of my chest that could not be operated on, and I knew that nothing could be done about it. I felt *perfectly all right* with it, and I started to get everything in order. I bequeathed part of *the interest* on my $10,000 to my family, and the rest I would give to you over a period of three to four years because I thought that you would need larger sums in the near future and that afterwards you would have a good job, anyway, or you would be married to a man with money. My mother wanted to know if it was for sure that it was a carcinoma and wanted to look at it, but I told her that I knew for sure and that I was too busy to dress and undress and that she could look at it in the evening when I would go to bed, anyway. So I took care of everything, and then I began to be interested in poison to end the affair once things would start to become painful.

Suicide in the face of painful death became a serious, and apparently recurrent, concern for Szilard; he would ponder it again when he did suffer from cancer in 1960. But here Szilard catches himself. "It was one of these typical dreams that are not worth analyzing because they are more interesting on the surface than underneath. Otherwise, I feel fine, and I am glad I came here during my lifetime." Then his introspection returns, along with a retreat to reason, as he continues:

It is characteristic that I am excited about every airplane that thunders over-head as a symbol of progress; in England I looked at airplanes with a frown. However, I am not sure if I want to live here myself, but if I were younger, I would certainly do it. I am in the mood to keep Oxford on a half-time basis, but this idea is still too vague to be discussed. Make sure that you have *a real* rest and get everything in order with the main [immigration] officer as well as you can; but do not take things too seriously! Have a very good time.

—Yours L. Sz.[10]

Szilard's life had become so unsettled by this time that he appears to have made no effort to attend the Washington meeting and sailed from New York aboard the RMS *Aquitania* on May 12, 1937.[11] In England he found it impossible to take up his full-time research but instead shifted between the Clarendon Hotel in Oxford and a room at Trude's address in London.

In June 1937, Szilard met Arno Brasch in Paris, trying to decide where their research on isotope separation might lead. Szilard considered, among other things, whether "induced radioactivity" might be used as "storage energy" to "drive aeroplane motors or rockets. . . ." And he made tentative calculations, but weeks later asked Brasch to double-check all work, including data from published articles by physicists Enrico Fermi and Ernest Lawrence. Szilard's mood wobbled, enthusiastic about an idea in one sentence, skeptical in the next.[12]

His career plans were shaky, too. Szilard annoyed Lindemann, his bene-factor, by suggesting a half-time research post at the Clarendon Labora-tory, then failing to keep the appointment set to discuss it. "I am not particularly enamored of your plan of working in Oxford half the year and the other half in America, especially if half of the time you should spend in England is spent in London," Lindemann complained. It was "quite useless" for Szilard "to endeavor to work here" unless he were to spend only summers in America.[13]

But that summer Szilard had other plans, and for the first time in months he seemed in a feisty mood. Aboard the cross-Channel ferry he had spotted Miss Singer, a physiologist he had met years earlier at lunches and country weekends with her uncle Charles Singer, and was miffed that she would not chat with him for the whole journey. He approached another woman, "who looked as if she came from Cambridge." She had, and was a medical student who knew Szilard's friend Maurice Goldhaber. Szilard invited her to come along to Switzer-land, and again he seemed disappointed when she refused. Was Szilard, who described these flirtations in a letter to Trude, being boastful or playfully naive?

"Do you understand?" he asked Trude before describing how in Swit-zerland he flirted with "a really beautiful girl on the bus." When he could find no language to speak with her, Szilard concluded: "Well, there are also some mountains to flirt with."[14]

His first morning in St. Moritz, Szilard climbed 700 meters "right away" and the next day took a 20-kilometer hike. During his two weeks there, Szilard was "momentarily disabled by a sunburn that nonetheless does not affect my friendly feelings toward the sun." He loved to brainstorm with friends and colleagues (alone, or with others, he called these mental excursions "botching") and reported to Trude that with Brasch he would "botch between breaks." Reading the newspaper every day, he wrote, convinced him that "everything happens, I am sad to say, according to my predictions, but that no longer concerns me. Everything is moving really quickly. . . ."[15]

Between hikes, Szilard wrote letters to coax Trude back from Vienna to London and on to America. In Vienna, meanwhile, Trude struggled with her emotions to leave home. None of her letters survive, but from Szilard's reactions it is clear that her psychiatrist's views annoyed him. "Your idea that I do not want to meet you on the Continent because I do not want to make it any easier for you to leave Vienna is, I am sorry to say, consumed by ideas about relationships and considerations," he complained. Szilard admonished Trude for making her departure "difficult for yourself" and, by implication, for him as well. "You are not making any decision on principle."[16]

A daylong train trip brought Szilard through the Swiss Alps to Lausanne, where he and Bela met to visit their sister, Rose, and her husband, the painter Roland Detre. A heavy rain prompted Leo to design a collapsible umbrella—full-sized when in use but many times smaller when folded—and he enlisted Rose, a textile and fashion designer, and Bela to help construct a model. In the Detres' small apartment all day, Leo and Bela crouched and crawled on the floor, cutting apart a large black umbrella. They struggled—by trial, error, and more error—to transform Leo's idea into a working model, but by early evening they had only managed to clutter the floor with shreds of black cloth and twisted, shiny struts. After hours of good-natured experimenting, many groans, and a few chuckles, they admitted defeat.[17]

During his trip to Switzerland that summer, Szilard had decided to emigrate to the United States, another reason for meeting Rose and Bela in September. Back in England that month, Szilard busied himself with plans to move with Trude to New York. But he also kept Lindemann informed about experiments to produce artificial radioactive elements, maintaining his hope that the key to nuclear chain reactions might at last be found.[18] Before sailing for New York, Szilard visited Maurice Goldhaber in Cambridge for a farewell dinner. At Magdalene College that evening he dined at the high table with Goldhaber and the literary critic I. A. Richards, and afterward they retired to the common room for port.

"I'll soon be going to the United States for a visit," Richards announced.

"You had better buy a one-way ticket," quipped Szilard.[19]

By late November, as he made plans to sail for New York, Szilard received a strangely appealing letter from Frédéric Joliot-Curie, an invitation to join his nuclear-research staff at the Collège de France in Paris. Joliot boasted about the power of his laboratory equipment and urged Szilard to conduct his experiments in artificial radioactivity there.[20] Szilard must have thought hard about this opportunity, delaying an answer— and his sailing—for nearly a month. But when he did reply, on Christmas Eve, 1937, Szilard was aboard the RMS *Franconia*. With his regrets Szilard sent Joliot a foil of indium, still his candidate to set off a chain reaction, and urged him to bombard it with photoneutrons to confirm an experiment Szilard had done recently at Oxford.[21] As Szilard sailed toward New York, Imperial Chemical Industries (ICI) wrote to announce an end to his Clarendon Laboratory grant. Beginning in January, support would be made just three months at a time, so the most he could expect to earn as a half-time researcher was $1,000 a year.[22]

After the *Franconia* docked in New York on January 2, 1938, Szilard later claimed, he "did nothing but loaf." This is misleading. He did not apply for a university or commercial research position, but his private records and correspondence reveal that he was hardly idle. Driving him on was his four-year-old "obsession" to create a nuclear chain reaction.

"You didn't know what he was up to," the Columbia physicist Isidor I. Rabi later complained. "He was always a bit mysterious."[23] At about this time, Szilard's friends the Polanyi brothers tried to help arrange a private research grant for him in England. A potential patron wanted a "detailed, objective opinion on the person of Szilard," something that may help "to sort out reliably the many contradictory facets of his character," Karl Polanyi wrote Michael. His insightful analysis of Szilard follows:

Enmity to him is partly based on formalities, such as curt behavior, which does not serve the purpose when official personalities are to be faced. What is needed is not a recommendation or a declaration of trust but a clarification. Nobody understands his motives, his interests, his attitude. His lack of self-interest evokes mistrust. Nobody understands his "essential" aims; in other words, one doubts the sincerity of his selflessness. So the following characteristics would have to be clarified; his curtness in view of his otherwise obvious worldliness, his organizing work [for the AAC], and the setting aside of his own ambitions. (Incidentally, I never knew that he is of a high rank as a scientist.) It would be entirely wrong, as I said before, to furnish anything in the way of panegyrics. What is needed is an analysis and explanation, keyed, so to speak, to his personality and to his total being. What is lacking in him, even short-comings, should be conceded and should be put into relationship to his total achievements rather than denied or "explained away." It would be desirable to submit this as an act of friendship. . . .

As far as I myself am concerned, I consider S. one of the rarest phenomena, to be judged in a positive way, a person whose qualities can be utilized only with difficulty in the present economic system. He is what he seems to be: an idealist devoted to the task. As his consciousness, however, is materialistic, leaning to experimenting, and agnostic, he fails to understand himself, same as the world fails to understand him. I am holding him in honor, and I value him.[24]

In New York, Szilard moved into the King's Crown, a cozy nine-story hotel between Amsterdam Avenue and Morningside Drive, at 420 West 116th Street. (Then a commercial hotel, the building is now an apartment house for Columbia University faculty.) Located just east of Columbia, this would become a haven for much of the rest of his life. There he quickly fell into a routine, walking to a corner pharmacy each morning to buy a newspaper or two and eat a breakfast (or two) while perched on a stool at the narrow Formica counter.

Trude had arrived three weeks before Leo and had found a sublet apartment on Riverside Drive, in the building where his friend Benjamin Liebowitz lived. By the following September she would become an intern and extern at Bellevue Hospital, riding ambulances and working for a while in the emergency room.[25] In his spare moments Szilard visited her or, more often, scribbled and mailed her short notes.

For the most part, his thoughts were elsewhere. In January and February, Szilard nagged Bela in Budapest with letters warning that the Nazis were about to conquer Europe, and in March, when Hitler announced *Anschluss,* a political union between Germany and Austria, Szilard cabled Bela: "NOW OR NEVER." By this time Bela had decided to leave, although his parents refused to budge. In fact, Bela couldn't budge, either, unless he found a way to arrange for an immigration quota. When three months passed without one, Leo tried shooting some "big cannon,"[26] Through Lewis L. Strauss, a Wall Street financier with whom he was negotiating isotope projects, Szilard reached a former under secretary of state, who made a few calls that yielded, by June, an allotment for the Szilards to immigrate through Montreal.[27]

To Lindemann, Szilard also wrote, in an almost cocky manner, but with no details, that his experiments at Oxford, comparing slow neutron and photoneutron absorption, "strongly threatens the current theory of the nucleus." Also with no other details, Szilard bragged that he and Brasch were in touch with persons who might finance an accelerator to make medical isotopes.[28] These no doubt included Strauss, whom Szilard first met "early in 1938" through a mutual acquaintance. With Brasch, Szilard began negotiating to build a "surge generator" to make isotopes for medical treatment. Strauss was eager to produce radioactive cobalt isotopes for cancer treatment, as a medical memorial to his parents, who had both recently died of cancer.[29]

Radium then cost about $50,000 a gram, but Brasch and Szilard convinced Strauss that their artificially irradiated cobalt, with a much shorter half-life, would cost only a few dollars a gram. Through a mutual friend, former president Herbert Hoover, Strauss approached Robert A. Millikan at the California Institute of Technology, who agreed to host the experiment.[30] And Strauss tried to interest executives at Westinghouse in Pittsburgh, General Motors in Detroit, and General Electric in Schenectady, taking Szilard along on his visits. At GE, Szilard met the same skepticism he had faced four years before with the company's British branch; nuclear energy, an official told him, was "for the science fiction fans."[31]

During 1938, Strauss and Szilard did manage to collaborate on research to investigate how radiation might eliminate budworms from high-quality cigar tobacco.[32] Their efforts to irradiate pork did not go as well. Without warning the public, meat packers feared the spread of trichinosis in 1938, and that summer, with Strauss's backing, the Szilard brothers conducted research to determine if X-ray doses were effective for treating pork: Leo, at the Strong Memorial Hospital in Rochester; Bela, at the Royal Victoria Hospital in Montreal. Leo also irradiated Canadian bacon, frankfurters, and salami at Columbia University and left them for months in the Liebowitz family's refrigerator, checking now and then to test their appearance and taste.[33] Leo and Bela met in Chicago with the National Live Stock and Meat Board to recommend an ambitious research program but scared them with an impromptu cost estimate for irradiation of $200 per hog carcass.[34]

Szilard's flighty independence eventually lost him his last reliable income, the part-time appointment at Oxford, while also straining his English friendships. Although the university's board of faculty knew that Szilard would be in America for several months a year to monitor nuclear research, it also expected him to lecture eight times a term for two of each year's three terms—ironically, the subject proposed, "High Tension Physics and Nuclear Stability," described Szilard's personal life and work aptly.[35]

In New York that summer, Szilard chanced to meet Fermi's colleague, Emilio Segrè. "Oh, what are you doing here?" asked Szilard in their surprise encounter.

"I'm going to Berkeley to look at the short-lived isotopes of element 43 [Technetium]," said Segrè. "I'll work there the summer, and then I'll go back to Palermo."

"You are not going back to Palermo?" asked Szilard. "By this fall, God knows what will happen! You can't go back."

"Well," Segrè replied, "I have a return ticket. Let's hope for the best."[36]

As usual, Szilard was expecting the worst. "I find to my amazement that the so-much feared summer heat of New York is quite comfortable to me," he wrote to Francis Simon. "Or do you think that the influence of my present political sympathies upon my attitude toward the heat could go so far as to let that heat appear to me as enjoyable [?]."[37]

When Maurice Goldhaber landed in New York, Szilard showed him around the city and brainstormed about continuing a collaboration they had begun in England. Goldhaber was off to become an assistant professor of physics at the University of Illinois. Szilard was clearly ambivalent about returning to Oxford because of his fears about the coming war, yet when Lindemann forced him to accept the lectureship's terms, he agreed.[38] Probably as frightening to Szilard as the coming war was the reality of a full-time job. To Szilard's friends, his uncertainties about returning to Oxford became annoying. As Francis Simon wrote:

> There is no gainsaying the fact that you do not make it easy for the people here to do something for you. One hardly hears anything from you, and if one does, then there are only a few lines which do not tell anything of importance, and nobody knows when you actually intend to come back. Then one hears suddenly from a visitor who has seen you on the other side that you have been engaged in psychology for some time—because that is much more important than neutrons. Of course, that doesn't spur on people, either, to make great efforts on your behalf. Now you write that this lectureship thing is not convenient for you if you have to lecture for two terms. . . .
>
> Once again I have to talk to you in all seriousness. All those to whom I have spoken are angry with you, whether this is [his AAC colleague] Miss [Esther] Simpson or [his former Oxford landlord James] Tuck or anybody else; and these are the people who wish you well! You do not get anywhere with your erratic actions, with your suddenly being unaccounted for; and it would be a pity if this would lead to failure. If you continue to act in that way, you cannot

expect ever to get sensible employment, and one cannot help to say that one couldn't blame anybody but you for this. At one time an American told me that there are quite a few people who are dying to converse with you for a few days but none who would like to offer you a job. I feel that this is, alas, quite correct.

Of course, you could say that you prefer to continue your lifestyle as before and not to give consideration to others and to their customs and habits. However, in that case, you have to accept the consequences and not ask such others for favors.

I don't say these things to you to annoy you, but I feel it is necessary, and many others feel the same way, but apparently nobody is telling you so. Perhaps it would be best that you marry, and preferably a woman who considers the realities somewhat more than you do. Of course, the responsibility for giving such advice is heavy, but you will not follow it anyway. . . . With heartfelt greetings, from my wife, too.[39]

Szilard's reply showed little regret:

Many thanks for your admonitions. I fear you are right.

The remedy [marriage], however which you suggest, appears to be somewhat too drastic. Anyway, why should a woman who has sense of reality mary [sic] a man who has none. Combinations the other way around seem to be much more common and appropriate.

Yours,
Leo Szilard[40]

This insouciance masked Szilard's darker political fears, which were aroused in late August. He had journeyed to Champaign, Illinois, to finish experiments begun with Goldhaber in England, and while there he sat in the living room of his apartment near campus, listening on his radio to news of the advancing Munich crisis.[41] Both now refugees, Goldhaber and Szilard met in Champaign with physicist R. D. Hill and together checked the calculations they had made months earlier when bombarding indium with fast neutrons at Oxford and Cambridge. The article about their work, which they sent to the *Physical Review* in October 1938, is curious for its mention of "chain reactions produced in cadmium by fast neutrons," a reference to a chemical, not Szilard's hoped-for nuclear, process.[42]

Into September, Goldhaber and Szilard hunched over the radio each night to learn what would happen next.[43] Years later, Szilard was proud of his prediction, made in a letter to Michael Polanyi from New York in 1935, that he would return to England until "one year before the war." Polanyi had passed the letter around at the time, and a few people chided Szilard about it when he returned to England that June. But when the Munich Pact was signed on September 30, 1938, allowing Germany to annex the Sudetenland in northern Czechoslovakia, Szilard decided to stay in America and cabled Lindemann—eleven months before the outbreak of war.[44]

Brainstorming with Goldhaber in Illinois had moved Szilard to try again to test his chain-reaction idea. From Champaign he went to Chicago to consult on hog-carcass irradiation, then visited the physics departments at Madison and Ann Arbor before stopping at the University of Rochester.

In June, Szilard had begun research at Rochester in a typical manner by strolling into Bausch and Lomb Hall, the main physics building, poking around, and asking directions from the janitor. To the next person Szilard met, he announced: "Some idiot is waxing your stone floors upstairs. If he doesn't stop, somebody is going to slip and kill himself!" Sidney Barnes, the quiet thirty-six-year-old physicist who received this greeting, ran the university's new 6-million-electron-volt cyclotron and soon learned that the flamboyant visitor with the curious accent had more than waxed floors on his mind; his fixation was indium. Szilard still considered it his candidate for producing a chain reaction and urged Barnes to run some experiments to test a "knock-on-proton reaction." Barnes had never heard of such a reaction because Szilard had just thought it up. A blend of curiosity and duty led Barnes to agree.

They conferred about experiments in the building's basement, with Szilard strolling nonstop around the cyclotron: a six-foot-square magnet set on two concrete pedestals connected to a wall of wires, lights, and switches. During his June visit, Szilard came in and out of Barnes's lab unexpectedly, vanishing and appearing as he had to Hans Bethe's amazement in London. So when Szilard reappeared in October, Barnes was hardly surprised and welcomed his visitor.

The "knock-on" reaction Szilard was looking for involved a proton striking an indium sample, "exciting" the nucleus to give it some energy but then flying off without actually penetrating the atom. Reactions like this had been reported at the Washington meeting that spring, and Szilard knew from his own research in Oxford that neutrons would not create this process in indium. But what about protons?[45] Barnes spent several weeks on the project, with Szilard appearing unpredictably, asking about results, suggesting new techniques, then strolling out the door. Barnes could never predict when Szilard would come or go, and as it turned out, during both the June and October visits he stayed at a downtown hotel and, usually at mealtime, called on his friends from Berlin, physicist Victor Weisskopf and his wife, Ellen.[46]

At the university, Szilard walked about constantly as he talked with Barnes and, after pacing the cyclotron room and the adjacent basement laboratories, entered the campus underground—a network of tunnels that carried steam pipes between buildings. Waving his arms, asking and answering his own questions, Szilard led Barnes from one building to another—upstairs to the ground floor, around the lobby and halls, then back into a new tunnel.

"I don't remember him ever sitting down," Barnes recalled years later. "If I had anything to say, I just waited until he stopped for breath, and I'd get it in. I generally didn't say much, though. I just asked questions." Once Szilard saw the results of their work, he insisted that Barnes alone should publish the papers. Szilard was "not interested in credit, just in getting things done," Barnes said.

Sidney Barnes never really understood the point of Szilard's indium experiments. But Szilard did. Indium would not give off extra neutrons whether bombarded with protons, electrons, or other neutrons. The prediction that indium would sustain a chain reaction was wrong, just as it had been for beryllium three years before. Unfortunately, one of Szilard's Rochester predictions proved correct: A few weeks after he first arrived, a senior faculty member had slipped on the physics building's waxed floor and cracked his skull.[47]

Back in New York that November, Szilard returned to the King's Crown Hotel and "for reference purposes" applied for a Columbia University Library card. This was a time for taking stock, for cleaning up loose ends, and in a contemplative mood Szilard sat by the hour in the cavernous library, thinking and scribbling notes and letters. To physicist Niels Bohr in Copenhagen he sent Barnes's papers on the radioactivity of indium.[48]

Around the campus in December 1938, Szilard had met physicist John Dunning, whose work he had cited in his own experiments at Oxford. It was an honor, then, when Dunning invited Szilard to "tell us informally about your experiments," at a Monday afternoon nuclear physics seminar. Szilard's severe honesty compelled him to acknowledge failure as readily as success, and this forum would allow him to describe his failed chain-reaction theory to understanding colleagues. Still, Szilard delayed accepting Dunning's invitation until after the Christmas holidays.[49] For although Szilard's five-year "obsession" with the nuclear chain reaction had reached a disappointing end, he couldn't quite decide how to admit it.

CHAPTER 13

Bumbling toward the Bomb

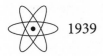 1939

The Christmas season in New York in 1938 was gloomy for Leo Szilard, who was convinced that his secret patent for a nuclear chain reaction would not work with the element indium. Five years of research, first with beryllium, then with indium, had failed and with it his hopes for a career in nuclear physics, the specialty he had once thought he could dominate by his visionary genius.

It seemed that Szilard could persuade no one to take seriously the commercial potential for the chain reaction, in part because he refused to describe in detail the natural mechanism that he envisioned—atoms hit and split by neutrons that, in turn, release more neutrons to split more and more atoms. He wanted to entice practical businessmen to buy into an alluring promise of cheap energy. But he also wanted to protect his discoveries by cloaking the whole topic in mystery. This made Szilard seem like an impractical visionary, a "character" with a hard accent and a string of improbable ideas.

In despair over these apparent failures, Szilard drew away from the few people he knew in New York, retreating to his small room at the King's Crown Hotel just east of the Columbia University campus. Szilard spent more time in his room, rummaging among papers and calculations or simply daydreaming in his bathtub.

That fall, Szilard's friend Trude Weiss had begun work as an extern in the pediatrics department of Bellevue Hospital in downtown Manhattan and so spent little time either in her sublet apartment or in the Columbia neighborhood. Szilard's brother, Bela, and his family were also living nearby, in a sublet apartment in Trude's building, but Leo visited them only every week or two for a meal and a chat. To them he seemed distracted, apparently upset by more than his finances.

Frustrated by his apparent failure, Szilard sat down in his paper-strewn hotel room on Wednesday, December 21, 1938, and wrote a short letter to the British Admiralty. In 1935, Szilard had assigned his chain-reaction patent to the Admiralty as a way to keep it secret from the Germans, whom he feared would use the device to make atomic bombs. He mentioned to the navy patent officer the "further experiments" conducted in Cambridge and Rochester, then admitted: "In view of this new work it does not seem necessary to maintain the patent. . . nor would the waiving of the secrecy of this patent serve any useful purpose. I beg therefore to suggest that the patent to [sic] withdrawn altogether. I am, Sir, Yours very truly, Leo Szilard,"[1]

Szilard could not know that this day would be a turning point in the history of modern science, a vindication of his hopes and fears. All he knew as he signed this letter was that he had conceived and explained an original process for releasing the binding energy that gives atoms their structure and it didn't work. Yet that same day, at the Kaiser Wilhelm Institute for Chemistry in Berlin, Otto Hahn and Fritz Strassmann bombarded uranium with neutrons, and it broke into two parts, or "fissioned," And in the process it released extra neutrons. Szilard, it turned out, had been right about the chain-reaction process all along. He had just tested the wrong elements for his demonstration.

Although Szilard would not learn about uranium fission for another month, once he did, the despair of that bleak December turned to exhilaration. Then to new fear. The coming year would bring a frenzy of scientific and political activity. Indeed, during 1939, Szilard would almost single-handedly lead the physics community and the US government to join forces in atomic energy research.

Two days into the New Year the SS *Franconia* eased into New York harbor with Enrico Fermi and his family aboard; life in Fascist Italy had finally become too dangerous for Fermi's Jewish wife, Laura. Fermi had left Rome a month earlier, received the Nobel Prize for physics in Stockholm, and there announced that he would accept a six-month teaching appointment at Columbia. Szilard had corresponded with Fermi since 1936, trying to interest him in experiments and control schemes for their nuclear patents, so the two physicists had much to talk about when they met, by accident, in the lobby of the King's Crown—the Fermis had checked into the hotel while apartment hunting.

Because of his recent international fame and his appointment as a university professor, Fermi arrived at Columbia a scientific celebrity. Fermi's Nobel Prize had been awarded for the research he conducted in Rome: bombarding different heavy atoms with neutrons and then identifying the new radioactive elements that were created. Although he had not realized it, Fermi had actually split uranium atoms with neutrons as early as 1934, the same year that Szilard had patented the chain-reaction concept. In 1938, Fermi had chosen Columbia from among six American universities, so his fateful meeting with Szilard that January brought together the two scientists in the world best able to advance the research that would produce nuclear power and nuclear bombs.

By contrast, Szilard was not well known, even to other physicists. He had worked to keep his pioneering nuclear research secret, and to Columbia's physicists, including Fermi, Szilard was just an unemployed visitor who poked around the department. There Szilard would appear unannounced in the laboratories, ask questions, make suggestions, then disappear. Frequently, he called on physicist Isidor I. Rabi, whom Szilard had first met in Berlin. Szilard and Rabi sensed an affinity, perhaps from their common Austro-Hungarian heritage, and the two tried to be friendly despite widely different professional interests. On each visit Szilard suggested new experiments for Rabi and his colleagues to perform, but after several weeks of this prodding, Rabi finally became impatient. "Please go away," he begged Szilard one day. "You are reinventing the field. You have too many ideas. Please, go away."[2]

But Szilard persisted, eventually imposing his concerns on not only Fermi and the Columbia physicists but also the president of the United States. What began as Szilard's personal crusade to harness the nuclear chain reaction would eventually become the federal government's $2 billion program to make atomic bombs: the Manhattan Project.

At first, Fermi and Szilard rarely met during working hours—Fermi's office was on the top floor of the new Michael Pupin Laboratory Building; Szilard worked in the basement or from his hotel room. To collaborate with Fermi, Szilard had to enlist the help of established faculty members, including Rabi and George B. Pegram, who was the physics department's chairman and dean of Columbia College.

Szilard needed to force a place for himself at Columbia because, as he tried to explain on January 13 in a long and uncertain letter to his mentor at Oxford, returning to England seemed doubtful.

Dear Professor Lindemann:

Three months have passed since, acting on impulse, I cabled you that I am postponing my sailing for an indefinite period on account of the international situation, and that I should be grateful if my further absence could be considered as a leave without pay. . . .

It seems to me that the Munich agreement created, or at the very least demonstrated, a state of international relations which now threatens Europe and in the long run will threaten the whole civilized world. This cannot fail to claim the attention of all of us, and, if the situation is to be improved, the active cooperation of many of us. I greatly envy those of my colleagues at Oxford who in these circumstances are able to give their full attention to the work which has been carried on at the Clarendon Laboratory and who are able to do so without offending their sense of proportions. To my great sorrow I am apparently quite incapable of following their example.

It seems to me that those who wish to continue to dedicate their work to the advancement of science would be well advised to move to America where they may hope for another ten or 15 years of undisturbed work. I myself find it very difficult, though, to elect such "individual salvation", and I may therefore return to England if I can see my way of being of use, not only in science, but also in connection with the general situation. It is hardly necessary to state that, if I shall be in England and if you want me to do so, I shall be most happy again to cooperate with those who work in the Clarendon Laboratory. It may be best, however, that I should not receive financial support from the Laboratory, as such financial support is bound to be linked with fixed obligations which I would rather avoid.

For the time being, I do not yet see my way of being of use in England in connection with the general situation, though I see certain potential possibilities in this respect. In view of these I am at present not looking for a "job" on this side of the Atlantic.

Perhaps I shall have an opportunity to talk to you about all this if I shall visit England in a not too distant future. . . .

Please excuse the three months' delay of this letter. Immediately after the Munich agreement it did not seem possible for me to have a sufficiently balanced view, and I had to allow some time to elapse before I was able to write without bitterness of this event.

With kind regards to all, I am

Yours very sincerely,
Leo Szilard[3]

In this strained and unusually revealing letter Szilard bared his personal frustrations and tensions, above all by citing the concern that controlled his life's emotional and moral decisions, a "sense of proportions." This sense was the instinct he most trusted at critical moments: a balance of physical laws, psychological expressions, and moral convictions. Now Szilard's emotions and motives were in turmoil. Unlike some colleagues, he could not avoid the war he saw coming and seek "individual salvation" in "the advancement of science." He had to play a part in solving the "general situation." But, being an alien, Szilard knew that he was not eligible to perform military research in England. Yet he might return there, anyway, provided he could still escape "fixed obligations." Or he might stay in the United States, become a citizen, and work against the coming war as best he could.

Three days after Szilard wrote that pained letter, on January 16, Enrico Fermi and his wife, Laura, were at the pier again, this time to meet the SS *Drottningholm* and to welcome Niels Bohr, the Danish physicist and Nobel laureate whom they had visited in Copenhagen the month before. Bohr landed in New York bearing the momentous news that the fission of uranium had been demonstrated by Hahn and Strassmann in Berlin.[4]

Bohr himself showed mixed emotions over his report. He was fatigued by worries that war in Europe was imminent and that Germany might use its discoveries in atomic science for military ends. Yet he was also clearly exhilarated by the news of nuclear fission and further excited by the cable he received shortly after landing in New York. The radiochemist Lise Meitner and her nephew Otto Frisch reported that they had confirmed the fission process by their own experiments in Sweden. Fermi, on the other hand, was unimpressed. He failed to recognize the importance of this news and failed even to mention what he heard to his Columbia colleagues. Fermi was so typically cautious, in fact, that as the grave consequences of fission became apparent to others around him, he repeatedly denied their significance.

On the day Bohr arrived in New York, an assistant went to Princeton and there—at an informal "journal club" where the latest scientific articles were discussed—announced the news about uranium fission.[5] Still unaware of this startling discovery, Szilard spoke that Monday afternoon at John Dunning's nuclear physics seminar, on the eighth floor of Columbia's Pupin Laboratory, about "Radioactive Isotopes Produced by Nuclear Excitation" and afterward attended an informal tea.

Szilard finally heard the news about uranium fission a few days later on a visit to physicist Eugene Wigner in Princeton. Wigner, his friend since their student days in Berlin, lay in the university infirmary with jaundice. As soon as Szilard entered Wigner's room, he heard Bohr's news that uranium could disintegrate into other elements. He was stunned. The element and the process he had searched for since 1933 had both been found. "When I heard this," he recalled later, "I saw immediately that these fragments, being heavier than corresponds to their charge, must emit neutrons, and if enough neutrons are emitted in this fission process, then it should be, of course, possible to sustain a chain reaction. All the things which H. G. Wells predicted appeared suddenly real to me."[6]

The political implications were also suddenly obvious. "At the time it was already clear," Szilard recalled, "not only to me but to many other people—certainly it was clear to Wigner—that we were at the threshold of another world war." Szilard vowed to test how many neutrons are emitted in fission and to keep his discoveries secret. "So I was very eager to contact Joliot [-Curie in Paris] and to contact Fermi, the two men who were most likely to think of this possibility."[7]

Szilard wanted to reach Fermi quickly. But the morning after his visit to Wigner's infirmary room, he woke to a heavy, cold rain. Throughout his life, and especially since his arrival in America a year earlier, Szilard invariably caught cold when he got wet, and the weather alone made him feel ill that morning. He worried that he had no rubbers or boots to wear and could find none in Wigner's apartment, where he was staying. But he went out, anyway, to send a telegram to Trude Weiss, "RETURNING KING'S CROWN WITH MODERATE COLD STOP YOU MIGHT VISIT ME WITH STETHOSCOP [sic] TONIGHT OR TOMORROW NIGHT = LEO."[8] But back in his hotel room he felt too weak to contact Fermi and lay sick in bed for days.

During this time, Szilard later learned, Fermi had also thought that an extra neutron might be emitted during nuclear fission, which raised in his mind the possibility of a chain reaction. But Fermi considered this notion so remote that he gave it little thought. Szilard, on the other hand, could think of little else. On January 25, Szilard wrote to Lewis L. Strauss, a Wall Street financier interested in the atom's commercial potential, about "a very sensational new development in nuclear physics," the fission of uranium—something far exceeding the medical-isotope schemes the two had discussed a year earlier. Szilard predicted that fission "might make it possible to produce power by means of nuclear energy" but that this was not very exciting and might be impractical. There were graver "potential possibilities," Szilard warned, that "might lead to a large-scale production of energy and radioactive elements, unfortunately also perhaps to atomic bombs." This news, he said, revived "all the hopes and fears" he had felt since 1934.

That night, in the Columbia physics labs, John Dunning and a young assistant to Fermi, Herbert Anderson, confirmed uranium fission by a procedure similar to that used by Meitner and Frisch. Dunning cabled the news to Fermi, who was in Washington for the Fifth Conference on Theoretical Physics.[9] The next day, January 26, Szilard dragged his aching body across the Columbia campus to the Western Union office on Broadway and sent an urgent cable of his own. Trying to intercept his December instruction to lift the chain-reaction patent's military secrecy, Szilard notified the director of navy contracts at the British Admiralty in London: "REFERRING TO CP10 PATENTS 8142/36 KINDLY DISREGARD MY RECENT LETTER STOP WRITING LEO SZILARD,"[10] He now believed that the chain-reaction process in his 1934 patent *would* create the "Violent explosions" he had warned the Admiralty about, and he wanted to be sure that the patent was kept a military secret.

The same day that Szilard cabled the Admiralty, Fermi and Bohr met at George Washington University, where they opened discussion about uranium's fission and its implications. But Fermi still made light of the discovery. A practical man, unlike the more theoretical Bohr or the more exuberant Szilard, Fermi saw progress in terms of experiments and mechanisms that could actually be performed. To him, scientific integrity meant being sure of your conclusions before considering their implications; to Bohr and Szilard it meant seeking the political and social implications of your work at each step.

Once Bohr and Fermi had discussed the possibility of uranium fission in Washington, researchers rushed to test the process. At the Carnegie Institution's Department of Terrestrial Magnetism (DTM) in Washington, on January 28, Richard Roberts and R. C. Meyers bombarded uranium with neutrons and saw unusually long spikes on their oscilloscope. They repeated the display that evening for Bohr, Fermi, University of Wisconsin physicist Gregory Breit, and Edward Teller, a Hungarian-born physicist Szilard had first met in Budapest and later befriended in London in 1934.[11] The Washington *Evening Star* carried a page 1 piece on the DTM conference headlined: 'Power of New Atomic Blast Greatest Achieved on Earth.' The article concluded, however, that "as a practical power source, the new finding has at present no significance." On January 30 the *Evening Star* reported on page 1 that uranium fission had been confirmed at Columbia, the DTM, Johns Hopkins University, and by researchers in Copenhagen.[12] The word was out. Unlike the "transformation" first seen when teams at the Cavendish Laboratory split light atoms, with uranium fission one of nature's heaviest atoms was easily burst by neutrons.

Recovering from his cold just as the DTM conference ended, Szilard caught a train to Washington and from the station called Teller. Mici Teller was weary from the parade of conference visitors who had come to their small wood-frame house on Garfield Street, so with Szilard's call she insisted to her husband that he *not* spend the night. They did agree to pick him up at the station, however, and once Szilard climbed into the car, Mici blurted out, "Will you stay with us?" He accepted promptly, then turned to Edward.

"You know what fission means," Szilard said as the car pulled into traffic and gained speed. "It means bombs." At the Tellers' house, Mici led Szilard to the small guest room. He dropped his bag and sat on the bed. "Too hard," he announced. "I'd prefer to stay in a hotel."[13]

When back in New York, Szilard called on Rabi, adding to the news about fission his own conviction that it might set off a nuclear chain reaction. Rabi told Szilard that Fermi, too, had made this connection but that he had expressed little desire to test the idea himself. "Fermi was not in," Szilard later recalled, "so I told Rabi to please talk to Fermi and say that these things ought to be kept secret because it's very likely that if neutrons are emitted, this may lead to a chain reaction, and this may lead to the construction of bombs."[14] A few days later, Rabi told Szilard what Fermi thought of his warning. Fermi enjoyed the punchy vernacular of

American English, and his reaction to Szilard's message was immediate: "Nuts!" Rabi couldn't explain what Fermi meant by this remark, so Szilard urged that they both call on Fermi in his office.

"Well . . ." Fermi told his visitors, "there is the remote possibility that neutrons may be emitted in the fission of uranium and then of course a chain reaction can be made."

"What do you mean by 'remote possibility'?" asked Rabi.

"Well, ten percent," Fermi said.

"Ten percent is not a remote possibility," Rabi said, "if it means that we may die of it."[15]

From this meeting on, Szilard realized how differently he and Fermi could view the same scientific evidence. "We both wanted to be conservative," Szilard recalled later, "but Fermi thought that the conservative thing was to play down the possibility that this may happen, and I thought the conservative thing was to assume that it would happen and take the necessary precautions." Bernard T. Feld, who became Szilard's research assistant at Columbia, characterized their difference in this way: "Fermi would not go from point A to point B until he knew all that he could about A and had reasonable assurances about B. Szilard would jump from point A to point D, then wonder why you were wasting your time with B and C."[16]

Szilard grew increasingly frustrated with Fermi's caution, which he only understood gradually as the two men worked together. In unpublished notes about the Manhattan Project, which he wrote in 1950, Szilard concluded that their differences stemmed from contrary views that each held about the connections between science and life. Put simply, science was Fermi's life, whereas for Szilard science was an endeavor ineluctably united with politics and personal sensitivities.[17]

After meeting Fermi on February 2, Szilard addressed the political consequences of fission in a letter to the British Admiralty, explaining vaguely his urgent cable. Indium, he wrote, "cannot be used for the process described in the patent," but "another element" (he did not name uranium) may create a "process" (he did not mention fission) that "might very well turn out to be similar to the process described in the patent assigned to you." Two months later, the Admiralty assured Szilard that his secret patent would be maintained.[18]

Keeping his chain-reaction patent a secret was one way for Szilard to prevent Germany from realizing fission's military potential. Another way was to urge fellow scientists to censor their own research. The most active research, Szilard knew, was under way in Paris, so he wrote to Joliot warning him not to publish the results of his neutron work. "Obviously, if more than one neutron were liberated, a sort of chain reaction would be possible," he wrote. "In certain circumstances this might then lead to the construction of bombs which would be extremely dangerous in general and particularly in the hands of certain governments." Szilard closed his letter, "In the hope that there will not be sufficient neutrons emitted by uranium, I am . . ." but then crossed out this conclusion, simply signed the letter, and mailed it.[19]

He may have hoped that extra neutrons would not be released, but Szilard could not rest until he knew for sure; the day he wrote to Joliot, he walked to the large apartment of his friend Benjamin Liebowitz on Riverside Drive. Liebowitz had studied physics at New York University, where Szilard met him during his first visit to America in 1931–32. As a successful inventor, he had a comfortable income from royalties. Szilard explained that the chain reaction might be possible and asked to borrow some money—he was nearly broke—for experiments to confirm this.

"How much money do you need?" Liebowitz asked.

"Well, I'd like to borrow $2,000," Szilard said. Liebowitz took out his checkbook. This was the first American money spent on the chain-reaction concept, and it is no exaggeration to say that this small loan helped change the fate of the world.[20]

Once he had cashed the check, Szilard rushed to Fermi's office to try to urge him to begin some fission experiments. Fermi was not interested. To him, fission was a curiosity, not a cause for concern. He was conducting a few experiments of his own, he said, but these were to answer basic questions about the behavior of neutrons, not to make the leap to chain reactions. Unlike Szilard, Fermi was not convinced that fission would lead to chain reactions or chain reactions to bombs. Still, Szilard had to be sure and called on physics chairman George Pegram to ask for permission to conduct experiments of his own. Pegram agreed, but only for three months, beginning on March 1.

To find out if a chain reaction was now possible, Szilard had to learn more about what happened when uranium is bombarded with neutrons. Seeking a reliable neutron source, he cabled Lindemann in Oxford, asking him to send a cylinder of beryllium—one that Szilard had ordered from Germany the year before, to begin just such bombardment experiments.[21] Then Szilard contracted to rent a gram of radium, which he planned to place within the beryllium to produce relatively weak "slow" neutrons.[22] Szilard suspected that uranium emitted the more energetic "fast" neutrons. If that were true, he reasoned, the difference between fast and slow neutrons would be apparent on monitors—a cloud chamber, for example, or an oscilloscope.

A clumsy and disinterested experimentalist himself, Szilard next needed equipment and a place to work. In mid-January he had looked in on Walter Zinn, a Canadian-born physicist skilled at rigging laboratory equipment for new uses. When Zinn agreed to collaborate, Szilard pursued uranium in his usually diligent way, hiring consulting physicist Semyon Krewer to seek industrial bids. Krewer also surveyed world uranium supplies and investigated processes for producing pure uranium metal and graphite, in this way making Szilard one of the best-informed people in America about the heavy metal.[23]

With fission "hope is revived that we may yet be able to harness the energy of the atom," the *New York Times* reported on the morning of February 5. With this and other newspaper stories about Bohr's appearances around the United States, word about uranium fission was slipping out. The article cited a "remotely possible atomic powerhouse" but made no mention of atomic weapons. Szilard clipped this article and mailed it to Strauss. At the same time, *Newsweek* reported on fission by recalling Einstein's guess about useful atomic energy: "It is like shooting birds in the dark in a country where there are not many birds." The magazine also quoted comedian Fred Allen's answer from a fictional professor about why he split atoms: "Well, someone may come in someday and want half an atom."[24]

In early 1939 the public read mixed messages about uranium fission. The new process was the greatest discovery since radium, the press reported, yet it also called into question Fermi's assumption that there are elements heavier than uranium—the so-called transuranic elements he and others had tried to create for several years.[25] Science Service, a respected news agency, played down the dangers of fission:

First of all, the physicists are anxious that there be no public alarm over the possibility of the world being blown to bits by their experiments. Writers and dramatists (H. G. Wells's scientific fantasies, the play "Wings Over Europe," and J. B. Priestley's current novel, "Doomsday Men") have overemphasized this idea. While they are proceeding with their experiments with proper caution, they feel that there is no real danger except perhaps in their own laboratories.[26]

Science Service also dismissed "forecasts of the near possibility of running giant ocean liners across the Atlantic on the energy contained within the atoms of a glass of water"; or of replacing steam and hydro-electric plants with "atom-motors"; "or the suggested possibility that the atomic energy may be used as some super-explosive, or as a military weapon. . . ." After all, uranium is very scarce and contaminated with nonfissionable elements. "This means that the release of atomic energy can only be achieved by direct intent, in the laboratory, and then only with considerable ingenuity of experiment."[27] Exactly what Szilard had in mind as he thought of Germany's moves toward war.

While awaiting his beryllium shipment from Oxford, Szilard tried to enlist scientists and financiers in a scheme to sponsor nuclear research and devised the Association for Scientific Collaboration, a legal entity for atomic research that he incorporated to raise money for experiments.[28] He rode the train to Washington with Strauss and there enlisted Teller as the association's representative. On his return, Szilard stopped in Princeton, where Wigner agreed to join the cause.

Despite earlier rebuffs, Szilard appealed to Strauss for money. In a letter written on February 13, Szilard shared his excitement over the pace of activities around him. "Almost every day," he wrote, "some new information about uranium became available, and whenever I decided to do something one day it appeared foolish in the light of the new information on the next day."[29] Teller added to Szilard's urgency on February 17 when he sent from Washington a handwritten note, in Hungarian, reporting that scientists at the DTM were eager to experiment with uranium. Bohr had told them that only scarce uranium[235] would fission, not uranium[238], the most plentiful form found in nature. But the DTM scientists had to defer any experiments because they lacked large amounts of uranium and were, in Teller's words, "cautious (perhaps a little too cautious)." Their chief concern was a reluctance to accept research money from the federal government. (Unlike conditions today, most scientific research before World War II was privately sponsored, by industrial investment or university endowment.) Teller closed by reporting "a chain-reaction mood" in Washington. "I only had to say 'uranium' and then could listen for two hours to their thoughts."

Finally, on February 18, Szilard walked into a US Customs Department office in downtown Manhattan, paid a fifteen-cent postal fee, and walked out with his beryllium cylinder.[30] It was time to dispatch Zinn to collect the rented gram of radium. Zinn rode the subway to the Eldorado Radium Corporation's office in Manhattan, expecting to carry away the tiny sample in his coat pocket. But when he presented Szilard's receipt, Zinn discovered that the gram of radium came stored in a 100-pound lead casket. "By taxi I dragged this load to Columbia," Zinn later recalled, and after "several days of concentrated effort" fitted the experimental neutron counter with Szilard's radium-beryllium source.[31]

Szilard, usually energetic and expressive, became even more outspoken, at times frenetic, as he thought about the news that uranium could fission. For weeks after, he rushed about to the Columbia labs, to the small post office on Amsterdam Avenue or the Western Union office on Broadway, to his cluttered hotel room, to colleagues' offices, bearing witness in any way possible to the great and dreadful events he foresaw. As he dressed hastily each morning, Szilard failed to notice that he kept putting on the same pair of shoes and that their heels had scraped down to the uppers. Even when Bela pointed this out, Szilard was too busy to change or repair them. Suddenly, Szilard was infatuated, and frightened, by the reality of fission and the thought that it might lead to chain reactions. More than any other person at the time, he was convinced that chain reactions were essential to winning or losing an impending world war: a fear he had lived with for five years.

As Bohr and Fermi appeared together at several scientific meetings during the spring of 1939, Bohr became more concerned, and outspoken, about the destructive potential for fission, while Fermi refused to predict any timely uses, commercial or military. After a discussion on February 24, *New York Times* science writer William L. Laurence buttonholed Fermi and asked him if he thought that fission could be used for weapons. At first, Fermi demurred. Then, slowly, he spoke. Such development was twenty-five to fifty years off, he said, if it proved practical at all.[32]

Remaining unknown to reporters (and to most other scientists), Szilard was increasingly anxious to have the United States begin studying fission for possible military uses—before the Germans did. In a February 22 "bulletin" he sent to Strauss, Szilard focused attention on the rare uranium[235] isotope. If separated and refined, he wrote, "a chain reaction could be set up in the concentrate." Szilard reported that at the time Fermi was convinced natural uranium (99.3 percent uranium[238] and 0.7 percent uranium[235]) was "no good" for producing chain reactions. Wigner wasn't sure. And Szilard's "own feeling is somewhere between Wigner's and Fermi's." As it turned out, the first chain reaction was created in 1942 by a Fermi-Szilard design that used natural uranium, but it was known by then that only the rare isotope sustained the fission process.

To certify when he thought up different ideas, Szilard liked to date them by mail, and on February 24 he posted himself an eight-page manuscript about uranium[235] separation. He mailed another manuscript on March 3, describing his inventions in isotope separation, including the Szilard-Chalmers effect. Separating uranium isotopes, he suspected, would soon be important.[33] Later that same day, Szilard and Zinn met at a two-room laboratory on the seventh floor of the Pupin building and worked into the evening, when, at last, their uranium-fission experiment was ready.

For this effort Szilard and Zinn planned to bombard uranium with slow neutrons, tracing their movements on a cathode-ray oscillograph screen—a device resembling a small television set. The neutrons would appear on the screen as gray streaks, their speed revealing whether they were fast or slow.

"If flashes of light appeared on the screen," Szilard explained later, "that would mean that neutrons were emitted in the fission process of uranium, and this in turn would mean that the large-scale liberation of atomic energy was just around the corner." As the picture tube heated up, Szilard and Zinn watched intently. No flashes appeared. They waited. Still nothing. Szilard felt relieved. Perhaps his chain-reaction theory was moonshine! Perhaps no one would make atomic bombs!

Then Zinn noticed a problem. The screen was not plugged in. The two men chuckled nervously, and with the device connected, they looked again. "We saw the flashes," Szilard recalled later. "We watched them for a little while and then we switched everything off and went home. That night there was very little doubt in my mind that the world was headed for grief."[34] Before retiring, Szilard placed a long-distance telephone call to Washington, interrupting a Mozart duet that Teller, on piano, and a friend, on violin, were practicing. In Hungarian, Szilard spoke a single sentence, "I have found the neutrons," and hung up.[35]

His fears confirmed, Szilard called on Fermi again to report the disturbing conclusions. Meanwhile, Fermi and his assistant, Herbert Anderson, had tried a similar uranium-fission experiment, but with inconclusive results.

"In your experiment," Szilard said, "you use a radon-beryllium source. That source, as you know, has rather energetic neutrons. How do you know that some of the neutrons are coming not from fission but from a direct (*n, 2n*) reaction?" In this way one neutron (*n*) hits a nucleus and knocks out two neutrons *(2n)* but does not actually split or fission the atom. Fermi conceded the point, while Szilard, restraining a powerful temptation to tease, showed irrepressible satisfaction.

"It just so happens that I have a radium-beryllium photoneutron source that produces neutrons of much lower energy," Szilard said. "With it, you won't have the problem of the (*n, 2n*) reaction." Fermi agreed to use it and in his next experiment unmistakably saw the fast neutrons.

"In our paper" reporting their results, Anderson recalled, "we acknowledged a curious organization called the Association for Scientific Collaboration, a Szilardian creation."[36] This entity, formed to solicit money for nuclear research, technically owned Szilard's radium-beryllium source. On the day after his decisive experiment with Zinn, Szilard tried to arrange a meeting to discuss fund-raising for his association. He wanted Fermi to attend and hoped that Strauss would invite his wealthy acquaintance, Lord Rothschild, then visiting New York. Another target of Szilard's fund-raising was W. T. Richards, brother-in-law of Harvard president James Conant.[37]

Szilard cabled Strauss on March 6 that the chances for a chain reaction were "NOW ABOVE 50%," telephoned to underscore his excitement,[38] and met with him for breakfast on March 8. The next day, assured where his research was leading, Szilard appeared before a patent attorney, raised his right hand, and swore a notarized oath. "I verily believe myself to be the original, first, and sole inventor of the improvements in apparatus for nuclear transmutation as described and claimed . . . ." he said, renewing his 1935 US patent application for the chain reaction. Just in case.[39] A day later, Szilard and Fermi met Rothschild over drinks at Strauss's apartment, but the two physicists failed to persuade the financier to invest in chain-reaction research.[40]

Germany had just annexed Czechoslovakia when Wigner visited Szilard in New York on March 16. They were upset by both the neutron experiments and the events in Europe and saw a link between the two: Czechoslovakia had Europe's richest uranium reserves. During a long and emotional meeting with Pegram and Fermi, Wigner argued that they must approach the government for support; Germany was on the way to building atomic bombs, he warned, and the United States must build them first or lose the war.[41] Szilard hesitated, afraid that government red tape might stifle his work.[42] To Pegram and Fermi chain reactions were only a theoretical possibility; atomic bombs, even more remote. They also saw war in Europe as unlikely, while Szilard and Wigner feared it was imminent.

After this Columbia meeting, Szilard and Wigner went to Princeton, where, in Wigner's office in Fine Hall, they met with Niels Bohr and other physicists. Bohr doubted that an A-bomb could be constructed because so much uranium[235] must first be separated. "It would take the entire efforts of a country to make a bomb," Bohr said. With Germany in mind, Szilard argued for self-censorship by all nuclear physicists outside that country.[43]

Fermi went by train to Washington after the Columbia meeting, and the next day, March 17, briefed a group of navy scientists. But Fermi's characteristic caution, his heavy accent, and his ambivalence about the meaning of uranium fission left his audience bemused. After Fermi's talk, Ross Gunn, a navy technical adviser working on submarine propulsion, telephoned Merle Tuve at the DTM, who had also sat in on the briefing. "Who is this man Fermi?" Gunn asked. "What kind of a man is he? Is he a Fascist or what?" Still, a power source that used no oxygen was a submariner's dream come true, and Gunn was intrigued enough by Fermi's ideas to offer Tuve $1,500 for uranium experiments. This might have been the first federal money to study fission, but for policy reasons— chiefly a reluctance to accept government funding for scientific research—the DTM declined the navy's grant.

Szilard and Teller met Fermi in Washington over the weekend and while there learned that Joliot and his colleagues in Paris had published in *Nature* their finding that uranium frees extra neutrons when it fissions.[44] Szilard and Fermi disagreed about how to report their own recent experiments; Szilard and Teller insisted that *all* uranium work must be kept secret, Fermi condemned censorship as unscientific. But, living in a democracy now, he proposed a vote, lost to the Hungarians two to one, and when back at Columbia, advocated censorship.[45]

"I invented secrecy," Szilard said after World War II.[46] At the time, he took credit for something he wanted abolished, the US military's monopoly on atomic-energy information. But from the moment on Southampton Row in 1933, when he first conceived the nuclear chain reaction, until 1945, Szilard schemed to keep all related work a secret. With Joliot's article, Szilard felt betrayed, although word about uranium was also leaking out in America: *Newsweek* reported the possibility that atomic energy might create "an explosion that would make the forces of T.N.T. or high-power bombs seem like firecrackers."[47]

In late March, Szilard, Wigner, and physicist Victor Weisskopf from Rochester sent cables and letters to their European colleagues, pleading for self-censorship and hoping that Joliot's article might be the only one published.[48] It appears "very likely," Szilard warned Weisskopf, that an A-bomb "will be too heavy to be carried by aeroplane" but could "probably easily be carried by boats" and with "engineering tricks" may cause an explosion whose destructive power "goes beyond imagination."[49]

One cable from Weisskopf to Joliot's group arrived in Paris on April 1, and they took it for an April Fools' Day joke. "To propose withdrawal of publication and communication by private letters," Joliot's colleague Lew Kowarski recalled years later, "between four countries, Europe being at that [time] what it was, was such a damn fool proposal . . . as a way to preserve secrecy." Besides, Kowarski thought the Nazis would take note if articles were not published.[50] And Joliot complained that Science Service had reported DTM's results in February, prompting Szilard to reply that these concerned "delayed neutrons" that are "harmless," not fast or slow neutrons. On April 6, Szilard asked Joliot, "KINDLY CABLE AS SOON AS POSSIBLE WHETHER INCLINED SIMILARLY TO DELAY YOUR PAPERS OR WHETHER YOU THINK THAT WE SHOULD NOW PUBLISH EVERYTHING." His answer arrived the next day: "QUESTION STUDIED MY OPINION IS TO PUBLISH NOW REGARDS JOLIOT."[51]

On the same day Szilard received Joliot's rebuff, Bela knocked on his hotel-room door with another cable. From an uncle in Budapest came the news that their mother had died of cancer. Szilard was grief stricken but strained to control his emotions with intense reason, hiding his sadness even from Bela. Leo had not seen his mother since his last visit home in 1936, yet the time and distance had not erased his fondness for her. When living in Berlin, he simply signed "Leo, too" at the bottom of Bela's letters, and in New York he also relied on Bela to convey messages to and from home. But by early in 1939, Szilard learned that his mother had cancer, and as a "gift" for her birthday in March he had written her a rare personal letter. The sorrow Szilard repressed must have hurt all the more a few days later when a letter arrived from his mother. Written in pencil in a frail hand from her Budapest hospital bed on April 1, the letter touched anew the deep personal and ethical bond that tied them together.

1st April 1939

My dear, good son!

I thank you for your birthday present. The fact that you offered that to me proves that we have the same ideas about life. It seems to me that you are apologizing and that you are looking for an explanation for the rareness of our correspondence. Be calm about that; I couldn't ever be angry at you, because in my innermost soul I sense that we always have been mutually honest and truthful and that therefore there could never be a misunderstanding between us. "Truthfulness" was, and always remained, what I was striving for. Thus, I would not like to fool either you or myself with respect to my illness. After all, it is not the most important thing how long one has lived but how and in what manner.

Looking backward, I have the feeling that I always did my duty, had some sorrow here and there, but was on the other hand blessed by plenty of joy that fate let me have; it would be ungrateful to ask for more. What I now still have to expect, I don't know.

What we experienced with Father remains ours. When my physical condition does not bother me, then we sit quietly holding hands and not speaking many words. For the time being, he accomplishes the wish to make my remaining life as easy for me as possible; and this effort supports him. What should become of him later is my greatest worry. You should try to diminish the great distance between him and you; the letters from all of you will be his only support for his life. Sure, these will be only a substitute, which, however, should not be underestimated, in contrast to a lack of communication, from the point of helping him over the difficult times that are to come.

Farewell, my good boy; the writing of long letters tires me out; so I will mostly write short answers.

I wish for peace for you; try to achieve that by keeping away from reproaching yourself.

I embrace you.

Mother[52]

Trying to extinguish his personal grief, Szilard confronted the crusade for scientific secrecy with a new intensity. But once the French reports on fission had been published, Szilard's Columbia colleagues insisted on releasing their own work. They resented this visitor's complaints, and some grumbled that his guest privileges in the department be ended. Rabi, the Columbia teacher closest to Szilard, took him aside and warned that his position in the laboratory was in jeopardy.[53] Reluctantly, Szilard went along. On April 15 the *Physical Review* published "Production of Neutrons in Uranium Bombarded by Neutrons," a March 1939 paper by Anderson, Fermi, and H. B. Hanstein, and the Szilard-Zinn paper on "Instantaneous Emission of Fast Neutrons in the Interaction of Slow Neutrons with Uranium."[54] Wigner urged Szilard that someone should alert the US government to the "possible sudden threat" of a German A-bomb, a threat made more real by Joliot's article in *Nature* estimating the "Number of Neutrons Liberated in the Nuclear Fission of Uranium."[55] The Paris group's conclusion that 3.5 neutrons were released later proved to be about one too many.[56]

Did a publication ban really slow the race to build an A-bomb, as Szilard had predicted? We now know how that race ended, but as Szilard feared, at its start the French disclosures did prompt several German actions when little private research and no government work were under way in America.

When Hahn and Strassmann bombarded uranium with neutrons in December 1938 in Berlin, they were unaware that the atom had split or "fissioned." That word was coined by Otto Frisch, who, working with Lise Meitner, repeated the Berlin experiment and alerted Niels Bohr as he sailed for New York at year's end. In early January, Hahn and Strassmann published their results under the misleading title "Concerning the Existence of Alkaline Earth Metals Resulting from Neutron Irradiation of Uranium," because at the time, they were fascinated by the new elements they created, not by the extra neutrons they released. So when Joliot and his Paris colleagues reported in *Nature* that March on "Liberation of Neutrons in the Nuclear Explosion of Uranium," their prestige highlighted the Hahn-Strassmann experiment.

Had all publications ceased then, as Szilard pleaded, the atom's military potential may still have been overlooked for several critical months. But once Joliot's "Number of Neutrons Liberated in the Nuclear Fission of Uranium" appeared in *Nature* in April, several governments outside the United States took direct actions.[57]

At London's Imperial College, Nobel physicist George P. Thomson saw in Joliot's second article a probable new power source, perhaps recalling his 1934 talks with Szilard. He thought a nuclear explosion unlikely, but with W. Lawrence Bragg, head of the Cavendish Laboratory since Ruther-ford's death in 1937, Thomson quickly alerted the British government. Four days after the *Nature* article appeared, on April 26, Britain's Ministry for the Co-ordination of Defence urged the Treasury and the Foreign Office to buy as much Belgian uranium as they could.

That spring, physicist Wilhelm Hanle described a "uranium burner" to a university physics colloquium in Göttingen, Germany. His superior wrote the Ministry of Education, which quickly appointed Abraham Esau to convene a conference on uranium. President of the German Bureau of Standards, Esau led an April meeting that urged buying all the uranium in Germany, banning exports, and negotiating radium contracts with the recently captured uranium mines at Jáchymov, Czechoslovakia.[58]

From Hamburg, physical chemists Paul Harteck and Wilhelm Groth alerted the War Office in Berlin to "the newest development in nuclear physics," which "will probably make it possible to produce an explosive many orders of magnitude more powerful than conventional ones.... That country which first makes use of it has an unsurpassable advantage over the others."[59] By summer, Siegfried Flügge at the Kaiser Wilhelm Institute in Berlin concluded that uranium fission might create an "exceedingly violent explosion."[60] With Kurt Diebner, an army physicist and ordnance expert, heading up Germany's uranium project at an office for nuclear research within the Army Ordnance Department, Germany was the first country with a military unit to study how nuclear fission might be used to make weapons.

In September 1939, Diebner was studying the rare isotope uranium[235] as the likely source of fission, and physicist Werner Heisenberg described how the metal might be made to explode. By month's end a handful of scientists working under Diebner and physicist Erich Bagge were studying two separate problems: how heavy water, which contains a double form of hydrogen, would act as a "moderator" to slow neutrons; and how uranium isotopes could be most efficiently separated.

In Russia during the spring of 1939, physicist Igor Tamm asked his students: "Do you know what this new discovery means? It means "a bomb can be built that will destroy a city out to a radius of maybe ten kilometers."[61] The French, too, realized where their research was leading. "Crude jokes were made," Kowarski later recalled, "about whether we will get the Nobel Prize for physics or for peace first because we made a war impossible by discovering nuclear explosives, which obviously would make war impossible."[62]

Leo Szilard at about age five, in a soldier suit (Bela Silard Collection)

Louis Szilard, Leo's father, at about age fifty. He changed the family name from Spitz to Szilard in 1900, when Leo was two years old. (Bela Silard Collection)

Leo Szilard's mother, Tekla Vidor Spitz, shown with one-year-old Leo (Bela Silard Collection)

Bela, Leo, and Rose Szilard in the Austrian resort town of Ischl, summer 1905 (Roland Detre Collection)

Rear view of the Vidor Villa, where Leo Szilard grew up, just before its completion in 1902 (*Magyar Pályázatok IV*; courtesy of Fővárosi Szabó Ervin Könyvtár, Budapest)

Leo Szilard in 1916, sporting a hat like the one he lost in a fight with an elephant (Egon Weiss Collection)

A Vidor family holiday in Aussee, Austria, summer 1915. Leo Szilard is fifth from the left in the back row. In the next row. his sister, Rose, is fourth and his brother, Bela, is sixth from the left. In the second row, Leo's father, Louis, is at the left and Leo's mother, Tekla, is fourth from the left. (Bela Silard Collection)

"Cooling the Passions" a humorous pose arranged in 1915 by Rose Szilard and photographed by Bela to tease their brother, Leo, about his infatuation with Mizzi Freund. From the left are Leo, Mizzi, and her husband, Emil Freund. In back of them, on the ladder, is Rose Szilard at the left, with the Szilards' cousin Otto Scheiber in the center (under the ladder) and Rezsin (Regine) Vidor, Emil's sister, on the right. (Photograph by Bela Silard/ Roland Detre Collection)

Leo Szilard's 1919 passport photo (Leo Szilard Papers Mandeville Department of Special Collections, University of California, San Diego, Library)

At a field artillery barracks in Budapest during World War I, 1917–1918, Leo Szilard leans on the wheel on the far right. (Bela Silard Collection)

Alice Eppinger, Leo Szilard's high school sweetheart, in Venice, about 1926 (Bela Silard Collection)

Leo Szilard's photo for a student ID card, Berlin, 1920s (Bela Silard Collection)

Leo Szilard asleep among friends at Lecco, Lake Como, Italy, in 1926. From left: Alice Eppinger; Elizabeth Fejer (later Mrs. Bela Silard); Mrs. Adele Eppinger, Alice's mother; Leona (Lola) Steiner; and Clara Erdelyi. (Bela Silard Collection)

Gertrud (Trude) Weiss in 1928, just before she met Leo Szilard in Berlin (Egon Weiss Collection)

Leo Szilard photographed in the spring of 1936 by Trude Weiss on a weekend trip in the English countryside (Egon Weiss Collection)

Leo Szilard and Ernest O. Lawrence at the 1935 American Physical Society meeting in Washington, DC. Szilard seems to urge secrecy, perhaps for his then-secret patent on nuclear chain reactions. Lawrence developed cyclotrons and would later join Szilard and other scientists in the Manhattan Project. (Leo Szilard Papers, Mandeville Department of Special Collections, University of California, San Diego, Library)

In the 1946 *March of Time* film "Atomic Power," Albert Einstein and Leo Szilard re-create the day in August 1939 when they drafted the letter that would alert President Franklin D. Roosevelt to the dangers of the German A-bomb program and lead to the creation of the Manhattan Project. (CriticalPast)

Eugene Wigner and Leo Szilard in Manhattan, in the late 1930s (Courtesy of *Fizikai Szemle*)

The creators of the world's first nuclear chain reaction pose in December 1946 on the steps of Eckhart Hall, their wartime office at the University of Chicago. *Back row* (left to right): Norman Hilberry, Samuel Allison, Thomas Brill, Robert G. Nobles, Warren Nyer, and Marvin Wilkening. *Middle row:* Harold Agnew, William Sturm, Harold Lichtenberger, Leona Woods Marshall, and Leo Szilard. *Front row:* Enrico Fermi, Walter Zinn, Albert Wattenberg, and Herbert Anderson. (University of Chicago)

Leo Szilard testifying before the House Military Affairs Committee in October 1945. Arguing for postwar civilian control of atomic energy, here he sketches a federal scheme for research and development. (AP/ Wide World Photos)

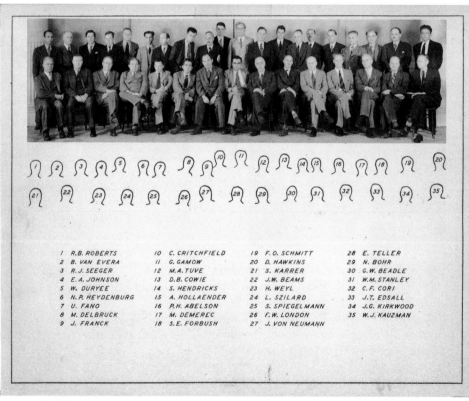

1 R.B. ROBERTS	10 C. CRITCHFIELD	19 F.O. SCHMITT	28 E. TELLER
2 B. VAN EVERA	11 G. GAMOW	20 D. HAWKINS	29 N. BOHR
3 R.J. SEEGER	12 M.A. TUVE	21 S. KARRER	30 G.W. BEADLE
4 E.A. JOHNSON	13 D.B. COWIE	22 J.W. BEAMS	31 W.M. STANLEY
5 W. DURYEE	14 S. HENDRICKS	23 H. WEYL	32 C.F. CORI
6 N.P. HEYDENBURG	15 A. HOLLAENDER	24 L. SZILARD	33 J.T. EDSALL
7 U. FANO	16 P.H. ABELSON	25 S. SPIEGELMANN	34 J.G. KIRKWOOD
8 M. DELBRUCK	17 M. DEMEREC	26 F.W. LONDON	35 W.J. KAUZMAN
9 J. FRANCK	18 S.E. FORBUSH	27 J. VON NEUMANN	

Szilard (fourth from the left in the front row) at the Carnegie Institution's annual theoretical physics conference in Washington, DC, spring 1946 (Egon Weiss Collection)

But in New York, Szilard feared just the opposite: that nuclear explosives would make war more possible and that Germany would make those explosives first. Niels Bohr said as much at the spring meeting of the American Physical Society when he surmised that a chain reaction or an atomic explosion could be created by bombarding a small amount of uranium[235] with neutrons. If uranium[235] could be separated and chain reactions begun, the *New York Times* reported in a science column, "the creation of a nuclear explosion which would wreck an area as large as New York City would be comparatively easy." The *Washington Post* on April 29 headlined an account of Bohr's speech "Physicists Here Debate Whether Experiments Will Blow Up 2 Miles of the Landscape."[63] *Times* science writer William L. Laurence called uranium[235] the "philosopher's stone" that would tap "the vast stores of atomic energy" and predicted that a tiny amount was enough to "blow a hole in the earth 100 miles in diameter. It would wipe out the entire City of New York, leaving a deep crater half way to Philadelphia and a third of the way to Albany and out to Long Island as far as Patchogue."[64]

CHAPTER 14

"I Haven't Thought of That at All"

 1939

In the United States during the summer of 1939, nuclear research ceased almost everywhere but in Leo Szilard's busy mind. Enrico Fermi went off to the University of Michigan at Ann Arbor to attend a theoretical physics conference and study cosmic rays, leaving Szilard in New York. "I still had no position at Columbia," Szilard remembered, his three-month appointment as a guest researcher ended, "but there were no experiments going on anyway and all I had to do was to think." Yet what Szilard thought about and shared with Fermi in four letters that summer became the basis for the world's first successful chain reaction.

His summer at Columbia was an auspicious time for Szilard because by that spring he and Fermi had a clear understanding of what would *not* produce a nuclear chain reaction. Szilard had persuaded Fermi to begin a large-scale experiment with 500 pounds of uranium, and borrowed this amount from commercial suppliers. For the decisive experiment Szilard and Fermi again teamed up with Herbert Anderson to test if a chain reaction might begin when the neutrons escaping from uranium fission were slowed down, or "moderated," by water. The physicists believed that if the neutrons moved slowly, they might have a better chance of hitting and splitting other uranium atoms. Working for several days around a circular pool of water, the three concluded that the uranium emitted more neutrons than it absorbed—a necessary step to start a chain reaction. But they could not make their uranium-water system create fission that was self-sustaining. Only later did they learn that many of the extra neutrons were not only moderated by the water but also absorbed by it, choking off the chain-reaction process.[1]

Working from his room at the King's Crown Hotel, Szilard sent Fermi a draft on their inconclusive water-moderated uranium work, which they submitted to the *Physical Review*. If water were not an effective neutron moderator, what would be? "I started to think about the possibility of using perhaps graphite instead of water," Szilard recalled. "This brought us to the end of June."[2]

The Szilard-Fermi correspondence over the next two weeks captures both the understandings and the tensions between them. After receiving the draft article, Fermi replied on June 26 with clarifying language, making his first admission that "if it will prove possible to slow down the neutrons . . . a chain reaction will probably be maintained."[3]

Fermi wrote Szilard on July 1. The cyclotron at Ann Arbor was out of order, he said, but when it was fixed, he wanted to repeat their last experiment on neutron absorption "because the results that we got seem to me rather crazy."[4] Writing to Fermi about "the trend of my ideas concerning chain reactions" on July 3, Szilard said, "It seems to me now that there is a good chance that carbon might be an excellent element to use [as a moderator] in place of hydrogen, and there is a strong temptation to gamble on this chance." Fermi, a cautious experimentalist, must have bristled when he read this and what followed: "I personally would be in favor of trying a large-scale experiment with a carbon-uranium-oxide mixture if we can get hold of the material," wrote Szilard. "I intend to plunge in the meantime into an experiment designed for measuring small capture cross sections for thermal [or slowed] neutrons," in other words seeking a material that would absorb the fewest neutrons. The higher an element's "cross section," the more neutrons it would absorb.

Already thinking beyond his own plans, Szilard speculated that "if carbon should fail, our next best guess might be heavy water. . . ." Ordinary, or "light," water, H_2O, has two parts hydrogen for each part oxygen. But hydrogen has a heavier isotope as well. Light water is composed of two lighter isotopes, H. Heavy water is composed of the heavier isotope, deuterium, whose symbol is D and is written D_2O. Deuterium, they thought, would capture fewer neutrons than hydrogen. Already Szilard had "taken steps" to "obtain a few tons of heavy water."[5] That day, Szilard also wrote to Strauss about his work with Fermi and Anderson, concluding: "There is also a fifty-fifty chance that the matter may be of great importance from the point of view of national defense."[6]

Two days later, Szilard wrote to Fermi again, sending a corrected value for his neutron-density calculations. His mind swung between algebraic calculations and practical business. Now "it seems that it will be possible to get sufficiently pure carbon at a reasonable price," Szilard reported. A chain reaction might be created "if layers of uranium oxide" are arranged between carbon layers. Still probing all possibilities, Szilard concluded: "Pending reliable information about carbon, we ought perhaps to consider heavy water as the 'favorite,' and I shall let you know as soon as I can how many tons could be obtained within reasonable time."[7] The same day he wrote this, Szilard visited the National Carbon Company offices on East Forty-second Street, seeking a sixteen-inch-square graphite block.[8]

In his third letter to Fermi, written on July 8, Szilard urged starting a "large-scale" experiment "right away," even before learning how well carbon absorbs neutrons. Szilard's style was impatient and irrepressible.

> Sorry to bombard you with so many letters about carbon. This is just to tell you that I have reached the conclusion that it would be the wisest policy to start a large-scale experiment with carbon right away without waiting for the outcome of the [neutron] absorption measurement which was discussed in my last two letters. The two experiments might be done simultaneously. The following can be said in favor of this procedure:
>
> A chain reaction with carbon is so much more convenient and so much more important from the point of view of applications than a chain reaction with heavy water or helium that we must know in the shortest possible time whether we can make it go.

Szilard urged experimenting with "perhaps fifty tons of carbon and five tons of uranium . . . as a start." The carbon and uranium should be "built up in layers" or stacked "in some canned form," making assembly and cleanup relatively easy. Then Szilard reported telling Pegram about this plan, "and he seemed to be not unwilling to take the necessary action. I wonder whether you think it wise to proceed as outlined in this letter."[9]

Szilard thought that Fermi's reply on July 9 seemed to be a debate with himself. Fermi proposed a "homogeneous" sixty-to-one carbon-to-uranium mixture, in effect blending everything in a pot. This only convinced Szilard that Fermi was not serious about a chain-reaction experiment "because it was the easiest to compute."[10] In a letter to Anderson more than a week later Fermi confessed to "not having understood" Szilard's proposal. "I think that the experiment is very important and should be performed."[11]

On his own, Szilard proposed a joint experiment between Columbia's physics department and his own Association for Scientific Collaboration[12] and visited the National Carbon Company to ask about the purity of commercial graphite.[13] "I have decided to prepare an experiment with all kinds of different materials . . . ." he reported to his friend Trude Weiss. "If things work out, I will be very busy."[14] But a disappointment followed within days when the navy, citing "restrictions" on government contracts, declined to support Szilard's ideas for nuclear research.[15]

Writing to Fermi "in a hurry" on July 11, Szilard was by now sure that he knew just how to make a chain reaction. "During the second week of July," Szilard later recalled, "I saw that by using a lattice of uranium spheres embedded in graphite, one would have a great advantage over using alternate layers of uranium and carbon." This, he said, convinced him "from then on that there is a good chance of maintaining a chain reaction in a uranium-graphite system."[16]

In fact, for nearly a year after the Anderson-Fermi-Szilard experiment with uranium and water, in the spring of 1939, there was almost no further work in the United States toward understanding fission or investigating the possibility of a nuclear chain reaction. University researchers were not accustomed to seeking federal support and scarcely tried, and the government scientists, though interested, failed to recognize the military significance of fission. It all just seemed too farfetched.

Not to Szilard, however, who still pursued private investors for his chain-reaction experiments. In the name of his Association for Scientific Collaboration, he met that month with a group of financiers in a suite at the Waldorf-Astoria Hotel, bringing along Bela "to restrain me when I begin to sound unrealistic." But Bela could scarcely say a word as Leo outlined to the skeptical money men a grand scheme that would bring electrical energy to poor nations yet cost "practically nothing." The machines to generate this power Szilard called "piles." Szilard urged the businessmen and bankers to invest their millions in his association but also insisted that a majority of shares be controlled by the physicists. Understandably, no one in the suite reached for a checkbook. "The physicists are too naive in practical matters, and the bankers are too businesslike," he complained to Trude. "I do not know how things will work out."[17]

In Germany, meanwhile, things were working out just as Szilard had feared. That summer, Siegfried Flügge at the Kaiser Wilhelm Institute in Berlin published "Can Nuclear Energy Be Utilized for Practical Purposes?" He, too, assumed "the energy liberation should thus assume the form of an exceedingly violent explosion."[18]

❉   ❉   ❉

When Eugene Wigner came to New York from Princeton in early July, Szilard showered him with the calculations he had made for the carbon-uranium lattice. Wigner was quick to see that this might work. They were also quick to link this approach—the closest yet to a workable chain reaction—with recent news from Europe that German military expansion could easily overrun Belgium, whose colony in the Congo was then the world's principal uranium source.

Wigner wanted to alert the Belgian government and suggested they seek advice from their former professor in Berlin, Albert Einstein.[19] Wigner occasionally saw Einstein around the Princeton campus; to Szilard, who had worked closely with him in the 1920s and early 1930s, Einstein had reverted from colleague and counselor to a famous but remote scientist. Einstein knew the Belgian monarchs well—in his unpretentious way calling her "Queen" and addressing the royal couple as "the Kings"—so perhaps, Szilard suggested, he might alert the queen of the Belgians about the perilous importance of the Congo's uranium. The two agreed that it was worth a try, and from his Princeton office they learned that Einstein was then at a cottage in Peconic, Long Island, owned by a friend named Dr. Moore.

Early on the morning of Wednesday, July 12, a clear and hot day, Wigner drove up to the King's Crown Hotel in his 1936 Dodge coupe, and Szilard climbed in. The two drove out of New York across the new Triborough Bridge, passing the New York World's Fair, whose theme "Building the World of Tomorrow" was symbolized by a 700-foot-high trylon, a tapered column rising to a point, representing "the finite," and a 200-foot perisphere globe, representing "the infinite."

Had Szilard and Wigner thought about it, their own drive that day had more to do with the "World of Tomorrow" than anything they passed on the fairgrounds. But their thoughts were fixed on finding Einstein's cottage, a task demanding all their attention. First the two Hungarians confused the Indian names in their directions and drove to Patchogue, on Long Island's south shore, instead of to Cutchogue, on the north. This detour cost them two hours, and once in Peconic, they drove around the tiny resort town asking vacationers in shorts and bathing suits the way to Dr. Moore's cottage. No one seemed to know.

"Let's give it up and go home," Szilard said impatiently. "Perhaps fate intended it. We should probably be making a frightful mistake by enlisting Einstein's help in applying to any public authorities in a matter like this. Once a government gets hold of something, it never lets go. . . ."

"But it's our duty to take this step" Wigner insisted, and he continued to drive slowly along the village's winding roads.

"How would it be if we simply asked where around here Einstein lives?" Szilard said. "After all, every child knows him." A sunburned boy of about seven was standing at a corner toying with a fishing rod when Szilard leaned out the car window and asked, "Do you know where Einstein lives?"

"Of course I do," said the lad, and he pointed the way.[20]

Szilard and Wigner were hot, tired, and impatient by the time they found the two-story white cottage. By contrast, the sixty-year-old Einstein was relaxed and genial; he had spent the early morning sailing in a small dinghy and now greeted his former colleagues wearing a white undershirt and rolled-up white trousers. Einstein bowed courteously as they met and led his visitors through the house to a cool screened porch that overlooked a lawn. There, speaking in German and sipping iced tea, Szilard and Wigner told Einstein about their recent calculations. They explained how neutrons behave, how uranium bombarded by neutrons can split or "fission," and how this process might create nuclear chain reactions and nuclear bombs.

"*Daran habe ich gar nicht gedacht*," Einstein said slowly, pondering what he had just heard. "I haven't thought of that at all."

Until that summer day, Einstein had believed that atomic energy would not be released "in my time," that it was only "theoretically possible." Einstein had not followed recent discoveries in nuclear research for years and sought only the "time for quiet thought and reflection" needed to unravel his unified field theory of the universe.[21] Einstein had published his famous equation $E = mc^2$ in 1905, but only now was that simple statement's ultimate significance clear. For even a small mass the potential energy released could be immense. Fission is the most efficient way to fulfill Einstein's equation because it releases the energy that gives matter its form—the binding energy holding the atomic nucleus together.

Einstein's next thought about the chain reaction was philosophical. If it works, he said, this would be the first source of energy that does not depend on the sun. Wind and solar energy are created by the sun's heat. And fossil fuels—oil, natural gas, coal—were once created from the carbon made by the sun's energy through photosynthesis. But releasing the binding energy of atoms was something new.

Einstein's third reaction was political. Although he was an avowed pacifist, he agreed to sound the alarm about atomic bombs, even if it proved to be a false one, in order to beat Nazi Germany to this awesome weapon. It took a scientist of Einstein's stature and personal conviction to take this risk, Szilard later noted. "The one thing most scientists are really afraid of is to make a fool of themselves," Szilard reflected on the day in 1955 when Einstein died. "Einstein was free from such a fear and this above all is what made his position unique on this occasion"[22]

When the three agreed that they should warn the Belgians, they sat around the dining-room table as Einstein dictated to Wigner in German a letter to the Belgian ambassador in Washington. Einstein warned that it might be possible to make bombs of unimaginable power from the uranium mined in the Belgian Congo and that Germany, which at first offered uranium for sale after taking over mines in Czechoslovakia, had recently banned all exports.

Wigner wondered whether the US government should also be notified, and into the afternoon Einstein and Szilard drafted a similar letter, also in German, to the secretary of state. That afternoon, they agreed to send the State Department a copy of Einstein's letter to the Belgian ambassador, giving the department two weeks to object if they opposed the letter.

After more iced tea and polite conversation, Szilard and Wigner took leave of their professor, and Einstein crossed the lawn and walked down some brick steps to a shaky wooden dock to go sailing again. The breeze that afternoon was fresh, the sky clear. Einstein steered the dinghy across the bay toward the sinking sun. That night, Wigner dropped Szilard at his hotel in Manhattan and returned to Princeton.[23]

Sitting in his room at the King's Crown, Szilard thought about the letters to the Belgian monarchs and the US government, but something didn't seem right. "We did not know our way around in America," he later recalled. "We did not know how to do business, and we certainly did not know how to deal with the government."[24] When an "uneasy feeling about this approach" led Szilard to "talk to somebody who knew a little bit better how things were done," he called on Dr. Gustav Stolper, a Viennese economist and publisher whom he first knew in Berlin. Stolper quickly understood Szilard's situation and suggested approaching a friend, Dr. Alexander Sachs, who was a vice-president of the Lehman Corporation, a large Wall Street investment bank. Sachs had worked privately since 1933 as an adviser to Roosevelt's New Deal and would surely know how to approach the government.

Szilard telephoned Sachs and soon called on him in his office at the corner of South William and Broad streets.[25] A serious-looking man with wavy hair and thick glasses, Sachs listened intently to what Szilard said. Sachs needed little persuading; he was familiar with popular reports about uranium fission and fearful of German aggression. Einstein's letter should not go to the Belgian royal family or a US government department, Sachs said; they wouldn't know what to do with it. It should go, instead, directly to President Roosevelt. Sachs boasted about his easy access to the White House and joined with Szilard in planning strategy. If Einstein would sign a letter, Sachs promised he would deliver it, in person, to the president. Szilard must have loved the idea.

But when Szilard "tried to draft a letter" from Einstein to the president the next day, it ran to more than four and a half typed pages in a convoluted and tentative style that was needlessly detailed and riddled with spelling errors—perhaps the result of a hasty collaboration with the orotund Dr. Sachs.[26] This draft told Roosevelt "it appears to be desirable" that "a man who has your confidence" ought to be in constant touch with the chain-reaction researchers. Either private or federal money might be needed for experiments "with several tons of material," and "a large stock of pitchblend" [sic] should be brought from Belgium or the Congo to the United States. Pitchblende is the ore for both radium and uranium.

Szilard mailed this rough draft to Einstein with a letter explaining Sachs's offer to approach the president and reassuring his mentor it "could not do any harm to try this way." Szilard asked if Einstein preferred to mark the draft with marginal notes or have him "come out to discuss the whole thing once more with you." If so, Szilard added, "I would like, if it is all right with you, to ask [Edward] Teller to take me, not only because I believe his advice is valuable but also because I think you might enjoy getting to know him. He is particularly nice."[27]

When Szilard telephoned Einstein about the draft letters, he learned that his mentor did prefer to meet again in Peconic. Wigner was then driving to California to teach a summer course at Cal Tech, so Szilard enlisted Teller as his chauffeur. Teller, who was teaching physics at Columbia for the summer, picked up Szilard in his 1935 Plymouth on Wednesday, August 2, and drove off to the cottage on Long Island—this time by a direct route. They found Einstein on the porch, wearing an old robe and slippers. There, over tea, Szilard urged him to send a revised letter to the president. Einstein quickly agreed. He had met Roosevelt personally, but for such an important matter Einstein thought his message should be communicated by letter.[28]

That settled, these brilliant physicists faced another tough decision: "We did not know just how many words one could put in a letter which a President is supposed to read," Szilard later recalled. "How many pages does the fission of uranium rate?" Einstein dictated in German a brief preliminary draft, leaving Szilard to write both a long and a short letter to the president.[29]

Back in New York, Szilard met again with Sachs and from him gained three other names as possible messengers to Roosevelt: financier Bernard M. Baruch, MIT president Karl T. Compton, and aviator Charles Lindbergh. Why this was necessary is unclear, but apparently, at their second meeting, Szilard and Einstein left undecided the choice as to who should deliver the letter.

Szilard telephoned Janet Coatesworth, a young stenographer who worked part-time for several Columbia departments. She came to his paper-cluttered room at the King's Crown and sat at his desk chair to take dictation while Szilard, excited and nervous about his task, alternately paced the floor or sat on the bed.

As Szilard began to dictate, in his crisp Hungarian-German accent, a letter to "F. D. Roosevelt, President of the United States," Coatesworth glanced up in disbelief. And when Szilard mentioned "extremely powerful bombs," she recalled, "that convinced me! I was sure I was working for a 'nutcase.'" Amused by her reaction, Szilard dictated more and more dramatically, his face beaming with mischief and merriment. He took special glee in closing the letter "Yours very truly, Albert Einstein." That convinced her that Szilard was deranged, a judgment he confirmed by dictating a second, even longer letter. To Roosevelt. From Einstein. Only years later did Coatesworth learn the truth about this historic session.[30]

Szilard mailed his long and short versions to Einstein, along with a letter in German about choosing the "middleman" to Roosevelt. Of the men Sachs had proposed, Lindbergh was Szilard's "favorite," so he also asked Einstein for a letter of introduction to Lindbergh. "The first version has the advantage of brevity," Szilard wrote about the two drafts, "but the second contains everything necessary to give the president a clear picture of what duties would have to be carried out by the person he would delegate." Szilard closed on a more personal note, saying, "If you liked Teller, I would like to come out to your place sometime or other with him."[31]

Szilard still meant to write a technical memorandum for the president's letter, but finishing these drafts for Einstein seemed to give him a sense of relief, at least temporarily. "We have a nice summer here, and New York actually is a summer resort," he wrote Trude on August 3. "There is such a breeze in my room that all papers flutter about."[32]

But on Friday, August 4, Szilard suddenly turned pensive. With the Einstein-letter drafts in the mail, he at last had a chance to pause. To think about what he was doing. To face the awesome and chilling consequences of his actions. This new and gravid collaboration with Einstein brought back Szilard's earlier reflections with his mentor: on the fate of humanity, on the question of a God. Now Szilard felt compelled to make sense of the dreadful future he was helping to bring about. In his contrary and rational way he turned on the experts in order to prove them wrong. What better "expert" could he pick on now than God himself?

Writing in German, Szilard's 10 *Gebote,* or "Ten Commandments," used the same form as the original but twisted and reversed the most familiar ideas. This is, in translation, what he wrote:

TEN COMMANDMENTS

1. Recognize the connections of things and the laws of conduct of men so that you may know what you are doing.
2. Let your acts be directed toward a worthy goal but do not ask if they will reach it; they are to be models and examples, not means to an end.
3. Speak to all men as you do to yourself, with no concern for the effect you make, so that you do not shut them out from your world, lest in isolation the meaning of life slips out of sight and you lose the belief in the perfection of the creation.
4. Do not destroy what you cannot create.
5. Touch no dish except that you are hungry.
6. Do not covet what you cannot have.
7. Do not lie without need.

8. Honor children. Listen reverently to their words and speak to them with infinite love.

9. Do your work for six years; but in the seventh, go into solitude or among strangers so that the memory of your friends does not hinder you from being what you have become.

10. Lead your life with a gentle hand and be ready to leave whenever you are called.

August 4, 1939[33]

Szilard's own spirit and moral certainty abound in this playful and profound exercise. In his view, as in Einstein's, obeying God required an understanding of nature, especially human nature. In nature's ways, Szilard seemed to hope, he could find that "sense of proportion" that guided his moral and ethical concerns. Moral example was a valuable end in itself. Blasphemy is not against God but against your own honesty, an echo of Szilard's deep bond with his mother's severe love of the truth. Szilard expanded the traditional "Thou shalt not kill" to all creation, not just life. For adultery, he substituted all appetites, linking them to need rather than desire. Stealing, in his view, was also prohibited except for need, as was lying.

"Honor children," Szilard commanded, rather than the traditional "father and mother," for in children he saw the bare truth that his own mind so fervently sought. This quest for truth was also a form of "love," and it is only here that the word is used.

The biblical admonition to work six days and rest on the seventh Szilard extended to years, and the regeneration that he sought was not physical but psychological—evidence of his own search for solitude and his quest for intellectual discoveries. Szilard's final thought, to lead life "with a gentle hand," must have been a frustrated dream, for it was rare that his mind ever let his spirit rest.

As a personal statement, these commandments meant a lot to Szilard, and he once told Trude that they should only be read in German. He would later joke about the meaning of the traditional commandments, but with humor that revealed a deeper moral understanding. Coming as it did, with war in Europe about to begin and with the horror of atomic bombs now a constant fear, Szilard's quest for moral guidance in these commandments was a likely comfort to a troubled soul.

At his cottage in Peconic that week, Einstein read the two letters sent by Szilard and agreed to his request for a letter to Lindbergh. But Einstein also cautioned his former student, nineteen years his junior, not to be too clever. "I signed the letters today right away," Einstein said in a handwritten German note to Szilard, adding that he, "too, would give preference to the more detailed one." Einstein said he hoped Szilard would finally "overcome your inner resistance" to being straightforward because "it always gives you pause for thought when a person wants to do something too smartly." Einstein, a thinker in abstract symbols who sought scientific truth in ultimate simplicity, felt uneasy with Szilard's many rational and complicated schemes. As he had done for more than a decade, Einstein again chided his younger colleague.[34]

The signed letters arrived at Szilard's hotel on August 9, along with the introduction to Lindbergh. After reading Einstein's note, Szilard penned a brief reply in German, thanking his mentor for the letters. "We will try to follow your advice and as far as possible overcome our inner resistances, which, admittedly, exist," he assured him. "Incidentally, we are surely not trying to be too clever and will be quite satisfied if we don't do things too foolishly"[35]

Compared with Szilard's first rambling draft and with the one-page shorter version, the longer letter to Roosevelt now seems a successful mix of the physicists' insights and concerns.

Albert Einstein
Old Grove Rd.
Nassau Point
Peconic, Long Island

August 2nd, 1939

F. D. Roosevelt,
President of the United States,
White House
Washington, D.C.

Sir,,

Some recent work by E. Fermi and L. Szilard, which has been communicated to me in manuscript, leads me to expect that the element uranium may be turned into a new and important source of energy in the immediate future. Certain aspects of the situation which has arisen seem to call for watchfulness and, if necessary, quick action on the part of the Administration. I believe therefore that it is my duty to bring to your attention the following facts and recommendations:

In the course of the last four months it has been made probable—through the work of Joliot in France as well as Fermi and Szilard in America—that it may become possible to set up a nuclear chain reaction in a large mass of uranium, by which vast amounts of power and large quantities of new radium-like elements would be generated. Now it appears almost certain that this could be achieved in the immediate future.

This new phenomenon would also lead to the construction of bombs, and it is conceivable—though much less certain—that extremely powerful bombs of a new type may thus be constructed. A single bomb of this type, carried by boat and exploded in a port, might very well destroy the whole port together with some of the surrounding territory. However, such bombs might very well prove to be too heavy for transportation by air.

The United States has only very poor ores of uranium in moderate quantities. There is some good ore in Canada and the former Czechoslovakia, while the most important source of uranium is Belgian Congo.

In view of this situation you may think it desirable to have some permanent contact maintained between the Administration and the group of physicists working on chain reactions in America. One possible way of achieving this might be for you to entrust with this task a person who has your confidence and who could perhaps serve in an inofficial, capacity. His task might comprise the following:

a) to approach Government Departments, keep them informed of the further development, and put forward recommendations for Government action, giving particular attention to the problem of securing a supply of uranium ore for the United States;

b) to speed up the experimental work, which is at present being carried on within the limits of the budgets of University laboratories, by providing funds, if such funds be required, through his contacts with private persons who are willing to make contributions for this cause, and perhaps also by obtaining the co-operation of industrial laboratories which have the necessary equipment.

I understand that Germany has actually stopped the sale of uranium from the Czechoslovakian mines which she has taken over. That she should have taken such early action might perhaps be understood on the ground that the son of the German Under-Secretary of State, von Weizsäcker, is attached to the Kaiser-Wilhelm-Institut in Berlin where some of the American work on uranium is now being repeated.

Yours very truly,

(Albert Einstein)

Although Einstein said later that he "really only acted as a mailbox" for Szilard, in popular history his famous equation $E = mc^2$ and his letter to President Roosevelt are credited with starting the American effort to build atomic weapons.[36]

With Einstein's letter in hand, Szilard then began work on a technical memorandum to accompany it: a review of research over the past five years that noted "one has to conclude that a nuclear chain reaction could be maintained under certain conditions in a large mass of uranium." For medicine, Szilard wrote, today's "quantities of grams" could be replaced by "quantities corresponding to tons of radium equivalents." Uranium might also serve "as fuel for driving boats or airplanes." He cautioned, however, that lead shielding to protect the pilot "might impede a development along this line," which, we now know, is just the problem that confounded nuclear airplane developers in the 1960s. Szilard also predicted "large quantities of energy" from a "stationary power plant."

Szilard advised that "it may be a question of national importance to secure an adequate supply of uranium." He referred to ore deposits in the United States, Canada, Czechoslovakia, and Russia but stressed that "the most important source" is the Belgian Congo. He suggested as a subterfuge arranging "a token reparation payment" to obtain a "large stock" of pitchblende from Belgium or the Congo, in this way concealing that "the uranium content of the ore is the point of interest."

Large-scale experiments are necessary to test the chain reaction, Szilard said. For these he urged strengthening existing research organizations or creating new ones. Another approach he suggested might be "the collaboration of the chemical or the electrical industry. . . ." So far, Szilard reported, extra neutrons—the precursor to chain reactions—have been created with slow neutrons. Whether fast neutrons would also work was not then clear. But "if fast neutrons could be used, it would be easy to construct extremely dangerous bombs" whose "destructive power" might "go far beyond all military conceptions." Szilard concluded that if chain reactions work, self-censorship by scientists would be advisable.[37]

When his technical memorandum was finished on August 15, Szilard delivered it and the longer Einstein letter to Sachs, along with a detailed letter of his own. In this letter, Szilard hinted broadly that he might be the man of "courage and imagination" who could "act with some measure of authority in this matter" between scientists and the government. Szilard reviewed his own actions since he learned about uranium fission in January and reminded Sachs that the Association for Scientific Collaboration offered a legal entity for research.

Sachs telephoned the White House for a private appointment with Roosevelt, but before it could be set up, German troops stormed into Poland at dawn on September 1, and the war Szilard had feared for years began. Immediately, Roosevelt's attention shifted to urgent—but more conventional—military matters. Roosevelt convened a special session of Congress and urged repeal of an arms-embargo provision in the US Neutrality Act. When passed on November 4, the amendment authorized the United States to extend "cash and carry" arms sales to countries in a state of war.

German military advances on the ground and in the air were matched by progress in the laboratory as teams of scientists worked on atomic weapons research. There was even a detailed "Preparatory Work Plan for Initiating Experiments on the Exploitation of Nuclear Fission"[38] In the United States, on the other hand, the only progress was in print, where two scientists from Washington—R. B. Roberts of the Department of Terrestrial Magnetism (DTM) and J. B. H. Kuper of the Washington Biophysical Institute—published "Uranium and Atomic Power," an article in the *Journal of Applied Physics* that summarized publicly what was then known about fission. They cited Einstein's $E = mc^2$ equation in the first sentence as the equivalence for "releasing atomic energy to furnish a new source of power," noted that fission was much more likely from slow neutrons than fast ones, and concluded "that the requisite conditions for a chain reaction were satisfied."

To Lindbergh, Szilard wrote that they had once met "at lunch about seven years ago at the Rockefeller Institute, but assume that you do not remember me, and I am therefore enclosing an introduction from Prof. Albert Einstein." Politely understated but forceful, Einstein's letter asked Lindbergh "to receive my friend Dr. Szilard and to consider carefully what he has to tell you." The subject, he warned, "may seem fantastic to a man not involved with science" but deserves attention "in the public interest."[39] Szilard's letter invited Lindbergh to meet with him to discuss how "large quantities of energy would be liberated" by a "nuclear chain reaction." Szilard also wanted to discuss how "to make an attempt to inform the administration. . . ."

But five weeks later, after Lindbergh, an isolationist, denounced Roosevelt's efforts to sell arms abroad, Szilard sent Einstein a caustic note. "I am afraid he is not our man," Szilard reported. Lindbergh's public discussion of the president's call for changes in the Neutrality Act "is on a pitiful level," Szilard wrote. "At that one becomes kindly disposed toward Lindbergh for he at least emits human sounds." In this note to Einstein, Szilard went on to predict "that Belgium will be overrun one of these days" and urged that they try to buy "at least 50 tons of uranium oxide" privately. "Whether I will be able to persuade a government agency to take such a step I do not know of course," Szilard concluded. "Perhaps one would have more luck with a smart speculator."[40]

At the same time, Wigner heard from Szilard that he wanted to approach Karl T. Compton at MIT, to act not only as a middleman to the president but also as a host for large-scale uranium experiments. Wigner was alarmed, fearing that Compton might find Szilard's style disconcerting. Put whatever you would tell Compton into a memo, Wigner urged Szilard, because "from a conversation with you" he "would probably obtain only a somewhat confused picture. . . ."[41]

Throughout the summer and fall Szilard pursued the separate mission of finding pure graphite for his "large-scale" experiment. He exchanged dozens of letters with chemical, carbon, and metallurgical companies, visited their offices every few days, and with the help of friends and consultants pressured manufacturers to bid on contracts for tons of fresh material. Szilard's correspondence for this period reveals that he held potential contractors to the most precise measurements and standards he could devise.[42] Already concerned that even slight impurities in the uranium and graphite might absorb neutrons, Szilard was, in fact, creating a decisive difference between US and German nuclear efforts.

Repeated calls to Roosevelt's secretary finally gained Sachs a meeting with the president on October 11. For this Sachs toted into the White House a note of his own, Einstein's letter, Szilard's technical memorandum, and an armload of scientific papers on nuclear fission. He may have expected to be ushered right in to see Roosevelt, but he first met Gen. Edwin M. ("Pa") Watson, the president's secretary, who asked Sachs to brief two ordnance experts, army colonel Keith F. Adamson and navy commander Gilbert C. Hoover. When they heard Sachs and saw his documents, the two nodded to Watson, who led him into the Oval Office.

"Alex, what are you up to?" Roosevelt asked, gazing at the papers cradled in his arms.[43] At first, Sachs didn't say, but sat across the large desk from Roosevelt and read from his own note, which warned about the possibility that a new energy source could be made into powerful bombs—a grandiose rendering of the Einstein letter. Then Sachs began reading from the technical papers heaped on his lap. When Roosevelt showed some discomfort and impatience, Sachs finally drew out Einstein's letter and read aloud the first and last paragraphs, which referred to uranium as a "new and important source of energy" but made no mention of a "nuclear chain reaction" or the "extremely powerful" bomb that might destroy a "whole port together with some of the surrounding territory." The paragraphs Sachs read did mention German research but not the request that "government departments" be alert to uranium ore supplies and follow experimental work by physicists.

Seeming preoccupied, Roosevelt asked Sachs to come back the next day. All this scientific talk seemed "premature" to the head of the US government.

Sachs left the Oval Office unsure about his effect and fearful he would have no second chance.[44] But the next morning Roosevelt was more chipper.

"What bright idea have you got now?" he asked, and "How much time would you like to explain it?" Again they talked about Einstein's warning, which Roosevelt seemed to have thought about overnight. Sachs still sounded vague and pompous, but Roosevelt was more attentive, listened quietly, and then interrupted.

"Alex," he said, "what you are after is to see that the Nazis don't blow us up."

"Precisely," said Sachs.

Roosevelt then leaned toward his desktop intercom and called his secretary. When Watson entered, Roosevelt waved to Sachs and his lapful of papers.

"Pa, this requires action," Roosevelt said. But action at that point only meant that Watson should create a government advisory committee to study the problem.[45] He telephoned Lyman J. Briggs, director of the National Bureau of Standards, then the government's principal physics laboratory, and asked him to chair an Advisory Committee on Uranium.

(Germany had done the same thing six months before.) Other committee members included Sachs, the ordnance experts Adamson and Hoover, and the physicists from the DTM and Bureau of Standards who had written the September summary article on "Uranium and Atomic Power."[46] Szilard, Wigner, and Teller were also invited. Finally, it seemed to the three Hungarians, things were starting to happen.

On Saturday morning, October 21, Szilard and Wigner joined Sachs at the Carlton Hotel in Washington for breakfast to review strategy. Then the three appeared at the Bureau of Standards office in the Commerce Department to attend the first meeting of Briggs's Advisory Committee on Uranium.[47] Almost from the beginning Szilard guided the discussion, explaining how a chain reaction might be created with uranium oxide and graphite. If graphite moderated the neutrons, making them "slow," Szilard said, then they might sustain a chain reaction; if too many neutrons were absorbed, nothing would happen because not enough of them would fly out to fission other atoms.

Colonel Adamson found Szilard's science fantastic. At the army's Aberdeen Proving Ground, he joked, "We have a goat tethered to a stick with a 10-foot rope, and we have promised a big prize to anyone who can kill the goat with a death ray. Nobody has claimed the prize yet."[48] Ignoring this, Szilard went on to tell the group that he and Fermi needed to conduct a large-scale experiment using uranium suspended in graphite to test how many neutrons the graphite absorbed—just the test he had proposed to Fermi in July.

Others at the meeting were skeptical, but Teller, pushing Szilard's optimism a step further, guessed that the graphite and uranium needed for a large-scale experiment might cost only $6,000. Colonel Adamson scoffed. It generally takes two wars to develop a new weapon, he said; besides, it was "morale," not research, that led to victory. Shifting in his chair, the formal and ever-polite Wigner could not contain his impatience.

"Perhaps," he told Adamson in a high-pitched but steady voice, enunciating every syllable, "it would be better if we did away with the War Department and spread the military funds among the civilian population. That would raise a lot of morale."[49]

"All right," Adamson snapped. "You'll get your money. . . . We do have money for this purpose."

Szilard was astounded by the offer. He had attended the meeting only expecting to gain the government's approval for their uranium research, which he assumed would be funded by industry, universities, or private investors. This promise of $6,000 for uranium research was the first commitment by the US government—a sum that would eventually swell to more than $2 billion before the first A-bomb was tested nearly six years later. But for the three Hungarians, all admittedly "green" in their dealings with American government, it must have seemed a glorious moment.

Indeed, that first step might never have been taken without Szilard's constant planning and meddling. While waiting for Sachs to see the president and even after the White House decision to form a government committee, Szilard continued to approach private investors for research money. It is clear from his letters and from his account of many conversations that Szilard thoroughly enjoyed his role as an organizer and manipulator. Szilard later wrote to Einstein that he had Teller invited to the meeting because he lived in Washington and could serve to "keep contact with Briggs in a workable fashion." After the first Uranium Committee meeting, Szilard returned to his room in the Wardman Park Hotel and wrote to Dean Pegram at Columbia. "It seems to me now," Szilard said of Sachs, "that he is performing his task efficiently and in the right spirit, and now I am in favor of giving him a fairly free hand, and see what he can achieve."[50]

Five days later, Szilard mailed Briggs a ten-page memo telling him what research was necessary to prove that uranium could produce chain reactions. He named laboratories where work could begin without attracting attention: Columbia, the DTM in Washington, the University of Virginia in Charlottesville, MIT, and Princeton. And Szilard urged that all research reports be withheld from publication, laying the groundwork for an atomic-secrets policy that in later years he would find a threat to both scientific advancement and civil liberties.

This memo to Briggs was Szilard's blueprint for a program to begin building a bomb and the first of its kind drafted in the United States. In it Szilard offered calculations about buying uranium and graphite, along with a strategy for cornering markets without revealing the extent of government interest. He described the large and small experiments that would show whether a chain reaction might work, stressing that to learn it *cannot* work would be just as important.[51]

On November 1, the Advisory Committee on Uranium reported to the president that a chain reaction was possible but not certain. If controllable, the chain reaction might become a power source for submarines. If explosive, "it would provide a possible source of bombs with a destructiveness vastly greater than anything now known." Briggs recommended buying what Szilard had asked for—fifty tons of uranium oxide and four tons of pure graphite. At last, Szilard thought, America was in the race for the bomb.

In some ways, Szilard ended the year 1939 much cheerier than he had begun it. Far from his gloom a year before over the apparent failure of his work on the chain reaction, Szilard could by November 1939 see that this "obsession" might soon work. He had designed a reactor— the uranium-graphite "pile"—during the summer, opened government channels for support to begin uranium research that fall, and in the process identified and united a circle of physicists who would become the core of the government's A-bomb work. By perseverance and pluck, Szilard had alerted other physicists, and finally the president of the United States, to the fears that had afflicted him since 1933.

Szilard must have been in high spirits as he hosted a dinner at the Men's Faculty Club of Columbia, on Morningside Drive and 117th Street, one Tuesday evening in November. Around the table were Fermi, now his intellectual admirer but still skeptical of Szilard's personal and scientific exuberance; Pegram, a patient sponsor turned active supporter; and Sachs, who would continue to help Szilard and Einstein influence President Roosevelt.

But Szilard's year also ended with familiar doubts and fears. He still lacked full-time employment, his affiliation with Columbia remained tenuous, and his Association for Scientific Collaboration existed only on paper and in his busy mind. Past insecurities also weighed on Szilard's mind as he recalled his early efforts at refugee settlement in London. "Though I finally succeeded in getting a number of things done by exerting myself up to the limit of my strength I learned a lesson," he wrote Sachs, "and now I am anxious to avoid a repetition of this experience."[52]

Yet as another Christmas season approached, Szilard's outlook remained bleak. To Benjamin Liebowitz he wrote that he had been unable to raise any private money for nuclear experiments and had to declare the $2,000 loan "a bad debt." "Unfortunately, I have not earned anything during this year, as I was tied up with this work on uranium. It looks as though I shall not be able to earn anything next year, either. . . ."[53]

CHAPTER 15

Fission + Fermi = Frustration

 1940–1941

For Leo Szilard, the first meeting of the federal government's Advisory Committee on Uranium, in October 1939, was followed by "the most curious period in my life." That meeting with military and civilian researchers—a direct result of Einstein's letter to Roosevelt—set in motion America's response to a possible German A-bomb. Or so Szilard thought. But for months "we heard nothing from Washington at all," he recalled, and by February 1940 this silence was unnerving. "I had assumed that once we had demonstrated that in the fission of uranium neutrons were emitted, there would be no difficulty in getting people interested; but I was wrong."

Physicist Enrico Fermi, Szilard's research collaborator at Columbia University, "didn't see any reason to do anything" as they awaited the Advisory Committee's promised $6,000 for graphite and uranium. So Szilard tried to "amuse" himself by making more detailed calculations for the graphite-uranium chain-reaction method they had codesigned. On January 27, 1940, Szilard mailed himself a twenty-two-page memo on a nuclear-chain-reaction system, then the most detailed design for a nuclear reactor ever made.[1]

Encouraged by what he read of the uranium research by Frédéric Joliot-Curie and his Paris colleagues, Szilard went to Fermi at Columbia and found he was about to leave for California to pursue cosmic-ray studies. Over lunch the two men caught up on their doings.

"Did you read Joliot's paper?" Szilard asked Fermi. He had.

"What did you think of it?" asked Szilard.

"Not much" replied Fermi. Szilard's frustration surged. "I saw no reason to continue the conversation and went home," he recalled.[2] But Szilard must have been further enraged and frustrated on February 1 when, in the *New York Times*, he saw a Science Service dispatch from Washington on Joliot's "announcement" that he had turned uranium into an atomic "firecracker" that might be "lighted" with neutrons.[3] This report must have frightened Szilard as well, since he lived in fear that German scientists would make the same discoveries. As he had done before when in need of help, Szilard turned to the one person he trusted for honest advice: Albert Einstein. Down to Princeton he went, to his white frame house on Mercer Street, and in Einstein's study suggested a newly aggressive approach. They had waited long enough for help from Washington. Szilard proposed "to go definitely on record that a graphite-uranium system would be chain-reacting, by writing a paper on the subject and submitting it for publication to the *Physical Review*." He and Einstein should "reopen the matter with the government," Szilard argued, with a new kind of pressure, one that amounted to political blackmail. Using this ploy, Szilard "was going to publish my results unless the government asked me not to do so and unless the government was willing to take some action in this matter."[4] Einstein thought about the plan and concurred.

In his New York hotel room, Szilard drafted a paper revealing just how chain reactions could be made to work: "Divergent Chain Reactions in Systems Composed of Uranium and Carbon." Joliot's reactions had been "convergent," as they shut down for lack of neutrons to extend the chain. Szilard's would be "divergent" by multiplying once the chain reaction began. He mailed his paper on this topic to the *Physical Review* on February 6 and followed it with a thirty-nine-page article on the fourteenth. The first source Szilard cited in the longer article was H. G. Wells, who had written about atomic bombs as early as 1913. Both papers, Szilard asked, should be withheld from publication until he had clearance from the government.[5]

Early in 1940, Szilard had new reasons for advocating secrecy among the atomic scientists; a warning about Germany's threat came when Peter Debye, a Dutch physical chemist and Nobel laureate, visited Columbia. Debye had headed the Institute of Physics at the Kaiser Wilhelm Institute in Berlin but was being forced out because he had refused to accept German citizenship under the Nazis. At Columbia, Debye first called on Fermi, who seemed unconcerned about reports of German A-bomb work. The German scientists were not at one location, Fermi said, and by being scattered around the country they would not be able to make a concerted effort. Szilard was not so assured by Debye's news; he was alarmed.[6]

In their February Princeton meeting, Szilard and Einstein agreed that a second letter to the president was needed to carry out their threat of publishing the chain-reaction manuscript. Accordingly, Szilard drafted an Einstein letter to Alexander Sachs, bearer of the first Einstein letter to Roosevelt, which relayed news that "since the outbreak of the war, interest in uranium has intensified in Germany [and] . . . that research there is being carried out in great secrecy and . . . has been extended to another of the Kaiser Wilhelm Institutes, the Institute of Physics." The research, Einstein warned in a March 7 letter, had been taken over by the German government. He mentioned that Szilard's manuscript "will appear in print" unless "held up" by a change in administration policy. Einstein promised a memo by Szilard on "the progress made since last October" on chain-reaction research. A March 15 letter by Sachs to Roosevelt alerted the president to Germany's advances and to Szilard's improvements on the French chain reaction. This prompted Roosevelt to ask his secretary, "Pa" Watson, to convene a meeting in Washington of Einstein, Advisory Committee chairman Lyman J. Briggs, and some military men.[7]

But more weeks passed with no results. Einstein declined to attend a meeting, and it was late April before Szilard had completed for Sachs the promised chain-reaction memo. In that memo (later called the "A-55 Report"), Szilard cited three chain-reaction "applications" that relate to national defense: power production for warships; a radiation device that might kill "human beings who are exposed to it within a radius of one kilometer"; and bombs whose "explosions of extraordinary intensity" may create tidal waves if detonated near ports. The bombs "would not be too heavy to be carried by small boats, but could hardly be carried by existing airplanes."[8]

In this memo Szilard also calculated relative speeds and ranges for oil- and nuclear-powered ships, based on advice from an Annapolis graduate and a quick read of *Jane's Fighting Ships*. "I should imagine," he wrote in April 1940, in his typical blend of geopolitics and physics, "that the combination of high speed [from lighter nuclear fuel] and a greatly increased cruising radius might be of decisive importance in case of a war with Japan."[9] Pearl Harbor was still twenty months away.[10]

Apparently, Einstein's second letter to the White House prompted release of the $6,000 promised to Fermi and Szilard, and they bought the graphite and uranium that Szilard had ordered from suppliers. Fermi and research assistant Herbert Anderson began measuring the way graphite absorbs neutrons and soon discovered that the pure material Szilard had specified absorbed very few: It would be a good "moderator," able to slow down neutrons for capture by uranium atoms without itself absorbing them. By contrast, in January 1940 the Germans had used impure graphite for similar measurements and had concluded that it absorbed too many neutrons to sustain a chain reaction.

Szilard went to Fermi's office and said that the neutron-absorption value should not be published. Fermi had honored Szilard's request for secrecy in 1939. But "this time Fermi really lost his temper; he really thought this was absurd" Szilard recalled. Szilard did not press his point but later asked, in a memo to physics department chairman George Pegram and Fermi, whether the "value for absorption cross sections of graphite" should be discussed within the lab, or should questions be evaded?[11] This led Pegram to pressure Fermi with the same question, and he reluctantly agreed to censor his work.

"From that point, the secrecy was on," Szilard recalled. And just in time. Had the Germans realized that their January calculations were off and that in graphite they had an abundant and inexpensive moderator, they might have pursued this research to make a reactor. Instead, acting on their erroneous conclusions, they used heavy water as a moderator—a choice that would doom their chances of making an A-bomb during the war.[12]

This second clash over censorship strained an already testy collaboration between Szilard and Fermi, a relationship that focused their different professional and personal styles. Ever the outsider, Szilard challenged conventions and authority with bursts of original ideas, while Fermi, a team player, enjoyed collaborating step-by-step on systematic research into the fundamental questions of nature. Szilard made intuitive leaps, while Fermi framed and resolved theoretical questions before designing and conducting experiments to test his ideas. Szilard's deep and concentrated thoughts led him to see through problems to sometimes outrageous, sometimes prescient, results, while Fermi's rigorous grasp of both theory and practice kept him centered on the task at hand. In personal style, too, Szilard was assertive, bumptious, and direct; Fermi was polite, quiet, and controlled.

"On matters scientific or technical there was rarely any disagreement between Fermi and myself or the rest of us whose opinions counted in those early days," Szilard recalled. "Nor was there much difference in our attitude toward those who 'directed' our work from Washington." Nevertheless, "Fermi thought we took too great chances in the [uranium] project," and "on this we thoroughly disagreed."

That fundamental disagreement Szilard traced to a very different view of the world and their lives as scientists.

> Fermi is a scientist pure and simple. This position is unassailable because it is all of one piece. . . . I doubt that he ever understood that some people live in two worlds like I do. A world, and science is a part of this one in which we have to predict what is going to happen, and another world in which we try to forget these predictions in order to be able to fight for what we would want to [have] happen. But then how many people are able to understand the coexistence of these two separate worlds? I certainly would not understand it were it not for certain accidents of my education. Fermi and I had disagreed from the very start of our collaboration about every issue that involved not science but principles of action in the face of the approaching war. If the nation owes us gratitude—and it may not—it does so for having stuck it out together as long as it was necessary.

Reflecting on their years of work together, Szilard once wrote that "of all the many occasions which I had to observe Fermi I liked him best on the rather rare occasions when he got mad (except, of course, when he got mad at me)." Szilard recalled that when a young physicist visited New York one summer and asked Fermi's help to use the library at Columbia, he personally talked to the librarian. She said the request must be decided by the central library. "If we admit every one," they told Fermi, "our libraries will be overcrowded." Fermi saw only two or three students loitering in the physics library. "Fermi got mad," Szilard recalled. "He got mad for the right reasons and in the right degree and in the right manner. . . . Whatever grudges I might still have against Fermi I am willing [to] forgive them for the sake of this one heartwarming memory."[13]

For his part, Fermi's "grudges" against Szilard were reciprocal, but so were his "heartwarming memories." A methodical worker, Fermi often rose before dawn to think and plan out his calculations for the day's experiments: reviewing the work that had come before and anticipating what might be learned from the tasks ahead. Szilard often slept late, then soaked in his bathtub for inspiration. So it angered Fermi when Szilard appeared at his office door, suggested new experiments or lines of inquiry, then sauntered on down the hall. Fermi had the status of a group leader among his collaborators in Rome, and he quickly attracted a circle of dedicated graduate students and colleagues at Columbia. Szilard's technique was to challenge every hierarchy he encountered.

Even more annoying to Fermi was Szilard's refusal to work with his hands. As the graphite began to arrive at Columbia in the spring of 1940, wrapped in paper packages the size of bricks, Fermi and his colleagues Anderson and George Weil unwrapped and stacked them in four-foot-square piles. Inside the piles they embedded powdered uranium in tin cans. Then they measured how many neutrons from the uranium reached different parts of the piles. Soon the physicists "started looking like coal miners" from the graphite dust, Fermi recalled. Szilard often peered in on the large rooms on the Pupin Laboratory's seventh floor, suggested new calculations and stacking methods, and then strolled off. This angered Fermi at first, especially when Szilard hired a burly undergraduate to do his share of the graphite-brick stacking. But Fermi soon mellowed and later admitted to colleagues that this was the best arrangement; because he lacked manual dexterity, the last thing Szilard should be doing is stacking graphite bricks.

Indeed, as more and more piles were created and activated, Pegram arranged to have the graphite stacked by members of Columbia's football team. Fermi later praised Szilard's "decisive and strong steps" to organize the pure graphite and uranium supplies that made the difference in their experimental success.

Reminiscing during a visit to Columbia in 1954, a few months before he died, Fermi remembered Szilard as "a very peculiar man, very intelligent." His audience—mostly physicists—roared with laughter. "I see that is an understatement," Fermi added. Again wild laughter. Fermi continued. "He is extremely brilliant, and he seems somewhat to enjoy, at least that is the impression that he gives to me, he seems to enjoy startling people."[14]

But Szilard himself was startled in May 1940 when he opened a letter from Louis A. Turner, a theoretical physicist at Princeton whom he had met casually. Before American scientists reimposed self-censorship on their publications, Turner had surveyed nearly a hundred articles written about uranium fission and in the *Reviews of Modern Physics* concluded that a "reasonable possibility" existed for "utilizing the enormous nuclear energy of heavy atoms" and that "the practical difficulties can undoubtedly be overcome in time."[15] Now Turner had a more pressing concern. Physicists knew by this time that only the rare isotope uranium^{235} could be split or fissioned by slow neutrons. At first, that made uranium^{238}, which is 140 times more abundant in nature, seem worthless to the chain-reaction enterprise. That spring, Szilard and Fermi had discovered in their experiments that when a uranium^{235} atom fissioned in natural uranium, at least one of the two escaping neutrons was absorbed by the nonfissionable uranium^{238}. This "neutron absorption" they first considered a nuisance, until Turner's letter gave it a frightening new importance.

Turner asked if he should publish an enclosed letter to the *Physical Review,* a letter that speculated how neutrons absorbed by uranium^{238} might transform it into a heavier element (later named plutonium^{239}) that might itself be fissionable. Since uranium^{238} was so much more plentiful than uranium^{235}, Turner saw this transformation as a way to create a new element for chain reactions.

Turner further suggested that a uranium-to-plutonium transformation might be truly beneficial if *more* plutonium were produced than uranium consumed. Turner asked if this notion should be withheld from publication "because of its possible military value." He thought not, for "it seems as if it was wild enough speculation so that it could do no possible harm. . . ."[16] But to Szilard, himself a master at wild speculation, the implications were stunning. "With this remark of Turner," Szilard said later, "a whole landscape of the future of atomic energy rose before our eyes in the spring of 1940, and from then on the struggle with ideas ceased and the struggle with the inertia of Man began."[17]

This letter, Szilard wrote to Turner, "will have to be delayed indefinitely in the same way as that of my own last paper," the A-55 Report on uranium fission. Szilard hoped to meet Turner to discuss his ideas, which "might eventually turn out to be a very important contribution." But Turner, still wondering why his "wild" speculation might need to be censored, challenged Szilard's decision to withhold publication, seeking a more orderly system for review than "the accident of our being acquainted. . . ."[18] In fact, Szilard now suspected, chain reactions—and perhaps bombs—might be made more easily with plutonium than with uranium, an intuition that proved correct.

Gregory Breit, the editor of the *Physical Review,* also wrote to Szilard, asking about Turner's letter and seeking some "official channels" to decide on withholding publication. But the issue was quickly resolved. On June 7, Lyman Briggs, director of the National Bureau of Standards and chairman of the Advisory Committee on Uranium, invited Szilard to become a member of a new "Advisory Committee on Nuclear Physics," with Nobel chemist Harold Urey, Breit, Pegram, the Department of Terrestrial Magnetism's Merle Tuve, Fermi, and Szilard's Hungarian friends Eugene Wigner and Edward Teller. Briggs offered Szilard five dollars a day for expenses, the first government support that he would personally receive for his chain-reaction work.[19]

Interest in chain reactions also gained attention within the government, largely because of concerns raised by Vannevar Bush, president of the Carnegie Institution of Washington and chairman of a policy panel at the National Academy of Sciences (NAS). On June 12, 1940, Bush and presidential adviser Harry Hopkins met with Roosevelt, who approved creation of a National Defense Research Committee (NDRC) that would direct scientific work for military uses. Bush's new committee quickly absorbed Briggs's Advisory Committee on Uranium, and with it the pioneering Columbia team of Fermi and Szilard.[20]

The day after Bush and Hopkins saw Roosevelt, Briggs's new nuclear physics committee met in Washington and agreed that papers on uranium should be subject to "censorship."[21] At a later meeting, the committee also decided that research on a uranium-carbon experiment should follow two lines: further measurement of the radioactive elements' nuclear constants and a search for a self-sustaining chain reaction.[22]

Back at Columbia, Szilard called on Pegram and urged that a "semi-large-scale experiment" using five tons of uranium metal should have priority over all other work. They also discussed Szilard's role in conducting research at Columbia. Szilard considered approaching Fermi to propose a range of experiments "under joint direction," and in a letter Szilard wrote to Fermi but did not send, he also took pains to review their uneasy relationship, concluding that if they had each worked alone with the necessary equipment, either one of them might have created a chain reaction by now.

"For us to work jointly in this matter has both its advantages and disadvantages, and we may at this juncture leave the question open whether the advantages outweigh the disadvantages from the point of view of obtaining speedy results," Szilard concluded. He sought "a satisfactory arrangement" because accepting one "that I would inwardly, rightly or wrongly, not consider as fair and just in the circumstances . . . would put a strain on our collaboration." Finally, Szilard said, if their effort taxed the Columbia department, then another laboratory—or work by other researchers—might be necessary. Clearly, Szilard was eager to create a chain reaction promptly and—though cautious—to enlist Fermi's mind and methods in this quest.[23]

When Szilard proposed collaborating, in a July 4 letter to Fermi, he was very tentative and left the decision to Pegram.[24] The compromise that evolved suited their different styles: Fermi took charge of all experimental work at Columbia, while Szilard generated fresh ideas and worked tirelessly to buy pure graphite and uranium supplies.[25] By the late summer, when Fermi, Zinn, Weil, and others were toiling in the heat to construct small graphite and uranium piles, Szilard appeared on campus every day or so to urge them on, convinced that they were racing with Germany for an A-bomb. One Friday, Szilard walked in and, hearing that his colleagues planned to take off for a holiday weekend, berated them for all the work that remained. Reluctantly, they agreed to cancel their plans and spend the weekend on campus, continuing their arduous calculations and construction.

"Good for you," Szilard said.

"Will you be around campus if we need your help?" asked one physicist.

"Oh, no," answered Szilard, turning for the door. "I'm spending the weekend in the country."[26]

In other ways, Szilard worked as hard, and as creatively, as anyone, although many of his insights came in unscientific settings. It was not in the lab but over lunch, at the Columbia Men's Faculty Club with Fermi and two men from the National Carbon Company, that Szilard first made one decisive discovery. He pushed his guests for more details about the impurities in commercial-grade graphite as, one by one, he named elements and compounds that might absorb neutrons. Then, "half jokingly," he mentioned an element that he knew gobbled neutrons.

"You wouldn't put boron into your graphite, or would you?" Szilard asked his guests. The two men looked at each other in embarrassed silence.

"As a matter of fact," one said, "samples of graphite which come from one of our factories contain boron, because it so happens that we manufacture in that factory graphite electrodes for electric arcs, into which boron is customarily put." Had the Columbia team relied on the National Bureau of Standards for graphite supplies instead of using Szilard's half-joking and eclectic procurement methods, it is likely that US researchers would have followed their German counterparts and rejected graphite as a moderator. Speaking of this discovery years later, physicist Hans Bethe noted that Szilard "finally persuaded one of the chief manufacturers to make some graphite without the use of boron and that graphite turned out, by Fermi's experiments, to be suitable as a reactor medium. So Szilard contributed in a very major way to the early success of perhaps the most important branch of the Manhattan Project."[27]

Still fearing the Germans' progress and not knowing about their failure with impure graphite moderators, Szilard warned other American researchers to keep silent about their work. To Ernest O. Lawrence, whose colleagues would be the first to create plutonium in their Berkeley cyclotron, Szilard wrote about Turner's letter, saying in July 1940 that "if element 239 [plutonium] shows fission for thermal [slow] neutrons, it would be highly advisable to keep this a closely guarded secret" and repeating in August that "information about it should not leak out in the newspapers or otherwise. . . ."[28]

When French physicist Bertrand Goldschmidt, a refugee from German occupation, came to Columbia in the summer of 1940 seeking to help the anti-Nazi cause, Szilard met him at the King's Crown Hotel and took him to lunch at the Men's Faculty Club. Szilard asked Goldschmidt to work with him on purifying uranium and tried through Pegram to arrange an appointment. But now that research at Columbia was funded by the government, the military was suspicious of "enemy aliens," and despite months of trying, Szilard could not maneuver around the new federal restrictions.[29]

The US military's security checks were not confined to foreign visitors, however, and in one of the most ironic episodes of the war effort, Fermi and Szilard were nearly barred from working on atomic research. Had this happened, it is safe to say there would have been no controlled nuclear chain reaction by December 1942 (the first "neutronic reactor" was codesigned by Fermi and Szilard) and no A-bomb by July 1945. An army report gave this information about Fermi and Szilard, then being considered for defense-contract work at Columbia:

(1) ENRICO FERMI. Department of Physics, Columbia University, New York City, is one of the most prominent scientists in the world in the field of physics. He is especially noted for breaking down the atom. He has been in the United States for about eighteen months. He is an Italian by birth and came here from Rome. He is supposed to have left Italy because of the fact that his wife is Jewish. He has been a Nobel Prize winner. His associates like him personally and

greatly admire his intellectual ability. He is undoubtedly a Fascist. It is suggested that, before employing him on matters of a secret nature, a much more careful investigation be made. Employment of this person on secret work is not recommended.

(2) MR. SZELARD. It is believed that this man's name is SZIL-LARD. He is not on the staff of Columbia University, nor is he connected with the Department of Physics in any official capacity. He is a Jewish refugee from Hungary. It is understood that his family were wealthy merchants in Hungary and were able to come to the United States with most of their money. He is an inventor, and is stated to be very pro-German, and to have remarked on many occasions that he thinks the Germans will win the war. It is suggested that, before employing him on matters of a secret nature, a much more careful investigation be made. Employment of this person on secret work is not recommended.[30]

Although said to be based on "highly reliable sources," the report is wrong on several points. Fermi was certainly not a Fascist, nor was Szilard a refugee from Hungary or his family wealthy merchants. These errors were minor compared to later ones that would be made about Szilard's maverick life, but on one point the report was prescient. Szilard did believe that Germany might win the war, a fear that drove him to work day and night to develop the chain reaction first.

On August 22, 1940, the War Department sent this report to Lt. Col. J. Edgar Hoover, director of the Federal Bureau of Investigation (FBI), requesting an FBI check of Fermi and Szilard "to verify their loyalty to the United States." The FBI agents repeated almost verbatim the army's details on Fermi and Szilard and continued filing reports about Szilard to Hoover after the two physicists began work for Bush's NDRC at Columbia in November.[31] Had the army investigators' advice been followed in the fall of 1940, neither Fermi nor Szilard would have been hired by the NDRC to work on the A-bomb. So in this case it appears that FBI surveillance, strongly influenced by a last-minute interview with Einstein, allowed the two physicists to continue building and designing their uranium-graphite piles. Ironically, when the army officers worried about "secret work" in 1940, they did not realize that the only secrets then worth protecting were not in government files but solely in the minds of "enemy aliens" Fermi and Szilard.

Szilard's mind centered on his secret work again in August 1940, after Germany began its bombing "blitz" of London. When Bela's wife, Elizabeth, lamented the destruction one evening, Leo suddenly stared at her intensely, thought for a pained moment, and said quietly: "Before this war is over there will be bombs thousands of times more powerful than those in the blitz."[32]

By joining the staff of the NDRC, Szilard accepted his first full-time job in more than a decade. But he complained that the $4,000 a year salary was such a "low figure" because he was "instrumental in inducing the government to assume expenditures" for testing chain reactions, a pursuit that "might come to nothing."[33] Still restless about the slow pace of their work at Columbia, Szilard turned to the John Simon Guggenheim Memorial Foundation for a fellowship. He wanted "to find out whether or not a nuclear chain reaction can be maintained by means of thermal [slow] neutrons or by means of fast neutrons in a system that contains uranium," but his application was rejected.[34]

When the work pace around Columbia quickened in the spring of 1941, Szilard realized he needed an assistant to conduct tedious but important calculations on the neutron absorption of various elements. He heard that a young graduate student, Bernard T. Feld, was interested. At Columbia, Feld later recalled, "Szilard was to us a mystery. He occasionally appeared at colloquia, and when he intervened, one could be sure that the comment or question would be particularly incisive." But Szilard remained for Feld "a remote and exotic figure" until one spring morning when the telephone rang in Isidor I. Rabi's laboratory. A call for Feld.

"This is Leo Szilard," a voice on the phone cracked. "Can you have lunch with me today at the King's Crown Hotel at twelve-thirty?" Feld accepted "with puzzled pleasure." Over lunch Szilard came right to the point. He needed help with his uranium-fission studies. Szilard explained what he and Fermi were doing. For their uranium-graphite pile, Szilard said, he concluded the uranium should be compacted into lumps. Then the neutrons released in fission would be slowed down by the surrounding graphite; as slow neutrons they would be most efficient in creating more fissions and more neutrons—thus maintaining the chain reaction.

"I listened with mounting excitement," recalled Feld, and when he realized that Szilard had anticipated a way to accommodate Feld's graduate studies, he accepted. After lunch, Szilard led Feld up to his room, dug out his notebooks, and spent that afternoon explaining in detail the calculations that covered the pages in his hasty scrawl. Szilard explained to Feld the theory and data behind these numbers. He wanted Feld to carry these calculations further. Their first session ended when Szilard handed Feld a fresh notebook and a twenty-inch slide rule, then the most reliable calculator available.

"Until I can make better arrangements," said Szilard, "you are free to work here. I will be away for the next few weeks." And he disappeared, in Feld's words "rushing up to Boston to cajole the carbon manufacturers into producing more, denser, and purer graphite at more reasonable prices; dashing down to Washington to extract greater support; stopping off at Princeton to consult with Einstein and the Wigner group." And as Szilard's "proxy hands at Columbia," Feld also got to stack graphite bricks.[35]

Szilard's rushing around seemed frantic to many colleagues, driven as it was by his fear of German A-bombs. That fear was renewed in April 1941 when physicist Rudolph Ladenburg at Princeton wrote to Lyman Briggs, reporting that a scientist had just come from Berlin via Lisbon with news that German physicists under Heisenberg were working on a bomb. Werner Heisenberg, the eminent German physicist, was trying to delay the effort, the traveler said, but all were under orders to make a weapon. Hurry up![36]

Eager to avoid bureaucratic delays, Szilard used $400 of his own money to rent fifty pounds of thorium, hoping to test if it fissioned. Later it was discovered to be the only natural element besides uranium that sustains a chain reaction.

In May 1941 came another stunning discovery, confirming Turner's speculation. In Berkeley, chemist Glenn T. Seaborg demonstrated that plutonium^{239}—the element made by bombarding uranium^{238} with neutrons—fissions as easily as uranium^{235}. Now it was known that slow neutrons could change natural uranium to fuel for possible chain reactions—and possible bombs. Unaware of Seaborg's work, the National Academy of Sciences' committee that Bush had convened to study fission reported on May 17 that three phenomena might result if a chain reaction could be maintained. The three were in just the order Szilard had presented to Sachs more than eighteen months before: power for ships, and possibly submarines; radioactive poisons; and, perhaps, bombs. The committee recommended extending NDRC funding for another six months. This work then included chain-reaction piles and gaseous diffusion experiments to separate uranium[235] at Columbia, and a gas centrifuge for isotope separation at the University of Virginia.

"We were in the Columbia labs, worried about how to stack bricks to maximize neutron moderation," recalled John Marshall, Jr., the young physicist Szilard hired from Rochester to work on the NDRC project, "and there was Szilard out in the halls. Pacing up and down. Worrying about reactor coolants and controls—problems we wouldn't confront for another two years. His mind was always racing ahead." Szilard's speculations by this time included both laboratory experiments and a cooling apparatus for huge industrial-sized nuclear plants. He was thinking hard and thinking big.[37]

The American public knew almost nothing about atomic energy at this time, and what it learned from the popular press was speculation. A July 22, 1941, issue of *PIC* magazine, a popular illustrated weekly, reported correctly about atomic research that *"this war will be won or lost in the laboratory."* The weapon *PIC* predicted, however, was not an atomic bomb but radioactive "death dust," and the use suggested for uranium[235] was not to explode but to fuel aircraft. "A lump of this U-235 the size of an ordinary pack of cigarettes would supply power enough to run the greatest bomber in the world for *three continuous years of unceasing flight,*" *PIC* reported. "It would mean that there'd be no point on this planet which a bomber could not reach starting from any other point on Earth."[38]

As farfetched as the death dust and three-year bomber seem today, they conform to ideas in the second report the NAS review committee handed to Bush in mid-July 1941. In its main report the committee was still reluctant to recommend a full-scale program. Ship propulsion, pollution by radioactive debris (*PIC's* "death dust"), and only possibly bombs still seemed the likely benefits of nuclear fission.

By the time Bush reconvened the NAS committee a third time, word of a British calculation had been spread among its members by Mark Oliphant, a Cambridge researcher who toured American universities in the summer of 1941. A "fission bomb" might be made with between 2 and 100 kilograms (4.5 and 220 pounds) of uranium[235], the third report to Bush noted. Working with Fermi in 1941, Szilard calculated roughly the amount of uranium needed to make a bomb. But it seemed enormous because of the arduous process needed to extract uranium[235] from natural uranium ore. Elsewhere on Columbia's campus, chemist Harold Urey worked with NDRC support to devise uranium-separation methods, and he soon realized that several pounds could be made if laboratory methods were expanded to industrial scale. But NDRC's contract specified that Urey was not to discuss his work with Fermi and Szilard, two "enemy aliens." As a result, Szilard later realized, "we were not able to put two and two together and come out with a simple statement that bombs could be made out of reasonable quantities of uranium-235."[39]

Oblivious to the strides made for separating uranium, Szilard continued to dash off memos on perfecting Columbia's chain-reacting piles. With help from Feld and Marshall, Szilard wrote on possible ways to extract plutonium from natural ores, on ways to create "explosive chainreacting bodies," on fast-neutron behavior in atomic piles, on how fission might be sustained by the neutrons that escape,[40] and on how neutrons are captured during fission. Teller, who knew that in Berlin Szilard had the nickname *Generaldirektor* for his imperious ways, joked at Columbia that he had raised his rank in published reports by the choice of his collaborators—to "*Feldmarschall* Szilard." But Wigner, who also enjoyed citing the *Generaldirektor* title, said after the war, "If the uranium project could have been run on ideas alone, no one but Leo Szilard would have been needed."[41]

Bush's NAS committee issued a third report in the fall of 1941 that concluded atomic bombs were "possible," perhaps within two to three years, at a cost of maybe $133 million. On November 27, 1941, Bush described in a memo to President Roosevelt an ambitious uranium program. He recommended turning over the work at the universities to the military.[42] This memo required no decision by Roosevelt, and Bush began to implement it on December 6, at a meeting of the NDRC. Gathered at the Cosmos Club, on Lafayette Square across from the White House, Bush's committee decided on an "all out" effort for military research on chain reactions.[43] The next morning, Japanese warplanes attacked the US naval base at Pearl Harbor. Roosevelt went to Capitol Hill to denounce the day that would "live in infamy." Congress declared war on Japan the same afternoon and on Germany and Italy three days later. More than two years after it had begun, the United States was in World War II. Consumed by his duties as commander in chief, Roosevelt did not respond formally to Bush's November memorandum until January 1942, when, in his own hand, he wrote on a tiny memo pad: "V.B. OK—returned—I think you had best keep this in your safe FDR."[44]

Bush and his colleagues had wasted no time awaiting formal approval for their plan. In December 1941, he designated a planning board to oversee the A-bomb project. University of Chicago physicist Arthur Holly

Compton became leader of a unified research and development effort. Increasingly the science leaders knew what they had to do. But with wartime secrecy in mind, they still had to decide where to do it.

At Columbia, Szilard continued to write uranium studies: with Zinn on December 12 on neutron behavior in uranium and other heavy elements; with Feld on the day after Christmas on ways to shape uranium to maximize fission.[45] Szilard's thoughts about the bomb project were irrepressible. In his cluttered room at the King's Crown he organized a survey of potential sites for the first chain-reacting pile, he devised new experiments, and he entertained schemes to organize not just the research and development but the entire project.[46]

CHAPTER 16

Chain Reaction Versus the Chain of Command

 1942–1944

Had Leo Szilard had his way, the world's first controlled, self-sustaining nuclear chain reaction would not have occurred in a squash court under a football stadium in Chicago. Instead, the historic site would have been an easy commute from Columbia University in Manhattan, where he and Enrico Fermi worked. With Fermi's assistant, Herbert Anderson, Szilard had picked seven sites by January 1942, including a polo field and a blimp hangar in New Jersey and a golf course in Yonkers. Szilard didn't have his way on this decision, or on many others, for his research to beat Germany to the A-bomb fell under federal patronage, then into the US Army's control. Yet despite his waning authority, Szilard continued brainstorming to find that "narrow margin of hope" he believed would aid not only the progress of the Manhattan Project but also the fate of the world once nuclear weapons became a reality.

Szilard lost the siting decision to intercampus politics, as physicist Arthur Holly Compton, named in January 1942 to direct US nuclear research, preferred his own domain at the University of Chicago. Szilard and Compton soon became friends, then allies against a mightier foe, Gen. Leslie R. Groves. Szilard's feisty rationality clashed repeatedly with the headstrong efficiency of this West Point graduate, who had just directed construction of the Pentagon (then the world's largest office building), ahead of schedule and under budget. Groves appeared as corpulent as Szilard, but with that their similarity ended. Before the war was over,

Groves would try to have Szilard jailed as an "enemy alien," order agents to follow him and to open his mail, and force him off the Manhattan Project payroll during a yearlong dispute over the chain-reaction patent.

Szilard officially joined the Advisory Council of the S-1 Physics Project, the code name for the Uranium Project, on New Year's Day, 1942, at the same time becoming a member of the scientific staff, Division of National Research, at Columbia University. He was now part of a bureaucracy that expanded daily, under the watchful and energetic direction of the Carnegie Institution's Vannevar Bush and James B. Conant, a chemist who at the time was also president of Harvard. As the S-1 Physics Project's new director, Compton came to Columbia in January 1942, met with Fermi, Szilard, and others, and after a tour of their graphite-uranium piles, sat down in the physics laboratories and spelled out his plans. "By July 1, 1942, to determine whether a chain reaction was possible. By January, 1943, to achieve the first chain reaction. By January, 1944, to extract the first element 94 [plutonium] from uranium. By January, 1945, to have a bomb."[1]

After their meeting, Szilard suggested a division of labor: Compton would be "official investigator," while Fermi and Szilard would be "in direct charge of laying out researches and overseeing them."[2] A search for larger work areas and a remote location led Herbert Anderson to look for a place to conduct their "egg-boiling experiment" within commuting distance of New York City. He found seven, and Szilard flew to Chicago to report to Compton. There he met with Compton, and Ernest O. Lawrence and Luis Alvarez from Berkeley, but after one meeting Compton simply announced that the uranium work would move to his laboratory in Chicago. Period. Szilard was miffed to learn that Compton had already told Conant where the site would be and drafted a protest letter complaining that he had abandoned orderly consultation among the scientists.[3] Gradually, the scientists who had begun chain-reaction research were being pushed aside, and Szilard resented it. But by the end of January, Szilard had returned to New York, packed up his bags and papers at the King's Crown Hotel, said farewell to Bela and his family and to his friend Trude Weiss, and moved to Chicago.

The University of Chicago campus reminded Szilard of Oxford. Gray stone neo-Gothic buildings spread in clusters and quadrangles along the broad Midway Plaisance, the main esplanade of the 1893 Chicago World's Columbian Exposition. There was no hotel close by, but Szilard quickly found the next best thing, the Quadrangle Club, the cozy English Tudor-style faculty club. There he could rent a small room with maid service, take all his meals in the large dining room, and generally enjoy the camaraderie of this new academic community. He was first appointed visiting research associate, later chief physicist, at the university's Metallurgical Laboratory, the code name for government nuclear operations on campus. The move for Szilard also meant a raise, from \$4,000 to \$6,600 a year.[4]

The Met Lab's mission was to prepare plutonium^{239} for A-bombs. But their first task was to prove that the nuclear chain reaction needed to produce that plutonium could actually work, a challenge Fermi and others still thought had little bearing on the outcome of the war.[5] The physicists moved into Eckhart Hall, a Gothic-style three-story building on the edge of the university's main quadrangle and in view of Szilard's room at the club. From the start, Szilard disliked Chicago's bitter-cold winter weather. And, increasingly, he resisted the bureaucracy imposed first by the university and then by the federal government. Soon there were armed guards at Eckhart Hall's doors. Passes. Scientific papers marked SECRET.

The tree-shaded, urbane Met Lab was destined to become just one of four secret sites in the S-1 program. Soon the largest factory building on earth would be laid out in the Clinch River valley of eastern Tennessee, near the tiny town of Clinton, by a rise called Black Oak Ridge. At that site, uranium^{235} would be separated from uranium^{238}; enough by July 1945 to make the single atomic bomb that destroyed Hiroshima. The Hanford Reservation, the third location, on a desert plateau in western Washington by the broad Columbia River near the town of Richland, replaced abandoned farms with squat, steaming power plants and soaring canyons of concrete. In the plants, which housed scaled-up uranium-graphite piles, uranium^{238} would be bombarded by neutrons to make plutonium^{239}, enough by July 1945 to make three bombs a month: one for the first test in New Mexico that month, another to destroy Nagasaki, and a third intended for Kokura on about August 20. The fourth site, a mesa near Santa Fe called Los Alamos, would later siphon brainpower from the Met Lab and collect the uranium and plutonium from Oak Ridge and Hanford to create—at first theoretically, then with elegant calculations and crude craftsmanship—the world's first atomic bombs.

The Met Lab scientists spent most of February 1942 getting settled and acquainted. Herbert Anderson offered to renovate the attic of a carriage house behind the Quadrangle Club in exchange for living there at low rent, and his aerie soon became a place for the scientists to meet and party. To Trude in New York, who had just begun an internship and residency at Willard Parker Hospital, Szilard scribbled a note reporting that he was

"Very, very busy"[6] But soon after Szilard began work in Chicago his enemy alien status forced Compton to remove him from the Met Lab payroll, although he continued to work, anyway.[7]

Compton raised three alternatives with the military for handling Szilard: transfer the navy's approval for his work at Columbia to Chicago; complete the clearance process from scratch; or consider "internment or otherwise keeping him under close surveillance. . . ." Compton reviewed Szilard's key role in starting the bomb project in 1939 and concluded:

> His work, while perhaps not indispensable, is really very helpful to us. If dropped from the project, he would have reason to be so dissatisfied that his loyalty to the country might be shaken. As it now stands, however, I have every reason to believe that he will work on this project with devotion to the welfare of the United States.[8]

But Szilard's devotion included an irrepressible desire to speak his mind, and as the Met Lab received more and more directives from Washington, he retaliated. Szilard wrote to Bush in May to complain "about the slowness of the work on unseparated uranium," a problem caused by a "division of authority along the wrong lines." Szilard complained that Bush's reorganization in the fall of 1941, creating four different divisions, compartmentalized a scientific effort that could thrive only by intellectual interchange. "We knew in August 1939 how to make a power plant with graphite and uranium," Szilard wrote. "By June 1940 we knew how to make 'copper' [plutonium] and bombs sufficiently light to be carried by airplane." Szilard sent along the memo he wrote with Einstein's first letter in August 1939 and a second memo (the A-55 Report) to the uranium committee chairman, Lyman J. Briggs, in October 1940. Both had proposed ways to organize the uranium research. "I wonder," he asked Bush, "whether, if you read the enclosed copies, you might not think that the war would be over by now if those recommendations had been acted upon."

With the first Einstein letter in mind, Szilard concluded:

> In 1939, the government of the United States was given a unique opportunity by Providence; this opportunity was lost. Nobody can tell now whether we shall be ready before German bombs wipe out American cities. Such scanty information as we have about work in Germany is not reassuring and all one can say with certainty is that we could move at least twice as fast if our difficulties were eliminated.[9]

Bush had assumed in 1941 that a large-scale bomb-making operation must involve the army, and with that in mind he proposed to reorganize the S-1 Committee in the spring of 1942. At first, money for research was siphoned informally by the budget director, but as plans swelled, a convenient cover was needed. Bush found one in the budget of the Army Corps of Engineers. So Szilard's ideas for reforms within the program came at a time when major changes were already under way. But the tone of Szilard's letter did not help his cause. Conant urged Bush to "acknowledge Dr. Szilard's letter and ask him to come and see you if you can take the time to listen to him"[10] Compton, too, wrote Bush from Chicago to praise Szilard's efforts.

> As you know Szilard was the first in this country, perhaps anywhere, to advocate trying to secure a chain fission reaction using unseparated tube alloy [uranium]. He has perhaps given more concentrated thought on the development of this project than has any other individual. As an experienced physicist and engineer and a man of unusual originality, his thoughts have been of great value in determining the direction of our work. He has likewise been from the beginning, actively concerned with the more far-reaching problems of organization and civil and military uses of the process. Even though not all of his ideas are practical, I consider him one of the most valuable members of our organization.

Compton described Szilard as "an independent individualist, vitally and I believe unselfishly concerned with the effective progress of our program. You will, if I am not mistaken, find a half-hour discussion with him to be time well spent."[11]

On June 1, the day Compton wrote that praise to Bush, Szilard thought up a way to monitor German nuclear activities: Use Swiss scientists as observers. He proposed to Compton that K. C. Cole, a physiologist at Columbia Medical School, act as a possible intelligence agent. Then came a grimmer thought. "It might be argued, of course, that if we are going full speed ahead, anyway, there is not much point in trying to find out what the Germans are doing, since there is no possibility of any defenses anyway."[12]

Indeed, Szilard's alarmist instincts kept anxiety high around the Met Lab. In June, Szilard warned Compton that the Germans might soon be capable of raining bombs or radioactive reactor debris on US cities. At one time during his work at Chicago he reported hearing that G. Dessauer had come from Germany to Switzerland with a warning about the Nazi nuclear program.[13] On another occasion he spread word about receiving a cable from Switzerland, sent by Dr. Fritz Houtermans, the brilliant physicist whom Szilard had first met in London in 1933. "Hurry up. We are on the track" Houtermans's cable warned. "Since it was sent to the Chicago project" Wigner said, "we also realized that they knew about our 'secret' work."[14]

On June 1, Szilard's formal status on the project was clarified when the navy wrote Compton with authorization for employment. But his informal status among project leaders remained ambiguous. Bush wrote to Compton on June 3 that he was trying to schedule a conference with Szilard. An internal routing slip from this time, undated and unsigned, tells a different story. "Following our conversation of yesterday, I understand you will see Szilard *after* the reorganization, say, the end of this month." At the project's highest levels Szilard was becoming a nuisance.[15] Briggs complained that Szilard was trying to usurp Compton's authority. "He is brilliant, enthusiastic, aggressive," Briggs told Bush, "but he is not a project leader."[16] Fermi recalled this uneasy time when he said that Szilard had done "a marvelous job" organizing uranium and plutonium procurement, "which later on was taken over by a more powerful organization than was Szilard himself. Although to match Szilard it takes a few able-bodied customers."[17]

Later in June, Szilard wrote a string of memos about cooling nuclear power plants by using helium and liquid bismuth.[18] And he filed a petition to become a naturalized US citizen. But by month's end his mind again scrambled physics and politics; he cited in a memo examples of what "we could do if we had an organization that could act with the freedom of an industrial corporation," then went on to report his negotiations with the Brush Beryllium Company in Cleveland about making "fused uranium ingots as a direct production of electrolysing uranium tetrafluoride."[19] Szilard was also perfecting an electromagnetic pump to cool a reactor with liquid bismuth.[20] His thinking and working hard on several projects meant that by evening he was "usually terribly awfully tired."[21]

Fears of a German A-bomb kept Szilard working hard, but they did not curb his wry humor. In a letter to Trude in July 1942, he urged her to consider entering pediatrics. "I hear that there are many babies being born these days. Probably because of a rubber shortage," and later that month, from a hotel in Cleveland, he reported that "at a certain place . . . it was written in pencil: 'Our aim is to keep this place clean; your aim will help.'"[22]

Szilard was in Cleveland with Edward Creutz and David Gurinsky to visit Brush Beryllium, where he wanted to investigate his idea for recovering uranium from its salts by using magnesium as a chemical reductant. In the factory they watched as technicians added chunks of magnesium metal to molten uranium salt. As the salt heated above magnesium's boiling point, Creutz remembered, "A minor explosion filled the lab with burning magnesium vapor and noise. But faster than the reaction was Szilard's exit. Possibly he had anticipated the result and vanished completely for several minutes. But the process worked."[23]

On the day of the beryllium explosion, August 20, 1942, two chemists back at the Met Lab were performing some stunning metallurgy of their own: Louis B. Werner and Burris Cunningham were isolating the world's first pure sample of plutonium.[24]

Szilard passed his first summer in Chicago working hard during the day on varied problems and most evenings returned to his cozy room exhausted, with barely the time to read a few pages of fiction or scribble a letter before nodding off. One letter reported, with amusement, a local crime. "The papers are full of a murder story," he wrote Trude. "A girl has shot her boyfriend, who was untrue to her. Today she gave an interview and said: sure I loved him; you do not go around shooting men whom you do not love. The rival, on the other hand, said: no man is worth so much."[25]

September 17 brought to the Met Lab scientists a new challenge as the army assumed command of the S-1 bomb project, which they still considered their own creation and domain. Herbert Anderson was the first to respond, in a memorandum that day on "Organization of the Metallurgical Project" in which he urged that a five-member executive committee be created to "have the final decision in all matters of actual procedure." Secrecy and bureaucracy have produced delays and difficulties, he said. Szilard's memorandum two days later ranged far beyond the management problems to consider "winning the peace after war with Germany." In this draft Szilard wrote that "at some future time, when the war is won or lost, the history of this chapter of *The Tragedy of Man* may perhaps be pieced together"—a reference to the Hungarian epic that had shaped his fears and hopes since childhood.[26] In two more days of thinking and editing Szilard drafted one of his most challenging documents on the war and the bomb: "What Is Wrong with Us?," an eleven-page report raising the problems at the Met Lab and warning about a postwar race for peace.

These lines are primarily addressed to those with whom I have shared for years the knowledge that it is within our power to construct atomic bombs. What the existence of these bombs will mean we all know. It will bring disaster upon the world even if we anticipate them and win the war, but lose the peace that will follow. . . . One has to visualize a world in which a lone airplane could appear over a big city like Chicago, drop his bomb, and thereby destroy the city in a single flash. Not one house may be left standing, and the radioactive substances scattered by the bomb may make the area uninhabitable for some time to come.

It will be for those whom the constitution has entrusted with determining the policy of this country to take determined action near the end of the war in order to safeguard us from such a "peace." . . .

Perhaps it would be well if we devoted more thought to the ultimate political necessities which will arise out of our present work. You may feel, however, that it is of more immediate concern to us that the work which is pursued at Chicago is not progressing as rapidly as it should.

Szilard went on to complain about the "compartmentalization of information" as the cause of delays and confusion.

I am, as a rule, rather outspoken, and if I do not call a spade a spade I find it rather difficult to find a suitable name for it. It may be that in talking to Compton I am overplaying a delicate instrument. This is, by the way, an opportunity to apologize to all members of our group for my outspokenness and to ask them to consider it as one of the inevitable hardships of war.

Next Szilard recounted the confusion and mixed signals over a decision about the reactor's cooling system; ultimately, a clash between physicists from the university and engineers hired from industry and conscripted by the Army Corps of Engineers. The scientists felt they were losing control of key decisions. And they were. Sarcastically, Szilard noted how "pleasant" life would be for those who sat back and simply followed orders—delegated from the president to Bush to Conant to Compton.

Compton delegates to each of us some particular task and we can lead a very pleasant life while we do our duty. We live in a pleasant part of a pleasant city, in the pleasant company of each other, and have in Dr. Compton the most pleasant "boss" we could wish to have. There is every reason why we should be happy and since there is a war on, we are even willing to work overtime.

Alternatively, we may take the stand that those who have originated the work on this terrible weapon and those who have materially contributed to its development, have, before God and the World, the duty to see to it that it should be ready to be used at the proper time and in the proper way.

I believe that each of us has now to decide where he feels that his responsibility lies.[27]

Two days after Szilard wrote this memorandum, on September 23, 1942, the recently promoted brigadier general Leslie R. Groves took command of the Manhattan Engineer District, the corps' code name for the S-1 work it had been performing for Bush and Conant since the spring. The name came from an office on Broadway in Manhattan where procurement orders were handled. At the war's end this guise would become the nickname for the entire bomb-making operation, the Manhattan Project. Groves had craved an overseas combat assignment and grudgingly accepted this desk-bound Washington appointment as a necessary but unpleasant duty that would do little to advance his career.

General Groves first visited the Met Lab on Monday, October 5, where, as he wrote in his history of the Manhattan Project, he met Compton and his assistant, Norman Hilberry, Fermi, James Franck, and "the brilliant Hungarian physicists Eugene Wigner and Leo Szilard. . . ." After the meeting, Groves and Compton "resumed a discussion we had begun earlier with Szilard on how to reduce the number of approaches which were being explored for cooling the pile. Four methods—using helium, air, water, and heavy water—were under active study."[28] This is the only mention Groves made of Szilard in his book, perhaps because it recounted the only time when they were not in direct conflict. The two men took a quick dislike to each other and personified the Manhattan Project's struggles between scientist and soldier.

But it was more personal than that: Groves's authoritarian, anti-Semitic views cast Szilard as a pushy Jewish busybody; Szilard's openness and glee at "baiting brass hats" cast Groves as a rigid militarist, who also seemed not too bright. Groves considered Szilard the only "villain" in the Manhattan Project, while Szilard considered Groves its biggest fool.[29]

Right away Groves warned the senior men on the project about the need to maintain tight security. Compton, Fermi, Szilard, and other scientists listened, nodded, and resumed their debates about reactor designs. After that first meeting, Charles D. Coryell, a Met Lab chemist, was walking toward the door with Groves. Glancing back at the large window in the conference room, Coryell wondered aloud about their own security.

"General, what would you think if someone threw a hand grenade through that window?"

"It'd be a damn good thing," Groves snapped. "There's too much hot air in here."[30]

Groves had chided the academics at their first meeting when he pointed out that while he didn't have a Ph.D, he had "ten years of formal education after I entered college. . . . That would be the equivalent of about two Ph.D.s, wouldn't it?" Besides, Groves said, his assistant, Colonel Nichols, did have one. The scientists listened in silence, exchanging a few embarrassed smiles. Szilard fumed, and as soon as Groves walked from the conference room, he turned to his colleagues.

"You see what I told you?" he said. "How can you work with people like that?"[31]

At his first meeting with Groves, Szilard pushed hard for his bismuth cooling scheme. Groves pushed back, and two days later Compton had a "directive" from the general. Szilard could build his bismuth-cooled reactor, but only if he acted as a "consultant" to the Met Lab's chief engineer, Thomas V. Moore. Moreover, the reactor must be ready for operation by June 15, 1943. If not, Compton said, it "will cease to be of the greatest urgency" for the project.[32]

This dispute over building a bismuth-cooled reactor involved much more than a technical challenge. It was the first head-on clash between Szilard and the evolving military-political-industrial complex that controlled the effort he alone had first led. Szilard, Wigner, Anderson, and even Fermi saw the choice of a reactor-cooling system as a dispute between physics and engineering, between their three-year-old Columbia team and the fresh consultants from Du Pont, Stone & Webster, and the Army Corps of Engineers. Here the physicists, led by Szilard, would draw the line.

"It is almost fall, but the weather is nice and clear, and there are a lot of fights that do not disturb me too much because basically I don't care which way it will turn out," Szilard wrote to Trude on October 11. He may not have cared which cooling system—water or bismuth—was used, for he was soon at work on reports about both. But he was under stress and complained about feeling "terribly awfully tired and [I] often go to bed around 8:00 p.m."[33] To Szilard, the fights with the engineers had more to do with authority than with technology.

Groves thought so, too. The few hours he had spent in conference with Szilard and the reports he had gleaned from others raised suspicions. "Groves thought Szilard was the perfect spy," said Samuel K. Allison, a

Met Lab colleague. "Not much was known about his background," and his movements were erratic.[34] Recalled John Marshall, Jr., Szilard's colleague at Columbia and Chicago, "He was always moving around . . . and he questioned army orders." Soon after they met, Groves became concerned about Szilard's "reliability." This enemy alien with the German accent and the very unmilitary bearing made Groves, the consummate military engineer, uneasy.

Szilard, too, was uneasy, about engineers in general. He complained to Compton one day that they knew nothing about designing and building a nuclear reactor. They were fine at putting up bridges, he said, but this was something special. Something new. Moreover, Szilard's own delight with tinkering made him think that he could design a reactor on his own.

"If you don't get rid of those engineers, I'm going to quit" said Szilard.

"You have just resigned" Compton told Szilard, venting the pressure he felt as mediator between his scientific colleagues and the army. Quietly, Szilard stood up and walked back to his office. There Compton found him a few minutes later.

"I didn't mean it, Leo," he said. "We still need you."[35]

To Trude, Szilard reported that "Wigner and I are fighting on the same side; if we are out of luck, we will get what we want."[36] Wigner also resented dealing with the engineers as equals.

"What do you want me to do?" Wigner asked when they called on him to cooperate.

"Well," one engineer said, "all we want you to do is to answer our questions."

"Oh," Wigner replied, his sarcasm barely masked by his polite manner, "if you know what questions to ask, you will find the answers to any question which you might ask and which I can answer in my files. All I have to do, then, is give you the key to my files, which I shall be very glad to do."[37]

On Monday morning, October 26, Szilard was in Compton's office again, this time at Compton's request. Szilard assumed they would discuss how to design and build a reactor cooling system, but Compton announced that no arrangement would work with Szilard involved. His very presence at Chicago made it difficult to organize a collaboration of physicists and engineers. Compton suggested that Szilard transfer back to Columbia, to work there in loose connection with the Met Lab. "Finally," Szilard recorded in notes he made, "he asked me to arrange matters so as to leave the Metallurgical Laboratory within forty-eight hours so as to make it easier for him to reorganize the technological division. . . ."

His work from New York would be of little value to the project, Szilard protested, and any salary would be little more than "a pension." Instead, Szilard insisted, he would prefer to apply for patents on his chain-reaction inventions and receive a royalty from the government.[38] Compton was surprised by this counterproposal. He cabled Groves that morning: "HAVE GIVEN SZILARD TILL WEDNESDAY TO REMOVE BASE OF OPERATIONS TO NEW YORK. ACTION BASED ON EFFICIENT OPERATION OF ORGANIZATION NOT ON RELIABILITY. ANTICIPATE PROBABLE RESIGNATION. SUGGEST ARMY FOLLOW HIS MOTIONS BUT NO DRASTIC ACTION NOW."[39]

Two days later, Compton offered a compromise. Szilard could form an independent organization to build a bismuth-cooled reactor at the Illinois Institute of Technology in Chicago. Again Szilard refused. If his presence caused trouble or led to friction, this would continue on the Met Lab's Technical Committee.[40] Fearing resignations by Wigner and others over Szilard's removal, Compton cabled Groves again: "SZILARD SITUA-TION STABILIZED WITH HIM REMAINING CHICAGO OUT OF CONTACT WITH ENGINEERS. SUGGEST YOU NOT ACT WITHOUT FURTHER CONSULTATION CONANT AND MYSELF."[41]

But Groves had ideas of his own and the same day drafted a secret letter to be signed by Secretary of War Henry L. Stimson.

WAR DEPARTMENT

October 28, 1942.

The Honorable,
  The Attorney General.

Dear Mr. Attorney General,
  The United States will be forced without delay to dispense with the services of Leo Szilard of Chicago, who is working on one of the most secret War Department projects.
  It is considered essential to the prosecution of the war that Mr. Szilard, who is an enemy alien, be interned for the duration of the war.
  It is requested that an order of internment be issued against Mr. Szilard and that he be apprehended and turned over to representatives of this department for internment.

Sincerely yours,

Secretary of War.[42]

Stimson refused to sign, which so angered Groves that he vowed to discover new reasons for Szilard's imprisonment and assigned a special agent to watch him. "Few enemies were causing us as much trouble . . ." as Szilard, Groves later recalled.[43]

At week's end, Szilard was eager to leave town, but as an enemy alien he could not travel without army permission. Compton had approved a brief vacation for Szilard, but now Groves had to decide if he could take it. Groves was to be in Chicago in three days.[44]

Waiting for the general to arrive, Szilard wrote Trude:

A lot has happened here in the meantime. On Monday they tried "to push me out of the project." It seemed, and for a while it was almost certain, that I would be "retired" and would be back in New York in one week. After that a "storm" broke out that raged for two days, so that everybody (including me) was flab-bergasted. Naturally I liked it very much! Now we have a terribly funny situation that I am quite comfortable with but which is not helpful to our work from a professional point of view. This, however, cannot be helped and will continue for some time. All the younger people who are important, no matter if they come from Columbia, Princeton, or Chicago, have behaved terribly decently and have given them "hell." I have cheered up accordingly; have resigned from all duties . . . and enjoy life. Also, I wanted to go on vacation right away, but I was asked to wait with that for another week.

The life of a horse is funny. What is new with you?[45]

Around the university campus, Szilard made friends with some biologists, whom he joined for their afternoon coffee break, drinking from glass beakers as they sat around a lab table. "They said there are only two kinds of scientist among the biologists," Szilard reported to Trude. "'Sons of a bitch' and 'bastards'; the former are those who write a paper and put in things that are not so, and the latter are those who point it out." (Szilard loved this distinction and cited it often once he became a biologist.)[46]

When Groves met with Compton, he must have complained again about Szilard's "reliability," because two days later Szilard turned over details about his life and work to help with "a more thorough investigation of my background in connection with the work of our project." He listed Einstein, Wigner, Benjamin Liebowitz, and Edward Teller as references.[47] To Trude, Szilard reported that Teller had problems of his own—a hernia—and asked for a medical book he knew about for nurses.

Meanwhile, Szilard reported, "Teller got hold of another popular book and immediately looked up 'hernia' and found this: 'With the latest advances in asepsis and anaesthesia and the influx into the nursing profession of vivacious and pretty girls an operation for hernia must be considered as one of the lighter forms of amusements.' No further anatomic details were given."[48]

Szilard was also amused to find himself writing to Compton and describing his own "attempts at secrecy from March 1939 to June 1940" He, the inventor of atomic secrecy, was now suspected of not being secretive enough. To Groves, Compton cited Szilard's early efforts to keep scientific secrets from Germany, his research in nuclear physics, his advancement of the US program, and his "important part in originating the ideas of the processes that we are now developing." Compton also felt obliged to remind Groves that Szilard had become a German citizen under the Weimar Republic and had first come to the United States on an immigrant visa before the Nazi takeover.[49]

Winter's chill gripped Chicago by mid-November, making the site for the first reactor especially inhospitable. Stagg Field was a medieval-style monument to Chicago's past glory as a football power, and this season the west stands stood mute against the gray sky, its ranks of empty seats overgrown with weeds. Below the sloping stands, the concrete walls of an abandoned squash court trapped and held the cold. There, bypassed and ignored by the university community, the Met Lab constructed its chain-reacting pile. (Another site, in the Argonne Forest west of the city, had been picked as a safer location, but a carpenters' strike interfered with construction there, and the existing structure was used. Once the physicists gave assurances that a chain reaction would not explode, Compton picked the closer site—right in the middle of the campus.)

His decision to build the first reactor at Stagg Field was not made until Saturday, November 14, and the following Monday morning, Walter Zinn, Herbert Anderson, and others began work by laying out a square balloon cloth—custom-made by Goodyear—in the squash court and setting down the first layer of graphite bricks for the nuclear pile. If the uranium-graphite pile did not work as planned, the physicists thought, then perhaps the neutrons could be encouraged to collide by surrounding the whole system with helium in the balloon. The scientists kept warm by the vigorous work of carrying and stacking graphite blocks, but the armed guards on duty danced from foot to foot in order to fight the damp chill. Zinn thought that graphite, a pure form of coal, might burn to heat the place. But when small fires were lit in steel oil drums, they fouled the dark room with smoke. Next, ornamental gas-fired imitation logs were hooked up, but these consumed the court's oxygen and produced eye-stinging fumes. Finally, when rummaging through the empty halls beneath the stands, Zinn found his solution in a dusty locker: a heap of full-length raccoon coats from the university's big-league football era. "For a time," remembers Zinn, "we had the best-dressed collegiate-style guards in the business."[50]

The brick stacking was slow and dirty work seven days a week. The scientists slipped and slid on the graphite dust, at times so blackened by the task that only their bright eyes and teeth gleamed in the dark chamber. More and more uranium spheres were set into the pile of graphite, with cadmium strips inserted to absorb any neutrons that might set off a chain reaction prematurely. As the pile grew, layer by dusty layer, Szilard sat at his desk in Eckhart Hall scribbling calculations for the next reactor— one designed to produce plutonium for the bomb. He refined his bismuth cooling system and devised a new way to assemble uranium fuel.[51] But no matter how absorbed he was in his work, Szilard hopped up from his desk every few hours and rushed out to buy the latest editions of the papers.

Finally, on the night of December 1, the fifty-seventh and last layer of graphite bricks was set.[52] Compton dismissed some colleagues' fears that the first nuclear reactor might somehow run wild, its atom-splitting neutrons exploding in a burst of heat and radioactivity. Szilard, too, doubted that this pile would misbehave, but he brooded nonetheless and seemed withdrawn from his friends.[53] That night, he walked to Culver Hall, a science laboratory on campus, where he knew that the physiological psychologist Heinrich Kluver often worked late.

"Come to dinner with me," Szilard said, and Kluver, who enjoyed Szilard's speculative conversations, accepted. As they walked through the bitter-cold night to a nearby restaurant, Szilard admitted that he had already eaten but would have a second dinner "just in case."

"Just in case what?" Kluver asked Szilard over dinner.

"In case an important experiment doesn't succeed," said Szilard with a typical air of mystery. He was not specific but did say that he and other scientists knew that something "might just fail to work; indeed, that seems to be the most likely outcome." However, another more remote possibility prompted Szilard's second meal that night. If their experiment "works too well," there might be an explosion. Did Szilard distrust the conclusions of his own colleagues? asked Kluver.

"Not at all," said Szilard. "But even the greatest theoretical physicists cannot be absolutely certain. So I felt that a second dinner was in order."[54]

Cold weather hung on that night, and by dawn on December 2 the temperature was 10 degrees Fahrenheit. A strong, raw wind rattled the city. The elevated trains and the trolleys were jammed, the result of gasoline rationing, which had begun the day before. Newspaper headlines reported an air battle over Tunisia. The US State Department announced on this day that 2 million Jews had already perished under Hitler and 5 million more were in danger.[55]

Still wondering if the Fermi-Szilard graphite pile would work, a group of visiting Manhattan Project executives, led by Warren K. Lewis of MIT, sat in Eckhart Hall that morning debating the merits of a heavy-water reactor. But under the stands at Stagg Field, the pile stood ready for the test. From a balcony overlooking the squash court Fermi ordered the cadmium control rods pulled from the black carbon pile. One by one. Until a single rod was left to absorb neutrons. Fermi eyed the neutron counter and flicked his slide rule. Everything was ready.

"I'm hungry," Fermi said. "Let's go to lunch," and back in went all the rods.

The break eased a spell of anxiety and anticipation, so that everyone seemed more relaxed as they filed back into the gloomy court that afternoon. By 2:20 P.M. Fermi's Columbia colleague George Weil stood before the balcony, facing the pile. When only one cadmium control rod kept the neutrons in check, Fermi called, "All right, George," to the floor below, and Weil pulled it to a predetermined point. The team watched in silence, checking dials. Fermi manipulated his slide rule. Szilard stood at his elbow, watching in silence. Thirty tense and busy minutes later, Fermi had Weil pull out the rod another foot, and the Geiger counters surged. Anderson remembered the final moments this way.

When the cadmium rod was pulled out to the position [Fermi] asked for next, the increase in neutron intensity was noticeably quickened. At first you could hear the sound of the neutron counter, clickety-clack, clickety-clack. Then the clicks came more and more rapidly, and after a while they began to merge into a roar; the counter couldn't follow any more. That was the moment to switch to a chart recorder. But when the switch was made, everyone watched in the sudden silence the mounting deflection of the recorder's pen. It was an awesome silence. Everyone realized the significance of that switch; we were in the high [neutron] intensity regime and the [Geiger] counters were unable to cope with the situation any more. Again and again, the scale of the recorder had to be changed to accommodate the neutron intensity which was increasing more and more rapidly. Suddenly Fermi raised his hand: "The pile has gone critical," he announced. No one present had any doubt about it. Then everyone began to wonder why he didn't shut the pile off. But Fermi was completely calm. He waited another minute, then another, and then when it seemed that the anxiety was too much to bear, he ordered, "Zip in!" Zinn released his rope [dropping a control rod], and there was a sigh of relief when the [neutron count] dropped abruptly and obediently. . . .[56]

The time was 3:53 P.M. No one cheered, but the excitement in that cold and shadowy room was felt and shared by all. With a rustle, Eugene Wigner stepped forward and pulled from a brown paper bag a large straw-bound bottle of Bertolli Chianti that he had clutched behind his back all afternoon. He handed it to Fermi, who pulled the cork and sent out for paper cups. Then the circle sipped quietly, solemnly, without toasts, as if partaking in a secret sacrament. Later, they penned their names on the bottle's straw wrapper.

Compton shared his colleagues' excitement over a long-distance call to his boss, James B. Conant, at Harvard.

"The Italian navigator has landed in the New World," Compton said, hoping to code his message but unable to conceal his pride.

"How were the natives?" asked Conant.

"Very friendly."[57]

The "New World" they had reached offered mankind the terrifying power of nature. From now on, a scientific speculation would be a wartime reality. And to Szilard, who had feared this moment for more than nine years—and had hoped all that time that it would not occur—his colleagues' exuberance must have seemed unnerving. He had first conceived the chain reaction in 1933 in a burst of defiant creativity. He had restrained his ego and energies for more than two years while collaborating with Fermi to design and build the first reactor. And now Szilard had witnessed his vision come true. As he had feared from the start, the world would never be the same again.

After the experiment and the silent sipping of wine, as their colleagues filed from the squash court into the cold evening light, Szilard and Fermi found themselves standing alone. "I shook hands with Fermi," Szilard remembered, "and I said I thought this day would go down as a black day in the history of mankind."[58]

CHAPTER 17

Visions of an "Armed Peace"

 1942–1944

A decade after the first nuclear chain reaction was created, Enrico Fermi, Leo Szilard, and their colleagues returned to Stagg Field to commemorate what they had achieved. Movie cameras rolled as Herbert Anderson paced the floor where the first pile had been built. Fermi and Szilard stood stiffly on the balcony: Fermi to describe the moment the reactor went critical; Szilard to recall the historic event with the crisp and smug assertion "I was here to say 'I told you so'" But in December 1942, in the days just after that historic event, Szilard and his world seemed anything but confident. For after nine years of fear and frustration, the chain reaction he had first envisioned in 1933 was a reality, and the war he had expected was then in its fourth grisly year; a race with Germany for the A-bomb seemed graver than ever.

Just as in the early 1930s Szilard had looked ahead to war when the world was at peace, in the early 1940s he looked ahead to peace from the gloom of war: to a peace that must be secured in new ways because of the new weapon he was helping to create; to a peace that might also be more stable and enriched because of the new power source the atom could provide. Szilard thought about the atom's peaceful uses the day after the Stagg Field experiment. He told Compton he wanted to "create a clear situation" with his chain-reaction patent rights and had doubts about whether to claim separate or joint patents with Fermi.[1]

But Szilard could rarely fret about just one problem at a time. He also worked to restore his security clearance.[2] And he worried about problems that were years away.

"Szilard was politically preoccupied at the time about the future of humanity and the organization of a proper type of republic," recalled nuclear physicist Pierre Auger, then a visitor to the Met Lab. Szilard especially enjoyed chatting with the tall, lean Frenchman, and the two men shared many brisk walks through the campus and along the Midway Plaisance. "His idea was to have a normally elected government and at the same time a 'shadow government' whose only task was to criticize the working government." In this, Szilard was seeking some balance between the critics and the workers, Auger explained; between power and reason. "And the important feature is that those who criticized could not take the power, but could only preside over new elections. He was always afraid of the trend to power, of people who were just after power and not after improving their world."

As Szilard and Auger walked—to Jackson Park on Lake Michigan, along the shore, and back up the Midway to campus—their brainstorming gave Auger the idea for the book *Discussions with Myself,* a collection of forty dialogues published in 1987. Szilard also took an interest in Auger's poetry and praised the power of English verse: Auger liked Blake; Szilard admired Shakespeare's sonnets. The two men enjoyed reciting limericks and clerihews about people they knew. There was one Szilard favorite that Auger remembers:

> *In a notable Family named Stein*
> *There was Gertrude, and Ep and then Ein.*
> *Gert's writings were hazy,*
> *Ep's statues were crazy,*
> *And nobody understood Ein.*

Auger had met Einstein in Paris in the 1920s and during the Chicago walks told Szilard about a tea party in the laboratory of the physicist and chemist Jean Baptiste Perrin. The poet Paul Valéry frequented these Monday afternoon affairs and asked Einstein, "What do you do when you have ideas? Do you write them down? I keep a notebook to write down my ideas," and drew from his pocket a small book. Einstein looked at Valéry and his book. "Oh," he said. "That's not necessary. An idea. It's so seldom I have one."[3]

Over the Christmas holidays Szilard saw *Bambi* and recommended the film to Trude. But despite the gaiety, he was disturbed by the new reality of the chain reaction and still tense about his ambiguous place in the project. "Although everybody is quite nice," he wrote to Trude the day after Christmas, "I am not too happy with the situation and I will take a vacation as soon as I can and think about the world." He ended 1942 writing to Trude, "Since it is New Year's Eve today, I will go to bed early. . . ."[4] Szilard's Met Lab contract expired at midnight.

Although off the payroll until his patent claim could be resolved[5] and clearly "impatient" with his plight, Szilard enjoyed a burst of creative energy. Within six weeks of witnessing that his chain-reaction concept worked, Szilard drafted the first of several proposals to Compton for a new kind of fast-neutron reactor.[6] In this ingenious design, fast neutrons from a radioactive core would bombard a surrounding blanket of uranium^{238}, turning it into plutonium^{239}. This plutonium would then be refined as new fuel for reactors or used as the explosive in A-bombs. Later that spring, Szilard was strolling by Eckhart Hall, in the university's main quadrangle, with Wigner and physicist Alvin Weinberg as they discussed his new machine.

"What should we call this process, this new reactor?" asked Wigner. Szilard thought for a few seconds and smiled with satisfaction.

"Let's call it 'breeding'" he said. "Let's call it a breeder." By the spring of 1945, Szilard would complete rough calculations and designs for three breeders.[7]

"I am again very busy (such a busy spell), and I am also a little bit tired because I wake up during the nights with strange ideas," Szilard wrote Trude in mid-January 1943. "Then I read in a book by [Baltasar] Gracian (300 years old, Spanish Jesuit). . . ." A satirist and epigrammatist, Gracian had the kind of biting wit that Szilard enjoyed. Gracian was a kindred spirit; like Szilard, he, too, often ran afoul of his superiors.[8]

Physicist Hans Bethe arrived in Chicago for two weeks' consultation on reactor designs, and during the visit he met with Szilard several times. The two friends discussed work at the Met Lab. "The things that were done and even more the things that were left undone disturbed me very much," Szilard later recalled, especially because he believed the Germans were ahead in the race for an A-bomb—a few days earlier he had proposed to Compton spying on the German nuclear program with Swedish and Danish scientists.[9]

"Bethe," Szilard said, "I am going to write down all that is going on these days in the project. I am just going to write down the facts—not for anyone to read, just for God."

"Don't you think God knows the facts?" Bethe asked.

"Maybe he does, but not *this* version of the facts."[10]

Military security agents, on orders from General Groves, were gathering their own version of the facts about Szilard, a suspicion he confirmed when a relief postman accidentally delivered surveillance instructions for opening Szilard's mail—and Fermi's.[11] After discovering that, Szilard openly joked about being watched, in one letter to Trude even chiding the "censor" for taking candy he had mailed to her.[12] All the agents read in Szilard's letters—if they could decipher his German scrawl—were comments about her health and complaints about his own, general remarks about his mood and fatigue, and accounts of his latest reading. At the time, Szilard was enjoying *Tom Jones*, whose short chapters he read in snatches during the day and just before nodding off at night. Trude knitted Szilard a long wool ski cap, which he could tug down over his eyebrows. Like the broad-brimmed felt hat of his adolescence, this cap made him feel "different," and he wore it with pride around the cold and windy campus. "We have a big snowfall today," Szilard wrote in March. "We had a few very cold days and the hood was admired by everybody. Somebody asked me 'did your mother knit it?' 'To the contrary' I told him. . . ."[13]

FBI agents tailed Szilard that March as he rode the Broadway Limited to New York and became a naturalized US citizen at the Southern District Court in Manhattan. They followed him in May to Washington, where he met with his former supervisor at Oxford, Frederick A. Lindemann, then Churchill's science adviser, with the title Lord Cherwell.[14]Although the war would not end for more than two years, Szilard argued to Cherwell about the need for strong international controls to prevent a postwar nuclear arms race. He also urged British intelligence to expand its spying on the German nuclear program.[15]

Furious when he learned about Szilard's travels and meetings, Groves urged even more spying. In June 1943 he informed a security officer that

> . . . the investigation of Szilard should continue despite the barrenness of the results. One letter or phone call once in three months would be sufficient for the passing of vital information and until we know for certain that he is 100% reliable we cannot entirely disregard this person.[16]

Szilard's return to Washington in June kept FBI agents hopping.[17] From their reports we know that an agent in the Pentagon read Szilard's security file, where he learned:

The surveillance reports indicate that the Subject is of Jewish extraction, has a fondness for delicacies and frequently makes purchases in delicatessen stores, usually eats his breakfast in drug stores and other meals in restaurants, walks a great deal when he cannot secure a taxi, usually is shaved in a barber shop, speaks occasionally in a foreign tongue, and associates mostly with people of Jewish extraction. He is inclined to be rather absent minded and eccentric, and will start out a door, turn around and come back, go out on the street without his coat or hat and frequently looks up and down the street as if he were watching for someone or did not know for sure where he wanted to go.[18]

In fact, Szilard usually knew exactly where he wanted to go, but he became so annoyed by his followers that he deliberately tried to trick them. The FBI agent's twelve-page report on Szilard's Washington visit reads like a script for the Marx Brothers. At 5:00 P.M. on Sunday, June 20, five FBI agents met at the Wardman Park Hotel in Washington to await Szilard's arrival from New York. They must have been conspicuous from the start because, as one agent admitted later, in this residential hotel there were few people in the lobby. When Szilard walked in at 8:30, "he had with him at the time one portable typewriter, one black leather suitcase, and one brown leather Gladstone Bag," the agents later reported.

Subject's description is as follows: Age, 35 or 40 years [at the time he was 45]; height, 5 '6"; weight 165 lbs; medium build; florid complexion; bushy brown hair combed straight back and inclined to be curly, slight limp in right leg causing droop in right shoulder and receding forehead. He was wearing brown suit, brown shoes, white shirt, red tie and no hat.

After twelve minutes in his room, Szilard was "observed leaving the elevator in the lobby . . . and carrying a *Newsweek* magazine, which he proceeded to read while walking through the lobby of the hotel." He "walked through the entire front lobby of the hotel," along an interconnected passage, around the tennis courts, "and paused a moment to observe the Shoreham Hotel" across the street, then returned to the lobby "and left by the main entrance, walking down the driveway to the drug store in one of the wings of the hotel."

Eight minutes later, "interested in finding a place to eat" but frustrated with the crowd in the drugstore, Szilard left and walked "on the various sidewalks in the grounds" and down the hill to Connecticut Avenue.

There he "paused momentarily" at the White Tower, Peoples Drug Store, and Chin's Chinese Restaurant, but chose a fourth, Arbaugh's Restaurant, where he ate dinner alone. Back at his hotel by 9:45, Szilard read his *Newsweek* in the lobby for twenty-five minutes before going to his room.

Four FBI agents were at the Wardman Park for Szilard's 8:00 A.M.wake-up call on Monday morning. He took a taxi to the Carnegie Institution of Washington and met with Captain Lavender, a Manhattan Project specialist on patents and legal affairs.[19] Three hours later, Szilard returned to the hotel, where he paced in the lobby, ate lunch in the drugstore, and after 2:00 P.M.met Eugene Wigner, who had just checked in. Unable to find an empty chair in the hotel barbershop, Szilard and Wigner (who was "approximately 40 years of age, medium build, bald head, Jewish features and was conservatively dressed") asked the hotel clerk how far Union Station was and took a taxi. "Due to adverse traffic conditions" the FBI agents lost Szilard's cab.

Although Szilard told the hotel's taxi starter he was headed for Union Station—and two FBI agents went there to head him off—he asked the driver to take them to the Supreme Court. But a few blocks from the court, Szilard asked for the nearest barbershop and was dropped with Wigner at the Plaza Hotel on First Street NW. FBI agents surmised that the two men went to the hotel barbershop, then walked to the Supreme Court, where they joined a public tour. The two were back at their hotel by 6:10 P.M., where seven FBI agents now awaited them. Then Szilard and Wigner walked to the Shoreham Hotel's Terrace Gardens for cocktails, ate dinner at Arbaugh's, and at 8:45 returned to the Wardman Park. After more wanderings they sat on a bench by the tennis courts, "where both pulled off their coats, rolled up their sleeves and talked in a foreign language for some time."

After Wigner retired at 9:30 P.M., Szilard walked to the hotel drugstore, where he read a newspaper and ordered grapefruit juice and a sandwich. Szilard retired at 10:55, but at 11:15 Isidor I. Rabi appeared at the front desk. Himself a hotel guest, Rabi called on Szilard in his room and talked for some time.

Six agents crowded into the hotel drugstore on Tuesday morning to watch Szilard eat breakfast. Next he visited the barbershop, took three aspirin at a drinking fountain, and caught a taxi to a temporary navy building at Seventeenth Street and Constitution Avenue NW. There Szilard "told one of the ladies that he wished to see Commander Lewis Strauss . . . and was interested in getting into a branch of the navy." Szilard and Strauss walked up Seventeenth Street to the Metropolitan Club for lunch and back to the navy building. Next, Szilard checked out of his hotel, cabled Trude "ARRIVING KINGS CROWN ABOUT 8:30 P.M.," and caught the Congressional Limited to New York.

The "Agent's note" that accompanies this surveillance report points out that Szilard "could easily leave the impression that he was conscious of being followed." He appeared to be "highly nervous and very absent-minded," the agent noted. "On one occasion he got off the elevator a short distance from his room, entered the room, came out in the hall about five minutes later and asked the maid where the elevator was located." Szilard's

> actions are very unpredictable and if there is more than one entrance or exit, he is just as apt to use the most inconvenient as not. It was found necessary to cover all possible exits to insure not losing him. It was also the observation of the Agents in Washington that he has very poor eyesight.

While the agent reporting this assumed that Szilard did not notice the surveillance, Szilard often joked with friends at how clumsily he was followed. On one trip to New York, he became so annoyed with his "tail" that he entered a building—probably the King's Crown or Trude's apartment—and stayed inside for three days. But on other occasions he pitied the agents and turned to invite them for a taxi ride or a cup of coffee.[20] He even used the army's security to hide his own lifelong fear of water. One stifling summer day, Szilard wore a heavy overcoat when he and physicist Katharine Way visited friends from the Met Lab on Chicago's north shore. For relief, the party walked to Glencoe Beach on Lake Michigan for a swim, but Szilard sat on the sand and refused to remove his coat because, he said, "it contains secret papers."[21]

Szilard managed to maintain his puckish sense of humor despite tensions at the lab. In July, two Met Lab scientists, John Marshall, Jr., and Leona Woods, were married, and when Szilard attended, he told the groom's mother that he was responsible for this match—because he had hired Marshall at Columbia.[22] And over the July Fourth weekend, at a lake in the country with a colleague's family, Szilard "flirted with his daughter." Szilard confessed this in a letter to Trude, adding, "She is a little younger than you, exactly 3½ years [old]. 'I *like* you,' she told me. 'You are funny.'"[23]

Szilard may have been amusing to children, but his behavior was not often appreciated by adults. In August 1943, Compton sent him a note dismissing him from the Met Lab.[24] Now other scientists were "no longer authorized to discuss our work and other secret matters with you," Compton said, and Szilard was to return all secret reports and notebooks and could not participate even as an unpaid consultant "until such time as the patent negotiations between yourself and the government have been completed. . . ."[25]

"Things are worse than semi-miserable (but not mousy), now," Szilard wrote Trude the evening he was dismissed. "Mousy" in their shared jargon meant depressed or dejected.[26] The next morning, Szilard began to fight back as he appeared at the downtown office of James Hume, a patent attorney whom the army had cleared to represent him. The two men formed a long and close association. "I was shocked when I learned what was going on," Hume remembered years later. "Groves's actions toward Leo were abominable. At that time we thought all the stiff-arm tactics were coming from the Nazis, but I soon discovered that Groves's actions fitted right into that pattern. That's the thing we're supposed to be fighting, I thought."[27]

Beating the Germans remained Szilard's chief concern. He wrote Compton that "since I am sure you would not expect me to sacrifice, for the sake of financial gain, my potential usefulness for our work, however small it may be, you will realize that I really have no choice . . ." but to assign his patents to the army.[28] The army offered Szilard $25,000 for all his nuclear inventions made before November 1940, when he was first employed at Columbia by the Office of Scientific Research and Development. This he refused, but agreed to negotiate for a higher fee.

Every few days, and sometimes more than once a day, Szilard dashed off notes to Trude, his constant support in this struggle. By mid-August, Szilard could write: "Things are still semimiserable here, and so I am ready for a vacation, but otherwise diversified and at times even funny."[29] He complained to her that "it is idiotic to live in a place where the weather is nice for only two months and otherwise it is too warm or too cold,"[30] and from 1943 on, Szilard wrote and talked about living in California, to him a seemingly ideal locale. But it would be two decades before he finally moved there.

"I am loafing, which, in due course, is getting on my nerves," he wrote Trude that fall. "If I would not be 'broke' at the moment, I would invite you for a visit."[31] He managed to pay his rent at the Quadrangle Club from savings and survived on the generosity of his colleagues, who often invited him home to dinner. Groves increased the pressure on Szilard in October when he demanded his signature to an Espionage Act pledge "that you will not give any information of any kind relating to the project to any unauthorized persons."[32] Szilard refused to sign.[33]

As negotiations for rejoining the Met Lab advanced, Szilard brought Hume to meet Compton.[34] Szilard also met Hume at his office in the Loop and several evenings called at his apartment on North State Street in the city's Gold Coast neighborhood. To maintain military secrecy, Szilard was never introduced to Mrs. Hume. Instead, the two men went right to Hume's bedroom at the back of the third-floor apartment and shut the door. Szilard sat on a French antique chair; Hume perched on the edge of a French Empire bed with rolled pillows. Then they talked patents. (From the window they looked out to the picturesque mansard roof of Isham House, then a private mansion and later publisher Hugh Hefner's Playboy Mansion.) When their negotiations ended Szilard left the apartment silently.[35]

Throughout the ordeal over his patents, Szilard offered Trude advice that also helped him maintain his wits. She was reluctant to visit a male psychoanalyst, but he urged her to try, as long as he has a "suitable sense of humor." This "is important in such cases," wrote Szilard, "because one wants to learn to laugh about oneself, at least."[36]

After many long meetings among Szilard, Hume, and Compton, Szilard limited the government's payment to his actual expenses, $15,416.00, plus a customary $1.00 patent fee. A settlement for the value of the chain-reaction patent itself, Szilard insisted, would be worth much more.[37]

Groves was coming to Chicago in early December and proposed a meeting to have Szilard sign over his patents. With Hume out of town, Szilard was reluctant to negotiate, but eventually agreed. As soon as they met, Szilard complained to Groves that he felt "under duress" by signing this contract and asked the general to state clearly whether he would be reemployed at the Met Lab if he failed to sign. Groves refused to answer, saying that such a statement *would* constitute duress and then their agreement would not be legally valid. After all, Groves replied, in any employment negotiation there is always some "duress." "Why, if Sears, Roebuck and Company offered me a million dollars a year as salary to be its general manager," Groves said, "that certainly would be 'duress' on me."[38]

But Szilard persisted, his anger veiled in sarcasm. He would only sign because in these circumstances he did not wish to be forced to desert his Met Lab work, Szilard said. If Groves said he had to sign to be rehired, Szilard explained, then "under these circumstances I would feel like a man who is accosted in the street by a man who draws a gun and asks for his money. In such a situation I would without hesitation surrender my money and report the case to the police. . . ." But today's choice was different, Szilard continued.

"In the absence of a clear statement by you, General, I feel like a man who is accosted in the street by somebody who demands my money while he keeps his hand in his pocket in such a manner that it is not quite clear whether he is holding a gun or just holding a pipe. Finding myself in a similar situation, I find it difficult to decide on a course of action."

Szilard told Groves that he would sign only because he was convinced, rightly or wrongly, that "the Germans have caught up with us and I do not wish to leave the project even though I can at present contribute very little to its success."[39] A year and a day after the first chain reaction had occurred, Szilard signed over his patents to the army and the next day wrote Hume with questions about the "fairness" of this agreement— questions that would haunt Szilard to the end of his life.[40]

As soon as he signed, Szilard was back to criticizing the project He wrote Bush with new "concern for the progress of our work" and reported "dissatisfaction" from most of the Manhattan Project laboratories.[41] Bush and Conant wondered how to deal with these renewed complaints. "I understand Szilard has now *signed* off on his patents," Bush wrote Conant. "You have been the buffer on this business once. Should Szilard see you?"[42] "My guess," Conant replied, "is Szilard will not be satisfied with seeing me, he thinks I am one of the evil ones. I suggest you offer me as a substitute, however, to take a load off your back. Say I shall be coming to Chicago in about a month or I will see him in Washington sooner."[43]

Just before Christmas, 1943, Szilard signed an employment contract with the University of Chicago Met Lab to be "Chief Physicist" at a salary of $950 a month, with back pay for the year, which left him feeling "temporarily very rich."[44] When he had money, Szilard carried bills in his vest and coat pockets, ready for any contingency or move. Physicist Frederick Seitz remembers that "on one occasion when a group of us were dining at the Quadrangle Club he came to us in a rather excited way, saying that he had to go to New York City and wondered if we could lend him enough money to buy a train ticket. We agreed to do what we could but asked him how much he needed. He started delving into his various pockets and turned up with several times more money than he needed for a round trip. . . ."[45]

At the Met Lab, Szilard and his colleagues had less to do than a year earlier. The physicists and engineers had agreed on a design for the huge reactors then being built at Hanford in the Washington desert. While Szilard was left behind, many of his colleagues were packing up for the train ride to Lamy, New Mexico—the nearest station to the secret Los Alamos Laboratory, where A-bombs were being designed and built under J. Robert Oppenheimer's direction. "Nobody could think straight in a place like that," Szilard warned as his colleagues departed. "Everybody who goes there will go crazy."[46] That left Szilard and others to ponder new power-plant designs, the "breeder" reactor among them, both to produce plutonium for bombs and, someday, to generate electricity. But now, Szilard complained, the atomic scientists "no longer consider the overall success of this [A-bomb] work as their responsibility." Lab morale "could almost be plotted on a graph by counting the number of lights burning after dinner in the offices of Eckhart Hall. At present the lights are out."[47]

For Szilard, too, physics gave way to the politics of atomic energy, which seemed more pressing than ever. In January 1944, when complaining to Bush about the Manhattan Project's compartmentalized operations, Szilard confessed why he worried so much. He predicted a postwar nuclear arms race unless some international control scheme could be created. And in what may be the first suggestion of a preemptive war to prevent the spread of nuclear weapons, Szilard proposed controlling all the world's uranium deposits, "if necessary by force, and it will hardly be possible to get political action along that line unless high efficiency atomic bombs have actually been used in this war and the fact of their destructive power has deeply penetrated the mind of the public. This for me personally is perhaps the main reason for being distressed by what I see happening around me."[48] Now, ironically, Szilard feared that the bomb he did not want to use to win the war must be used to win the peace.

In March 1944, Szilard met Bush in Washington for "a long conference" that Bush told Conant "lasted practically all day" and at later encounters proposed various postwar control schemes for atomic energy.[49] He also wrote about diverse nuclear-physics projects.[50] But by 1944, Szilard's scientific curiosity veered toward biology, as it had more than a decade before. In April he bought a college zoology text, but found the reading "discouraging." "A protozoan is almost as complicated as a human being," he wrote Trude, "an analogy which carries very far: protozoans do not learn from experience either."[51] Reading Macleod's *Physiology,* Szilard noted several significant differences from an earlier edition he had seen.[52] Szilard later speculated to Trude about the chemical makeup of individual cells and about genetic mutations.[53] And that fall he attended "a nice presentation on genetics" by Sewall Wright, who founded at Chicago the mathematical discipline now called population genetics. Wright's talk, Szilard lamented to Trude, "made me homesick for genuine science."[54]

Looking to world government as a way to control the spread of the A-bomb after the war, Szilard believed that "in spite of this hope being slim, it would I believe be necessary at least to make an attempt in this direction." He could not forsake his "narrow margin of hope." At the same time, Szilard predicted an "armed peace" if several countries acquired nuclear weapons, the stalemate that became the cold war.[55]

Szilard's imagination led him to predict problems that few contemporaries could see: some, like the cold war, were serious; others were silly. To allow Met Lab scientists in Ryerson Hall access to a library of secret documents in the adjacent Eckhart Hall, the army built a passageway connecting the two. His colleague Charles Coryell recalled that this worried Szilard, because he "was afraid that somebody would get lost in the passage ways connecting various secret parts with others. He had heard that if you were ever lost in the mountains, if you could find a porcupine and kill it, you could have food. So he proposed that the army put some porcupines around there in case somebody got lost."[56]

Knowing Szilard, he was probably serious about this danger, although no one mistook what Fermi said at one seminar for anything but fun. "The universe is vast," Fermi began, "containing many stars like our own sun . . ." and he spun a shaggy-dog story about the probability that air and water and chemical compounds elsewhere might resemble those found on earth. On and on he conjectured, about self-reproducing biota and primitive life-forms and evolution more advanced than our own. The overly intelligent creatures would surely explore other galaxies, he concluded, and "could hardly overlook such a beautiful place as our earth, with its ample supply of water and organic compounds, its favorable temperature range and all its other advantages."

Fermi paused. "And so," he said, at last posing his question to the group, "if all this has been happening, they should have arrived here by now, *so where are they?*"

"They are among us," Szilard declared, "but they call themselves Hungarians."[57]

This joke, in many variations, regenerated throughout the Manhattan Project. In a popular retelling, the Hungarians were Martians. After all, four Hungarians—Szilard, Wigner, Teller, and von Neumann—were among the brightest scientists in a stellar gathering. Like Martians, the Hungarians were said to have superhuman intelligence and to speak an unearthly language. Were Hungarians really Martians? Whenever Szilard was asked, he smiled, and answered: "Perhaps."

As he was devising ways to make international control of the atom more palatable to both the American public and foreign statesmen, Szilard proposed sharing the atom's peaceful benefits. He suggested irradiating cobalt disks in the Hanford reactors and offering them "as a present to other nations for hospital use" to replace X-ray machines. This might signal US intentions "to refrain from monopolizing the humanitarian applications of atomic power." Compton liked this "good idea" and urged Szilard to send it on to Bush. Szilard did and in the same letter raised a theme that would drive his thinking for the rest of his life.

> Clearly, the existence of atomic bombs and their effect on the organization of the world will overshadow every other practical application for the next twenty-five or fifty years to come, but one might in the meantime try to keep up the faith in the ultimate peaceful applications of atomic power by putting into effect one by one such applications even though they are of minor importance.[58]

Around the Met Lab at that time, a small group headed by Zay Jeffries of General Electric drafted a "Prospectus on Nucleonics" that looked for peaceful applications not twenty-five or fifty years into the future but five or ten. The group argued for "setting up an international admin-istration with police powers which can effectively control at least the means of nucleonic warfare."[59] Szilard had no direct role with Jeffries's group. But what he overheard about it both convinced him that a postwar nuclear arms race was likely and scared him about who might start it. Szilard heard that Jeffries believed the army engineers wanted to keep the A-bomb a secret not only from the Russians but also from the American public. If US citizens learned about the bomb, the engineers were said to believe, then public opinion would prevent its use against Japan. And the engineers wanted to bomb Japan not just to help end the war but to test and improve their new weapon for use against their next enemy—Russia. Heard in 1944, this eerie rumor convinced Szilard that the most urgent task ahead was to improve US-Soviet relations.[60]

CHAPTER 18

Three Attempts to Stop
the Bomb . . .

 1945

By early in 1945, Leo Szilard had a place in the history of nuclear physics for his work in creating the world's first chain reaction. He was not sure, however, what role he could play in deciding how his invention would be used. Those decisions were being made by military and political leaders in Washington, not at the Metallurgical Laboratory at the University of Chicago, where Szilard worked, or at the secret Los Alamos site in New Mexico, where A-bombs were then being designed and built.

Isolated from both political and practical deliberations, Szilard had time to reflect on the admix of science, politics, and personality that had brought the United States into the atomic age. A dozen years earlier, the A-bomb had been little more than Szilard's private obsession. Now he was part of a $2 billion nationwide enterprise and found it necessary to argue, in a series of memoranda, for a research and development program for the "breeder" reactor he had invented "and the part that, given favorable conditions, I might be able to play in it." Believing that the breeder could yield abundant nuclear fuel, Szilard speculated about plentiful atomic energy in a postwar world and on ways to assure its control. As he later recalled:

> Initially we were strongly motivated to produce the bomb because we feared the Germans would get ahead of us and the only way to prevent them from drop-

ping bombs on us was to have bombs in readiness ourselves. But now, with the war won, it was not clear what we were working for.[1]

Arthur Compton, the Met Lab director, shared Szilard's concern but, despite his access to the project's leaders in Washington, had no say about the A-bomb's use. Because the Manhattan Project's strict secrecy left no intermediate level in the government to consider these issues, Szilard concluded that "the only man with whom we were sure we were entitled to communicate was the president." Why not write to him?

In a memo on "Atomic Bombs and the Postwar Position of the United States in the World," Szilard warned Pres. Franklin D. Roosevelt that the choice facing US strategists was starkly simple: Strike an arms control agreement with the Russians or be forced to beat them in a nuclear arms race. With no direct knowledge of bomb-design work at Los Alamos, Szilard warned that "six years from now Russia may have accumulated enough [fissionable material] to make atomic bombs . . ." that would be so small they could be hidden in US cities for later detonation. Worse, he said that ". . . after this war it is conceivable that it will become possible to drop atomic bombs on the cities of the United States from very great distances by means of rockets."

In a US-Soviet nuclear arms race, Szilard wrote, the "greatest danger" is "the possibility of the outbreak of *a preventive war*. Such a war might be the outcome of the fear that the other country might strike first, and no amount of good will on the part of both nations might be sufficient to prevent the outbreak of a war if such an explosive situation were allowed to develop." Only a worldwide system of controls could avert this danger, Szilard warned: a system involving both Great Britain and the Soviet Union. By diluting and denaturing fissionable uranium and plutonium, the world powers could develop nuclear energy peacefully, free from the danger that this fuel might be diverted to weapons making.[2]

Finishing this visionary document on Monday, March 12, Szilard faced a practical problem. "Since I didn't suppose that he would know who I was, I needed a letter of introduction." He turned to his mentor and friend Albert Einstein.[3] His letters on Szilard's behalf had reached FDR in 1939 to initiate the bomb program and in 1940 to help move it along. So, on March 15, Szilard drafted a letter to the president that he hoped Einstein would sign and a memo to the Met Lab's associate director, Walter Bartky, about postwar controls of the A-bomb.

In the spring of 1945, Szilard was not the only person actively trying to influence the president's thinking about such postwar controls, but strict "compartmentalization" within the project kept like-minded individuals from sharing their thoughts. The same day that Szilard drafted the letter for Einstein, Secretary of War Henry L. Stimson met with FDR at the White House for lunch to review the need for policy decisions about international control of atomic energy.[4] A respected cabinet member to four presidents, Stimson posed two control strategies: a secret pact by the United States and Britain to share nuclear technology or an open international exchange of information. This issue had to be settled before the bomb was used, Stimson said, and he left the White House convinced that the president agreed with him.[5] Later in March, Danish physicist Niels Bohr, a consultant at Los Alamos, also sent FDR a follow-up letter to their August 1944 discussions on postwar controls.[6]

To prepare his approach to the president, Szilard arrived by train in Princeton on March 23 and walked the few blocks to the white clapboard house on Mercer Street where Einstein lived. The two friends met in the study, a sunny room that looked onto a back garden and lawn, and once Szilard explained his plight, Einstein agreed to sign the letter of introduction.

112 Mercer Street
Princeton, New Jersey
March 25, 1945

The Honorable Franklin Delano Roosevelt
The President of the United States
The White House
Washington, D.C.

Sir:

I am writing you to introduce Dr. L. Szilard who proposes to submit to you certain considerations and recommendations. Unusual circumstances which I shall describe further below induce me to take this action in spite of the fact that I do not know the substance of the considerations and recommendations which Dr. Szilard proposes to submit to you.

In the summer of 1939 Dr. Szilard put before me his views concerning the potential importance of uranium for national defense. He was greatly disturbed by the potentialities involved and anxious that the United States Government be advised of them as soon as possible. Dr. Szilard, who is one of the discoverers of the neutron emission of uranium on which all present work on uranium is based, described to me a specific system which he devised and which he thought would make it possible to set up a chain reaction in un-separated uranium in the immediate future. Having known him for over twenty years both from his scientific work and personally, I have much confidence in his judgment and it was on the basis of his judgment as well as my own that I took the liberty to approach you in connection with this subject. You responded to my letter dated August 2, 1939 by the appointment of a committee under the chairmanship of Dr. Briggs and thus started the Government's activity in this field.

The terms of secrecy under which Dr. Szilard is working at present do not permit him to give me information about his work; however, I understand that he now is greatly concerned about the lack of adequate contact between scientists who are doing this work and those members of your Cabinet who are responsible for formulating policy. In the circumstances I consider it my duty to give Dr. Szilard this introduction and I wish to express the hope that you will be able to give his presentation of the case your personal attention.

Very truly yours,

(A. Einstein)

After the two men had chatted awhile, Szilard's thoughts rushed on to his next task: drafting a new memorandum to send with Einstein's letter. In it Szilard warned FDR that "our 'demonstration' of atomic bombs will precipitate a race in the production of these devices between the United States and Russia and that if we continue to pursue the present course, our initial advantage may be lost very quickly in such a race." Szilard suggested delaying the A-bomb's use, called for a system of international controls, and asked that a cabinet-level committee meet to hear the scientists' views on atomic-energy issues.[7]

On the other hand, Szilard's second memo to FDR failed to mention the prospect of rockets delivering A-bombs, and most significantly, it omitted "the outbreak of a preventive war" as the "greatest danger" in a postwar world, substituting instead a nuclear arms race itself.

Szilard recalled that in July 1943 an assistant to Eugene Wigner had managed to put complaints about the Hanford reactor project before the president by going through First Lady Eleanor Roosevelt; he now thought this approach worth a try and, with a note to her, enclosed a copy of Einstein's letter.[8]

Szilard could not know that on March 27, two days after he met with Einstein and drafted his memorandum, General Groves wrote a memo of his own, declaring that he expected the A-bomb to end the war in the Pacific, thereby justifying the project. Eleanor Roosevelt replied to Szilard in early April, proposing a meeting at her Manhattan apartment on May 8. Excited by this break, Szilard rushed to Compton's office and told him what he had done. Szilard was nervous as Compton slowly read the memo, expecting to be scolded for again working outside official channels. "I hope that you will get the president to read this," Compton said, and Szilard left the office elated. He was at his desk about five minutes later when he heard a knock on his office door. In walked Norman Hilberry, Compton's assistant, with news he had just heard on the radio. President Roosevelt had died.[9] Szilard's first attempt to stop the bomb had ended.

That afternoon, Vice-President Harry S. Truman was sworn in as the thirty-third president of the United States, and after a brief meeting with the grief-stricken cabinet, Secretary of War Stimson told him in vague but ominous terms about a new explosive of unbelievable power. The next day, Truman learned more details from James F. Byrnes, a friend from the Senate who had served until recently as FDR's director of war mobilization.

"There I was now with my memorandum, and no way to get it anywhere," Szilard recalled. "At this point I knew I was in need of advice."[10] The Met Lab's associate director, Walter Bartky, suggested a talk with university chancellor Robert M. Hutchins. A man of independent views, Hutchins listened intently to Szilard, then asked what this situation might lead to. Szilard's answer was abrupt: A world under one government.

"Yes," Hutchins said, "I believe you are right." Despite their quick rapport, Hutchins was no help to Szilard at this, their first meeting. "I do not know Mr. Truman," he said.

Anxious days and nights passed as Szilard paced his room and the campus, groping for some way to reach Truman. The Met Lab was a large project, he reasoned. *Someone* there must come from Kansas City, Truman's political base. At Bartky's office, Szilard flipped through the lab's personnel files and found Albert Cahn, a young mathematician from Kansas City. Cahn agreed to help, telephoned home, and in two days—through the political machine of Tom Pendergast—Szilard had an appointment at the White House. But not until May 25, more than a month away.[11]

Truman learned more about the bomb by April 25, through a briefing by Groves and Stimson. Groves insisted that Japan had always been the target, and Stimson raised the long-range implications of the bomb, mentioned a possible nuclear arms race, warned about the horrors of a nuclear war, and urged international control. Szilard, the outsider, was struggling to give Truman similar advice, but "official" events were outpacing him. On April 27, Groves's Target Committee, which included four military men and nine scientists (Oppenheimer among them), held its first meeting. To measure the new bomb's effectiveness, the committee sought targets that had suffered little damage from conventional weapons and picked seventeen cities, including Hiroshima and Nagasaki.

Stimson convened an Interim Committee of the bomb program "to study and report on the whole problem of temporary war controls and later publicity, and to survey and make recommendations on post war research, development and controls, as well as legislation necessary to effectuate them." The war in Europe ended on May 8, and the next day, the Interim Committee met for the first time. At the table with Stimson were navy under secretary Ralph A. Bard, Assistant Secretary of State William L. Clayton, bomb-program executives Vannevar Bush and James Conant, Byrnes (Truman's representative), and Karl T. Compton, Arthur Compton's brother from MIT. At Conant's suggestion, the Interim Committee appointed a Scientific Panel, which included Oppenheimer, Ernest O. Lawrence, Enrico Fermi, and Arthur Compton.[12]

For his White House appointment on May 25, Szilard took along Bartky and Cahn. Matthew J. Connelly, Truman's appointments secretary, read Einstein's letter and Szilard's memo carefully. "I see now," Connelly said, "this is a serious matter. At first, I was a little suspicious, because this appointment came through Kansas City. The president thought that your concern would be about this matter, and he has asked me to make an appointment for you with James Byrnes, if you are willing to go down to see him in Spartanburg, South Carolina." Szilard and Bartky were surprised but said they were glad to go wherever the president directed them. Szilard asked if he might also bring Harold Urey, a Manhattan Project chemist and Nobel laureate, and Connelly agreed.

As their overnight train rolled south through the Virginia hills, Szilard, Bartky, and Urey wondered aloud why the president had sent them to Byrnes. He had been a US representative and senator from South Carolina, becoming a budgetary expert during Roosevelt's New Deal legislative reforms. Named to the Supreme Court in 1941, Byrnes had resigned a year later to direct economic stabilization and, later, war mobilization. Now in his first weeks of retirement from government service, he seemed to have the president's ear. But why? Szilard suspected that Truman must be planning to appoint Byrnes to a new government post, perhaps to head atomic work after the war.[13]

During the train ride the three agreed to raise two points: First, they wanted to explore how the A-bomb would affect world affairs after the war and how America's role would change if it used the bomb to end the Pacific war. Second, they worried about the future of atomic energy and the need to plan postwar research.

Set in the mountains of northwestern South Carolina, Spartanburg was a small market and university town. From the tiny station the three walked past flat-front stores and beneath arching trees to Byrnes's red-brick colonial and once seated in the living room, Szilard handed him Einstein's letter to Roosevelt and his own memorandum. Byrnes glanced at the letter and studied the memo, but before he could finish reading, Szilard began a forceful lecture about the dangers of Russia's becoming an atomic power if the United States demonstrated the A-bomb's power and used it against Japan.

"General Groves tells me there is no uranium in Russia," Byrnes interrupted.

Wrong, said Szilard. There are rich ore deposits in Czechoslovakia, which Russia can obtain, and their own vast territory must contain some uranium.

Szilard then argued that the United States should not reveal the A-bomb's existence until the government had decided its postwar policy. Indeed, the United States should not even test the bomb, since its existence is its greatest secret.

"How would you get Congress to appropriate money for atomic energy research if you do not show results for the money which has been spent already?" Byrnes the politician replied.[14] He thought the war would be over in about six months, and his worry was Russia's postwar behavior. Russia had invaded Hungary and Rumania and wouldn't be persuaded to withdraw troops unless the United States demonstrated its military might. The bomb, Byrnes said, would make the Russians "more manageable."

Szilard was "flabbergasted by the assumption that rattling the bomb might make Russia more manageable."

"Well," Byrnes added, "you come from Hungary—you would not want Russia to stay in Hungary indefinitely." This made Szilard furious. Byrnes had assailed Szilard's chief moral guide, his "sense of proportion," and his anger persists in an account written fifteen years later: "I was concerned at this point that . . . we might start an atomic arms race between America and Russia which might end with the destruction of both countries. I was *not* disposed at this point to worry about what would happen to Hungary."[15]

Szilard's anger turned to astonishment as the four men began to discuss the future of the Manhattan Project and Byrnes seemed indifferent. Only later did Szilard learn that Byrnes was not in line to head the project but to become Truman's secretary of state.

Byrnes later wrote that Szilard's "general demeanor and his desire to participate in policy-making made an unfavorable impression on me. . . ." At the same time, Szilard became convinced that Byrnes did not grasp the importance of atomic energy, or much else.[16] The encounter left him frightened that if Byrnes had his way, a US Soviet nuclear arms race would be inevitable.

Byrnes tucked Szilard's letter and memo into his suit-coat pocket as he rose to bid his three visitors farewell, and Szilard later imagined that his memorandum stayed there—all the way to the dry cleaner. But an unsigned memorandum was found by historians among Byrnes's papers. In retrospect, Szilard thought their Spartanburg visit "fittingly naive," while Bartky concluded that it was "purely academic" and accomplished nothing.[17] As far as Szilard knew, his second attempt to stop the bomb had been as futile as his first.

"I was rarely as depressed as when we left Byrnes' house and walked toward the station," Szilard wrote later. "I thought to myself how much better off the world might be had I been born in America and become influential in American politics, and had Byrnes been born in Hungary and studied physics. In all probability there would have been no atomic bomb, and no danger of an arms race between America and Russia." At the station, Szilard, Urey, and Bartky boarded the next train for Washington.

Eager to influence the Interim Committee, Szilard audaciously telephoned General Groves's office in the New War Building on May 30 to arrange a meeting there with Oppenheimer for later that morning.[18] But almost as soon as the two scientists sat down, they disagreed: first over plans to use the bomb, then over its postwar control.[19] It would be a serious mistake to use the bomb against Japanese cities, Szilard said.

"The atomic bomb is shit," Oppenheimer replied, surprising Szilard.

"What do you mean by that?" Szilard asked.

"Well, this is a weapon which has no military significance," said Oppenheimer. "It will make a big bang—a very big bang—but it is not a weapon which is useful in war."

Yet Oppenheimer did think it important to tell the Russians that we had an A-bomb and intended to use it on Japan's cities rather than taking them by surprise. Szilard knew Secretary of War Stimson shared this view, but he complained that while warning Russia was necessary, it was not sufficient.

"Well," said Oppenheimer, "don't you think that if we tell the Russians what we intend to do and then use the bomb in Japan, the Russians will understand it?"

"They'll understand it only too well," Szilard replied, no doubt with Byrnes's intentions in mind.[20]

When the Interim Committee met at the Pentagon on May 31, the four Scientific Panel members—Oppenheimer, Lawrence, Fermi, and Compton—attended, along with Gen. George C. Marshall, there to hear firsthand the scientists' concerns. Oppenheimer echoed some of Szilard's ideas as he proposed an exchange of atomic information, with emphasis on peaceful uses. The goal of atomic energy, he said, should be the enlargement of human welfare, and America's moral position would be greatly strengthened if information were offered before the bomb were used.[21] What's more, Oppenheimer said, Russia had always been very friendly to science, and the United States should not prejudice its attitude toward cooperation. They would not have to await such cooperation, however, for Soviet spies had already informed Stalin of American progress.[22]

The Interim Committee agreed later that day that "the most desirable target [for the A-bomb] would be a vital war plant employing a large number of workers and closely surrounded by workers' houses." While Szilard could not attend this meeting, he was there in spirit, thanks to the anger of General Groves. Complaining about the "handling of undesirable scientists," Groves said that "the program has been plagued since its inception by the presence of certain scientists of doubtful discretion and uncertain loyalty." After the bomb becomes public, he vowed, there should be "a general weeding out of personnel no longer needed."[23]

Back in Chicago, Compton assured his Met Lab colleagues that Interim Committee members were receptive to the scientists' suggestions.[24] And, he said, they would welcome more advice at their next meeting at Los Alamos in two weeks. Compton offered to convey to the Scientific Panel any recommendations prepared by the time he left Chicago on June 14, and by Monday, June 4, the Met Lab scientists had seized this opportunity, forming committees on organization, research programs, education and security, production, and social and political implications.

Szilard was named chairman of the production group but declined in order to focus his energies as a member of the social and political committee that was headed by physicist James Franck.[25] Szilard quickly became its catalytic member and its conscience.[26] Eugene Rabinowitch, who drafted the Franck Committee report, recalled "many hours spent walking up and down the Midway with Leo Szilard arguing about these questions [and] sleepless nights when I asked myself whether perhaps we should break through the walls of secrecy and get to the American people the feeling of what was to be done by their government and whether we approved it. . . ." Rabinowitch credited Szilard with "the whole emphasis on the problem of the use of the bomb which really gave the report its historical significance—the attempt to prevent the use of the bomb on Japan." The report's "fundamental orientation," he said, was "due above all to Leo Szilard and James Franck. . . ."[27]

While Szilard and Rabinowitch were arguing along the green expanse of the Midway, Compton found himself again mediating between his scientists and the army. After Groves's assistant, Col. Kenneth D. Nichols, telephoned to ask about Szilard's visit to Byrnes, Compton admitted, "I have never been able to control Szilard's actions in matters such as this. . . ." Nevertheless, he wrote a thoughtful memorandum that stated eloquently the motives that would compel his colleagues to political action in the coming months.[28]

"The scientists have a very strong feeling of responsibility to society regarding the use of the new powers they have released," Compton wrote on June 4. "They first saw the possibility of making this new power available to human use . . ." and "have perhaps felt more keenly than others the enormous possibilities that would thus be opened for man's welfare or destruction." He continued:

> The scientists will be held responsible, both by the public and by their own consciences, for having faced the world with the existence of the new powers.

The fact that the control has been taken out of their hands makes it necessary for them to plead the need for careful consideration and wise action to
someone with authority to act. There is no other way in which they can meet
their responsibility to society.

Compton noted that two approaches by the scientists, through official
channels, had both failed to reach policymakers: Zay Jeffries's "Prospectus
on Nucleonics" report and a November 1944 memo by Compton to Groves
that summarized his colleagues' concerns. And Compton echoed Szilard's
fear that US negotiators needed to know about the A-bomb as they drafted
the new UN Charter. Groves had told several Met Lab scientists that he
discussed the bomb with "members of the State Department," but based
on comments by Secretary of State Edward R. Stettinius after a briefing
on the new weapons, Compton thought "his appreciation of the problem
was so limited as possibly to serve as a hazard to the country's welfare."[29]
The Franck Committee report—with a recommendation that the
A-bomb be demonstrated to Japan before use against its civilians—was
finished in time for Compton to take it along as he boarded the train for
Los Alamos on June 14, and when he arrived, he passed copies to Fermi,
Lawrence, and Oppenheimer. But in a hasty session on Saturday, June 16,
the four scientists concluded: "We can propose no technical demonstration likely to bring an end to the war; we see no acceptable alternative to
direct military use."[30] With this "expert" advice by the Manhattan Project's
respected Scientific Panel, the Interim Committee confidently rejected the
Franck Committee's recommendations on June 21. Although he would not
know it for several days, Szilard's third attempt to stop the bomb had just
been shattered.

General Groves was angry with Szilard long before he read Compton's
memo about the Byrnes visit and became even more angry about
Szilard's other travels. "I understand that at frequently recurring
intervals Dr. Szilard is absent from his assigned place of work at the
Metallurgical Laboratory in Chicago," Groves wrote to Compton on
June 29, "and further, that he travels extensively between Chicago, New
York and Washington, D.C." He asked Compton for details on whether
Szilard traveled on leave or on duty, what he was paid, whether the
US government financed these trips, what business he transacted, and
whether his travels were cleared in advance. Demanding a "complete
report" on all Szilard's activities for the project in the last six months,

Groves also wanted to know "what positive contribution, if any, to the project he has made since 1 July 1943." Groves warned in closing that "these inquiries must not of course be discussed with Dr. Szilard either directly or indirectly."[31]

Groves even wrote to Lord Cherwell, Winston Churchill's science adviser and Szilard's former mentor at Oxford, seeking details of his meeting with Szilard more than a year earlier. "Frankly, Dr. Szilard has not, in our opinion, evidenced wholehearted cooperation in the maintenance of security," Groves said. "In order to prevent any unjustified action, I am examining all of the facts which can be collected on Dr. Szilard...."[32]

Groves would have been livid had he known what Szilard was actually doing then: bucking the army's chain of command to reach the commander in chief. "I knew by this time that it would not be possible to dissuade the government from using the bomb against the cities of Japan," Szilard wrote later. Having lost in the Interim Committee, "all that remained was for scientists to go unmistakably on record that they were against such an action."[33] Beginning on the first day of July, Szilard drafted and circulated a petition to the president, first around the Met Lab, then by colleagues who traveled to Oak Ridge and Los Alamos. "However small the chance might be that our petition may influence the course of events," Szilard wrote on July 4 to an Oak Ridge colleague, "I personally feel that it would be a matter of importance if a large number of scientists who have worked in this field went clearly and unmistakably on record as to their opposition on moral grounds to the use of these bombs in the present phase of the war."[34]

At Los Alamos, Ed Creutz delivered several copies to Edward Teller. But Teller was then striving ambitiously to promote a "super"—a "fusion" hydrogen bomb, expected to be hundreds of times more powerful than the "fission" bombs then nearing completion—and used Szilard's petition to gain political favor within the tight-knit and competitive Los Alamos community. Teller took the petition to Oppenheimer, knowing that he advocated immediate use of the bomb. Years later, Teller would write that he thought he could not circulate the petition without Oppenheimer's permission. But he also admitted to having "considerable respect for his opinion," and "I sincerely wanted to be on friendly terms with Oppie."

In Oppenheimer's office, Teller handed over the petition, saying it had come from a scientist "near Pa Franck." At once Oppenheimer criticized the Chicago scientists in general and Szilard by name. Scientists had no right to use their prestige to influence political decisions, Oppenheimer complained. In fact, while denying Szilard the right to influence policy, at the Interim Committee, Oppenheimer had advocated immediate bombing and had won the concurrence of his three reluctant colleagues by asserting his own status as lab director.[35]

Relieved that a decision about the petition had been taken from him, Teller nonetheless felt he owed his friend Szilard an answer. But the letter he wrote, which may have been drafted even before meeting Oppenheimer, was convoluted and disingenuous.[36] Teller complained that "no amount of protesting or fiddling with politics will save our souls" or clear their consciences for working on the bomb, although he knew that Szilard's motives had been purely defensive against Germany. Teller suspected that the A-bomb's "actual combat use might even be the best thing" to frighten the public about the weapon's horrors. But Teller's "main point," he said, siding with Oppenheimer, was that "the accident that we worked out this dreadful thing should not give us the responsibility of having a voice in how it is to be used."[37] Teller wrote this knowing that all mail was censored and that Oppenheimer was sure to see it, so, independently, he asked for his permission to send the letter to Szilard.[38]

Using physicist Ralph Lapp as a courier, Szilard sent eight sets of the petition to Los Alamos. "Of course, you will find only a few people on your project who are willing to sign such a petition," he wrote to Oppenheimer's brother Frank, to Teller, and to physicists Philip Morrison and Robert Wilson. "I am sure you will find many boys confused as to what kind of a thing a moral issue is." Admitting the futility of his petition to stop the bombing of Japan, Szilard said that

> from a point of view of the standing of the scientists in the eyes of the general public one or two years from now it is a good thing that a minority of scientists should have gone on record in favor of giving greater weight to moral arguments and should have exercised their right given to them by the Constitution to petition the President. . . ."[39]

When Groves learned about Szilard's secret approach to the Los Alamos scientists and Teller's reply, the general at once feared that Szilard would somehow try to publicize the bomb. Suspecting the worst from Szilard, the lieutenant who reported on Szilard to Groves warned that army resistance to the petition might backfire: "Dr. Teller's attitude is rather interesting and might furnish Szilard with a new approach, i.e., to attempt to get fellow scientist [sic] to stop work."[40]

Szilard, for his part, had no time to think about organizing a scientists' boycott.[41] Independently, Groves's assistant, Colonel Nichols, asked Compton to check on the petition by confirming his colleagues' attitudes, and Compton asked Met Lab director Farrington Daniels to poll the scientists. On July 13, Daniels reported to Compton that a majority (83 percent) favored a military demonstration in Japan or a demonstration in the United States with Japanese representatives present.[42]

Szilard redrafted the petition during the second week in July but knew his time was running out when told that they were no longer permitted to telephone from the Met Lab to Los Alamos. "This could mean only one thing: Los Alamos must get ready to test the bomb, and the Army tried by this ingenious method to keep the news from the Chicago project."[43]

Another ingenious method to keep Szilard from public mischief came on July 15 when a Manhattan Project security officer arrived at the Met Lab with two copies of a secret document strapped to his body. With armed guards at the door, the captain sat Szilard and others in a university classroom to read galley proofs of chapters from a report by Henry DeWolf Smyth, chairman of the physics department at Princeton. To be released at the war's end, Smyth's "Atomic Energy for Military Purposes" described in a straightforward, nonscientific style the basic concepts in nuclear physics and how A-bombs were conceived, designed, and developed. The report credited Szilard and Fermi with devising the world's first reactor; mentioned Szilard and others for restricting their publications; credited Szilard and Wigner with prompting Einstein's letter to Roosevelt, with participating in the Uranium Committee, and with arranging government-sponsored research; and cited Szilard and Fermi for codirecting uranium research at Columbia University before the Federal Office of Scientific Research and Development (OSRD) and then the Manhattan Project were created. Smyth's report also revealed the project's scale and immense costs.

Asked to sign a receipt stating that he had read and approved the report, Szilard balked. The report gave away the important secret of "the general ideas and the knowledge of the methods that actually worked." This posed a problem beyond the war's end, Szilard feared, for once the bomb's secret was released, other powers would have no incentive to heed US calls for international control. Refusing to agree with the report's release, Szilard scratched out "approved" and signed that he had "read" it.[44]

At 5:29.45 mountain war time on July 16, on a desert artillery range designated Trinity, near Alamogordo, New Mexico, the world's first nuclear explosive was tested successfully. A few hours later and unaware that the test had taken place, Szilard redrafted the petition, his last attempt to go on record against the weapon he had worked for years to create and for months to control. Szilard's final petition gained 155 signatures at Chicago and Oak Ridge. It read:

A PETITION TO THE PRESIDENT
OF THE UNITED STATES
July 17, 1945

Discoveries of which the people of the United States are not aware may affect the welfare of this nation in the near future. The liberation of atomic power which has been achieved places atomic bombs in the hands of the Army. It places in your hands, as Commander-in-Chief, the fateful decision whether or not to sanction the use of such bombs in the present phase of the war against Japan.

We, the undersigned scientists, have been working in the field of atomic power. Until recently we have had to fear that the United States might be attacked by atomic bombs during this war and that her only defense might lie in a counterattack by the same means. Today, with the defeat of Germany, this danger is averted and we feel impelled to say what follows:

The war has to be brought speedily to a successful conclusion and attacks by atomic bombs may very well be an effective method of warfare. We feel, however, that such attacks on Japan could not be justified, at least not until the terms which will be imposed after the war on Japan were made public in detail and Japan were given an opportunity to surrender.

If such public announcement gave assurance to the Japanese that they could look forward to a life devoted to peaceful pursuits in their homeland and if Japan still refused to surrender our nation might then, in certain circumstances, find itself forced to resort to the use of atomic bombs. Such a step, however, ought not to be made at any time without seriously considering the moral responsibilities which are involved.

The development of atomic power will provide the nations with new means of destruction. The atomic bombs at our disposal represent only the first step in this direction, and there is almost no limit to the destructive power which will become available in the course of their future development. Thus a nation which sets the precedent of using these newly liberated forces of nature for purposes of destruction may have to bear the responsibility of opening the door to an era of devastation on an unimaginable scale.

If after the war a situation is allowed to develop in the world which permits rival powers to be in uncontrolled possession of these new means of destruction, the cities of the United States as well as the cities of other nations will be in continuous danger of sudden annihilation. All the resources of the United States, moral and material, may have to be mobilized to prevent the advent of such a world situation. Its prevention is at present the solemn responsibility of the United States—singled out by virtue of her lead in the field of atomic power.

The added material strength which this lead gives to the United States brings with it the obligation of restraint and if we were to violate this obligation our moral position would be weakened in the eyes of the world and in our own eyes. It would then be more difficult for us to live up to our responsibility of bringing the unloosened forces of destruction under control.

In view of the foregoing, we, the undersigned, respectfully petition: first, that you exercise your power as Commander-in-Chief to rule that the United States shall not resort to the use of atomic bombs in this war unless the terms which will be imposed upon Japan have been made public in detail and Japan knowing these terms has refused to surrender; second, that in such an event the question whether or not to use atomic bombs be decided by you in the light of the consideration presented in this petition as well as all other moral responsibilities which are involved.[45]

Oppenheimer had banned an earlier draft of Szilard's petition at Los Alamos, but just as Teller had used it to curry favor with Oppenheimer, so Oppenheimer used this version to please Groves.[46]

Franck persuaded several colleagues to sign Szilard's petition on condition that it be sent to Truman by official channels only. Szilard at first mistrusted this approach, fearing that Groves or his allies would intervene, but he finally agreed and handed the petition to Compton on July 19. He asked Compton to place the petition in an envelope addressed to the president and to seal the envelope before it left his office, keeping the signers' names secret. Compton agreed.[47]

After checking with Groves, Compton sent the petition to Nichols, on July 24, noting that "since the matter presented in the petition is of immediate concern, the petitioners desire the transmission occur as promptly as possible." Compton also reported on Daniels's July 12 poll that the "strongly favored procedure" is to "give a military demonstration in Japan. . . ." This is Compton's own preference, he said, "and is, as nearly as I can judge, the procedure that has found most favor in all informed groups where the subject has been discussed."[48]

Except, of course, at the highest levels of government, for the next day—in Potsdam with Truman, Stalin, and Churchill—Secretary of War Stimson approved the final orders to drop the A-bomb "after about 3 August. . . ."[49]

When Nichols received Compton's memo and Szilard's petition on July 25, he dispatched them to Groves by military police courier. Groves held the petition until August 1, when a telex from Tinian Island in the Pacific assured him that the A-bomb was ready for use.[50]

Having tried to reach the president, then having gone on record against use of the A-bomb, Szilard could only await nervously the disaster he was certain would occur. But for mischievous distraction—and to annoy the FBI agents who stalked him—Szilard joined Albert Cahn aboard a Braniff flight to Kansas City, met Cahn's parents, spent a night in the Phillips Hotel, and (according to an agent's report) the next morning returned to Chicago aboard a Santa Fe express train. Back at the Quadrangle Club, Szilard faced a new problem: eviction. His personal habits had annoyed the housekeeping staff during his years there: He seldom drained the bathtub and didn't always flush the toilet; that was "maid's work," he claimed, asserting his bourgeois European mentality in one of the few ways he could. The club's management had complained, and Szilard had refused to change his habits. Now he was asked to leave.[51]

President Truman's Potsdam Declaration on July 26 offered Japan "an opportunity to end this war" by urging the government "to proclaim now the unconditional surrender of all Japanese armed forces. . . ." The alternative for Japan, Truman and his allies said, is "prompt and utter destruction."[52] When Japan rejected the declaration two days later, Szilard grew more anxious than ever that the A-bomb would soon be used. To friends in Chicago he seemed disheartened and gloomy, as he had on the eve of the first chain-reaction experiment.

On August 1, worried that the bomb might soon be used and annoyed about his expulsion from the faculty club, Szilard stuffed his suitcases with papers and rumpled clothes and moved five blocks east, where he rented two back rooms from Dr. and Mrs. Paul A. Weiss in a three-story brick apartment house on South Blackstone Avenue. The view there was much less picturesque than Szilard enjoyed at the club; now he looked from large windows onto a back alley and garages instead of seeing through small leaded-glass panes the neo-Gothic spires of the Quadrangle and the divinity schools nearby.[53]

The day of Szilard's move, Groves at last forwarded Szilard's petition to Stimson's office. But still in Potsdam with Truman, Stimson would not see the petition until his return later in August. Three US B-29s lifted from the Tinian airstrip the morning of August 6, one bearing the uranium bomb nicknamed "Little Boy," one armed for aerial protection, and one along to photograph and film the mission. (Four cities had been spared conventional bombings so an A-bomb's effects could be seen clearly: Hiroshima, Kokura, Niigata, and Nagasaki.) Over Hiroshima the single bomb fell to explode above the waking city at 8:15 A.M. A shock of light, a blast, heat, whirlwind firestorms. By sudden incineration and lingering death some 200,000 people died.[54] In a flash, the villains of the Pacific war became its greatest victims.

News of the single blast reverberated around the world. At a small cabin on Saranac Lake, in the Adirondack Mountains of New York, Albert Einstein's secretary heard a radio news item about the war in the Pacific that told of a new kind of bomb dropped on Japan. 'Then I knew what it was,' she recalled later, "because I knew about the Szilard thing in a vague way. . . . As Professor Einstein came down to tea, I told him, and he said, 'Oh, Weh' ['Oh, woe'] and that's that." To a newspaper reporter that day Einstein could only say, "Ach! The world is not ready for it."[55]

At Farm Hall, a seventeenth-century country house in England, ten eminent German scientists who the Allies thought had worked to build an A-bomb for the Führer were under house arrest: among them Otto Hahn, Werner Heisenberg, and Carl von Weizsäcker. Hahn, who was first to hear, was shattered by the news, said he felt responsible for the deaths of hundreds of thousands of people, and gulped several drinks to brace himself. Hahn's colleagues were incredulous, then suspicious.

"If the Americans have an uranium bomb," Hahn said to the others, "then you're all second-raters. Poor old Heisenberg." At first, Heisenberg thought the news a hoax. Their conversation, secretly taped by British intelligence, turned to moralizing.

After another radio announcement convinced the Germans that the bomb was real, they contemplated the magnitude of America's effort and their own failure. Heisenberg recalled a time, in early 1942, when "we had absolutely definite proof that it could be done." But Weizsäcker's excuse— though not shared by everyone—closed that speculation: "I believe the reason we didn't do it was because all the physicists didn't want to do it, on principles. If we had all wanted Germany to win the war we could have succeeded."

"I don't believe that," the gloomy Hahn replied, "but I am thankful we didn't succeed."[56]

In Chicago, Szilard "knew that the bomb would be dropped, that we had lost the fight," he recalled later. "And when it was actually dropped, my overall feeling was a feeling of relief. . . . Suddenly the secrecy was dropped and it was possible to tell people what this was about and what we were facing in this century."[57]

But first he had to tell someone special, and he sat down to write Trude Weiss. When she heard about the Hiroshima bombing on the radio, she later recalled, "Then I knew. Like a shock I knew. I was at my desk in New York, and I had to go home. Everything was suddenly out of proportion."[58] Szilard knew she was not prepared for the news that the A-bomb was her Leo's doing and needed to explain. Nervously he shifted from the matter-of-fact report that he had moved to a casual query about a party to rare self-confession. In English and German and English again. From past to present to future.

THE QUADRANGLE CLUB

CHICAGO

Monday, Aug. 6, 1945

Dear Ch,

I report my new address: 5816 Blackstone Ave., c/o Weiss. Telephone: MIDWAY 0545.

How are things with you? How was the Calderon party?

I suppose you have seen to-day's newspapers. Using atomic bombs against Japan is one of the greatest blunders of history.

—Both from a practical point of view on a ten-years scale and from the point of view of our moral position.—I went out of my way (and very much so) in order to prevent it but, as today's papers show, without success. It is very difficult to see what wise course of action is possible from here on. Maybe it is best to say nothing; this is what I suggest you do.

I hope you do not feel like a little mouse anymore. [This sentence is in German.] Maybe I'll come East for a visit soon, now that the cat is out of the bag.

Yours L.[59]

"I always thought it was his way of apologizing," Trude said after Leo's death. "It was one of the most important letters he ever wrote to me."[60]

Relief at being free to discuss the bomb was tempered by Szilard's horror at early reports from Hiroshima, a horror that roused him to new thinking, writing, telephoning, buttonholing and berating his colleagues—and anyone else who would listen. For months he rushed around the campus, around the city, around the country, in a frenzy of activity.

His first stop on the day he heard about Hiroshima was Robert Hutchins's office because he needed a sympathetic person to talk to about the tragic news. He asked if the Met Lab staff might wear black mourning bands on their arms, but Hutchins thought the gesture "a little Hungarian" and suggested Szilard find some less dramatic way for the scientists to demonstrate their grief. A few days later, Szilard enlisted Hutchins to head a Chicago scientists' group to meet with President Truman, but this White House visit was never arranged.[61]

American B-29s took off from Tinian Island again on the morning of August 9, this time to drop a single plutonium bomb, nicknamed "Fat Man," on Kokura. But bad weather there sent the bomber to Nagasaki. The bomb killed only 70,000 because hills deflected the blast and radiation. "Dropping the bomb on Hiroshima was a tragic mistake," Szilard quoted his colleague Samuel K. Allison as saying. "Dropping the bomb on Nagasaki was an atrocity."[62]

On Saturday, August 11, Szilard asked a University of Chicago chaplain to hold a special prayer service for the dead of Hiroshima and Nagasaki and offered to transmit the prayer to the Japanese survivors.[63] Szilard's petition to the president three weeks earlier had been based on moral principles, and he was upset that political and religious leaders seemed to be silent about these concerns. Szilard enlisted philosophy professor Charles Hartshorne for a taxi ride uptown to call on Chicago's Roman Catholic cardinal. When ushered into the churchman's presence, Szilard began at once to spout his views about the import and dangers of atomic energy. The cardinal listened patiently as Szilard demanded that the church confront the morality of the A-bomb, then replied briefly.

"God had locked up the energy in question so securely that only after thousands of years has it been unlocked," he intoned. "Surely there was a reason for this long delay."

"What *was* the reason?" Szilard demanded.

"The church will consider the matter and in due time will make a statement about it," the cardinal announced. End of discussion. Szilard and Hartshorne left disappointed and, riding back to campus, agreed they now wished the atom's energy had been locked up still more securely. Szilard called on other religious leaders but found them just as unconcerned.[64]

When a prayer service was held at the university's Rockefeller Chapel, Szilard attended with Met Lab physicist Alexander Langsdorf and his wife, Martyl—the artist who would design the "minutes to midnight" clock for the *Bulletin of the Atomic Scientists*. As the three walked out, Szilard noticed that he had forgotten his hat in the pew, and Langsdorf offered to fetch it.

"No, no," Szilard said. "General Groves will know where I am and it will get back to me."[65]

Groves may not have been watching Szilard quite so closely, but he did worry about any publicity he might create. Once Groves learned that Manhattan Project officers had agreed to allow Szilard to publicize the scientists' petition to Truman, the general ordered it classified "Secret," blocking Szilard from releasing it to *Science* magazine. An army officer explicitly forbade Szilard from publishing the petition anywhere and threatened to fire him from the Met Lab if he did.[66]

Quoting a letter sent to justify classification, Szilard complained to Hutchins how "the Manhattan District's definition of 'Secret' includes 'information that might be injurious to the prestige of any government activity,' which is, of course, very different from the definition adopted by Congress in the Espionage Act."[67]

He also complained directly to a Met Lab officer about what a "big mistake" it would be to publish the Smyth report. When it appeared, Szilard said, he would be known to the world as a "war criminal." Szilard insisted that for his own protection the army should furnish him with "a personal bodyguard and an automobile." This was ignored by the army but reported matter-of-factly by the FBI.[68]

In a memorandum for discussions with other Met Lab scientists, Szilard predicted that if A-bombs spread, US cities "will be threatened by sudden annihilation within ten years. The outbreak of a preventive war will then hang over the world as a constant threat." He also fretted about other threats "for the large-scale extermination of human beings" from "biological methods," conceding that his latest speculation—city dispersal schemes—would be useless against this warfare. Only "moral inhibition" might work. And, he grieved, "Hiroshima shows that moral inhibitions can no longer be counted upon. . . ."[69]

The Soviet Union declared war on Japan on August 8 and this, with the two A-bomb attacks, convinced Japan to surrender on August 14, a US national holiday called V-J day. Autos with their horns blaring paraded through cities and towns across the country, and the movie newsreels showed jitterbug dancing and a whirl of embraces. But Szilard appeared depressed that night when he visited Hartshorne's wood-frame house on East Fifty-seventh Street, near the university, and his spirits were little improved the next day, after the Smyth report's release. The press described Szilard's role in the early development of the bomb and pictured him along with Einstein, Fermi, and other well-known scientists. Good company, Szilard probably thought, but a sorry way to gain celebrity. Back at Hartshorne's for the family's traditional Friday afternoon tea, Szilard was still in a sour mood.

"Well," a young philosopher said to Szilard, referring to the Smyth report, "I hear you have become a great man."

"I've always been a great man," Szilard replied.[70]

Later that month, Szilard boarded an overnight train for Buffalo (tailed, as usual, by the FBI) and there met with his cousin, city planner Laszlo Segoe. Szilard wanted to know what dispersing urban centers—to defend against atomic attack—would cost: in land area, transportation, capital investment. Thanks to Segoe's advice, Szilard was able to spout elaborate statistics about city dispersal, although this line of pragmatic brainstorming seemed more fantastic with each telling.[71]

Japan formally signed documents of surrender on September 2, and to most Americans peace had come about because of the A-bomb. Without this weapon, they thought, the war would have dragged on for months. But recent scholarship suggests that Japan was about to surrender before the A-bombs were used and would have before the US invasion planned for November 1945 and March 1946.[72]

The United States was victorious and seemed invincible with its new weapon. Now the atom that had ended the war could also enrich the peace. Just after V-J day a Scripps-Howard news service feature on the "Era of Atomic Energy" predicted that "no baseball game will be called off on account of rain" because atom-generated heat would dispel bad weather. Artificial "suns" would assure clear skies at resorts and heat "indoor farms," using uranium cores. "There is no reason why an internal combustion engine cannot be developed" using "tiny explosions of uranium^{235}."[73]

Such fantastic optimism was just what Groves and his engineers had hoped for during the weeks that followed Hiroshima and Nagasaki. But for the atomic scientists, who knew better and feared worse, the new technology they had created was still principally a weapon of mass destruction. Somehow, Szilard believed, something had to be done to stop the army from using it again. By the fall of 1945, the only solution he could imagine was to disarm the military and establish civilian control of the atom.

CHAPTER 19

... And Two to Stop the Army

 1945

At a press luncheon on September 1, 1945, at the Shoreland Hotel on Lake Michigan, University of Chicago chancellor Robert M. Hutchins announced that physicist Samuel K. Allison—the Met Lab's associate director—would head a new Institute of Nuclear Studies. Allison gave the reporters something more stirring to write about. "Scientist Drops A-Bomb: Blasts Army Shackles," the *Chicago Tribune* reported about his remarks. "We are determined to return to free research, as before the war," he said, warning that if military regulations hampered the free exchange of scientific information, researchers in America "would leave the field of atomic energy and devote themselves to studying the color of butterfly wings."[1]

Incensed by this, Col. Kenneth Nichols, assistant to Manhattan Project director Leslie R. Groves, flew from Oak Ridge to Chicago the next day for a swiftly arranged luncheon of his own at the Shoreland, where he told Allison, Enrico Fermi, Harold Urey, and chemist Thorfin Hogness to halt all banter about butterflies. The War Department was preparing a bill for Congress to restructure postwar control of atomic research, Nichols said, and cracks like Allison's might hurt the chance of passage. Besides, he said, there would be ample opportunity for the scientists to speak when Congress held hearings on the bill.

Leo Szilard had not attended either luncheon[2] but on his own kept quiet about the bomb in public because Manhattan Project officials had asked the scientists to remain silent while important international negotiations were under way. Szilard worried about more than restrictions on research. Once his three attempts to prevent the A-bomb's use on Japan had failed, his next task was to stop its use entirely by placing this weapon under civilian control at home and international control abroad.

Behind the scenes, however, Szilard was typically plotting political tactics and always eager to speak to politicians personally. His September 7 memo on "An Attempt to Define the Platform for Our Conversations with Members of the US Senate and the House of Representatives" became a seminal document for the atomic scientists' lobbying efforts as well as a basis for their public statements and for atomic-energy policy in general. An Atomic Power Commission, perhaps with a permanent committee of its own, should be created, which Congress should supervise. But recalling his own troubles with the chain of command, Szilard insisted that "the scientists ought to be free" to communicate to members of the congressional committee.[3]

Military secrecy was sure to thwart scientific work, as it had during the war, Szilard warned, creating the "intolerable situation" in which scientists might be intimidated by the officials who controlled their research funds. Congressmen should also be warned, Szilard wrote, that the A-bomb creates the danger of a preventive war if an arms race develops. Only international inspections could keep such mistrust from occurring.[4]

Szilard urged Hutchins to convene a small, private conference on atomic-energy policy at the university and helped to attract influential participants. At this time, Sen. Brien McMahon of Connecticut proposed that the UN Security Council be licensed to conduct nuclear research, and physicist James Franck and sixty-four other scientists and academics from the university sent President Truman a petition urging him to share atomic secrets with other nations as a way to avoid a nuclear arms race.[5] Also on campus, the Atomic Scientists of Chicago was formally established to work for international control of the atom.[6]

Within the Truman administration, Secretary of War Stimson urged the president to contact the Russians for an agreement to limit the use of atomic weapons, arguing that "if we fail to approach them now and merely continue to negotiate with them, having this weapon rather ostentatiously on our hip," they would undertake "an all-out effort to solve the problem," making any agreement dubious.[7]

Unable to muzzle the atomic scientists who were trying to debate this issue in public, the War Department's Bureau of Public Relations sent the press a "Confidential—Not for Publication Note to Editors" in the president's name. "In the interest of the highest national security," the memo warned, editors and broadcasters were requested to consult the War Department before printing any atomic information, except for official releases. Groves also tried to pressure Hutchins directly, warning about "grave security hazards" at the planned atomic-energy conference and enclosing the War Department's note to editors. Groves also reminded Hutchins of the War Department's secret contract with the university. As far as he was concerned, Groves said, "the limits of discussion are the Smyth Report and other authorized public releases."[8]

Insulted and angered by this tactic, Hutchins replied to Groves that the conference was in response to the Smyth report's call for public discussion of atomic energy. He had told the Manhattan Project office on campus "some days ago" that the conference would be private, with no public or press at the discussions, so the War Department's note "has, of course, no application to a conference at which no editors or broadcasters will be present."[9]

The Atomic Energy Control Conference on September 19 and 20 attracted economists and political scientists from several universities. Beardsley Ruml, treasurer of the R. H. Macy department store and chairman of the Federal Reserve Bank of New York, offered the scientists public relations advice. Commerce Secretary Henry A. Wallace and Census Bureau director Philip Hauser attended, unofficially, as did David Lilienthal, chairman of the Tennessee Valley Authority (and by 1947 the first chairman of the new US Atomic Energy Commission). Also present was Chester Barnard, president of the New Jersey Bell Telephone Co., who would soon draft the State Department's proposal for international control of atomic energy.[10]

Addressing the conference on opening day, Szilard predicted that Russia would have A-bombs in a short time, suggested control arrangements with Moscow, and urged relocation of city populations as the only defense. And he discussed plans for an eventual world government to secure atomic-energy control. William Benton, who had just become Secretary of State Byrnes's assistant for atomic-energy affairs, asked Szilard for a memo on his views, and the two corresponded throughout the fall, with Benton using some of Szilard's ideas on Byrnes.[11]

As usual, Szilard worked more freely out of the formal sessions, buttonholing conferees, expounding his views, and sometimes even listening to theirs. After the Smyth report, Szilard believed, the only remaining secrets concerned the next stage of development, the hydrogen bomb, or the "super." He warned that "we cannot rely on more and better bombs for more than a few years. . . . An armaments race in atomic weapons may well become the greatest single cause of a future war."[12]

Szilard had the last word at the conference—and the first on record about the touchy topic of "verification"—when he said a necessary first step would be to "guarantee immunity to scientists and engineers everywhere in the world in case they should report violations of the [arms-control] arrangements agreed upon. . . ." With "a Bill of Rights for scientists and engineers . . ." they would become "the guardians of the international arrangements relating to the control of atomic energy." Typically, Szilard carried his idea to its too logical conclusion, advocating that all countries revoke their espionage laws for scientific and engineering secrets.

The secrecy issue divided the Atomic Scientists of Chicago, and during an emotional debate about a declaration, Szilard finally persuaded the group to change the phrase that "secrecy is not advisable" and substitute that "secrecy is not possible."[13]

The Chicago scientists also argued about the War Department's bill to keep the atom's control with the army, which Colonel Nichols had alluded to at his Shoreland luncheon. In the House, the bill would go to the sympathetic Committee on Military Affairs, but by the end of September, the Senate adopted a resolution by Senator McMahon to create a new committee to write all atomic-energy legislation.[14]

At about that time, Szilard was in New York to appear on "Round Table," the University of Chicago's weekly radio program. Discussing "The Atom and World Politics" with Norman Cousins, editor of the *Saturday Review of Literature,* William Fox, a research associate at the Institute of International Studies at Yale, and William Hocking, professor emeritus of philosophy at Harvard, Szilard said it was unlikely that there would be world government within three years but added that "this is the only solution for permanent peace." One alternative, a "durable peace," would require an agreement among nations not to stockpile atomic weapons, an accord verified by a widespread international detection system run by the scientists themselves.

If both inspections and "some international authority" fail, Szilard said, then, logically, the United States should consider a ten-year plan to relocate 30–60 million people. The cost of $15 billion a year was the price that must be paid if nuclear arms are not controlled, he said. Half a century later, Szilard's relocation plans for the US urban population seems naive, but it does capture his mental and emotional spirit at the time. Frightened by Hiroshima, he longed to give the world a grand scheme that would counteract the A-bomb's growing perils. He wanted to alert policymakers and the American public to the fundamental change that had occurred, and in his friendly way of shocking listeners this plan posed the jarring alternatives to the arms race he saw coming. He sometimes admitted not knowing for sure when he was kidding. But this time he seemed serious, logical, and to most who heard him, a slightly mad scientist.[15]

Around the time of his "Round Table" broadcast, Szilard stopped in Princeton to visit Albert Einstein—their first encounter since March. "Our conversation turned back six years to the visit on Long Island when we discussed the letter he might write to the president," Szilard recalled.

"You see now," Einstein said to him, "that the ancient Chinese were right. It is not possible to foresee the results of what you do. The only wise thing to do is to take no action—to take absolutely no action."[16]

But Szilard disagreed and left the white frame house on Mercer Street determined to continue his arms-control crusade. In Washington he met William Benton at the State Department and attended a dinner at his house with some of the department's top desk officers. Physicist Edward Condon was also invited and matched Szilard's puckish humor with good-natured repartee of his own. The two men enjoyed each other's company and together that fall enlivened any gathering they attended. They were a match in another important way: As a vice-president of the American Physical Society, Condon was a scientific insider; Szilard, still a creative and quirky outsider.[17]

In October, President Truman proposed to Congress that total control over "the use and development of atomic energy" be vested in a new Atomic Energy Commission (AEC), empowered to operate all existing facilities, acquire minerals, and conduct research for both peaceful and military uses. The AEC would also license other researchers "under appropriate safeguards" and establish security regulations for "the handling of all information, material and equipment under its jurisdiction." International control, Truman proposed, should begin with talks by the United States, Britain, and Canada, "and then with other nations," seeking an "agreement on the conditions under which cooperation might replace rivalry in the field of atomic energy."[18]

The same day Truman's message was sent to Capitol Hill, the army's allies introduced identical versions of their bill: in the House by Andrew Jackson May, a Kentucky Democrat who was chairman of the House Military Affairs Committee; in the Senate by Edwin C. Johnson, a Colorado

Democrat and ranking member of the Senate Military Affairs Committee. The May-Johnson bill proposed a nine-member, part-time AEC, empowered to select a full-time administrator. The House bill was referred to May's committee, but in the Senate a two-day fight erupted over jurisdiction between the Military Affairs and Foreign Relations committees.[19] Still in Washington after the Benton dinner, Szilard "picked up more or less accidentally" a copy of the May-Johnson bill and took it with him to Chicago.[20]

Back on campus, Szilard walked the bill across the university's tree-shaded Quadrangle to the law school and handed it to Edward Levi, an acquaintance who had been advising the Chicago scientists. In particular, both men were alarmed that national and international policies for atomic energy would be set and supervised by a part-time board.[21] At an Atomic Scientists of Chicago meeting on campus, Condon and Szilard stressed the bill's security restrictions. "If this bill passes, we have no choice but to get out of this work," Szilard said.[22] When Arthur Holly Compton, the Met Lab's director, returned to Chicago the next morning, he told his fellow scientists that the War Department had asked them to keep silent not because of delicate international talks but to pass an atomic-energy bill in Congress without "unnecessary discussions."

"I got mad at this point," Szilard remembered. He rose to his feet and declared that no bill would be passed without discussion if he could possibly help it. It was their "duty," he said, to fight any attempt to "smuggle" a bill through Congress.

Szilard was even more angry the next day when he read in the newspapers that while Compton was talking with them, May's committee had held a five-hour, closed-door hearing on the bill and was about to report it to the full House for passage. A chance telephone call from Hutchins gave Szilard the opening he needed: a *Chicago Sun* request for an interview.[23] Szilard's complaint about the army made the front page, and the *Washington Post* reported the scientists' call for a joint congressional committee and public debate on atomic policy, alerting politicians in the capital to a growing struggle.[24] The fight was on.

Szilard enlisted two allies when he persuaded Hutchins to free Dean Robert Redfield from some university duties to concentrate on the "political implications of the atomic bomb," and asked Redfield to "persuade" sociologist Edward Shils to drop some classes and work against the May-Johnson bill. Then he left Chicago for Washington aboard the Liberty Limited, tailed—as usual—by the FBI.

The Chicago scientists enlisted their Manhattan Project colleagues around the country, hoping to bring pressure on Congress from all directions. At Oak Ridge, support was strong; 90 percent of those working on the bomb agreed that no hearing on atomic energy should be held until a bipartisan committee was created to consider the matter. But at Los Alamos, where the A-bomb was designed, built, and tested and where work continued on the "super," support was minimal. Seeking to influence his distant colleagues from Chicago, Herbert Anderson, Fermi's assistant at the Met Lab and at Los Alamos, criticized the bill in a letter to William Higinbotham, a founder of the Atomic Scientists of Los Alamos. The bill's security provisions "are frightening," he warned. "They place every scientist in jeopardy of a jail sentence or a large fine." But more disturbing, said Anderson, was his conclusion that their "leaders" on the Interim Committee—Oppenheimer, Lawrence, Compton, and Fermi—"were duped" when they urged the scientists to keep silent about the army's bill. "Let us beware of any breach of our rights as men and citizens. The war is won, let us be free again!"[25]

The Chicago scientists were eager to educate the American public as well as Congress, and Katharine Way, a chemist at the Met Lab who became the group's publications director, urged a few colleagues to write a book. In October she journeyed to New York with an outline, where she sold the idea to editors at McGraw-Hill. Compton would write the introduction, Niels Bohr the foreword. Physicist Philip Morrison, who was among the first Americans to survey Hiroshima, would compare that city's destruction to a similar attack on New York. Wigner would write a history of the atomic age. Oppenheimer would describe the new weapon. Physicist Louis Ridenour would emphasize that there is no defense against the A-bomb, while physicists Frederick Seitz and Hans Bethe would describe how other countries might build a bomb. Urey would analyze the politics of arms control. Szilard would give a plan for international inspection. Columnist Walter Lippmann would propose an international-control scheme. And Einstein would offer a rationale for nuclear disarmament.[26]

Arriving in Washington on Friday, October 12, Szilard and Condon checked into the elegant Mayflower Hotel, on Connecticut Avenue near the White House. The Mayflower was too expensive for their budgets, but Anderson's brother, who worked for Schenley distillers, had interested his company's owner, Louis Rosensteil, in the cause of the atomic scientists. To help, Rosensteil had offered Anderson and his colleagues use of his company suite. Once in the spacious rooms, Szilard and Condon began to place telephone calls. For several minutes they called congressional offices, dictated telegrams to their allies in Chicago and Oak Ridge, and arranged meetings of scientists for the following week. Assuming their phones might be bugged, they ended each call, sarcastically, with "And God bless General Groves!"[27]

From his calls Szilard learned that May's committee planned to meet the following Tuesday, vote out a bill on Wednesday, and bring it to the House floor for passage by Thursday.

As Szilard and Condon chatted away, a key clicked in the door, and a well-dressed man stepped into the suite. Surprised to find the two men there, he bustled into the room and announced: "I'm Mr. Strauss, president of Schenley! And who are you?"

"I'm the bastard son of Louie Rosensteil," answered Condon.[28] His real explanation made little difference after that quip, and the two physicists picked up their papers and bags and moved out, this time up Connecticut Avenue to the Wardman Park, a stately brick-and-white-columned structure near Rock Creek Park. This was Szilard's favorite hotel, and his room there quickly became a center for planning the scientists' assault on Capitol Hill. Colleagues recall it as strewn with clothes and papers; the hotel operators remember the bursts of calls, day and night, plugged through the switchboard to all parts of the country.

From the Wardman Park, Szilard taxied about the city impulsively, his erratic movements an endless challenge for the FBI agents on his trail. Caught by a downpour on one outing, Szilard snapped up an umbrella he was toting, paused, turned, and with a graceful thrust of the arm offered shelter to the agent at his heels. The agent declined and was soaked during the rest of the walk.[29] His followers must have been confused by Szilard's frequent calls at a town house on Vermont Avenue near McPherson Square: the office of the Independent Citizens Committee for the Arts, Sciences, and Professions. In fact, this committee had offered two small rooms there to physicist John Simpson as temporary quarters for the Atomic Scientists of Chicago.[30]

Condon and Szilard persuaded Representatives Chet Holifield, a member of the Military Affairs Committee, and his close and respected fellow Californian George Outland to oppose the May-Johnson bill. Holifield arranged for the two physicists to talk with May, but he was not impressed and on Saturday told newspaper reporters that his hearings would remain closed.[31] By now, however, telegrams protesting the bill were reaching congressional and Senate offices from scientists' groups in Chicago and Oak Ridge. The scientists' views attracted congressional interest quickly, and those who ventured to Capitol Hill found it easy to meet their representatives and senators. "Mention to a senator's secretary at the door that you're a 'nuclear physicist' and you come from 'Los Alamos,'" recalled Szilard's assistant and collaborator Bernard T. Feld, "and you were ushered right in to see the senator. We were celebrities, and the lawmakers wanted to learn about the bomb—right from the horse's mouth."[32]

Szilard had his own weird ideas about lobbying and was "flabbergasted" when he met chemist Charles Coryell on Capitol Hill one day and learned he had appointments with Senators William Knowland of California and Kenneth McKellar of Tennessee. "How did you get this?" Szilard asked. "You don't meet congressmen this way. You go to a friend who has a friend who knows someone to take you there." Szilard could not believe that it was possible to arrange a meeting just by asking.[33]

When riding in a taxi near the Capitol one day, Szilard noticed Seitz walking along the sidewalk, stopped the cab, and asked him to join with a few other scientists on their way to call on a senator.

"What's the subject of your meeting?" Seitz asked, and when told, protested that he knew little about this topic.

"No matter," Szilard insisted. "I understand that when you call on a senator, it is a good idea to have a tall person in the group."[34] Seitz was tall, and that was enough. When Szilard met with James Newman, the Truman administration's adviser on atomic-energy legislation, he joked that he had brought Condon along because he had an honest, farm-boy face—a reassurance to those uneasy with Szilard's pudgy features, Hungarian accent, and blunt speaking style.[35] In fact, Szilard and Condon made an amiable and successful team—so successful that they had to schedule their time: "We would keep cabinet members waiting one day, senators for two days, and congressmen for three days before we'd give them an appointment," Szilard recalled.

Busy and bumbling he may have been, but Szilard managed to collect and use whatever details he gleaned from his Washington contacts. Commerce secretary Henry Wallace, an internationalist who liked Szilard and Condon, introduced them around Washington. Interior secretary Harold Ickes complained that he had not read the May-Johnson bill because the War Department had loaned him a copy for only a few hours. From Rear Adm. Lewis Strauss, an associate interested in atomic energy since the late 1930s, Szilard learned that the Navy Department had no views of its own about the bill, nor did the president's top assistants. Clearly, this was not an administration bill but a War Department bill. President Truman had no designated adviser on atomic-energy legislation until, in mid-October, he named Newman head of the science section in the Office of War Mobilization and Reconversion. Szilard, Condon, and Newman all agreed that the May-Johnson bill did little more than extend General Groves's wartime powers and practices and worked together to defeat it.[36]

Pressured by Holifield and other colleagues, May reopened hearings on his bill for one day. Szilard was invited, with less than two days' time to prepare. He urged that Anderson speak against the bill for Oak Ridge and Chicago scientists. The War Department asked Oppenheimer and Compton to endorse the bill.

All the day before, Szilard scribbled notes for his testimony, stopping only in the early evening to attend a dinner for a few senators with Fermi, Condon, Urey, and Oppenheimer, who had resigned that day as director of Los Alamos.[37] This was the first of several meals, arranged by Watson Davis of the Science Service news agency, for "educating" senators, and that night's guests included Brien McMahon, the vigorous and astute Connecticut Democrat who was reported to have said the A-bomb was "the greatest event since the birth of Jesus Christ."[38] From the administration came two Szilard allies, James Newman and Henry A. Wallace. Republican Charles W. Tobey of New Hampshire set the tone for his colleagues when he glanced around the table and said, "It looks as if we have a nonpartisan issue."

But as the meal progressed, it became clear that the scientists themselves were at odds. Oppenheimer and Szilard both supported international control of atomic energy but disagreed about how to bring it about. Now representing different factions in American science and government—Oppenheimer the agile political insider, Szilard the restless agitator from outside—the two men must have sensed that their differences might destroy the common goal. They agreed to talk after dinner, and with Urey as their mediator and Newman as adviser, the two met first at Szilard's room at the Wardman Park, then downtown at Oppenheimer's at the Statler Hotel. But neither man would yield, and Szilard walked out, taking Urey and Newman back to the Wardman Park to help draft his testimony.

Just after midnight, Szilard received a four-page telegram from Chicago, with support from the Oak Ridge and Met Lab scientists. Now certain of his colleagues' backing, Szilard worked through the night, dictating his statement to a hotel stenographer who had to operate a switchboard as she took shorthand.[39] When he arrived on Capitol Hill on Thursday morning, October 18, Szilard was tired and uneasy about the state of his prepared text but eager to declare his views.

Below the tall ceiling of the committee's hearing room were walls trimmed with marble columns. Flags, plaques, and other military gear decorated the chamber. The spectators who packed the room gazed past the tables for the press and the witnesses to a raised, semicircular dais and to a wall behind where three flags were splayed.[40] The committee members appeared pompous and remote as they peered from the dais. A showdown was at hand.

Promptly at 10:00 A.M. Chairman May rapped a wooden gavel. The hearing, he said in a voice smoothed with a Kentucky drawl and trimmed with sarcasm, was to allow "a group of interested people, known as scientists." to state their views. May denied that his "committee was trying to rush things" and vowed to give "patient consideration" to the witnesses. He stared across the press table at Szilard, who seemed stiff as he sat in a leather chair, his dark suit buttoned tightly around his belly. "We have as our first witness," May said, "a Dr. Sighland."[41]

"My name is Leo Szilard," he began, pointedly correcting the chairman. Cameras flashed as Szilard recited his background, saying twice he was a naturalized American citizen. Szilard thanked the committee for this moment. Then, matter-of-factly, he explained that the government's atomic facilities can make two substances, uranium[235] and plutonium, which can be used for generating electricity or "for manufacturing bombs." Szilard declared "the hope of all physicists" that plutonium would be used for power rather than bombs. Now out of physics and into politics, he was ready to reorganize the US government along the way.

Szilard strode to a portable blackboard to make his next point: how to administer the postwar nuclear enterprise. He sketched three large circles representing new government-owned corporations—one for uranium and plutonium production; another for scientific research and development; a third for the manufacture, research, and development of bombs. Typically, Szilard spelled out details for selecting the three-member boards for each corporation. He also proposed a policy-making commission to coordinate national and foreign policy, with the secretaries of state, commerce, interior, and war as members.

The policy choice ahead, Szilard declared, was between short-term nuclear power development on a small scale, with plants operating in perhaps three to five years, and long-term development on a large industrial scale in a decade. Szilard also proposed clever ways to divide research among five or six laboratories, dotted around the country near large universities and institutes of technology.[42]

Several committee members seemed eager to embarrass Szilard in the question-and-answer session that followed his testimony. Rep. R. Ewing Thomason of Texas asked about Szilard's nationality, and that of Wigner, Teller, and Fermi, but changed the subject when he learned that all were naturalized US citizens. Thomason also asked about Szilard's dispute with the army over patent rights. Szilard held fast, responding with a calm precision that soon showed Thomason he had little to gain. When Rep. Leslie C. Arends, an Illinois Republican, asked Szilard about defense against the A-bomb, his answer was direct.

"There is no military defense," Szilard said. "I think our vulnerability will be much less if we relocate 30 million to 60 million of our population. It may be necessary to do that unless the international picture improves."

Can atomic energy be developed for peaceful uses? Rep. Charles H. Elston of Ohio wondered: "Can it be used to run locomotives and steamships and automobiles?"

"When you produce atomic power, you also produce radiations," Szilard explained. "You have to protect the driver and the passenger against those radiations. That might mean that an automobile would have to carry 50 tons of shielding material, and that would be rather on the heavy side." Ships might be atom powered, but locomotives are doubtful, said Szilard. For now, he said, "stationary power plants" are the first likely application, although very large airplanes "might be a possibility."

After more than an hour on the stand Szilard began showing signs of his sleepless night as he became testy. Admonished by Chairman May for failing to answer a question directly, Szilard asked permission "to give a correct answer rather than a short one."[43]

Holifield asked questions to put on the public record Szilard's opposition to the Smyth report's release, his complaints about wartime compartmentalization, and his charge that penalties in the May-Johnson bill for violating secrecy would delay research but would not deter spies.[44] But while Holifield and Szilard enjoyed their colloquy, a uniformed aide to Groves huddled with Representative Thomason behind the dais. Suddenly, Thomason broke in to ask again about Szilard's nationality, to question his refusal to sign over eight patents, and to challenge his delay in taking an oath about these patents. The hesitation about the oath, Szilard explained, was over questions of "sole" or "joint" invention, an especially delicate point of law. At this point, Chairman May complained that "Dr. Sighland" had "consumed" an hour and forty minutes and ended his testimony.[45]

Herbert Anderson spoke for the younger scientists in the Manhattan Project (he was thirty-one), repeating Szilard's criticisms about security driving away talented researchers. And he read the telegram from the Chicago and Oak Ridge atomic scientists, which criticized eight points in the May-Johnson bill. The morning session complete, the atomic scientists walked across Capitol Hill to Union Station, to eat at the Savarin Restaurant and to plan for the afternoon.[46]

After lunch, the May committee heard Compton and Oppenheimer speak for the bill. Compton, who by this time was sympathetic to the Chicago scientists' views, criticized the civil-liberties problems that the bill raised and suggested amendments but would not ask for the bill's defeat. Oppenheimer followed, asking that the bill be passed in order to permit continued scientific work.

"Oppenheimer's testimony was a masterpiece," Szilard recalled later. "He talked in such a manner that the congressmen present thought he was for the bill but the physicists present all thought that he was against the bill." Asked if he thought it was a good bill, Oppenheimer said that Bush and Conant thought it was a good bill and he had very high regard for them. "To the congressmen this might mean that Oppenheimer thinks this is a good bill," Szilard recalled, "but no physicist believes that Oppenheimer will form an opinion on the basis of his good opinion of somebody else's opinion."[47]

Ignoring protests from Urey, a scheduled witness who was not called, the Military Affairs Committee ended its hearings after Oppenheimer's testimony. Still, Szilard and Condon knew by now that May's committee would not abandon the War Department's bill and could not amend it enough to correct the basic military bias. But they also knew from their many visits around town that almost no one *but* the War Department liked the bill. "We did not have very much more to do than tell everybody what everybody else thought of the bill" in order to kill it, Szilard recalled. Szilard complained later that for many on May's committee the only decision seemed to be "whether to make the bombs and blast hell out of Russia before Russia blasts hell out of us."[48]

That weekend, Szilard, Condon, and Urey left Union Station by train, followed by FBI agents. In New York, Urey attended a luncheon for Nobel laureates at the Waldorf-Astoria Hotel, where he called for international control of atomic energy and attacked the May-Johnson bill. Szilard and Condon probably met with Einstein, as he soon joined associates in denouncing the May-Johnson bill in a telegram to President Truman.[49]

Besides lobbying for Senator McMahon's resolution to create a special atomic-energy committee, Szilard followed votes in the May committee, which split seventeen to ten when it reported its bill to the full House.[50] Szilard attended another dinner for scientists and senators and joined in the organizing meeting for the Federation of Atomic Scientists, later to become the Federation of American Scientists, a research and lobbying group active in public policy debates ever since.

After McMahon became chairman of the Senate's nine-member Special Committee on Atomic Energy, he hired Newman as an assistant and prepared to hold hearings on legislation. Condon ended the frantic lobbying pace he had maintained with Szilard when President Truman nominated him to succeed Lyman Briggs as director of the National Bureau of Standards, but before accepting that post he served briefly as a scientific adviser to McMahon's committee.[51] With these two allies working in the Senate, Szilard could turn his attention back to the House, where eleven of the May committee's members now opposed their chairman's bill and a floor fight was certain. Szilard's first attempt to stop the army seemed to be succeeding.

Through California Democrat Jerry Voorhis, Szilard organized a meeting for House members to hear the atomic scientists' views. More than seventy representatives filed into the spacious Caucus Room of the Old House Office Building (now the Cannon Building) on Thursday afternoon, November 8, and latecomers, congressional staff, and the press packed the long, high-ceilinged chamber. With practice, Szilard's delivery was becoming more personal, more precise, and more persuasive. He plumped for international cooperation to avoid an arm race and said the only defense against it was massive city relocation. Szilard's remarks to the caucus appeared in the *Congressional Record* and were reported by the *New York Times*.[52]

But for Szilard and his scientific colleagues the most memorable event that fall was an Armistice Day tea sponsored by former Pennsylvania governor Gifford Pinchot and his wife, Cornelia, on the grounds of their fifty-four-room mansion on Scott Circle, a few blocks from the White

House. An ardent naturalist, conservationist, and advocate of public power who had founded the Progressive party with Theodore Roosevelt, Pinchot actively supported federal control of atomic energy. Mrs. Pinchot considered that the scientists' youth and political inexperience made them "ideally inefficient" for the lobbying task at hand.[53]

Szilard and Lyle Borst from Oak Ridge were featured speakers at the tea, and other colleagues who circulated around the elegant grounds included Condon, Higinbotham (by now a May-Johnson opponent), and Daniel Koshland, an Oak Ridge biochemist who had signed Szilard's petition to Truman in July. The scientists chatted with members of Congress, among them Representatives Jerry Voorhis and Franck Havenner of California, and Clare Boothe Luce, a Connecticut Republican and Military Affairs Committee member who opposed the May-Johnson bill. As a maid moved around the lawn with a tea tray, she approached a group that included Higinbotham and Szilard.

"Would you like cream or sugar in your tea?" she asked.

"Cream and sugar," Szilard replied, "and *no* tea."[54]

Conversation that afternoon focused on congressional politics; on various international-control schemes; and on the recent proposal by navy captain Harold Stassen, a US delegate to the UN conference in San Francisco, to place twenty-five American-made A-bombs under the control of a UN air force. Talk was so lively that it continued throughout the mansion until 9:00 P.M.[55]

That same Armistice Day afternoon, talk about control of atomic energy was taking place on a higher level as President Truman and Secretary Byrnes, with British Prime Minister Clement Attlee, Canadian Ambassador Lester Pearson, and others, sailed down the Potomac past Mount Vernon on the elegantly old-fashioned yacht *Sequoia*. After rounds of negotiations on shore, the following week, Truman, Attlee, and Canadian Prime Minister W. L. Mackenzie King issued a Three-Nation Declaration on Atomic Energy. Acknowledging that "there can be no adequate military defence" from atomic weapons, and that "no single nation can in fact have a monopoly," the three leaders pledged to exchange basic scientific information for peaceful uses with any nation that would reciprocate. They also asked that a UN commission draft proposals for ways to assure peaceful uses of atomic energy and to eliminate nuclear weapons from all military stockpiles.[56]

By November, Szilard had decided to stay in Washington and wrote the Met Lab in Chicago for a leave of absence without pay.[57] Although impatient (and inept) with the details of legislation, Szilard did enjoy brainstorming with Newman, Thomas I. Emerson, general counsel to the Office of War Mobilization and Reconversion, and Emerson's assistant Byron Miller. And Miller engaged Szilard in more practical problems. Szilard and his friend Eugene Wigner were at Miller's house for dinner one night when the furnace shut off. Miller had no idea what to do, but the two Hungarian physicists doffed their jackets and poked around in the basement to examine the situation. There they fiddled with some dials and valves and stoked the coal pile in the furnace, relit the fire, tended it until sure the flame was going again, and then proudly returned upstairs. "Very high grade combustion engineering help," Miller later boasted.[58]

Back in New York City the third week of November, Szilard appeared at a WQXR radio forum on "What Would You Do with the Atomic Bomb?" with *New York Times* science reporter William L. Laurence, who had written the Manhattan Project's official press releases about the bomb. On the air Szilard criticized as futile "international bargaining" to forestall a nuclear arms race if it involved making and storing A-bombs in the United States.[59] Celebrity continued for Szilard during that week as the *New York Post* featured him in its "Closeup" column, describing his 1933 escape from Germany to England and the 1939 approach to Roosevelt through Einstein's letter. The article described Szilard as a "somewhat rotund man of five feet six, weighing 170 pounds," and the large photograph showed him looking younger than his forty-seven years, with hair short but with lips—obviously touched up by an artist—that seemed about to purse into a kiss. "I am satisfied I could reduce if I wanted to eat less." he was quoted as saying, "but I have never put it to a test."

In the *Post* interview Szilard first revealed two points he would make in later statements and writings. He had no hobbies, he said, "except possibly baiting brass hats." And he mentioned that *The Tragedy of Man* by Madách had "influenced my whole life." The moral he recalled from it was that no matter how gloomy the human condition, we must maintain a "narrow margin of hope" and take action.[60]

Senator McMahon's hearings on atomic energy opened on Tuesday, November 27, with Alexander Sachs as the first witness. He recounted, in tedious detail, his 1939 approach to President Roosevelt with Einstein's letter that led to the Manhattan Project.[61] The next day, the McMahon committee heard Urey's familiar call for international inspection and control and Groves's rejoinder that international inspections could endanger sovereignty, the sanctity of the home, and private commercial enterprise.[62]

That evening, in New York, Szilard watched as McMahon publicly challenged the army's supremacy in atomic-energy control at an "Atomic Age Dinner" at the Waldorf sponsored by the Americans United for World Organization, a group working for legal restrictions on the atom through the UN Charter. Other speakers included toastmaster Raymond Swing, whose popular national radio programs that fall had publicized the efforts of Szilard and Einstein; Henry DeWolf Smyth, author of the official report on the Manhattan Project; physicist Ernest O. Lawrence, Gen. Carl Spaatz of the Army Air Force, who had directed the bombing of Japan; and Col. Paul W. Tibbetts, Sr., the pilot "who dropped the bomb on Hiroshima."[63]

In New York a few days later, Szilard attended a gathering that would be one of his best public events of the year: the Nation Associates' three-day forum on "The Challenge of the Atomic Bomb" at the Astor Hotel in Times Square. This well-publicized event celebrated the liberal magazine's eightieth anniversary and the six hundred participants included many people Szilard already knew or would work with, including Smyth; Rep. Helen Gahagan Douglas; physicists Louis Ridenour and Victor Weisskopf; and Harold Laski, an acquaintance from his work on the Academic Assistance Council, a politics professor at the London School of Economics, and chairman of Britain's Labour party.

At the concluding dinner, Szilard made a stirring speech that he had revised and rehearsed for days. After brief remarks by Mrs. Franklin D. Roosevelt, Szilard began by recalling the moment in 1939 when he and Walter Zinn had first detected extra neutrons from uranium fission: "That night there was very little doubt in my mind the world was headed for grief."[64]

From this dramatic opening Szilard recounted the bomb's development, his July 1945 petition to Truman, the army's effort to muzzle the scientists and pass the May-Johnson bill, and the scientists' lobbying against the army. He proposed educating public officials about the A-bomb by staging "a demonstration" for them.

To avoid a nuclear arms race, Szilard called for a treaty prohibiting atomic bombs—backed up by inspection under the United Nations Organization (UNO). Then, with a Szilardian idea that must have baffled many listeners, he proposed amending the Espionage Act to allow scientists and engineers to vacation for four weeks a year as guests of the UNO. "Those vacations abroad would give an opportunity to all those who wish to report [on their own country's] secret [arms-control] violations to secure immunity by staying abroad rather than returning home after delivering their report." This scheme Szilard would propose in years to come, with added inducements and $1 million rewards. Although farfetched, the scientists' vacation proposal struck at the question of verification that still confounds arms-control negotiators.[65]

Early in December Metro-Goldwyn-Mayer and Paramount both announced plans to make feature-length films about the A-bomb. And Time-Life planned a film in its *March of Time* series in which the atomic scientists—Bush, Conant, Einstein, Szilard, and others—would play themselves in a dramatization of the bomb's development.

In Washington representatives from the many Manhattan Project laboratories came together at George Washington University to officially create the Federation of Atomic Scientists (FAS). Under Willie Higinbotham's direction, the younger men were forging an organization that would supplement, and eventually replace, the early lobbying efforts by Condon and Szilard. For his part, Szilard met with many of these scientists and always seemed to be plotting some new scheme on his own. "He was a great objector to other people's statements," Higinbotham recalled, "but he seldom consulted us about his own. It became a byword around the [FAS] office that to reduce the Russians to helpless confusion it would only be necessary to parachute Szilard into Moscow."[66]

Szilard fashioned a rambling six-part statement for the McMahon committee that surveyed most of the day's issues: fissionable material production, weapons development and manufacture, "preparedness" for nuclear war, the time left until another country develops the bomb, international control schemes, and remarks on secrecy.[67] He was still editing this on Monday morning, December 10, as he arrived at the Senate Office Building just before 10:00 A.M.—marking paragraphs he might omit when reading the text, drawing slashes under sentences for phrasing—eager to begin his second attempt to stop the army.

In the large caucus room, he met Newman and Condon, who encouraged his viewpoints. Unlike his hostile reception by the May committee, Szilard received a gracious welcome from McMahon, who introduced him as "one of the most eminent of the pioneers in the science of uranium fission."[68] Szilard began by describing the production of uranium and plutonium, including his wildly optimistic plans for the breeder reactor, as he wove together some possible economic, political, military, and diplomatic consequences of the atom's development. He even speculated that atomic energy might help to stabilize the US economy: "When a depression threatens, electrification of our railroads, based on atomic power plants, may be pushed with the support of the federal government, whereas in boom periods an expansion of atomic-energy power projects might be discouraged. . . ."

But in answer to a question, Szilard said that if an international security system could be set up by giving up peaceful uses of atomic energy, he would "gladly renounce" it. Besides, he said, making bigger and bigger bombs holds little interest "from a scientific point of view and will be pursued only if it is necessary for political reasons."

"The bombs already made are big enough, aren't they?" asked Sen. Eugene D. Millikin, a Colorado Republican.

"They are big enough for my taste," Szilard replied.[69]

Speaking about what "preparedness" in a nuclear arms race might entail, Szilard posed this ironic analysis: "We are afraid of Russia, not because she has atomic bombs; we are afraid of Russia because we have atomic bombs." But in such a race the United States "would lose ground steadily," Szilard argued, citing city-dispersal studies by economists Jacob Marschak and Lawrence Klein at Chicago to demonstrate that those who talk to the American people about "preparedness" should mention a peacetime expenditure of more than $20 billion a year. With these calculations Szilard was trying to dramatize that the alternatives to effective international arms control involve unacceptable financial, political, and social costs. Szilard's rare ability to analyze a problem rationally, then pursue in interlocking detail the results and options, led his inquisitive mind to conclude that being "prepared" for a nuclear arms race demands profound (even absurd) upheavals. Clearly—at least to him—the best way to be prepared is to prevent the arms race itself.

The *New York Times* the next day called his presentation "astonishing and challenging." Responding to his answers, the senators asked him how many suitcases would be needed to smuggle a disassembled A-bomb into the country; how much time the United States had before Russia could build a bomb; how an international-control treaty would punish a transgressor; and why the Germans failed to build a bomb. Szilard had lucid and accurate answers for each but in the end returned to his complaint that the May-Johnson bill would perpetuate compartmentalization to maintain secrecy—and as a result stifle science. It was self-censorship by scientists, not the result of compartmentalization, he said, that kept details about plutonium production a wartime secret. And self-censorship would continue to work as long as scientists saw a need for it.

Through the morning hearing Szilard made critical remarks about the army, General Groves's limited scientific understanding, and "the incompetence of military intelligence. . . ."

Edwin C. Johnson of Colorado, the Senate sponsor of the May-Johnson bill, showed no hostility toward Szilard, although he did complain to another hearing witness, "You scientists have got the world in a mess and now you want the politicians to straighten it out."[70] Johnson seemed delighted with Szilard's city-dispersal ideas, wondering how to relocate Connecticut's rich industries to Colorado.

"Let us get away from the horrible hypothesis that you propose," said McMahon of Connecticut.

"It is horrible, I suppose," admitted Johnson, "horribly good."

Indeed, as the long hearing drew to a close that morning, the senators seemed to be mulling over Szilard's perverse ideas about "preparedness" and its international alternatives.

The CHAIRMAN. Doctor, I assume that if you had a voice in the election you would view the dispersal of our cities with the consequent cost and general transferring from one part of the United States to another of thirty to forty million people to be far less preferable than some sensible international agreement, for the control of this thing, on which we can rely?

Dr. SZILARD. It is certainly less sensible; yes, provided we can get the arrangement. I am not advocating dispersal, but it is a necessary step within the framework of a policy which is merely based on "preparedness." No preparedness makes any sense without it. If we have to anticipate an attack of this sort, we have to disperse.

Senator TYDINGS. If we get into a war of that type, just having a big navy, army, or air force, while they are essential, as compared to previous wars, without the ability to carry on and supply them by the dispersal of plants all over the country, preparedness is only an illusion?

Dr. SZILARD. Exactly.

But Szilard and his fellow scientists still had a long struggle ahead before civilian control of the atom could be assured, and their crusade for international control would persist for years. Szilard himself, despite his impressive blend of science and social engineering, left some listeners bewildered. After the session, Sen. Arthur Vandenberg of Michigan, the ranking Republican on the Foreign Relations and the Atomic Energy committees, asked Condon: "Who was that guy Lizard you had in here yesterday?"[71]

Throughout 1945, Szilard had come from quiet discontents about the future of the postwar world, through three failed initiatives to stop the bombing of Japan, to finally play a leading role in shifting control of the atom from military to civilian hands. This was achieved first by opposing the army's continued management of the atom, then by supporting creation of a civilian Atomic Energy Commission (AEC). McMahon introduced his bill to create the AEC on December 20, about the time Szilard's imagination rushed on to another idea.[72] He approached his friend Chancellor Hutchins at the University of Chicago and through him proposed to Assistant Secretary Benton in the State Department an arms-control technique that would come to dominate Szilard's life: Let the scientists themselves solve the problem they had created. At the time, Byrnes of the United States, Ernest Bevin of Great Britain, and Vyacheslav Molotov of the Soviet Union were in Moscow, where, two days after Christmas, they would issue a communiqué recommending that the UN General Assembly create a commission on atomic energy.[73] During that conference, Benton in Washington cabled Conant, a member of the US delegation in Moscow:

22 December 1945

Hutchins telephoned today querying the advisability of inviting five or ten Russian physicists to the United States, bringing them in under private auspices to visit Harvard, University of Chicago, and other institutions. He felt such joint discussions of atomic physics between scientists of the two countries might promote a basis for international cooperation and control. Such discussions would of course be private and unofficial without governmental comments. I told him I would cable you this suggestion for possible exploration by you in Moscow.

"Politics" has been defined as "the art of the possible." Szilard had told his distinguished audience at the celebration dinner for *The Nation* magazine in December. "Science might be defined as the art of the impossible. The crisis which is upon us may not find its ultimate solution until the statesmen catch up with the scientists and politics, too, becomes the art of the impossible.

"This, I believe, might be achieved when statesmen will be more afraid of the atomic bomb than they are afraid of using their imagination, because imagination is the tool which has to be used if the impossible is to be accomplished."[74]

PART THREE

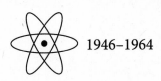 1946–1964

CHAPTER 20

A Last Fight
with the General

 1946–1964

In *The Ascent of Man,* a popular television series and book in the 1970s, the mathematician and social scientist Jacob Bronowski praised Leo Szilard's "integrity" in renouncing nuclear physics after Hiroshima and adopting a new career in biology.[1] Historian Max Lerner also praised the decision as "rejecting death and embracing life."[2] Noble sentiments by two of Szilard's friends but not completely true. Szilard's career shift in the spring of 1946 was in part a virtue made of necessity.

Although biology had appealed to him for more than a decade, Szilard had no desire to quit his research in nuclear physics. In the 1940s and 1950s he continued to file patents for reactor designs and even sketched plans for a nuclear airplane. He relished the publicity when he and the late Enrico Fermi were awarded the joint US patent on the first nuclear reactor in 1955. And he enjoyed devising government-funded export schemes to spread the "peaceful" atom to developing countries. In fact, Szilard's shift to biology was occasioned by his nemesis, Gen. Leslie R. Groves, the military leader of the Manhattan Project.

The "baiting of brass hats," Szilard's self-professed hobby, had set him and Groves at odds ever since the fall of 1942, when the Army Corps of Engineers took command of the nuclear-research project that Szilard and Fermi had begun at Columbia more than three years before.[3] From their first meeting, the straitlaced and steady Groves had found Szilard impetuous and rude: He openly questioned the general's orders, joked with his colleagues and superiors, and criticized and debated decisions. This conduct seemed downright subversive to an army engineer like Groves, and within a month he had ordered Szilard transferred back to New York from the Met Lab in Chicago. When Szilard's colleagues defended him, Groves withdrew the order but secretly tried to have Szilard jailed and deported as an enemy alien, a move ultimately rejected by the secretary of war.

From that first conflict, Groves's "intuition" had led him to suspect Szilard. "I just didn't trust him," Groves later explained. "I knew he was a detriment to the project."[4] Before long, Szilard had become an obsession for Groves.

Groves's animus toward Szilard grew from several causes. Groves was an all-American boy, active in sports and devoutly Christian, a patriotic militarist and engineer, while Szilard was an Eastern European immigrant, active in science and irreligious, a vagabond with no nationalistic sympathies who in his youth had spurned both the military and engineering. Groves's authoritarian rectitude and anti-intellectual swagger clashed with Szilard's austere reason and playful erudition. Unlike his colleagues, Szilard also seemed pushy and arrogant to Groves, outspoken on any subject, from physics to politics. In addition, Groves's anti-Semitism was focused and personified in Szilard, in all making him a perfect villain.

Groves's personal dislike and distrust of Szilard was amplified when he challenged Groves at two critical points in his army career: The scientists' petition had questioned wartime use of the A-bomb, Groves's pivotal professional achievement; and the scientists' double-barreled attack on the May-Johnson bill and support of the McMahon bill had turned the congressional hearings from a moment of glory for Groves to a public confrontation that put him on the defensive. The general was irate, but his aide, Kenneth Nichols, urged him "to refrain from making any comment on Dr. Szilard's testimony unless requested to do so," warning Groves that "when dealing with one who juggles the truth and warps facts as this individual does, it is not advisable to argue with him publicly."[5] Groves kept an angry silence about Szilard, but privately the general fumed. Besides their direct disagreements, Groves detested Szilard's call for international control of atomic secrets and his idea that a world government was necessary to enforce peaceful uses of nuclear power.

By the spring of 1946, Groves's view that the atom should remain a military secret was commonly accepted, even proclaimed on March 5 when Winston Churchill, Britain's wartime prime minister, journeyed with Truman to Westminster College in Fulton, Missouri, to deliver his stirring Iron Curtain speech against Soviet aggression. Three days after that speech heralded the cold war, the army announced that it would ban Communists from all "sensitive" positions. Also on March 8, J. Robert Oppenheimer, the former director of the Manhattan Project's Los Alamos laboratory and the most famous of the atomic scientists, was quoted in a newspaper as opposing the May-Johnson bill—a shift from his position the previous fall.[6] And in Washington that day, *Time* magazine reporter Frances Henderson interviewed both Groves and Szilard, separately, for an article about the Manhattan Project scientists and their fight for the McMahon bill.

Sen. Brien McMahon had led his colleagues in forming a special committee to deal with atomic energy. He was widely respected as a thoughtful and creative lawmaker, but in the interview with *Time*, Groves called McMahon a "fool" and sniped at Szilard, criticizing his influence on other scientists, questioning his loyalty, and disputing his motives.[7] Groves was gruff and candid with Henderson throughout the long and rambling interview, then caught himself and ordered her not to quote his remarks about "any of the scientists." But the general's anger was irrepressible, and he kept talking, obsessively, about one scientist in particular.

"Do you know his background?" Groves asked, and not waiting for her reply, said, "Well, Szilard was born in Hungary" and "served in the German army—or rather the Austrian army." Groves continued:

> Anyway, after the [First World] War he studied—didn't teach, or so to speak ever earn his way. . . . In this country he was at Columbia, here and there, never teaching; never did anything really you might say but learn. Everywhere he went, from what I hear, he was hard to work with. The kind of man that any employer would have fired as a troublemaker—in the days before the Wagner Act [a 1935 law granting minimal employee rights].

At that thought, the general grinned. Henderson said that it was Szilard who had initiated research that led to the Manhattan Project, and Groves had to say: "Yes, as a matter of fact, I might even go so far as to say that if it hadn't been for Szilard, it would never have reached the president." Then, realizing he had complimented Szilard, Groves added: "Only a man with his brass would have pushed through to the president. Take Wigner or Fermi—they're not Jewish—they're quiet, shy, modest, just interested in learning."[8]

Then "why was Szilard kept on the project?" Henderson wondered.

"Well, he was already on it," Groves snorted, "transferred from Columbia out to Chicago when we came in on it. Frankly, we would have let him go except we didn't trust him loose."

Still unable to stop thinking about Szilard, Groves rambled on and on.

> He made a lot of security breaks. Nothing important, but he violated security a dozen times or more. . . . Oh, we've had quite a time with him—he keeps the young men all stirred up. And I wouldn't have a bit of trouble with Wigner (and the Princeton people) or Chicago [scientists] if it weren't for Szilard. If there were to be any villain of this piece, I'd say it was Szilard. . . .

The scientists were "not practical," Groves complained. Szilard, for example, wanted the scientists to manage construction of the huge plutonium-production reactors at Hanford. "Said they would have gotten it done sooner," Groves intoned in a mocking way. He also complained about the "trouble we had" with those reactors, trouble that "a bunch of impractical scientists" would only have compounded. But he failed to note that the scientists quickly recognized the reactor problems once they occurred.[9]

By contrast, Groves praised Oppenheimer as "a real genius" who "knows about everything—he can talk to you about anything you bring up. Well, not exactly, I guess there are a few things he doesn't know about. He doesn't know anything about sports."[10]

The general also admitted to approving Oppenheimer's security clearance, even though Groves knew he was "reddish-pink in his youth," just to get the job done. "I just had to say—after all, it was wartime—they're cleared." Security would have been much stricter, Groves said, if only he had had his way. "If this were a country like Germany, I should say there were a dozen [scientists] we should have shot right off. And another dozen we could have shot for suspicion or carelessness."

Answering the charges Szilard had made before Congress and in the press that compartmentalizing information had slowed the scientists' work on the bomb, Groves insisted, "It was the only way to get the thing done. Otherwise [the scientists] would have spent all their time talking. . . . This way we made them work. . . ." With compartmentalization, Groves said smugly, "only one person knows everything," and "I was more interested in accomplishment than learning." But he admitted that by continuing to compartmentalize atomic-energy information well after the war "we're having trouble keeping the scientists."

His own "idea of an atomic commission," Groves said, would have four military members, four businessmen, and engineers; only two scientists and one lawyer-statesman—all serving part-time. McMahon "thinks serving on the commission would be like being on the Supreme Court . . ." Groves said, adding, "Incidentally, you might want to write that he [McMahon] has never worn a uniform himself."

Mention of McMahon sparked a familiar anger. "Do you know who wrote the McMahon bill?" asked Groves. "I couldn't swear to it, but I'll bet it was Szilard. It's badly drafted by somebody who knows nothing about it but yet has legal knowledge." Groves's angry monologue continued.

> Of course, most of his ideas are bad, but he has so many . . . you know no firm wants him for a consultant. Why, he's the kind of guy that advises a company one way and after they're half way through that says, "no, let's try this way." Of course, *he* isn't paying the bills.
>
> And I'm not prejudiced. I don't like certain Jews, and I don't like certain well-known characteristics of theirs, but I'm not prejudiced. . . .

After Henderson thanked Groves for his time, she left to join Szilard, finding "a short, chubby man who likes to meet his luncheon companions outside of restaurants so he can enjoy an extra moment of fresh air." Szilard, too, was angry and candid that day, but unlike Groves, he was eager to be quoted.

"We're working according to the methods of 1940," Szilard complained, slighting Groves's management. Uranium and plutonium production methods are now "obsolete" and far too costly. In the United States, Szilard said, the first-rate scientists are leaving the project, stressing the point just made to her by Groves.[11]

Szilard used this interview to complain about *Time*'s earlier description of the atomic scientists as "befuddled." It is unique for thousands of intellectuals from diverse cultural backgrounds and political views to find themselves in practically unanimous agreement about the atom's civilian control, he said. And over lunch he restated the points that the scientists were then making to senators, congressmen, and anyone else who would listen: that compartmentalization had cost the United States at least a year's time building the bomb; that the Smyth report about how the bomb was made should have been withheld "until a policy was set" for international control of the atom; that secrecy works because the scientists want it to work, not because the military enforces it. But scientists who continued government work could be dismissed without a stated cause, Szilard warned, leaving them defenseless against the military bureaucracy. "All a scientist has is his reputation," Szilard said, sounding vulnerable and no doubt reflecting on his own encounters with Groves's hostility. "Destroy that and you destroy him."

In the interview that morning, Groves had told Henderson he knew what some of the scientists were saying about him, "but I won't throw mud." Publicly, he kept his silence by preventing her from using his quotes. Privately, however, Groves was quick to seek revenge. Within two weeks, Groves traveled to Chicago, where he made a show of presenting medals of merit to "key figures" in the Manhattan Project. Harold Urey, Enrico Fermi, Samuel Allison, metallurgist Cyril Smith, and the project's health director, Robert S. Stone, were present. Szilard was not.[12] Later that spring, Groves struck again. On May 7, the nationally syndicated columnist Leonard Lyons wrote in the *New York Post* about Groves's anger with Szilard—the source, apparently, was the general himself. In part, the column read:

Maj. Gen. Groves, head of the Atomic Bomb project, in his private discussions of the Army sponsored [May-Johnson] bill and of the opposition to it by the scientists, makes no secret of his dislike for Dr. Szilard, who first interested Roosevelt in the bomb. If the Army bill passes, Szilard—because he was born in Hungary and served in the German Army in the first World War—wouldn't be allowed to work on the project. . . .[13]

A few days before this column appeared, Farrington Daniels, director of the Met Lab in Chicago, had praised Szilard's contribution to the Manhattan Project in a letter to the army, urging that he receive a commendation. And on May 27, responding to Daniels's letter, the army's Decorations Board recommended "that a Certificate of Appreciation for civilian war service be awarded to Dr. Leo Szilard. . . ."[14]

But before the paperwork for a certificate was complete, Groves learned about the award and sent the district engineer at Oak Ridge a secret memo with "certain facts with which the [Decorations] Board was not familiar and which were unknown to Dr. Daniels that prevent the approval of the Board's recommendation" of a certificate. Groves mentioned difficulty over Szilard's patents "and also his failure to devote his full energies to the work that he was assigned to do." Groves also wrote that "it was quite evident" Szilard "showed a lack of support, even approaching disloyalty, to his superiors, particularly with respect to Dr. Compton, director of the Metallurgical Laboratory," although Compton and Szilard were friendly and sympathetic colleagues during and after the war. No doubt aware that these falsehoods might be traced to him, Groves cautioned that "this disapproval will be kept secret, but Dr. Daniels will be informed verbally by General Nichols."[15]

Still not content to snub Szilard, Groves plotted behind the scenes to bar him from further work on the government's nuclear research. To lobby in Washington for civilian control of atomic energy, Szilard had taken an unpaid leave from the Met Lab in November 1945, and when he returned to Chicago in late April 1946, he heard from friends on campus that Groves was trying to block his rehiring, whether or not the army's May-Johnson bill passed. The Lyons column in May had only confirmed Groves's animosity, and two days after it appeared, Szilard confronted his fate by asking Daniels in a letter "whether it was suggested to you by some representative of the Manhattan District that my contract shall not be renewed. . . ." Szilard asked Daniels directly if he would be receiving a reemployment offer. "It is my continued desire," Szilard wrote, "to work in the field of atomic energy and to do this, if possible, for the United States Government rather than for some private corporation." Szilard also praised Walter Zinn's decision to direct the new Argonne National Laboratory near Chicago, the entity that would replace the Met Lab for federal nuclear research programs.[16]

From this and other correspondence, from his later enthusiasm for the peaceful uses of nuclear power, and from his occasional work as a consultant, it is clear that Szilard did not renounce nuclear physics after Hiroshima. Instead, he was barred from it by Groves, as Daniels confirmed in a letter on May 10. "I feel it is only fair to tell you now that the Area Engineer of the Manhattan District . . . has requested me not to offer you a position in the new [Argonne] laboratory," Daniels wrote, adding praise for Szilard's "valuable contributions" to the bomb project's success and paraphrasing the recommendation he had written a week before to request the army's Certificate of Appreciation:

> Your foresight and initiative were largely responsible for obtaining support for the original atomic energy program and your work on piles [reactors] and your vision for new types of piles have been important in the development of the research program of the Laboratory. You have made important contributions to

the patent structure of the Manhattan District and you have been vigorous in pointing out the political and social implications of the atomic bomb.[17]

Angry and sad but not surprised, Szilard accepted the rejection stoically, for he could rarely admit pained feelings or disappointments to others. In a letter to his friend Trude Weiss, Szilard reported it was now "certain" that his contract would not be extended. "I shall hang around for a couple of months more here to determine if other possibilities will materialize," he wrote in a style more formal than the punchy prose he usually dashed off to her. But in three weeks nothing did materialize, and on May 31, a month before his Met Lab contract expired, Szilard sent his resignation letter to Daniels, effective the next day.[18]

For Szilard, his last fight with Groves was over. But Groves's obsession with Szilard persisted until after his death. When testifying at a security-clearance hearing on Oppenheimer in 1954, Groves dwelled on Szilard's behavior. Groves recalled his efforts in 1942 "to intern a particular foreign scientist" because "intuition" convinced him this "alien" was a spy. On the witness stand Groves recounted in detail a visit to Secretary of War Stimson's office. "I didn't accuse [Szilard] of disloyalty or treason, but simply that he was a disrupting force and the best way out of it was to intern him," Groves recalled.

> I was told that this man [on Stimson's staff] didn't want to take it up with the Secretary. I insisted on it. He came back and said, "General, the Secretary said we can't do that. General Groves ought to know that. I told the Secretary, of course General Groves knew that would be your answer. He just still wanted to make a try." I think that is essential to realize.

When the transcript of Oppenheimer's hearing was published in 1954, Szilard bought a copy and marked this passage, no doubt amused by the efforts the general had made against him.[19]

Those efforts continued even after Szilard's death in 1964. Responding to an obituary of Szilard by Eugene Rabinowitch, Groves dictated a memo-to-file that begrudged Szilard's "brassiness" as the reason for early government funding of atomic research, but complained that his approaches to the White House and to Byrnes "were violations of his oath of secrecy." Szilard was "such a disruptive influence" at the Met Lab, Groves recalled, that after their strained meeting over patents in December 1943, Groves thought of a way to rid the project of Szilard. Groves asked Conant if he would offer Szilard a faculty position at Harvard, for which Groves would cover his salary and expenses. Groves thought "this would be perfectly proper use of Government funds as he was such a deterrent to the successful operations of the Manhattan Project." But Conant refused, Groves recalled, "even if I paid him 1000% profit on the deal."

In a critique of atomic-energy entries in the *Encyclopedia Americana* in 1965, Groves attacked Szilard again by saying the biographical sketch about him was "unnecessarily long" and "overemphasizes Szilard's importance." Groves complained that Szilard "was not particularly interested in the saving of American lives in the war against Japan" and said it was "untrue that Szilard played any prominent part in the development of the atomic energy legislation which led to the creation of the AEC."[20]

When Groves read historian Margaret Gowing's *Britain and Atomic Energy, 1939–1945*, he added angry comments to the text almost everywhere he saw Szilard's name in print. "Szilard was not the leader" of American scientists who tried to withhold publications on uranium fission, Groves wrote, "nor did he lead any other activity. He was a parasite living on the brains of others."[21] Groves could not believe that Szilard and Wigner had drafted the 1939 Einstein letter to FDR, because "it was too cleverly written and indicated a knowledge not only of international affairs but also of what would appeal to President Roosevelt." Groves thought that Alexander Sachs had written it.[22]

"Szilard was *not* the most distinguished scientist," Groves noted later in Gowing's text. "Certainly he contributed nothing, in fact much less than nothing to our success. He had seized an opportunity to push government aid for atomic research while at Columbia. He was an entrepeneur [sic] not a scientist."[23] Around the description of the scientists' petition, Groves noted: "From the moral standpoint [of dropping the A-bomb], Szilard had never displayed any evidence of having any."[24]

Finally, Groves belittled Szilard in marginal notes made in his copy *of The Decision to Drop the Bomb*, a book based on a 1965 NBC News White Paper. "He was completely unprincipled, amoral, and immoral," Groves wrote by Szilard's name. He "was an inveterate troublemaker and not a great physicist."[25]

CHAPTER 21

A New Life, an Old Problem

 1946–1959

In 1946 an investigation was conducted by [the Federal Bureau of Investigation] concerning Dr. Szilard based on his association with known "liberals," his activities as a member of the Atomic Scientists of Chicago, his outspoken support of internationalization of the atomic program, and his constant influence on other scientists concerning the support of this international program.[1]

With that report, FBI director J. Edgar Hoover summarized Leo Szilard's ambitions and actions in the years following World War II. When barred from working on the government nuclear program that he had helped to create, Szilard took up biology. As a novice in a new field, his scientific whimsy proved useful, although he never dominated the discoveries as he had with the nuclear chain reaction and its consequences. For the first time in a dozen years, Szilard found himself without a single focus and purpose for his life, without a fear to drive and define his furious mental energies.

Instead, he turned to the kinds of experiments he always enjoyed: the fusion of science and politics. His work to gain civilian control of atomic energy, thought Szilard, was the logical first step toward the more critical task of creating an atomic-control scheme for the whole planet; what the scientists called humanity's fateful choice between "one world or none."

But Hoover and his aides found Szilard's actions so threatening to America's "national security" that they and other federal agents kept him under surveillance long after he left the Manhattan Project in June 1946. Annoyed by this, Szilard dared to make it amusing, sometimes jumping in and out of taxis or skipping from restaurants and drugstores through a side door when he sensed he was being followed. Szilard once told physicist Harold Agnew that he wanted all records of his fingerprints returned from the Manhattan Project, because, Agnew recalled, "he saw no reason why he should be handicapped if he decided to lead a life of crime."[2] But as cold-war paranoia spread through the land and onto college campuses, as "liberals" and "internationalization" became more suspect, Szilard's response—humor and ridicule—only prompted more FBI attention. Being naturally contrary, Szilard continued to fight for the issues he cherished: nuclear arms control and regard for the planet's population and resources abroad, individual freedom and economic equity at home.

Szilard's first struggle after leaving the Manhattan Project was more mundane, however. He had to find a job. He thought of the University of Chicago's new Institute of Physics, but his relationship with its director, Enrico Fermi, had not been easy before or during the war and had been strained further by disagreements that spring over the scientists' lobbying. Privately, Fermi let his university colleagues know that he did not want Szilard at the institute,[3] and it was September before help came—from his friend and admirer university chancellor Robert M. Hutchins. A maverick himself, Hutchins was eager to keep Szilard on campus and created a joint appointment ideally suited to his eclectic interests. Szilard would join the faculty as a half-time professor of biophysics in the new Institute of Radiobiology and Biophysics and a half-time adviser to the Office of Inquiry into the Social Aspects of Atomic Energy, a new interdisciplinary project in the Division of Social Sciences. With no teaching responsibilities for his above-average $6,000-a-year salary, this was, Szilard later said, "one of the best positions that exist at any university in the United States." Yet he would soon come to find even this arrangement too confining.[4]

His newfound financial security freed Szilard's mind for other wonders—and other worries. Besides his arms control and biology, Szilard brainstormed and wrote about whatever topics or problems seemed ripe at the moment—racial prejudice, high school education, global inflation, high-fat cheese. And yet this new freedom to think spontaneously raised an old problem: He could concentrate on anything for a while but on nothing for very long. The mental agility that led Szilard to make surprise connections among a dozen disciplines also lured him from making sustained contributions to any one.

"Theoretically I am supposed to divide my time between finding what life is and trying to preserve it by saving the world," he wrote to physicist Niels Bohr in 1950. "At present the world seems to be beyond saving,

321

and that leaves me more time free for biology."[5] For Szilard the postwar years were both giddy and frightful, focusing and scattering his energies, exciting and exhausting his gallivanting mind. He moved about impulsively by train and plane, often on the road more than at "home" in his room at the University of Chicago's Quadrangle Club, sometimes renting hotel rooms in two cities at once in his fervent quest to find new people and places that might somehow embody his rush of fresh ideas. "Things meant little to him," recalled law professor Hans Zeisel, a friend since the 1920s who was his neighbor at the Quadrangle Club in the 1950s. "Home was anyplace where his intellectual interests were at the moment."[6]

An itinerant scholar who had for years kept his bags packed for hasty escape, Szilard cared little about his surroundings. His university office was as gray and spare as the boxy science building that housed it. A few books leaned and lay on the shelves. A few files slumped in cabinet drawers. No pictures, photos, or parchment degrees graced the walls. And all the records, important letters, newspaper clips, and other "nuggets" he wished to preserve Szilard mailed to Gertrud (Trude) Weiss, his longtime friend in New York.

Impulsive travels and his very diverse half-time appointments kept Szilard from having a full-time secretary, so he hired or cajoled help any way he could. When in Chicago, Szilard turned most often to Norene Mann, a lively and patient veteran of the Met Lab who had been physicist James Franck's secretary since 1941. She opened Szilard's mail when he was away and took his dictation in her free moments. Some evenings Szilard walked to Mann's apartment on Maryland Avenue to dictate letters and papers; some days he sat at her desk or at Franck's. For this assistance Szilard paid her with sporadic checks and bribed her with flowers, ice cream cones, or—when her work became exhausting—use of his room at the nearby Quadrangle Club for a midday nap. She remembers Szilard as "extraordinarily generous" but also "exasperating" for his frantic dictation and urgently scribbled editing on papers, draft after draft.

Once annoyed that Szilard had ignored her advice on punctuation, Mann sighed and grimaced. He paused, sensing her frustration. "If you stop for a minute. I'd like to tell you a funny story," he said in a grinning attempt to break the rising tension. Szilard also enjoyed performing small favors for Mann, such as taking her letters to the mailbox. And when her seven-year-old granddaughter sloshed into the office from school one day, soaked from a spring downpour, Szilard took the girl's hand. "You come with me," he said, flashing an avuncular grin as he led her to his biology laboratory and stood her in front of a hot-air vent to dry.[7]

Away from Chicago, Szilard hired stenographers wherever he stopped: scribbling notes on planes and trains, dictating letters and article drafts in paper-strewn hotel rooms, then "filing" his papers in small suitcases that he bought in transit and stashed in friends' and relatives' closets.[8] Yet from this frantic thought and motion came some original results. Szilard helped to frame the emerging field of molecular biology by arranging informal fortnightly seminars around the Midwest with Joshua Lederberg, James Watson, and Salvador Luria as a way to speed information exchange in a rapidly developing field. And he drafted plans for a research center that addressed both scientific and social concerns (eventually realized as the Salk Institute for Biological Studies), his way to foster the blend of reason and imagination that energized his own restless life. Also, Szilard helped create new forums for arms control, such as the Pugwash Conferences on Science and World Affairs.

As in the past, Szilard pursued his urgent activities by working mainly alone; publicly in the shadows of science and world affairs, privately in the contours of his potent and playful mind. And, as in the past, this fumbling genius amused, annoyed, and bewildered the very people he tried to help and who, in turn, might have helped him stabilize his vagabond life. He enjoyed generating ideas that people in power might use and spouted advice to anyone who would listen. But he shunned the commitments and perseverance needed to join and flourish in the scientific and foreign policy "establishment."

Indeed, Szilard's quirky and creative life left many who knew him wondering if he were serious—a question he sometimes had to ask himself. On a live, nationally broadcast radio discussion about the hydrogen bomb in 1950, he stunned fellow scientists Hans Bethe, Harrison Brown, and Frederick Seitz by proposing that the new weapon—if built—should be made so dreadful that no nation would dare use it. His proposal led to the idea of the cobalt bomb, later a model for the doomsday machine in the film *Dr. Strangelove*.[9]

Often lonely and forlorn, Szilard gradually began to appreciate after the war the uses and pleasures of sociability. Shy behind his bombastic quips and wisecracks, Szilard savored the friendly lunchtime conversations at the round tables in the Quadrangle Club's dining room, yet often returned there to eat his dinner alone: reading, scribbling notes, or simply staring across the huge, dark room. Even when dining with friends, he could be alone with his thoughts. After working late at the *Bulletin of the Atomic Scientists* one night, Szilard invited editor Katharine Way to join him for dinner at the club, then added: "But bring a book. . . . I have to do some thinking."[10]

Szilard cared about the lives of his colleagues, showering them with heartfelt advice even as he hid his own pleasures and pains. In this turbulent period he married Trude Weiss, although they would not live together for another decade. And while a nomad himself, Szilard helped create structures and systems that gave permanence to the work of others, institutions that endured even as he pushed on to other ripe ideas. For after years of trying to master the effects of his genius, Szilard finally discovered in the postwar years that solitary struggles, no matter how brilliantly conceived or cleverly executed, are seldom as effective as enterprises shared.

To appreciate Szilard's frantic creativity, consider what he did in the summer of 1946, after he returned to Chicago from lobbying for the McMahon bill in Washington. First he drafted an essay on racial security, suggesting ways to guarantee the rights of racial minorities under a world government. Next he drafted a plan to organize the proposed National Science Foundation (NSF). The NSF should sponsor research on "unrecognized problems," he wrote, ignoring those already known; and it should pay scientists $12,000 a year to study whatever interests them, assuring the "leisure" needed for creative science.[11] Then he boarded a train for his first visit to California, where he rewrote scenes portraying himself and Albert Einstein in an M-G-M film about the A-bomb, *The Beginning or the End*. After three days around the movie lot in Culver City, Szilard flew to New York to lobby for Bernard Baruch's American proposal for international control of atomic energy at the UN Atomic Energy Commission.[12] Then he returned to Chicago and flew off to a conference on the atom's international control at Estes Park, Colorado, a visit that sparked his love of the Rocky Mountains.

Back at the University of Chicago, Szilard turned his mind to a sweeping philosophical review of the A-bomb's creation in a public lecture entitled "Creative Intelligence and Society: The Case of Atomic Research, the Background in Fundamental Science." In bursts of thought, with each paragraph just a sentence or two, he dramatized the findings and mistakes that had led nuclear science to its present, fateful condition.[13]

Beginning with Becquerel's 1896 discovery of radioactivity, Szilard described how Madame Curie isolated radium. His speech was exuberant, his delivery pithy and direct.

Transmuting one chemical element to another chemical element was, as you know, the unsolved problem of the alchemists.

324

> But Madame Curie, who isolated radium, could not pride herself to be a successful alchemist.
>
> She did not produce radium.
>
> She merely separated it chemically. . . .
>
> So, in spite of this new discovery, God remained the first and only successful alchemist.

Simply, abruptly, dramatically, Szilard spoke on, leading his audience through vivid descriptions of alpha particles and on to Ernest Lawrence and his cyclotron. Szilard first made public his recollection of thinking up the nuclear chain reaction while crossing London's Southampton Row in 1933. He credited H. G. Wells's 1914 science-fiction story *The World Set Free* as the source that prompted this discovery, dramatizing the connection between literature and science. And he praised Princeton physicist Louis A. Turner for recognizing that nonfissionable uranium^{238} might be converted to fissionable plutonium. Then came this stirring conclusion.

> The first use of plutonium, as you know, was in the form of a bomb which destroyed a city.
>
> The next use of plutonium might be the same again.
>
> With the production of plutonium carried out on an industrial scale during the war, the dream of the alchemists came true and now we can change, at will, one element into another.
>
> This is more than Madame Curie could do.
>
> But while the first successful alchemist was undoubtedly God, I sometimes wonder whether the second successful alchemist may not have been the Devil himself.

Szilard's return to Chicago in the summer of 1946 and his move back to the Quadrangle Club restored his favored morning routine of soaking for an hour or two in his bathtub to dream up ideas and schemes that he might ponder and proclaim throughout the day. His next-door neighbor at the club, Hans Zeisel, knew Szilard's routine well and noticed after a day or two that he was not coming down for meals. Yet he heard Szilard's bath running each morning for those long and thoughtful soaks. A week passed. Then two. Finally, Szilard appeared at breakfast, smiling.

"Where were you, friend?" Zeisel asked.

"Working on a problem," he said with obvious delight. "I had a theory. But it was all wrong. It didn't lead anywhere."[14]

Many of Szilard's proposals didn't lead anywhere, but he found such pleasure in thought itself—in chasing ideas to their limits, and beyond—that he seldom cared if his speculations were right or wrong. To him, disproving a hypothesis was just as important as proving it. Szilard "was as generous with his ideas as a Maori chief with his wives," said microbiologist Jacques Monod, who acknowledged winning a Nobel Prize with a concept Szilard had pressed on him. Szilard "was too rich in ideas, whether scientific or political, too joyfully familiar with all of them; he derived too much sheer pleasure in playing with them as a child with his toys ever uniquely to pursue only one of them, aggressively reiterating, illustrating, and defending it, as most of us do."[15]

Theories and ideas were made in such earnest, as so many new ways to save the world, that they claimed Szilard's serious attention even when they seemed farfetched. In September 1946, Szilard turned his mind to economics, writing "Market Economy Free from Trade Cycles," his plan to diminish inflation and reconcile the different values of national currencies. Curbing trade cycles was an essential step, he argued, toward establishing the world government needed to control the atom. His scheme called for a two-currency system (red and green dollars) to stabilize prices for different kinds of transactions.[16]

University of Chicago economist Milton Friedman often played tennis on courts behind the Quadrangle Club, and there Szilard assaulted him with his ideas. Friedman came to expect a "sophisticated economic proposition" whenever they met: by the courts, over lunch in the club, or elsewhere around the campus. He found Szilard's brainstorming "fascinating," often "clever," sometimes "insightful," always amusing, and "most surprising, it was usually correct. . . ." Friedman saw that Szilard focused intuitively on issues and problems then at the heart of the profession's debates, often reworking contemporary details with arcane statistical techniques.[17] Szilard also buttonholed economist Jacob Marschak, a friend from Berlin and Oxford who thought that Szilard was "prophetically aware" of economic trends without following the field's literature.[18]

The two-currency scheme was proposed to eliminate boom-and-bust trade cycles in market economies. Money serves two different purposes, Szilard reasoned: spending and saving. By having green dollars for spending and red dollars for saving and enforcing a floating exchange rate between them, the economy would have more flexibility than with a single currency, making it easier for both governments and markets to respond to trade cycles. Szilard reworked this scheme in 1948 and again in 1949, by then just for fun because there was no longer much hope for a world government.

Colleagues and friends soon learned that even Szilard was not always sure when his ideas were serious and when they were playful. Often they were both. "My Trial as a War Criminal," written in 1947, was Szilard's bitter reaction when the Justice Department censured his efforts to start arms-control discussions with Soviet scientists and to urge US-Soviet peace talks in an open letter to Stalin. Yet in the story, Szilard also revealed his playful-serious spirit by rejecting the chance to move to Russia: "How many years would it take me to get a sufficient command of Russian," he wondered, "to be able to turn a phrase and to be slightly malicious without being outright offensive?"[19]

Away from his Quadrangle Club routine, Szilard found a bracing outlet for his many ideas in the Colorado Rockies, where, he said, "the lack of air stimulated thought."[20] Szilard's friend Trude moved from New York City to Denver in 1950 to teach public health at the University of Colorado Medical School, and about that time he also hoped for a post there.

Once in the mountains, Szilard shed his customary dark suit for a tan gabardine zippered jacket. Sometimes he even tugged off his necktie and rolled up his shirtsleeves. On one outing he wrapped his head in a red bandanna and sat, Buddha-like, on a flat rock in a field of wildflowers. But he never relaxed enough to change his black dress shoes from the city, and on bright days he donned a floppy canvas hat, dark glasses, and a tan raincoat kept buttoned to his neck—all to protect his skin from the sun. Now much heavier than when he enjoyed daylong hikes in the Swiss Alps in the 1930s, Szilard only strolled on the wooded paths around his lodging or rode by automobile through the steep passes of the Rocky Mountain National Park along the Continental Divide. When he left the car, it was to step into a meadow, sit on a boulder or tree stump, or settle in a folding beach chair at the edge of the dusty road.

Much of Szilard's time in Colorado centered around a log cabin that he and Trude rented on the Stead Ranch near Estes Park or in the ranch's large inn, where the guests dined. Often Szilard sat in a corner of the inn's lobby or in the small reading room, there to peruse books and magazines, to scribble rapidly on a yellow legal pad, or to stare away in thought, oblivious to the people strolling and chatting about him. Sometimes his eyes closed as his mind played and pondered, giving the appearance he was napping when he was really only "botching."

In the dining room of the inn one day, Leo and Trude met Walter Volbach and his wife, both German refugees. Volbach, a theatrical producer in Europe who then taught theater arts and ballet in Texas, knew Szilard's reputation as a scientist but was surprised by his other interests. In their conversations, Szilard predicted that Russia would have an A-bomb within a year or two. (It did, thirteen months later.) But Volbach recalled that Szilard "seemed to be in the best mood" when talking about the arts: not only films, which he and Trude enjoyed, but drama and the theater; concerts and opera; and the philosophy of art, painting, and sculpture. "He had not just a fine memory but a tremendous knowledge in all the fields—and a very delicate feeling for the complex aesthetics of various styles."

The Volbachs invited Leo and Trude for a ride into the mountains, and off they drove toward Bear Lake. Parking the car, the Volbachs and Trude decided to hike for an hour to a higher lake. Not Leo. He quickly found a large, flat rock and sat down to botch.[21] In fact, all the while Trude chatted with guests, read, or took studied black-and-white photographs of the landscape, Leo lost himself in writing, usually short pieces of fiction and satire. Oblivious to the surroundings, he scribbled thoughts and plots on paper that he spread on picnic tables, balanced on the log railing of his rustic cabin porch, juggled on his lap, or held on the arm of his chair.

During a visit to the Stead Ranch in July 1948, Szilard wrote satires on medicine and politics, "The Mark Gable Foundation" and "The Diary of Dr. Davis." The first described how a person escaped social and political problems by being preserved and revivified in the distant future. The second questioned how to finance productive scientific research. Szilard and organic chemist Aaron Novick had opened a biology research laboratory at the University of Chicago that spring, so blending the theory and practice of science had become a daily challenge. When puzzled, Szilard often found answers in his fiction. But this sometimes confused readers who could only take science—and life—very seriously.

Alone for a week at the Rancho Del Monte near Santa Fe, New Mexico, in September 1948, Szilard wrote "Science Is My Racket," which urged academics to be more active politically. A few months later, he began work on one of his best-remembered tales: "Report on 'Grand Central Terminal,'" a satire on archaeology, politics, and more. When invaders from outer space visit New York City after a nuclear war, they analyze the terminal's pay toilets, trying to understand the values of the vanished society. The invaders are led by Xram (Marx spelled backward) and produce a curious socioeconomic explanation for the disks (coins) found in the door slots. Because they had found no disks in other "depositories" around the city, the visitors conclude that those in the pay toilets "had been placed there as a ceremonial act" connected with "deposition *in public places* and in public places only," the images on the coins being tributes to their leaders.[22]

With mirth and anger, Szilard also used satire to confront the cold war. At the Stead Ranch in the summer of 1949 he wrote "Calling All Stars," a science-fiction satire about cybernetics, biology, and nuclear annihilation. What worries superhuman minds on a distant star is that creatures on Earth were both smart enough to separate uranium[235] yet dumb enough to use this knowledge to make weapons. The distant minds conclude, as Szilard himself had by then, that if these Earth organisms are "engaged in co-operative enterprises which are not subject to the laws of reason, then our society is in danger."[23]

In his fiction, Szilard satirized the US-Soviet nuclear arms race by looking behind the military confrontation to the science and politics at play in the two countries. His "Nicolai Machiavellnikow," a comedy skit finished in July 1949, mocked the congressional hearings then being led by Sen. Bourke B. Hickenlooper, an Iowa Republican eager to discredit Atomic Energy Commission (AEC) chairman David Lilienthal. That spring, Szilard had advised witnesses appearing before the committee but did not testify himself. The skit treats these hearings as part of a Soviet plot to undermine America's nuclear power research. Lilienthal wants to combine US nuclear reactor development with British efforts, which are more advanced. Oust him and America's program will remain "a complete mess," a failure, chides Szilard, because "they did not even get back any of their good men who had worked on [atomic energy] during the war."[24]

Szilard's direct—and daring—solution to the anti-Communist hysteria in America was simple. "If someone is accused of being a communist," he said even before Sen. Joseph McCarthy made red-baiting a nationally televised sport, "we should all take him out to lunch. If someone is charged with subscribing to the *Daily Worker*, we should all subscribe."[25]

When Trude returned from Colorado to New York in August 1949, Szilard moved from the Stead Ranch to the white-columned Stanley Hotel in Estes Park, a larger and more formal resort, where he stayed alone to botch. There he quickly established an easy routine, reported by letter to Trude as "breakfast in town (steak). Lunch in the cocktail room at the hotel at 11:30 a.m. (melon, bread, butter + honey, milk). Dinner in the evening in town." He shaved every day, walked into town for meals, and slept from 10:00 P.M. to 6:00 A.M. One day in town he stepped onto a scale to read his weight and fortune. "Weight: 181 lbs. Fortune: You are very attractive to the opposite sex." He wondered if the scale would give the same fortune at 185 pounds, a weight he expected to reach "probably very soon."[26]

Botching went well at the Stanley, especially the essay on red and green money, and one evening economist Jacob Marschak stopped by to pick up Szilard for dinner and a chat about the paper.[27] Marschak mentioned Orwell's just-published satirical novel *1984*, but Szilard refused to read it, saying the book's horrors conformed too closely to his own.[28] Physicist Hans Bethe visited the Stanley, too, and drove Szilard to the Fall River Pass near the Continental Divide. Aaron Novick stopped by for two days of talks on his way to a biology conference in California. But mostly Szilard kept company with his many thoughts. "Not talking to anyone here about anything," he wrote Trude, "which is very good for writing."[29]

Szilard's demeanor was often formal, and his thinking was at times dead serious. Yet in many ways he was still a child at heart and enjoyed the company of young children. At the Stanley Hotel that August, he wrote a delightful fable about a child's fears and fascinations. The tale is based on his encounters with the daughter of Dr. Gertrude Hausmann, a friend of Trude's who lived in Denver. At the Stead Ranch earlier that summer, Szilard and Kathy had enjoyed wadding tissue paper to plug holes in the screen door of the Hausmanns' cabin, and at the Broadmoor Hotel in Colorado Springs the two had dined at a table of their own. With Kathy, Szilard saw his mountain retreat through the eyes of another child, one more open with feelings and fears.[30]

KATHY AND THE BEAR

This is a beautiful spot to stay in the summer. The hotel is well run. The food is good. Only the service at meals is too slow for my taste; it is intolerably slow for a child of four.

Kathy and her mother spent the last weekend here. At lunch Kathy sat next to me.

"You want some sugar in your milk, Kathy?" I asked.

"She never takes any sugar," her mother said.

"Yes, please, I want sugar," said Kathy.

I gave her lots of sugar and she started to drink her milk. Halfway through she put her glass down.

"Mother," she said, "can I go to see the bear?"

"What bear?" I asked.

"She means that huge bear skin on the wall in the lobby," the mother said. "Would you like to take her there? I shall get you when the meat comes."

Kathy led me to the bear, but stopped at a respectful distance.

"He is dead," she said. "He was shot. You can put your hand into his mouth, he can't do anything to you." And after a moment of silence: "I won't put my hand into his mouth; I am scared." She stood there fascinated and looked at the bear, but kept her distance.

"All right," I said, after a while. "You have seen him now. Let us go back to lunch."

The meat was not there yet when we got back to our table, and Kathy's mother was across the dining room talking to some friends.

"Does a bear like to be shot?" Kathy asked.

"I don't know, Kathy," I said. "I do not think so."

"Doesn't a bear go to heaven when he dies?" Kathy wanted to know. I wasn't going to compromise with truth, not even for the sake of a girl four years old.

"I do not know, Kathy," I said. "But I am sure your mother can tell you. Why don't you ask her after lunch, when she puts you to bed?"

"My grandfather is dead and he is in heaven," said Kathy. "And my grandmother is dead and she is in heaven. God is in heaven too, and he is not dead. How is that?" asked Kathy. This was beginning to get difficult.

"Why don't you drink your milk, Kathy," I said. "It isn't too sweet for you—or is it?"

"It *is* too sweet," said Kathy. I accepted the verdict in silence.

"But I like it too sweet," said Kathy, and with that she picked up her glass and started to drink.

The next day at lunch we had again to wait for the meat.

"May I go to see bearsy-wearsy?" said Kathy.

"All right, Kathy," I said. "Come along." When we got into the lobby, Kathy said:

"He was a bad bear. He killed all the chickens. And the farmer took his gun and shot him. And then they saw that he was beautiful, and they put him on the wall so that little girls can look at him." With that she started to stroke tenderly the one paw that she was able to reach.

"Lift me up, please," she said. "I want to put my hand into his mouth." I lifted her up and she had her wish.

"You want to come back to the dining room now?" I asked.

"No," she said. "I want to stay with the bear."

331

"All right," I said, "but don't stay too long." Kathy settled down on the couch and began to pet the beast. I went back to the dining room. After a while Kathy appeared and sat down in her chair.

"Honeybear," she murmured and picked up her milk.

Soon after writing "Kathy and the Bear," Szilard confronted his own fears when he reread Thucydides' *History of the Peloponnesian War*. The work "considerably frightened" him because "neither Sparta nor Athens wanted war, yet they went to war with each other" for thirty years. Only half-joking, Szilard thought his situation was more dangerous than the ancients' because "in many respects these Greek city-states were politically more mature" and "both Sparta and Athens were much more democratic than are Russia or the United States." Szilard feared that just as in 431 B.C., when the conflict had "started as a war between an ally of Sparta and an ally of Athens," so any country in the North Atlantic Treaty Organization (NATO) or the Warsaw Pact might prompt a nuclear war that both superpowers dreaded.[31] Three days after Szilard wrote this, on September 23, 1949, President Truman announced that the Soviet Union had exploded an atomic bomb. For several months, Szilard carried a copy of Thucydides in his pocket and warned about its ominous message repeatedly until the book became dog-eared and fell apart.[32]

In the 1940s and 1950s, Szilard worried about nuclear war incessantly, but he also feared the planet's extinction from other causes, especially poverty and overpopulation. To the advisory committee of the newly created Ford Foundation, Szilard offered, in January 1949, his thoughts on how their money might be used "for the betterment of mankind." He urged analyzing ways to create alternative forms of democratic government that are more flexible than the Westminster and Washington models. His key concern for eliminating world poverty focused on finding effective birth-control methods, especially for India and China. He also urged the foundation to sponsor a "Voice of Europe" radio system to increase international understanding by beaming programs to the United States.[33]

With nuclear chemist Harrison Brown, a colleague at the Met Lab and in later arms-control groups, Szilard brainstormed about the connections between governance and biology, advancing the idea that the earth has a natural "carrying capacity" that must not be exceeded. This theme Brown developed in seminars at the University of Chicago, which Szilard attended in 1950 and 1951, and in the 1954 book *Challenge of Man's Future*. Brown and Szilard discussed the buildup of carbon dioxide in the atmosphere, population pressures, and alternative food supplies, issues raised by the Club of Rome in the 1970s. In the late 1940s, Szilard believed that nuclear power would someday be cheaper than coal or oil and would become a significant energy source worldwide, especially in developing countries.[34] By 1956, however, Szilard wrote that Pres. Dwight D. Eisenhower's 1953 "Atoms for Peace" proposal to export nuclear reactors had "offended my sense of proportion" because "to establish a secure peace when the bombs stare us in the face is a tall order. If we want peace, we have to make peace and not atoms."[35]

Fearing that a nuclear war with H-bombs might occur soon, Szilard worried about how children would survive to rebuild the shattered planet. He proposed founding a private boarding school in Mexico City and urged his University of Chicago colleagues to enroll their children. In June 1951 he drew up plans for the school, "Where will your children spend the war?" Children would enter at age nine and could graduate at eighteen or twenty-one, receiving "a solid body of knowledge" in "a climate physical, intellectual, and moral that will make them fit to live later as they may choose either in the United States or in South America." In conversations around the Quadrangle Club, Szilard claimed he had talked about the school with the Mexican president and had arranged for a high-quality, bilingual education. Some faculty members resented the fact that Szilard, who had no children, would be so insistent about how theirs should be educated and protected. For his part, Szilard was angry when friends shunned his rational advice but gladdened when Chancellor Robert Hutchins—an admirer given to his own quirky schemes—smiled on the plan.[36]

In another flight of rational fancy Szilard proposed a way to assure fairer presidential elections when one party's candidate is overwhelmingly popular. President Dwight Eisenhower had soundly defeated Illinois's Democratic governor Adlai E. Stevenson in the 1952 presidential election and was a popular favorite for reelection in the winter of 1955 when Szilard strolled into the K Street office of Washington lawyer Joseph L. Rauh, Jr., swept past his secretary, and stood before his desk.

"Joe, I've got it figured out," Szilard announced. The two had known each other since they had lobbied for the McMahon bill in 1946, when Rauh was chief counsel of the National Committee for Civilian Control of Atomic Energy. Now Rauh was active in Democratic party affairs and a respected civil rights lawyer.

"Yeah?" Rauh said, then recalling they had recently discussed electoral reforms.

"This is what we do next year," said Szilard. "We run Ike on both tickets, Stevenson for vice-president on the Democratic ticket, Nixon for vice-president on the Republican ticket. Then the voters can decide who they want to succeed Ike."

"Who's in charge of getting these nominations on the two tickets?" Rauh asked.

"Oh, you are!" Szilard said.[37]

Like most of Szilard's schemes, plans for a Mexican school and a two-party ticket were soon forgotten. But his concerns about education and politics thrived, especially in conversations with historian Max Lerner, a professor at the new Brandeis University. In January 1952, Lerner invited Szilard to speak to seniors about his "experiences and working philosophy," and a month later he flew to Boston and rode by taxi to the hilly campus in the western suburb of Waltham. There Szilard spent three joyful days chatting with students and faculty, savoring his role as a welcome celebrity. Since childhood Szilard enjoyed designing social and political institutions, and Brandeis was an appealing place to apply his musings. Soon he had proposed organizing a tutorial system there by drawing on specialists at the better-known universities in the Boston area.[38]

The vitality at Brandeis was a welcome contrast to Chicago, where Szilard was about to close his biology laboratory because his indispensable partner, Aaron Novick, planned to leave for a sabbatical in Paris. In fact, as early as the previous fall Szilard had considered quitting biology research altogether, preferring to join the university's central administration "to work on the improvement of the finances of the University . . ." by brainstorming investment schemes that would profit from the school's tax-exempt status.[39] In the fall of 1952, Szilard took an unpaid leave from Chicago to become visiting professor of physics at Brandeis. The university's press release announced that this "nuclear physics pioneer who was largely responsible for the development of the atomic bomb . . ." would assist in developing the expanding graduate and undergraduate science program, conduct seminar courses in "Frontiers of Science" for advanced students, and teach in the graduate school.[40]

Szilard played his new role as teacher with the same gusto he brought to his science. "Mass murderers have always commanded the attention of the public, and physicists are no exception to this rule," he said to shock his audience at a talk "On Education," but quickly added that ". . . the most important thing to remember about science is the fact that it is supposed to be fun. . . . Doing nothing—in a pleasant sort of way—was always considered in Europe a perfectly respectable way of spending one's time," he said, voicing his own predilection. "Here in America you are expected to keep busy all the time—it does not matter so much what you are doing as long as you are doing it fast."[41]

In another "Talk on Education" Szilard urged Brandeis students to study the humanities as a preparation for public life. "The skills and knowledge you acquire determine what you can do," he said. "Your knowledge and wisdom determine who you are. In our society, there is a market for skills and knowledge. But I have some doubts if there is much of a market for wisdom." In Szilard's view, "the great educational problem in our society does not lie at the college level but at the level of the high school," and as a remedy he proposed a national high-school reading program keyed to college study.[42]

At Brandeis two days a week, Szilard met with students and faculty in seminars but soon adopted a role that he considered far more important. "Leo said that his function was to head up the Happiness Committee," Lerner recalled. "He said a university ran on the happiness of the faculty, and he wanted to be the one to think up ways of keeping them happy." See that they are well paid, Szilard said, that their offices are comfortable, their graduate assistants are bright and eager, and that the faculty club food is appetizing. Then you will have a first-rate university![43]

But a growing involvement at Brandeis only complicated Szilard's already hectic life, adding a new stop on his bumblebee's itinerary. With the King's Crown Hotel in New York as his base, he wandered impulsively to the Quadrangle Club in Chicago and the Somerset or Statler hotels in Boston—also making quick visits to his ailing father, then ninety-two and living in a nursing home in Yonkers. Szilard visited the home every week or two and by mail advised the staff on his father's diet and treatment. When in New York, Szilard also began work as a consultant to the president of the International Latex Corporation, Abram N. Spanel, and for the next few years spent many nights and weekends at Drumthwacket, Spanel's mansion in Princeton.[44]

Szilard had not lived in a house since he left his family's villa in 1919, and except for Drumthwacket he had no idea what domestic life was like. So one afternoon, when at Brandeis, he decided to visit Trude's brother, Egon Weiss, and his wife, Renée, at their home in the nearby suburb of Cochituate. "I'm on a diet," Szilard announced, walking in with a paper bag in his hand. From the bag he nibbled grapes throughout the visit, popping them into his mouth between rapid questions and quips. Egon was then studying library science at Simmons College, and Szilard was fascinated by the arrangement of knowledge into rational categories and the possibility that information might be ordered and retrieved by computers. But Szilard's real focus that day was the Weiss's house itself. "He was interested in what it cost, how it was heated, when it was painted," recalled Renée Weiss. "He wanted to know how the toaster worked. What was in the basement. How we washed our clothes."[45]

To friends and family, Szilard's mention of a "diet" usually prompted grins, for his sweet tooth always won over his good intentions. In the ornate red plush Russian Tea Room in New York one day, Szilard was eating a *nusstorte,* a cake made with nuts, eggs, and sugar, when Efraim Racker and his wife, Franziska (Frances), Trude Weiss's sister, walked in.

"How is your *nusstorte?*" asked Frances.

"This is no *nusstorte,*" Szilard said, a grin wrinkling his round face. "This is a *genusstorte,*" *genuss* being the German word for enjoyment.[46]

Yet, periodically, he vowed to lose weight and boasted about his efforts, if not the results. Walking around the Hyde Park neighborhood with Milton Weiner, a student at his biology laboratory who was Enrico Fermi's son-in-law, Szilard passed a haberdashery and remembered he had to buy some underwear. "I'm on a diet," he said as they entered. "I need to get them smaller than before." When Szilard picked a size and handed it to the clerk, the cleric suggested, politely, that they might not fit.

"But I'm on a diet," Szilard insisted, to which the clerk replied: "The underwear don't know that."[47]

An itinerant life continued for Szilard even after he returned to the University of Chicago's payroll in 1954 as professor of biophysics. Morton Grodzins, dean of the Division of the Social Sciences, described Szilard's "responsibilities as a full-time member of the faculty" with wry amusement, for he had almost none. "I take pleasure in contemplating that a great physical scientist joins Chicago's social science group in order initially to devote himself to biology," he said of the arrangement. Later, Grodzins told Szilard that "the University will be satisfied with three months of annual residence provided that the University considers your activity while not in residence as being of substantial service to the University, to scholarship or to the public interest."[48]

Understandably, Grodzins had trouble knowing what Szilard would do, because even he wasn't sure. Szilard did know that he disliked Chicago's harsh winters and hellish summers, and with Novick gone, he missed the daily thinking-and-tinkering routine that had made their biology laboratory so enjoyable. But where else could Szilard work? By the summer of 1954, Theodore Puck had told Szilard that a full-time biology appointment in Denver was "unlikely," first citing a lack of funds but later admitting his deeper fear that Szilard might dominate the research.[49] Szilard considered a research post at the Albert Einstein College of Medicine in New York and even had Einstein write a letter of introduction, but then barely pursued this opportunity. Instead, Szilard devised a "roving research professorship" for himself to permit occasional work at six institutions.[50] But this plan was abandoned when he refused even to meet a potential funder's minimal request for a work schedule.

By 1956, Szilard devised an even grander scheme when he enlisted New York University biologist Bernard Davis and others in sponsoring a "fellowship for life," but he dropped the idea before it could be arranged and proposed, instead, a course of research at NYU on cancer of the prostate gland—to be funded, Szilard hoped, by a cancer-research foundation headed by his friend Lewis L. Strauss. At NYU's Medical Center, Dr. Lewis Thomas offered laboratory facilities for the cancer research, but then Szilard balked because he feared no suitable assistants could be found.[51] In 1956, Szilard asked for a research position at the Rockefeller Institute in New York but could do no better than accept the offer to become an unpaid "affiliate." At the time, Szilard also drafted plans for a research institute at Brandeis, but he never followed through on this idea, either.

Szilard even wondered anew about atomic energy, looking beyond his tireless nuclear arms control efforts to the possibility that reactors might produce industrial heat and electricity. A few weeks after President Eisenhower announced his "Atoms for Peace" program, an ambitious plan to spread nuclear technology around the world, Szilard contacted John Menke, a Manhattan Project veteran who had founded Nuclear Development Associates in White Plains, New York, to export reactors. In 1955, Szilard gained new attention for his pioneering role in atomic research when his joint patent with Fermi for the first nuclear reactor was declassified and published. In 1956, Szilard wrote a memo on a proposed federal financing scheme for exporting nuclear power plants. And in a draft letter to the *New York Times* that summer, he proposed a full exchange of information on fusion energy with the Russians—

an idea signed in 1992 by the US government. With fusion energy, Szilard believed, the large number of neutrons released could be used to convert thorium to fissionable uranium—one of the two "breeder" reactor schemes he had first pondered during the war.[52]

Early in 1957, two former colleagues wooed Szilard to reconsider nuclear physics. First, his thesis adviser in Berlin, Max von Laue, invited Szilard to head a new physics institute there. Szilard declined, and when he visited Berlin that October, it was to speak about molecular biology.[53] But Szilard accepted when Edward Creutz, director of research for General Atomics (GA) in California, asked for "novel ideas" about new power reactors and "the possibility of your helping us get going in the atomic energy business." Szilard admired Creutz, loved California, and agreed to become a consultant.[54] But at the same time, Szilard's enthusiasms also ran with a scheme he had just proposed to create a biology study center that would combine science and social problems: what became the Salk Institute for Biological Studies, now just down the road from GA in La Jolla.[55]

When Szilard arrived in La Jolla, he spent little time at GA's futuristic headquarters, preferring to sit by his motel's pool with a yellow pad on his lap, just botching. He began by making calculations on heat transfers in different alloys but quickly turned his thoughts to generating electricity directly from uranium fission, without employing steam to drive a turbine generator, as today's nuclear power plants do. Szilard refused to sign an agreement that his inventions were GA's property, complaining, "That's not how my brain works." He could not keep track of which concepts were his and which were the company's, he said. "I have no idea where my thoughts come from and no control over where they go."[56] While in La Jolla, Szilard made crude and overly complicated drawings for a thermoelectric generator and later completed patent plans for the device. "It was clear that by this time Leo was no longer at the cutting edge of nuclear physics; was no longer a pioneer," Creutz recalled. "He was stimulating to have around, but didn't really contribute much to our work designing new reactors."[57]

Another field Szilard had once pioneered was information theory, or "cybernetics," which by the 1950s had developed under Claude E. Shannon at the Bell Telephone Laboratories in New Jersey. Szilard had first linked the concepts of information and entropy in the 1920s. At the time, Szilard and his friend John von Neumann were teaching courses together at the University of Berlin, and in 1947, von Neumann reconsidered Szilard's ideas and later urged Shannon to use the term "entropy" in his work.[58] By the early 1950s, around Columbia University, Szilard also discussed his information-entropy ideas with physicist Leon Brillouin, and in 1952, at Brillouin's suggestion, Szilard was invited to speak at an American Physical Society symposium on "Entropy and Information" during the society's spring meeting in Washington. But here Szilard admitted his limitations and declined, saying: "All I could do is to present—as an introduction to the topic—the original considerations which I published in 1927. I have not done any further work in the field since I wrote this one paper."[59]

For Szilard the joy of discovery was first having the idea, not applying it afterward. Szilard did not seek to tie his early insights to later commercial developments, although he and von Neumann talked about computers from time to time and in the 1950s von Neumann himself wanted to develop Szilard's information-entropy ideas as part of his pioneering work in computer design. Not until the 1970s and 1980s, however, in two *Scientific American* articles about Maxwell's demon and the second law of thermodynamics, did Szilard's role in information theory gain the attention of a wider scientific community.[60]

In his wandering life in the 1950s, Szilard thought up dozens of inventions and even drew patent applications for a few: a pocket calculator to compute calories, a phonograph with both sonic and supersonic frequencies, a method to desalt seawater by freezing, and decades ahead of the cholesterol and fat scares of the 1980s, a line of cheeses and other dairy products with high-iodine vegetable oil substituted for more than 80 percent of the milk fats.[61]

But Szilard's most memorable—even notorious—concoction was his extract for "rum tea," and just recalling it makes some friends smile, and wince. The project began in the early 1950s when Hans Zeisel was research director of the Tea Board, an industry promotion group. Talking about tea one day, Szilard recalled from his Budapest childhood drinking rum tea, which was served like a hot toddy. Zeisel also remembered the drink from his youth in Vienna. Grinning with anticipation, Szilard vowed to invent an extract or pill that would allow them to mass-produce and market this delicacy. With help from Maurice Fox, a graduate student in his biology laboratory, Szilard brewed a rum extract and drained the thick brown liquid into a small medicine bottle, which he slipped into his suitcoat pocket. For months Szilard produced the bottle at restaurants, called for hot water, and carefully poured in a tot of his rum-tea extract.

"One can use water from the hot-water tap and does not need to boil it," he told Trude. "It tastes better than Nestea," a powdered instant tea then on the market, and boasted, "I have a full bottle in the bathroom and make myself tea every morning." But as the extract took on a greenish tinge, Szilard reported to Zeisel that the concoction would not keep at room temperature. Szilard abandoned his hope for making instant rum tea only when he could devise no way around the stiff taxes on alcohol products. "All his thoughts, from inventing a rum-tea pill to controlling the bomb, centered around making this a better place to live," Zeisel recalled fondly. "He just wanted to improve the whole world, and everything in it."[62]

CHAPTER 22

Marriage on the Run

 1951–1959

On a fall day in 1951, Leo Szilard strolled into the smoky chatter of the large dining room in the Quadrangle Club, surveyed the lively scene, and took a seat at the "physics table," one of several round tables under the tall leaded-glass windows where different faculty specialties converged.

"Leo," said Herbert Anderson, raising his voice across the noise as Szilard sat down. Anderson had known Szilard since 1939, when they collaborated with fellow physicist Enrico Fermi at Columbia University.

"Leo," Anderson repeated, smiling. "I hear you got married."

"Where'd you hear that?" Szilard replied, seeming annoyed.

"Ohhh . . . I read it in the newspapers."

"Do you believe everything you read in the papers, Herb?"

"Should I believe this?" Anderson asked.

"If you want to," Szilard said, and dropped his eyes to the menu.[1]

Aaron Novick worked almost daily with Szilard at the biology laboratory they had founded on campus, yet he only heard about the marriage when Harold Urey's wife, Frieda, asked at a party if rumors of a secret wedding were true. University chancellor Robert Hutchins was also surprised when Anderson brought him the news.

"Who would marry Szilard?" Hutchins wondered. "It must have something to do with taxes."[2] Indeed, Walter Blum, a friend of Szilard's at the law school who was legal counsel to the *Bulletin of the Atomic Scientists*, only learned about the marriage when Szilard raised the subject of taxes. They brainstormed often about tax laws and loopholes, seeking ways for the university to profit from its tax-exempt status and as an intellectual game. But that fall, for the first time in their years of discussion, Szilard asked Blum about medical deductions, and when Blum answered, Szilard replied: "That figure is for a single person."

"But the figure for married people doesn't concern you."

"Yes, it does," said Szilard.

"Why?" Blum wondered.

"Because I *am* married."

"I knew it would take something to do with taxes to persuade you to get married," Blum teased. In response, Szilard just smiled,[3] although with another friend he became flustered. "It was almost impossible to embarrass Szilard," fellow Hungarian Edward Teller recalled. Teller saw Szilard and his girlfriend of many years together at a motel in Santa Fe and later heard they were married. But when Teller saw Szilard in Chicago a little later and congratulated him, he blushed. "Some people would have . . . blushed when found with their girlfriend, but that was natural for Szilard. . . . To get married was not natural for Szilard."[4]

Szilard was secretive about his marriage to Gertrud (Trude) Weiss because he could scarcely admit it to himself. "I am a bachelor by birth," he told a reporter six months after the wedding.[5] For despite a serious relationship with Trude that had evolved over more than two decades, Szilard still nurtured an adolescent vow to shun emotional involvement—his way to focus and sharpen the clarity of his mind. He could be affectionate and solicitous with small children and childishly flirtatious with older women, but rarely had he been comfortable with mature passions. Above all, he valued a freewheeling life of the mind and enjoyed the eccentric freedoms that came with being a refugee savant. Love and marriage seemed common threats to this uncommon life on the run.

Szilard had met Trude in Berlin in 1929, and they corresponded and visited each other after she returned to Vienna to study medicine in 1930. During the next few years, she was engaged to two other men but both times ended the relationship in favor of her distant and sporadic friendship with Szilard.[6] He seemed to care about her, persuaded her to leave Vienna and join him in England in 1936, and helped her arrange medical study in London and a hospital job in New York.

From the mid-1950s on, when Szilard was in his thirties and Trude nine years younger, he treated her in an avuncular manner. Still, beginning with their time together in London, Leo and Trude grew closer. In letters about illness and medical treatments, Leo advised Trude to take hormones as a way to alleviate her anxiety and pain with menstruation—he called it her "brainstorm"—and often explained her mood changes by the strength and frequency of this medication.[7]

Work kept them separated when Leo and Trude both lived in New York City, from 1938 to 1942, but he came to value their friendship and made a habit of writing her quick notes and calling to check on her. When Szilard moved to Chicago for the Manhattan Project in 1942, his notes and calls increased. He also visited her in New York several times during the war and, whenever he traveled by plane, took out life insurance that named her as beneficiary. Gradually, Trude had become someone Leo could turn to, often at day's end, to relate his activities, amusements, and anxieties. She, in turn, valued his advice about an array of worries: about how to care for her ailing mother, about decisions in her public-health career, about whether to visit friends for a weekend in the country, about what he thought of a new hat or dress.

As lovers do, they also made up a vocabulary of their own for shared emotions. A "brainstorm" was certainly her period, and no doubt with sex in mind, Szilard at times planned his visits around it. This planning also seemed to involve his concern about emotional outbursts of any kind. As he advised in a July 1943 letter, "If I were in your place I would finish the period of brainstorming with a physiological brainstorm and then adjourn other brainstorms (with one exception) until September. You can say to yourself: 'God, will I be excited in September!'"[8] In their private language, things "ici-pici" were quick, rushed, or slight,[9] "kuc-kuc" was weariness or anxiety or a problem,[10] and "shush-kush" were more personal, upbeat, and gossipy feelings.[11]

After the war, Szilard sometimes visited Trude's family in New York: a bumptious free spirit who appeared at holidays, asked about their lives and problems, dispensed advice, sometimes stashed a suitcase of papers in their closets, and then disappeared. Trude's family treated her long and intimate friendship lightly, and her sister Frances's husband, Efraim Racker, called Leo his "brother-not-in-law."[12] Outwardly, Leo and Trude seemed to be good friends, and, privately, a strong dependence had evolved.

In 1946, Szilard's illustrious work in nuclear physics was cut short by disputes with Gen. Leslie Groves, head of the Manhattan Project. But he stayed on in Chicago, maintaining a long-distance relationship with Trude by telephone and letter. He depended on her to collect clippings and other historical documents about him, to visit his father in New York, and to concur in the opinions he held about his own health. He and Trude took vacations together in the Rockies and planned other trips—not taken—to Hollywood and Bermuda.[13] Szilard recounted his dreams with Trude and interpreted hers. At the same time, while prizing his freedom, Szilard could be "terribly lonely," according to Victor Weisskopf, a longtime friend of both Leo's and Trude's. During some visits to Chicago, Weisskopf avoided staying at the Quadrangle Club because he "wanted to sleep" but knew that Szilard "only wanted to sit up talking, in the lobby or in my room."[14]

Unhappy with Chicago's climate and uneasy about the fitful progress he was making in biology, Szilard craved a permanent appointment at the University of Colorado's new biophysics department and negotiated for months with its director, geneticist Theodore Puck. The two had first met at the University of Chicago, and in 1948, when Puck left to found the new department, he had invited Szilard to Denver as a visiting lecturer. This prospect coincided with Trude's move from New York to Denver, in April 1950, to teach public health at the University of Colorado's Medical School. Beginning in November 1950, Szilard visited Trude regularly in Denver, and for weeks at a time, through the next spring and summer, when he came to town to lecture, he often stayed at Trude's apartment— the first time in their twenty-one-year friendship that they spent more than a few days together.

Friends and relatives who visited them in those days remembered Trude and Leo as a study in contrasts: she bustling about, picking up and rearranging her papers and knickknacks, puttering in the kitchen, and tending the phonograph as it played and replayed Mozart; he sitting in an armchair, reading the newspapers, scribbling on a pad, or fluttering his eyelids in a trance that led him in and out of sleep. Szilard enjoyed the Mozart when Trude played it and appreciated the attention she gave him, at least for a while.[15]

"Trude complained that Leo never stayed long because he said there was no comfortable chair," recalled Dr. Gertrude Hausmann, her medical-school classmate in Vienna and a close friend in Denver. But when a friend at the medical school gave Trude a big armchair, Leo still made his visits brief. For him it was more than the chair that made him uncomfortable. It was the whole situation and his cozy place in it.[16]

Trude was delighted to have Leo around but also felt uneasy because of her friends' and neighbors' conservative attitudes about marriage. "It was a small community, and we were seen together a lot," Trude recalled in a 1978 interview. "It was not like now, where it's more accepted not to be married." During the summer of 1951, Trude became embarrassed that one of her students had seen her and Leo staying together at a cabin in Estes Park, and that fall, with humorous discomfort, Szilard told a friend in Chicago that Trude might lose her university job because his visits to her Denver apartment contributed to her "moral turpitude."[17]

Despite the jokes, Szilard felt uneasy about their new intimacy for another reason: It threatened his notions of freedom and privacy. Before accepting the visiting lectureship, he considered Trude his only confidant and routinely sent her newspaper clips and photographs for "the family collection,"[18] He called her several hours a week and wrote to her every few days—sometimes twice a day. But Szilard discovered that being together for a week or two at a time in Denver was much more demanding than dashing off notes from planes or hotel rooms and chatting at bedtime on the telephone.

Szilard was uneasy about his career as well: His writing and research and conversations on biology stretched from consulting at the Conservation Foundation in New York to working in the laboratory in Chicago to lecturing in Denver. He became depressed and uneasy about his work and his future in February 1951 when a "snag" blocked a promising research appointment at the Rockefeller Institute.[19] "I do not believe that the atmosphere in which one can conduct serious science can last much longer," he complained from Chicago in April 1951, already sensing that his colleague Novick would be leaving. "What next I do not know, unfortunately."[20]

Trude tried to comfort Leo and urged him to visit a psychoanalyst, as she had done for years. But he resisted, only complaining from afar that his "soul" was not "in order."[21] They were together in Colorado all summer in 1951, and during that time, after conversations that bared Szilard's need for freedom with Trude's for love and attention, they finally agreed to wed.[22] But once Szilard had accepted the idea, he delayed setting a date. In Denver that fall, he and Trude once got as far as the waiting room of a justice of the peace, but Leo grew fidgety and walked out. Finally, on October 10, Szilard and Trude checked into the King's Crown Hotel in New York and from there visited her brother-in-law, Efraim Racker, a physician, for the required blood tests. On Saturday, October 13, Leo and Trude appeared before a city clerk and in a quick civil ceremony were married.[23]

No relatives were there, no celebration or honeymoon followed, and only one telegram marked the occasion: From girlfriends Trude had known from Vienna, it read "HOW IS THIS NIGHT DIFFERENT FROM ANY OTHER NIGHT?" Their reference was to the question asked at the seder during the Jewish Passover. Trude's friends had in mind that the only difference this night was that she and Leo were officially married.[24]

Friends and relatives had trouble understanding the long-term, long-distance relationship that had persisted between Leo and Trude, with its crosscurrents of freedom and dependence. Szilard had resisted marriage to the day he took the vow, and during that weekend he suffered fresh anxieties. The couple spent their wedding night at the King's Crown Hotel. On Sunday, Trude returned to Denver; and on Monday, Leo visited the New York City marriage-license bureau, retrieved a copy of their certificate, and mailed it to Trude. "Enclosed a photostat, congratulations!" he penned in a curt letter. As always, he addressed her "Dear Ch" and signed with a cryptic "Yours, L."[25]

On the Tuesday after their marriage, Szilard wrote Trude again, this time about wedding announcements they planned to mail sometime in early November, "if we so decide." As a precaution against further delays, the announcement gave no date; it simply read:

Gertrud Weiss	Leo Szilard
Married	
New York City	
October nineteen hundred and fifty-one	

Leo knew that Trude planned a party for her friends but made it clear the celebration should be held "before I come to Denver or after I have left."[26]

On Thursday, Szilard flew to Chicago and the next day met Anderson's query at the physics table, which only made him more anxious about his new status. On Friday, Szilard was back in New York for a conference on World Population Problems and Birth Control at the New York Academy of Sciences. In Chicago again by Sunday evening, Szilard complained in a letter to Trude that he was upset after just a week of wedlock. His "ego" was "very rebellious," he reported. That week he had already inquired "in detail" about divorce procedures in several states. "But then I lost all hope of 'freedom' and I felt terrible; almost incapable to work at the lab., absent-minded, sweats, and high pulse rate. This has been going on for three days." He worried anew about sending the wedding announcements and complained that he found it impossible to work in Chicago or

Denver. Again he shunned Trude's repeated suggestion that week to see a psychoanalyst.[27]

Szilard and Trude struggled with his anxieties by phone and by letter, he insisting on self-help efforts and aggressive distractions (he called them "constructions" or "constitutionals") to improve his attitude, she urging analysis by a marriage counselor or psychiatrist. Only his side of their correspondence survives, but it is pained by both apologies and anger. He is not "mad" at her, did not want to "punish" her, worries often about their shared and separate problems.[28] "How things will continue in the end depends on you," he wrote. "In the meantime, I do not want to talk about it."[29] Trude, also self-absorbed and anxious, became annoyed with Leo's stubborn, almost childish need for independence. Szilard enjoyed chatting with the children of his friends and associates and praised the "terrible and inexorable logic of the child" while sometimes thinking and talking as impulsively as they do. His host in Denver, Theodore Puck, found Szilard's rapid and impertinent ideas disruptive to the routines of the biophysics laboratory. "Leo's real problem," Trude once told Puck, "is that he doesn't know what he wants to be when he grows up." At times, Leo thought the same of Trude.[30]

Forsaking his work in Chicago early in 1952, yet unable to secure a full-time post that would allow them to live together in Denver, Szilard migrated to New York, where he felt most "at home" living in the King's Crown Hotel.[31] From there he accepted historian Max Lerner's invitation to lecture at Brandeis University and also began part-time work as a consultant to industrialist Abram Spanel, president of the International Latex Corporation, and often stayed at his mansion in Princeton.

Lerner saw Szilard as "painfully and profoundly lonely" in the first years of his marriage and tried to engage him in a circle of friends around New York City and Long Island that included socialite Kitty Lehman. During lively dinner parties at her Long Island estate, Szilard's moods and manners were erratic. One minute he was silly and winsome, the next, sullen and withdrawn. "Had dinner with Kitty L. and Co. (at a restaurant)," Szilard wrote Trude from New York, "they drank a little much (me not at all) and afterwards she put her arms around me and said: 'Poor Leo, all your friends are drunk.'" Lehman liked the quip that Hungarians are really superintelligent Martians and, in one mock-serious presentation after a dinner at her Park Avenue apartment, dubbed Szilard an "honorary mortal."[32]

Szilard's alienation from human contact betrayed his own struggles between independence and intimacy, between the mind and the heart. In one impatient letter to Trude he complained that she was "too involved to really understand what is the matter with me" and should merely listen to learn "what I can do and what I cannot do." You "could have spared yourself a lot if you had listened more closely last summer and autumn," when they decided to marry.[33] He visited Trude in Denver for brief holidays, clearly cared about her well-being, and shared his thoughts with her in hasty notes and telephone calls. But in the first three years of his marriage, Szilard's immediate "family" became Abram and Margaret Spanel, whom he had first met through physicist Edward Condon.[34]

Abe Spanel was a self-made millionaire, a Russian-Jewish immigrant with a knack for business and a flair for posing original solutions to the world's ills, often in editorial advertisements he ran in the *New York Times* and the *New York Herald Tribune*. Like Szilard, Spanel was impulsive, often mixing practical detail and pure whimsy. They also shared an interest in patents, with Spanel urging Szilard to file applications for just about any idea that came to mind.[35] Spanel and Szilard even looked alike: both short and round, both paying lip service to diets as they nibbled and noshed with gusto. For a while, Szilard kept a notebook in his vest pocket and at day's end tallied the calories that each had consumed. Totaling these amounts, he then divided by two and declared that number to be their ideal diet. Weight watching with Spanel also led Szilard to design (and patent) a pencil-shaped pocket calorie counter. Frequently, the two men met for lunch in Manhattan, and they rode together by train to Princeton, where Szilard spent many nights and weekends in Spanel's Greek Revival–style twenty-one-room mansion, Drumthwacket, since 1981 the New Jersey governor's official residence.

Spanel and Szilard chatted constantly about world affairs and in the fall of 1952 even dabbled in a political crisis; impulsively the two men flew to La Paz trying to prevent nationalization of the tin mines. There to visit Spanel's good friend Enrique de Lozada, Bolivia's ambassador to Washington, they called on government officials and labor leaders, lunched with the president, and met mine owners and diplomats. On his first and only trip to South America, Szilard found La Paz "unbelievably beautifully situated," the Indians "very friendly," the Bolivians "very bright and educated." For a few days he and Spanel had to share a room at the Hotel Sucre, and "feeling a little inhibited about snoring," Szilard did not sleep well. Although he also became winded in the high altitude and suffered nosebleeds, he found the whole adventure "refreshing."[36] At breakfast in the hotel Szilard encountered John and Leona Marshall, his former Manhattan Project colleagues whose wedding he had attended during the war.

"Good morning," Szilard said casually, as if the three old friends met there often.

"What brings you to La Paz?" John Marshall asked.

"Just traveling," Szilard said in a mildly conspiratorial tone, and later added that his purpose was to "buy fabric" for a coat.[37] Despite the many meetings with all participants in the dispute, Spanel and Szilard had no influence on the situation, and Bolivia's tin mines were nationalized on October 30, 1952.

In Princeton, Szilard also relished the company of Spanel's vivacious wife, Margaret, and spent many evenings with her before the living-room fireplace at Drumthwacket, talking out the "inner conflict" of his own marriage. Although Szilard had rejected psychoanalysis, in these rambling, soulful conversations he became unusually confessional as he probed the nature of personal attachments and struggled to understand his own long-distance relationship with Trude. Yet while he was finding some personal comfort living at Drumthwacket, he still had trouble empathizing with Trude's loneliness. In July 1952, Szilard wrote her that "it is sad that you feel bad and that in addition to all the real troubles you also plague yourself with chimeras."[38] While still living at the Spanels' more than a year later, Szilard advised Trude that "to invite somebody over every evening is nice, but it is not a solution for your problem (to live alone) because when you need it most the initiative is lacking to arrange it. It is too bad that you always get so irritated about Princeton (and also very silly)."[39]

Except for the private conversations with Margaret Spanel, Szilard refused to discuss his personal affairs or even reveal that he was married. During one of the Spanels' festive Sunday lunches in the large, sunny dining room, Szilard fell silent and tried to ignore the conversation. Charlotte Howell, the wife of a Princeton English professor, was seated at Szilard's right, and the more she tried to engage him, the more he resisted.

"You are monumentally quiet, Dr. Szilard," she told him, but he continued to eat and said nothing. When the conversation spun from Asian culture to marriage and to Abe Spanel's essay "The Plea for Monogamy," Mrs. Howell looked again to Szilard.

"How do you feel about monogamy?" she asked him directly.

Now unable to avoid her question, he kept his face in the plate and answered. "Madam," he declared, "I am against rationing in any form," and scraped a forkful of food.

At another lunch, Szilard met Dr. William and Louise Welch; she was high priestess of the Gurdjieff Movement, a religious group devoted to a Russian-Greek mystic whom many considered a pretentious fraud. Szilard was quiet, even sullen, and clearly annoyed by the guest couple's conceit and their boasts about attaining physical perfection through yoga. After lunch, in the living room for coffee, Szilard entered to find Dr. Welch sipping his cup while seated in the lotus position by the fireplace. Welch looked up from the floor, eyed Szilard's girth, and said loudly:

"I bet you can't do this, Dr. Szilard."

Glowering, Szilard stared down at Welch. "There are those who think the important thing about the Buddha is the way he sat," said Szilard. "There are others who think the important thing about the Buddha is the way he thought."[40]

When he wished to be, however, Szilard was engaging and lively. "You were so interested in fishing and the country," one luncheon guest at the Spanels' reminded Szilard years later.[41] And when Szilard's longtime friends Eugene Wigner and John von Neumann (both Princeton residents) came to lunch, the three Hungarians delighted in a conversation so brisk that no one could complete a sentence before his thought was augmented by another's idea, round and round in bursts of intelligent agitation.

Along with the rousing conversations, Drumthwacket could also be a menacing place for Szilard. One fall day in 1954, with a hurricane predicted, Szilard and Abe Spanel took a midafternoon train from New York in order to be home before the storm struck. After dinner that night, Margaret Spanel produced a copy of *A High Wind in Jamaica*, a Richard Hughes novel with a suspenseful account of a hurricane. Seated by a blazing fire, she began to read aloud about the tense calm before the storm, and when glancing up at a dramatic pause, she noticed pain and fear in Szilard's face. He seemed pale and so frightened that she stopped reading.[42]

In less dramatic settings, too, Szilard suffered anxiety and gloom during the 1950s, including a spell of self-diagnosed "psychological depression" and a "psychosomatic" rash on his face.[43] In the fall of 1954, Szilard complained about double vision and began wearing a black eye patch.[44] Then his back and shoulder ached, and an uncertain ailment even led him to consider hospital treatment. For the first time in his life, Szilard drafted a will.[45]

By letter and telephone Szilard repeatedly told Trude to "cheer up" as she complained about her own depression and mood swings. Still, their letters from this period seem preoccupied with illnesses, with both physical and mental pain.[46]

By 1955, when alternatives to his Chicago professorship failed to open up, Szilard turned harshly dour and began to worry about his own fate as well as mankind's. Physicist Victor Weisskopf learned of Szilard's poor health and despair and arranged some pleasant publicity to cheer up his longtime friend: On May 18, 1955, the US Patent Office awarded the late Enrico Fermi and Szilard a patent for their codesign of the first nuclear reactor.[47]

But despite public acclaim, Szilard's own sense of gloom only deepened in 1955 with the illness and death of the two men most important in his life: Albert Einstein and his father. In a sense, Einstein was Szilard's intellectual father, a mentor and authority figure he had turned to for advice and aid throughout his adult life. Beginning in 1921, Einstein had endorsed Szilard's novel doctoral thesis, devised and shared patents, wrote recommendations for immigration and employment, vouched for the Bund and other altruistic schemes, three times signed letters to President Roosevelt about the A-bomb, and served as figurehead for the Emergency Committee of Atomic Scientists, which Szilard helped create. In 1954, Szilard had asked Einstein to promote his application for a research post at the Albert Einstein College of Medicine, and as late as the spring of 1955, Szilard had drafted for Einstein a cover letter to Indian Prime Minister Jawaharlal Nehru, urging him to read Szilard's views on the US-Chinese dispute over the islands of Quemoy and Matsu—at the time, Szilard feared the United States was prepared to use nuclear weapons against China.[48]

On April 11, five days after sending Szilard's letter to Nehru, Einstein signed the Russell-Einstein Manifesto, a joint statement with British philosopher Bertrand Russell calling on the world's scientists to unite in their efforts for nuclear disarmament and peace. It was Einstein's last correspondence. The next day severe pains and cramps interrupted his work; then he collapsed from a chronic heart ailment. On April 18, at the age of seventy-six, he died at a hospital in Princeton.[49]

Interviewed about Einstein on national television that day, Szilard was grim when recalling a visit shortly after Hiroshima. "'You see now,' Einstein said to me, 'that the ancient Chinese were right. It is not possible to foresee the results of what you do. The only wise thing to do is to take no action—to take absolutely no action.'"[50]

At the time of Einstein's death, Szilard's ninety-four-year-old father, Louis, lay ill at the Saw Mill River Convalescent Home in Yonkers, New York, near where his son Bela and family lived. Louis Szilard had maintained robust health for most of his life, but by 1953 had developed vitamin deficiencies and a heart condition. Leo's visits were always brief but apparently intense, and in his final years Louis Szilard wrote that he was pleased with his life, proud of his sons' achievements, and grateful that the United States had offered them all such opportunities. Leo and his father were always formal with each other, but when he died, Leo became upset and emotional, then left the room to hide his feelings from his brother, Bela, and his sister, Rose.[51]

Szilard could not conceal his depression from his good friend Eva Zeisel, niece of the Polanyi brothers, wife of a University of Chicago law professor, and a confidant since they first met in Berlin in 1927. A sculptor and ceramic designer, in 1955 she lived next to her mother, Laura Polanyi Striker, in a small fifth-floor apartment at 431 Riverside Drive in New York City. Mrs. Striker had run an avant-garde kindergarten in Budapest early in the century (writer Arthur Koestler was a pupil) and caused a scandal when she had the children dance naked to Beethoven. In later life she had researched and written an account of Capt. John Smith's expedition to Hungary, made a few years before his celebrated journey to Virginia. Szilard often attended tea at Mrs. Striker's, a rambling apartment with Biedermeier furnishings, but he usually just sat in an armchair amid the lively conversations, stared into the distance, and said nothing. To lift Szilard's spirits, Eva sometimes walked across the Columbia campus to the King's Crown Hotel, where he lived, and sat in the lobby with him, chatting in the eerie light cast by a large fishbowl. Once or twice a week, at about suppertime, Szilard turned up at her door. Eva Zeisel always invited him in, and after insisting, "I'm not hungry," Szilard always agreed to stay. Over dinner Szilard said little, appeared lonely and preoccupied, and usually left early.

"You look so bourgeois," she once teased him at the door as he pulled on his huge, double-breasted, blue vicuña topcoat. "It must give you a feeling of luxury."

"Still, I like it," he said, wrapping the thick collar around his neck. "It makes me feel warm."[52]

Szilard was restrained with Zeisel but could be more gregarious and "found great solace" talking to children (among them her daughter Jean). Szilard also enjoyed talking with Eva Zeisel's German-born friend Inge, then a twenty-two-year-old photojournalist, later the wife of the Italian publisher and political activist Giangiacomo Feltrinelli. At the Zeisel's country house in New City, New York, near the Hudson River town of

Nyack, Szilard seemed awkward as he sat about in "crumpled, sober city clothes," Feltrinelli recalled. But back in the city Szilard perked up in Inge's company, teased her in an avuncular way, and entertained her in his own quirky style: one night she remembers his taking her to a Mexican restaurant and a Japanese film. (Or was it the other way around?) Mostly she found Szilard "very depressed in these times, having no particular job," but remembered "one charming moment" when, at the airport as she was leaving for Europe, "he gave me fifty dollars, 'just for drinks, and you never know!' So nice of him, as he didn't have much money himself."[53]

In fact, in 1955 money was not one of Szilard's worries. He had about $21,000 in his checking account at the Chase Bank, was drawing a $10,500-a-year salary from the University of Chicago, and earned consulting fees and generous travel expenses from Spanel and the University of Colorado. What troubled him deeply, however, was his *future* earnings: He anticipated only a modest pension because he had joined the Chicago faculty at the age of forty-eight and could scarcely work for twenty years before retirement. By year's end, Theodore Puck finally admitted that he could not offer Szilard the full-time appointment in Colorado he had hoped for.[54] Puck's rejection upset both Leo and Trude. To him it meant loss of a full-time research job; to her, a missed opportunity to live in the same place with her husband.

Independently, Szilard's situation and behavior had come to anger even his fondest friends. Eva Zeisel thought Szilard mistreated Trude by not living with her in Denver, and she disliked the fact that he never mentioned to her friends that he was married. At one party in New York, which Leo and Trude both attended, he meant to introduce her as "Dr. Weiss" but slipped and called her "Dr. Wife." Biologist Maurice Fox, a longtime friend, recalled a situation in New York in the late 1950s when Leo and Trude attended the same medical conference separately. "When I met Trude, I proposed that she come to our house for dinner," Fox remembered, "and she said, 'Yes, if Leo's agreeable.' She searched around the conference to find him and only accepted our invitation when he had consented." As the Szilards' longtime friend Victor Weisskopf quipped about Leo, "It was always said that he has a wife in Denver, an office in Chicago, and lives in a hotel in New York."[55]

Most of the time, Szilard found it easier to dispense rational advice about marriage to others than to confront the emotional dimensions of his own life. Biologist Bernard Davis complained that he was confounded by loneliness and so insecure that he fell hopelessly in love with the first young woman he met. As a result, Davis said, he was unable to make a thoughtful decision about whom to marry. "I think," Szilard answered, "that if you can't stand to be lonely and can't wait to make the right choice, you should go to a place where statistics are in your favor. Get a fellowship and go to Denmark!"[56]

When trying to fathom his own peculiar marriage, Szilard theorized that his emotions and Trude's were somehow interdependent. "Maybe you felt so bad lately because of complementarity," he wrote her in 1957, referring to the theory in physics that electrons or photons can behave as either particles or waves but not as both at the same time. He reported that as he recently felt a need to channel his thoughts positively, in what he called "a massive constitutional," perhaps at the time she was having a pleasant time eating lobster. "I am gaining weight, and since the sum remains constant, you will probably have lost weight. And your depression is probably also because I enjoyed myself in Heidelberg."

In the same letter Szilard reported that *Brighter Than a Thousand Suns,* a new book about the Manhattan Project, had made him a "great man" in Germany. "I have sent you a book by airmail, today, for you to see what a fine husband you have. ('Better a fine husband far away than a horrid one close by.')"[57]

And yet for all his anguish and protest at being a husband, Szilard also found ways to be supportive, more in sickness than in health. When Trude was hospitalized in Denver after a serious automobile accident in 1957, he visited and called her often. Illness drew them together again, in 1958, when Szilard stayed in Denver while recuperating from a mild heart attack that he suffered during his last trip to Europe.[58]

In June 1959, Szilard brought Trude along to a Pugwash conference for the first time—the Fourth Pugwash Conference, on "Arms Control and World Security," held in Baden, Austria. And that fall a medical emergency finally united Leo and Trude for the rest of their lives. In Vienna after the conference, Szilard noticed traces of blood in his urine, had tests performed there and in Stockholm, and returned to New York, where doctors ordered him hospitalized. Immediately, Trude took a leave from her teaching in Denver, flew to New York, and moved into a graduate students' dormitory at Rockefeller University, near Leo's hospital. Illness, it seemed, was the best tonic for their ailing marriage, for it allowed Leo and Trude to realize how desperately they needed each other.

CHAPTER 23

Oppenheimer and Teller

 1946–1959

Leo Szilard's many trips to the Colorado Rockies freed him from Chicago's oppressive summer climate, but he could not escape the anticommunism that spread to college campuses and throughout the government in the postwar years. In Boulder on one visit he met Walter Orr Roberts, a solar astronomer who had been investigated by the House Un-American Activities Committee (HUAC) for his professional contacts with Soviet scientists. Szilard wrote letters supporting Roberts and probably had him in mind when penning "Security and Equality," an essay that argued that universities should continue the salaries of scientists removed from their jobs as security risks until they find new work.[1] These salaries might be subsidized, Szilard proposed, by deducting 1 percent from all academics' paychecks. At the time, Szilard also attacked the loyalty declaration required for Atomic Energy Commission (AEC) fellowships.[2]

In September and October 1950, Szilard himself became the target of CIA and FBI surveillance when his name was linked with physicists Philip Morrison and Katharine Way, whom he had known at the Met Lab during the war. Incorrectly, Szilard was also said to be an "associate" of George N. Perazich, who had recently been named as a member of a Soviet espionage ring. Szilard had no contact with Perazich or the Communist party, but the FBI still tried for years to uncover damaging information about him.[3]

"Szilard was outraged that nobody spoke up at the German universities against Hitler in the 1930s," recalled Roberts, "and he was outraged that nobody spoke up against HUAC and the AEC in the 1950s."[4]

In 1952, Szilard urged his colleagues to take "some collective action" when the AEC denied a security clearance to David M. Bonner, a biologist he had met at the Cold Spring Harbor biology symposium the previous summer.[5] And he tried to raise money for University of Colorado philosopher David Hawkins (a wartime aide to J. Robert Oppenheimer who had belonged to the US Communist party at Berkeley in the 1930s) and coached Hawkins on testifying before HUAC.

"Call your friends, and all agree to name each other and not to talk about each other," advised Szilard. "That way you're cooperative but you're not an informer." Growing furious at the thought of HUAC's heavy-handed tactics, Szilard urged Hawkins to be even more defiant.

"Say you'll testify but don't want to talk about your friends. Then answer their questions freely. Then announce at the end of your statement, 'Now I wish to rejoin the Communist party. It is no different from your behavior, and it has better aims.'"[6]

"Are you serious?" Hawkins wondered. Szilard said he wasn't really sure.

As he had done for refugee scholars from Nazi Germany in 1933, Szilard raised money to support the victims of anti-Communist discrimination. He urged University of Chicago chancellor Robert M. Hutchins to seek Ford Foundation support for researchers denied National Institutes of Health (NIH) grants because of problems with a federal loyalty oath, and he appealed for funds to Warren Weaver, director of the Rockefeller Foundation's natural-science programs.[7]

The AEC was the logical focus for America's anti-Communist fervor: It alone protected the atomic secrets behind the accelerating nuclear arms race, and through its network of national laboratories the agency dominated federally funded science. With the Eisenhower administration in 1953 came a new intensity to the cold war and new precautions to protect the nation's common defense and security.[8] Soon after financier Lewis L. Strauss became AEC chairman that July, the commission was embroiled in a bitter scandal over physicist Oppenheimer's security clearance. Wartime director of the Manhattan Project's Los Alamos laboratory that had designed and tested the first A-bombs, Oppenheimer then headed the Institute for Advanced Study at Princeton and served as a consultant to the AEC.

It all started with solicited charges made to FBI agents the year before by Edward Teller, ambitious promoter of the "super," or H-bomb, and an obsessive anti-Communist. Teller saw Oppenheimer as an opponent of his cherished weapon and in two interviews raised doubts about his

Communist sympathies. When FBI director J. Edgar Hoover received charges against Oppenheimer's loyalty in a letter by William L. Borden, former executive director of the congressional Joint Committee on Atomic Energy, Hoover promptly reported Borden's allegations to the White House. President Eisenhower quickly ordered a "blank wall" erected between Oppenheimer and the AEC's secrets.[9] The AEC revoked Oppenheimer's clearance on December 23, 1953, and when he protested, a special Personnel Security Board was set up to review the threat he posed to "national security."

As the incident became public, Szilard deplored the AEC's use of clearances for what seemed to be political and ideological ends; Oppenheimer's public questioning of the H-bomb was at odds with the administration's enthusiasm to test and deploy the new weapon. Typically, Szilard struck back first with political satire. In Colorado during the Christmas holidays, he wrote "Security Risk," a fictional tale of anti-Communist paranoia and fears of blackmail in the State Department.[10] Szilard even wrote himself into the story, as a character advising the department to "simply publish once a month a list of known homosexuals on your payroll," for "clearly those whose names you have made public can no longer be blackmailed." The story's narrator confessed that "with Szilard, I never know when he is serious and when he is joking, and I suspect that often he does not know himself."[11] But the outcome in Szilard's satire was no joke: Based on details that were later proven to be false, the State Department imposed rigid security restrictions that forced a diplomat of unquestioned loyalty to lose his security clearance and his job.

On New Year's Day, 1954, at the stately Broadmoor Hotel in Colorado Springs, Szilard penned a letter to the editor criticizing security checks but apparently never mailed it.[12] The next day, however, he drafted "Cyclotron," a satire on academic politics that included the anti-Communist antics of Wisconsin's Republican senator Joseph McCarthy and the cumbersome procedures for obtaining security clearances.[13] That spring, Szilard drafted a "statement" about Oppenheimer, again half sarcastic, half serious. "We are now rapidly approaching a state of affairs when scientists will say to each other, 'Some of my best friends are security risks.'" Of the many scientists that Szilard spoke with about the case, none thought Oppenheimer would leak secret information to Russia. And if real grounds for suspicion did exist, Szilard wrote, then with all that Oppenheimer knew, "wouldn't arresting him and shooting him without trial be the only prudent course of action from the point of view of 'National Security'?"

Szilard's caustic humor brewed to outrage in his statement's final line: "Classing Oppenheimer as a Security Risk and subjecting him to a formal hearing is regarded by scientists in this country as an indignity and an affront to all; it is regarded by our friends abroad as a sign of insanity—which it probably is."[14]

Szilard and Oppenheimer had first met in 1945 during the Manhattan Project. In March of that year Szilard had telephoned Oppenheimer to seek his advice and help on an attempt to educate the cabinet about the postwar problems that would be created by the A-bomb. In May, just before his visit to the White House and to James Byrnes, Szilard wrote Oppenheimer to argue that a nuclear arms race might be forestalled if the United States decided not to use the A-bomb against Japan. Szilard assumed, correctly, that Oppenheimer favored using the new weapon but still invited his views, enclosing at the same time the memo he had prepared for President Roosevelt on the need for postwar arrangements. When Szilard and Oppenheimer met in Washington at the end of May, they disagreed over using the A-bomb on Japanese cities, and that summer, at Los Alamos, Oppenheimer barred the circulation of Szilard's petition to President Truman.

After the war, both men favored the atom's civilian control at home and its international control abroad, although with differing priorities: Szilard thought one would lead to the other, while Oppenheimer thought, correctly, that the fight for a civilian atomic agency in the United States would deflect the scientists' efforts to gain international control. As a result, Oppenheimer had at first supported continued army control of the bomb and only backed the McMahon bill when passage seemed assured. Yet in 1946 they each wrote a chapter in *One World or None,* the scientists' book on nuclear arms control and disarmament, and they conferred in 1947 when Szilard's passport was revoked by the State Department to prevent him from attending an international peace conference.[15]

In 1948, Szilard approached Oppenheimer to ask if the Institute for Advanced Study might be interested in "applying scientific methods of investigation to the problem of racial discrimination and race relations." Szilard had read that under Oppenheimer the institute was interested in "taking up problems to which scientific methods have not been applied in the past" and reported that he knew of possible funders for such a project.[16] In 1950, Szilard and University of Chicago chancellor Hutchins tried to enlist Oppenheimer in a fund-raising project for a study of ways to maintain peace, but at this Oppenheimer demurred.

Szilard disliked Oppenheimer personally, especially for his hypocrisy with fellow scientists, saying during the Manhattan Project that they should have no role in setting policy when he pressed his own views on politicians and government officials freely. But Szilard saw a threat to all scientists when the AEC revoked Oppenheimer's security clearance, and he tried to influence the review board in several ways. Szilard wrote to scientists who might be called to testify, urging them to support Oppenheimer.[17] Szilard paraphrased his draft "statement" for newspaper reporters, and he was quoted at length about the AEC hearing in the American Communist party's *Daily Worker*, a fact not missed by FBI agents at the time and duly noted in Szilard's file.[18] Szilard also helped draft an editorial for the *Bulletin of the Atomic Scientists* that declared that charges Oppenheimer is a security risk are "contrary to both decency and common sense," although he later complained when his name was used with the statement.[19]

But most importantly, Szilard tried to prevent Edward Teller, Oppenheimer's most influential accuser, from appearing before the AEC review board. For, unlike Teller, Szilard did not consider Oppenheimer a national security threat. This put Szilard on the spot: opposing someone he liked but disagreed with while supporting someone he agreed with but disliked.

Beyond Teller's anti-Communist paranoia was a vindictive hatred of Oppenheimer for opposing development of the H-bomb. Teller had criticized Oppenheimer before the Joint Committee on Atomic Energy as early as 1950, and in two FBI interviews in 1952 he made charges that would later form the basis of the AEC's case.[20] But Teller was also Szilard's friend, despite their fierce differences over national security issues, such as building the H-bomb and negotiating with the Russians. According to Szilard's widow, he wanted to save his friend Teller from his own "worst instincts."[21]

Szilard and Teller had first met casually in Budapest and became friends in London in 1934–35, when both were refugees from Nazi Germany. At the time, Szilard explained his chain-reaction theory to Teller along with his fearful prediction that it might lead to a nuclear explosion. While impressed, Teller later recalled, he "did not then expect that it would actually be realized."[22] They also worked together to enlist the US government's support for A-bomb research in 1939, through Einstein's letter to President Roosevelt and their participation in the Advisory Committee on Uranium that the letter prompted. Just after the war, they socialized when both men taught at the University of Chicago, and in 1952, Szilard and Teller were among the scientists who attended physicist Joseph Mayer's popular Thursday afternoon seminar at the Institute for Nuclear Studies.[23] The same year, Szilard invited Teller to join a group of scientists and academics to meet with Adlai E. Stevenson, later the Democratic candidate for president,[24] but the meeting was canceled.

When Teller arrived in Washington for the Oppenheimer hearing in April 1954, he was eager to speak with AEC chairman Lewis L. Strauss.[25] Teller had earlier sought Strauss's help to arrange the release of his mother, Elona Teller, and his sister, Emma Kirz, from Hungary. It is possible that Teller's concern about his family increased his eagerness to please Strauss by publicly criticizing Oppenheimer, although he had strong personal and professional motives for doing so as well.[26]

According to Trude Szilard, her husband came to Washington on the eve of Teller's scheduled testimony and set out from his hotel to find his friend. Szilard rode taxis to restaurants and clubs, walked to other hotels, but after hours of searching, finally returned dispirited. "If Teller attacks Oppenheimer," Szilard grumbled to Trude, "I will have to defend Oppenheimer for the rest of my life. . . ." Half joke, half gibe. Nevertheless, the search was futile. Teller was sequestered that night with Roger Robb, the AEC prosecutor, and with Hans Bethe and his wife in their hotel room.[27]

On the witness stand the next day, Teller was asked directly: "Do you or do you not believe that Dr. Oppenheimer is a security risk?" Teller answered, in part:

> . . . I thoroughly disagreed with him in numerous issues and his actions frankly appeared to me confused and complicated. To this extent I feel that I would like to see the vital interests of this country in hands which I understand better, and therefore trust more. In this very limited sense I would like to express a feeling that I would feel personally more secure if public matters would rest in other hands.[28]

By these words, Teller drove a wedge between himself and many of the country's leading scientists. In May 1954, like the character in Szilard's "Security Risk" satire, Oppenheimer was certified "a loyal citizen" by the AEC board but still denied his clearance.[29] Later that year, Szilard proposed to Oppenheimer that he defend Teller publicly. Szilard added, however, that he could not tell if this proposal was serious or not.[30] Teller later lamented that from his testimony "I lost practically all my friends." But he was pleased that although "Szilard disapproved very much," they could still like one another.[31]

A few weeks after the Oppenheimer hearing, Strauss contacted CIA director Allen W. Dulles to seek help in releasing Teller's family—an approach that ultimately failed.[32] Szilard remained friendly with Teller after the hearing and later urged him to defend himself before his scientific peers, but the two continued to disagree about most military and political matters. Yet Szilard advised Teller about his right to a German pension[33] and urged him to participate in several scientific conferences at which the nuclear arms race was discussed. Teller always refused.[34]

For his part, Teller liked and defended Szilard and proposed his name for the annual Atoms for Peace Award. Citing both Szilard's "genius" and the way he "disregards some social conventions," Teller acknowledged that Szilard's odd behavior had "cost him dearly" during his career.[35] Their friendship remained "strong," Teller would later recall, "but strangely unemotional" in its manner.

By contrast, intellectual arguments between the two friends were often emotional and intense. Over dinner in Washington in the spring of 1958, Szilard urged Teller to visit the Soviet Union as a way to soften his hostility to the country. Teller's response was curt: "I have no intention of visiting Russia."

"Why, why?" Szilard asked. "Don't be unreasonable."

"Look," said Teller. "There are many reasons, but one of them is this."

"Is what?"

"My mother and sister live in Hungary. Once I'm in Russia I don't know what the Communists might do to force me into a situation which I don't like. . . ."

Szilard listened, shaking his head. "Teller, you are completely wrong. The Russians would never stoop to such methods. But I understand how you feel. Let me see what I can do about it."[36]

Szilard considered this reply wrongheaded and typical of Teller's anti-Soviet paranoia, but from that conversation he also realized a personal reason for "the emotions which manifest themselves in some of Teller's public statements, as well as the emotions he had displayed in private conversations." The next day, Szilard wrote to the third secretary of the Soviet embassy in Ottawa, seeking to arrange the release of Teller's mother and sister.[37]

Nothing happened until August, when Szilard was at the ski resort of Kitzbühel, Austria, for a Pugwash conference. Szilard and chemist Harrison Brown met with Soviet academician Alexander Topchiev privately to discuss plans for a US-Soviet scientists' conference. A Moscow meeting had been proposed for that summer but was postponed when some American scientists—among them Teller and Eugene Wigner—followed AEC chairman Strauss's advice and refused to participate. Topchiev promised knowledgeable and influential Soviet participants. "If you bring Teller," he said, "we will produce the Russian Teller,"[38] no doubt having in mind physicist Andrei Sakharov, the Soviet Union's leading H-bomb designer.

At the Kitzbühel meeting, Szilard again told Topchiev that one reason for Teller's bitterness against the Russians was the fate of his mother and sister. Topchiev promised to help. After the conference, on the train down the mountain passes to Vienna, Szilard was sitting in a compartment next to the Hungarian physicist Lajos Jánossy.

"I understand you want to talk to me," Jánossy said. Topchiev had mentioned Teller's problem, and during the ride Jánossy, Szilard, and Brown discussed ways to arrange the women's release. Jánossy was the stepson of Marxist philosopher George Lukács and himself a well-known Communist, so it was possible that he might have influential friends in Budapest and Moscow who could help.

The three men talked as the train descended from the jagged Alps and cut through broad valleys between Kufstein and Salzburg, then through farm fields punctuated with fanciful, vegetable-shaped steeples—so many baroque turnips, pears, figs, and radishes atop pastel churches. They rode by the cliff-top Benedictine abbey at Melk and glided through the shady Vienna woods and into the city. At his favorite Vienna hotel, the Regina, Szilard telephoned Teller in Berkeley, asking for the names and addresses of his mother and sister.[39] Szilard also kept in touch with Janossy before his return to Budapest, and the two had a snack in the Regina Hotel before going off to an American-Canadian reception organized for the Pugwash participants.[40] But after this August encounter, Szilard heard nothing more from Jánossy until after Christmas, when a letter arrived in Chicago. "I am very glad to be able to tell you that the mother and sister of Dr. Teller have received permission to leave the country."[41]

Soon after Teller's mother and sister arrived in San Francisco, in January 1959, Szilard had occasion to telephone Teller. When he heard the good news, he said: "Give my condolences to Mici," Teller's wife. Teller didn't understand the quip, but Mici replied, "Leo, you are the only completely honest man I know."[42]

CHAPTER 24

Arms Control

 1946–1959

In January 1946, *Life* magazine's readers saw Leo Szilard in a typically impish pose. Under the headline "Scientists Scare Congress," Szilard wearing his favorite tan trench coat, his eyes roving intently, but his round jaw seeming to stifle a broad grin, stood in a shadowy attic room with two colleagues. Capturing the "ironically humble Washington office" of the Federation of Atomic Scientists (FAS), the pictures showed "a suite of borrowed rooms on [the] top floor of an old, ill-heated brown-stone building." These scientists were the "League of Frightened Men," applying their "coldly intellectual kind of pressure" as they lobbied members of Congress. This pressure, *Life* concluded, had already contributed to a Big Three conference of the United States, Britain, and the Soviet Union in Moscow, in December, to seek postwar controls for the A-bomb.[1]

Although the US conferees in Moscow had ignored Szilard's suggestion to arrange meetings among Soviet and American scientists,[2] his writings, speeches, and congressional testimony on inspection and verification schemes to forestall a US-Soviet arms race had already identified what would become the nuclear era's trickiest political problem. With his FAS colleagues in Washington, with other groups he helped to found in Chicago and Princeton, but mostly working on his own, Szilard struggled after the war to devise—and sell to political leaders—new ways to lessen mistrust and increase accord between the US and Soviet governments.

Szilard's ardent faith in the powers of human reason, and his view that individuals and elites run the world and only need to be educated to assure peace, led him repeatedly to propose high-level discussions among statesmen and scientists, including his attempts to reach Soviet premiers Joseph Stalin and Nikita Khrushchev directly. After the war, Szilard also took up biology and worked fitfully in that expanding field. But in daydreams and nightmares, the fear that nagged him most was the bomb.

Unfortunately, most of Szilard's attempts to control the weapon he had helped create were too visionary. Too rational. Too clever. Too impatient. And too quixotic to deal with the world's newly complicated and dangerous postwar situation. "He was always ahead of his time," concluded Hans Bethe, a physicist who worked with Szilard on many arms-control projects. "He had many good ideas, but seldom carried them to completion."[3] To do that, Szilard would have to manipulate expanding military and diplomatic bureaucracies in the US government and new forums at the United Nations rather than the secret contacts he knew before and during the war.

Szilard's first mission in 1946 was to shift federal nuclear programs to civilian hands, and he spent all spring lobbying with the FAS and other groups to defeat the May-Johnson bill, by which the army would have retained control over atomic energy. Szilard and most FAS colleagues favored the McMahon bill to create an independent Atomic Energy Commission (AEC), but when Pres. Harry S. Truman endorsed it that spring, the army's allies retaliated with an amendment by Sen. Arthur Vandenberg to give the AEC a permanent Military Liaison Committee with access to all commission business.[4] Vandenberg's amendment was the issue that finally split the atomic scientists: J. Robert Oppenheimer, Ernest O. Lawrence, and Enrico Fermi sided with the army. Oppenheimer, in particular, argued that the May-Johnson bill was a lesser evil that should be passed quickly so they could lobby for the atom's international control. These three colleagues had gathered at a downtown Washington hotel room when physicist Edward Condon and Szilard swept in dramatically, both eager to defend McMahon's approach. Fermi, who believed that Szilard exaggerated military restrictions on atomic science, stared for a moment.

"Leo," he said, shaking his index finger, "you don't always tell the truth." Szilard glared at Fermi, and without saying a word, he and Condon spun on their heels and stalked out.[5]

Vandenberg's amendment prompted the FAS scientists to join a coalition of citizens' groups called the National Committee for Atomic Information (NCAI), which organized a lobbying campaign that hit the McMahon committee with 42,189 pieces of mail. But the army exploited news of an atom spy ring's arrest in Ottawa by announcing plans to bar Communists from all "sensitive" positions.[6] These polarizing events, bitter divisions within the scientists' own ranks, and a majority conservatism of McMahon's committee all merged to pass the amendment 6 to 1, perpetuating the military's strong voice in atomic policy.[7]

Glum and dispirited by the army's continued influence, the scientists found what relief they could in humor. Los Alamos physicist Philip Morrison got a laugh whenever he described the member of McMahon's committee who groused, "Why is it that witnesses we have here, these scientists, have such difficult names, like this man 'Zillard' or whatever it is? Can't we have some scientists with real American names?" The senator asking these questions, Morrison said, "was named Bourke B. Hickenlooper."[8] Szilard joked about the army's obsessive secrecy and at a dinner party in Washington amused senators and representatives with tales about heavy-handed tactics. In one account, which Drew Pearson later featured in his nationally syndicated "Washington Merry-Go-Round" column, Szilard recalled walking into his office at the Met Lab during the war to find a bookcase turned toward the wall—an army officer's way to conceal its secret contents. Szilard later complained that "the most powerful weapon of this war was not the A-bomb, but the secrecy stamp."[9]

For a time, the atomic scientists were celebrities, their views widely reported, their suggestions respected. One Sunday in March 1946, book reviews in both the *New York Times* and the *New York Herald Tribune* gave front-page acclaim to *One World or None,* the scientists' "Report to the Public on the Full Meaning of the Atomic Bomb." This collection of essays included articles by physicists Arthur Compton, Niels Bohr, Eugene Wigner, Oppenheimer, Louis Ridenour, Condon, Morrison, Frederick Seitz, Bethe, Szilard, and Einstein; by chemist Harold Urey; and by the distinguished political columnist Walter Lippmann. Szilard's chapter, "Can we Avert an Arms Race by an Inspection System?" identified just the issue that has bedeviled arms-control negotiators to this day, although his rational solution—teams of inspectors to monitor scientists—seemed extreme. Szilard admitted this surveillance would only buy time, to avert an arms race while a "world community" creates "permanent peace."[10]

But the *Trib*'s reviewer, science editor John J. O'Neill, found even these temporary steps too intrusive, complaining that Szilard "struggles, apparently without much happiness in his effort, to justify an international Gestapo snooping into laboratories, a procedure utterly repugnant to every scientist and destructive to the foundation of free scientific inquiry."[11]

The *Times* review, by reporter Rufus L. Duffus, was printed around an ominous pen-and-ink sketch of the Manhattan skyline exploding. "Leo Szilard outlines an inspection system which might be effective if every country, including the United States and Russia, would open its mines and factories to UNO agents," he wrote.[12] Duffus and other readers were also impressed and frightened by the conjecture of Morrison—an early US observer at Hiroshima—about how an A-bomb might destroy mid-town Manhattan. "This book does not leave one in a hopeful state of mind," Duffus wrote. "The beautiful optimism that has buoyed Americans up during most of their history is in for a shock."[13]

The day after these reviews appeared, President Truman named financier Bernard Baruch to be US representative to the United Nations Atomic Energy Commission (UNAEC).[14] A self-styled "adviser to presidents" and a wartime counselor to Secretary of State James Byrnes, Baruch had written an influential report on postwar economic conversion but seemed to know little about scientific or international affairs.

Ten days after Baruch's appointment, the State Department released its Acheson-Lilienthal Report, which proposed international control of atomic energy under the United Nations. A small group headed by Under Secretary of State Dean Acheson and David Lilienthal, chairman of the Tennessee Valley Authority, had drafted the report since January. This report pleased the scientists, who thought that their calls for international control of the atom were finally answered.

In April, McMahon thanked Szilard for his "splendid cooperation" on the bill's Senate passage, but by this time he was back in Chicago thinking more broadly about the fate of the postwar world.[15] Szilard gained national attention in May when the Associated Press published a twenty-eight-page supplement for its newspaper subscribers by science writer Howard W. Blakeslee on "The Atomic Future." Pictured with Einstein and Alexander Sachs as one of "Those Who 'Sold' the Atom to America,"[16] Szilard now had to sell its control to the world.

Szilard called on Einstein in May 1946 to propose creating an Emergency Committee of Atomic Scientists (ECAS) to raise money for public education on atomic energy, for the citizens groups in the NCAI. Just after Hiroshima, Einstein had told Szilard that he wished he had done nothing to develop atomic energy, but soon he wrote a chapter for *One World or None*, dictated an *Atlantic Monthly* article advocating worldwide control of atomic energy, signed a petition to

President Truman against the May-Johnson bill, and publicly endorsed world government.

Einstein trusted Szilard, usually followed his suggestions, and in this spirit he signed a May 23 fund-raising telegram that announced formation of the ECAS. (Other founders included Bethe, Condon, Urey, and Szilard.) "The unleashed power of the atom has changed everything save our mode of thinking and we thus drift toward unparalleled disaster." Einstein warned.[17] This statement, which has been widely quoted ever since, was actually drafted by Szilard and Harold Oram, a press agent and fund-raiser for liberal political causes.[18]

In June 1946, as the House Military Affairs Committee amended the Senate-passed McMahon bill, as Baruch delivered the US plan for the atom's control to the UNAEC, and as Soviet delegate Andrei Gromyko began months of debate—and delay—with a counterproposal,[19] Szilard bobbed from Washington to New York, via California. He traveled west to assure that his role and Einstein's were portrayed accurately in *The Beginning or the End*.[20] On M-G-M's movie lot in Culver City, producer Samuel Marx handed the script to Szilard, who went off, read it, and returned to declare: "It's lousy." He told Marx just how Einstein had written to FDR—not after rapturous inspiration, as the script portrayed, but after a visit by Szilard and Wigner.[21] Eager to keep the character of Einstein in his film, Marx agreed to correct the story, gave Szilard an office, and urged him to rewrite this important scene; Szilard did, and several other episodes as well.[22] Then he boarded a plane for New York to talk up the Baruch Plan.

That month, Szilard also journeyed to Colorado, where he spoke at a conference on the "Regional Consequences of the International Control and Utilization of Atomic Energy,"[23] continuing a frantic life of public appearances and private negotiations aimed at saving the world from his own invention. Szilard and his fellow scientists gained fresh notoriety in August when *Life* magazine touted *Atomic Power*, a film in the monthly *March of Time* series: "While Hollywood is laboring over a ponderous epic of the atomic bomb, the *March of Time* has scooped them with a remarkable piece of living history."[24] *Atomic Power* "tells how the bomb was made, but instead of actors [has] the real scientists and directors of the Manhattan Project to re-enact their own parts." One of the scenes was captioned "ALBERT EINSTEIN is asked by Physicist Leo Szilard to urge US atomic-research backing. Then Einstein wrote famous letter to Roosevelt." Reenacted on the back porch of Einstein's Princeton home, the scene showed a pensive professor, in shirtsleeves, sucking a churchwarden pipe; a formal Szilard, as always wearing a necktie, was seated at his side to study the letter. In another scene, scientists, including Oppenheimer in his familiar porkpie hat, fiddled with electronic "controls which set off the first atomic bomb ever exploded, on July 16, 1945."[25]

While Oppenheimer relished his screen debut, Szilard seemed stiff and self-conscious in his four scenes. In the Columbia laboratory with Dean Pegram, Szilard eyed an instrument nervously, pointing with a pencil to some calculations on paper. Over the letter to FDR, Szilard pointed stiffly as Einstein slouched forward. In a scene where Einstein approved the fund-raising telegram for the ECAS, Szilard sat upright and glanced down at the paper, as if afraid to nod his head. Szilard could be flamboyant and funny when speaking to small groups or when he chatted into a radio microphone, but before a camera he could look rigid and uncomfortable, as he did when reenacting his "art of the impossible" speech for the film.

The "ponderous epic" that M-G-M was then producing displeased not only Szilard but also syndicated columnist Walter Lippmann, who complained to Marx after a preview that "the basic theme of the film is not the problem of the atomic bomb in the world, but the success story of the Americans, particularly General Groves, in making the bomb." He protested "melodramatic simplifications" and "falsifications" of historic events, including a fictional meeting between Groves and Truman to decide on the bomb's use. From this and other scenes, Lippmann concluded that "serious people abroad are bound to say that if that is the way we made that kind of decision, we are not to be trusted with such a powerful weapon."[26]

Szilard arrived at Princeton for a weekend meeting of the ECAS board in November, where Einstein announced a $1 million drive to educate the public about atomic weapons.[27] Szilard warned the group that a nuclear arms buildup would move the United States "along a road that leads to war." In December 1946, the University of Chicago hosted a fourth-anniversary commemoration of the first chain reaction at Stagg Field. Szilard moved uneasily among his former colleagues, no doubt sensing that many of them disliked his extroverted political actions. On the steps of Eckhart Hall, in the university's neo-Gothic Quadrangle, Szilard posed for a photograph with Fermi, Zinn, Herb Anderson, and other Met Lab colleagues. The group had stepped outside for the photo, but ever cautious, Szilard alone wore a coat. He stood off to the side, now both physically and emotionally on the fringes of nuclear science.

Other scientists committed to international control joined that December at a meeting that united the Federation of Atomic Scientists with organizations from rocket, radar, and related fields to form the Federation of American Scientists. Szilard helped organize an Atomic Energy Conference to educate important persons about the bomb, which was held at Chicago's Shoreland Hotel in mid-December.[28] The FBI agents who followed him at the time described Szilard as walking "with his head back and stomach protruding; rarely wears a hat; has traits of nervousness and absent-minded [sic]."[29]

Absentminded he surely must have seemed, for Szilard's thoughts seldom focused on the university and the city about him; he was simply too preoccupied with saving the whole world. In January 1947, Szilard addressed the Foreign Policy Association's Cincinnati chapter, saying that no one knows much about foreign policy in an atomic age, so scientists—who are at least rational—might as well be allowed to offer their views.[30] He pleaded "not guilty" for his role in creating the bomb.

Judgment of another sort came to Szilard that January when he joined Manhattan Project chemist Harrison Brown to drive uptown in his sedan to an office building in Chicago's Loop, there to attend a screening of *The Beginning or the End* for scientists and journalists. At first, the scientists roared with laughter as the actor playing Fermi spoke his lines in a thick Italian accent. And they guffawed as Hume Cronyn portrayed Oppenheimer's nervous mannerisms. But through most of the screening the scientists were silent, bemused, or angered at the story's elusive message, distracted by the excessive flashing lights and electronic sound effects. As the screening ended and the lights flickered on, M-G-M's representative looked around nervously, seeking in the scientists' faces some reaction. Most sat quietly, volunteering nothing, but Szilard was more direct. Before journalists could turn to ask his opinion, he slipped out the door, rode the elevator to the lobby, climbed into Brown's car, and crouched on the floor by the backseat. Days later he would only mention the film with pained humor. "If our sin as scientists was to make and use the atomic bomb," he declared, in mock-theological tones, "then our punishment was to watch *The Beginning or the End*."[31]

In March 1947, Gromyko rejected America's proposal at the United Nations for international control of atomic energy, and a week later the president told a joint session of Congress that his Truman Doctrine on aiding allies now bound the United States to contain Soviet aggression. Szilard reacted to these cold-war tensions by hitting the road. After his article "Calling for a Crusade" appeared in the *Bulletin of the Atomic Scientists* that spring, he addressed a university audience in Bloomington, Indiana, as a scientist speaking about peace. To bring about a transition to world-government control of atomic energy, Szilard proposed that the United States sell "peace bonds," as it had sold war bonds during World War II, and called for a constitutional amendment to redefine US sovereignty. "The Constitution was twice amended in this century over the issue of Prohibition," he said, "and if we are willing to go out of our way for the sake of being permitted to drink or for the sake of preventing others from drinking, maybe we shall be willing to go out of our way for the sake of remaining alive."[32]

Szilard ended his hour onstage with a peroration that linked arms control to politics. "We cannot look for our salvation to the Eightieth Congress," he said. "But this country is a democracy. We are the masters of our destiny. There will be elections in '48 and again in '52," In Portland, Oregon, Szilard repeated this point, but in Spokane, Washington, he turned philosophical. When speaking about "Atomic Energy—A Source of Power or a Source of Trouble," he made his first public admission that the atom's peaceful benefits may not be as grand as he had first hoped.[33]

In May, Szilard's "Calling for a Crusade" speech gained him national attention as a cover story: "The Physicist Invades Politics," in the *Saturday Review*.[34] And in June, Szilard and Harrison Brown met with publicist Harold Oram, who urged that as a fund-raising gimmick the ECAS should issue a public declaration after the forthcoming atomic scientists' conference at Lake Geneva, Wisconsin. "I was a negative member" of the ECAS, Bethe recalled, "trying to restrain Szilard from making extravagant statements."[35]

Einstein warned Szilard and his other younger colleagues that the ECAS should not strive to become a mass movement. "You must not use razor blades for chopping wood," he said.[36] As scientists, you should work in more subtle ways.[37] Einstein was right, for after the Lake Geneva conference the ECAS board haggled for more than a week before it agreed on a draft statement for Oram—a statement so bland that it only criticized the UNAEC for failing to reach accord on international control.[38] "Usually we agreed with the ideas that Szilard had," Bethe recalled. "But most of us didn't think we should make so many statements."[39]

Groping for new ways to forestall a nuclear arms race, Szilard thought it worth a try to contact Kremlin leaders directly and turned to University of Chicago law professor Edward Levi for legal advice. "Why not go right to the top?" Szilard asked Levi with a touch of his good-natured mischief. And he did, by drafting a letter to Soviet Premier Joseph Stalin. In it Szilard weaved concerns about Truman's "containment of Russia" with US fears about Soviet retaliation. He urged that Stalin speak often by radio to the American public, as Truman should to the Russians. And he proposed informal discussions by people from all walks of life to devise a Soviet-American peace agreement.[40]

A schemer who preferred working behind the scenes, Szilard tried to sidestep Foreign Minister Andrei Vishinsky, UN delegate Gromyko, and the "layers of Marxists" in the Kremlin by calling on Marshall MacDuffie, an international lawyer and former State Department official who had helped negotiate lend-lease aid contracts with Stalin during World War II. In October 1947, Szilard met with MacDuffie at his New York apartment, at Columbia's Men's Faculty Club, and on a train trip to Washington, hoping to contact the Soviet premier directly. When MacDuffie warned about the Logan Act, a 1799 law that prohibits US citizens from dealing personally with foreign governments, Szilard was undeterred, claiming grandly that he would seek permission from Secretary of State George C. Marshall, perhaps by way of Eleanor Roosevelt.[41] Szilard asked MacDuffie to find "some special person" to penetrate the State Department: Under Secretary Dean Acheson, perhaps; or George Kennan, then director of the department's Policy Planning Staff. And Szilard used his favorite technique—a letter from Albert Einstein. Einstein wrote Marshall about Szilard's letter "to support his request that he be permitted to transmit this letter to Mr. Stalin through channels chosen by him."[42]

When MacDuffie's meeting with State Department lawyers failed to win permission for Szilard's letter, he sent copies of the "Letter to Stalin" and Einstein's letter to Marshall to Attorney General Thomas C. Clark. But this only prompted Clark to ask FBI director J. Edgar Hoover for "any derogatory information" about Szilard and later to circulate his FBI file to the State Department and the AEC.[43]

Finally, after Under Secretary of State Robert A. Lovett refused Szilard's request, he published the letter in the *Bulletin,* to a flurry of newspaper articles. Press coverage was mostly positive, in which the "father of the atom bomb" now wrote to slow the arms race he helped start by telling Stalin that "peace can yet be saved by you, yourself." But a *Washington Post and Times Herald* editorial that praised Szilard's "good will" and "great energy, imagination and patriotism" also criticized his "naivete" and compared the open letter with the "peace ship" that industrialist Henry Ford had sponsored during World War I.[44] Szilard may have loved thinking up clever ways to control the bomb, but he lacked both the patience and the subtlety necessary to influence the defense and foreign policy establishments that would carry out his schemes.

To muddle his efforts, Szilard complained that he could not always tell whether his own ideas were serious, and in October 1947, when busy plotting approaches to Stalin, Szilard's mind twisted one grim thought into satirical fantasy. He imagined himself a condemned "war criminal" after Russians had invaded New York City in World War III. The Russians had bombarded the East Coast with a deadly virus (fortunately, Szilard added, the attack was limited to New Jersey), and once the president surrendered, the occupying forces rounded up all the scientists who had worked on the A-bomb. In "My Trial as a War Criminal" the Russians spurn members of the American Communist party because they have no sense of humor and enlist Britain's government—now neutral—to conduct a Nuremberg-style tribunal. But before the accused Americans could be sentenced, Russia suddenly appealed to the United States for help. The vaccine against the virus had been manufactured incorrectly at a plant in Omsk, and an epidemic had killed more than half the city's children. Riots broke out. And in the unrest, a postwar settlement "was in every respect very favorable to the United States and also put an end to all war crimes trials."[45]

In the real world, no such settlement was even being discussed by the spring of 1948, but to take a first step, Szilard asked University of Chicago chancellor Robert M. Hutchins to organize and lead a new committee to devise and offer Russia fresh proposals for a stable peace. Szilard suggested that this "peace mongering organization" enlist such distinguished members as Einstein, Felix Frankfurter, Nelson Rockefeller, lawyer Abe Fortas, poet Archibald MacLeish, geographer Gilbert White, *New Republic* publisher Michael Straight, diplomat and politician Chester Bowles, Institute for Advanced Study director Frank Aydelotte, and Hutchins himself. But Szilard as much as conceded its futility as he urged Hutchins that "we necessarily have to operate within the narrow margin of small possibilities." From his childhood, Szilard tried to summon faith in a "narrow margin of hope," and now he struggled to be hopeful that his own efforts were useful, no matter how marginal.

"I am trying to figure out whether this business of saving the world is being left undone because it is too difficult for most of us," Szilard admitted to Hutchins, "or rather because of the doubt in our minds whether the world is worth saving."[46] Like so many of Szilard's schemes to enlist bright and influential men (rarely women) to solve the world's problems, this committee soon stalled over how to include US government officials; Soviet scientists could work only with their government's cooperation, so a private citizens' approach from the United States would always confuse Moscow's leaders.

Confusion about US-Soviet intentions was already abundant. The Soviets imposed a yearlong traffic blockade on Berlin in June 1948 and conducted their first A-bomb test in August 1949—an event that prompted Szilard to urge the United States to scrap its newly formed North Atlantic Treaty Organization (NATO) and negotiate a treaty with the Soviets for the "elimination of atomic bombs." *Newsweek* gave a typical press reaction to Szilard's idea when it reported, "To most Americans, this apparent combination of scientific erudition and political naivete was hard to understand."[47]

The arms race that Szilard feared advanced by January 1950 when President Truman announced that the United States would build a hydrogen, or "super," bomb. And that February, Sen. Joseph R. McCarthy launched an anti-Communist crusade that would, by 1954, lead to his condemnation in the Senate but also incite fears of Moscow and its agents throughout the land. Szilard created a scare of his own that month when he told a national radio audience that by adding cobalt to H-bombs the weapons might create enough intense radioactive fallout to annihilate the human race. This astonished physicists Hans Bethe and Frederick Seitz as they sat by Szilard in the radio studio.[48]

"It was terrible but typical," Bethe recalled. "Szilard was his own worst enemy.... The H-bomb was bad enough. Why go beyond it? Why devalue the H-bomb in a way by that?" Here, said Bethe, was "one of the occasions where I thought Szilard hurt his cause by going too far."[49] This cobalt-bomb suggestion was made to oppose H-bomb development by showing where it might lead, but Szilard's ultimate weapon soon had a life of its own. Other scientists—and the AEC—dismissed the idea, opening a debate that ran for months. In a science column headlined "Hydrogen Hysteria," *Time* magazine compared Szilard to the nurserytale character Chicken Little, warning that "the alarmists, however well-intentioned they may be, are helping to frighten the US public into forcing dangerous concessions to Russia."[50] The *Bulletin* commissioned University of Chicago physicist James Arnold to critique the cobalt bomb's practicality, and after several months of study, he reported that Szilard's ideas were feasible.[51]

Still hoping that influential persons might help their government find its way in the world, Szilard proposed to Einstein in March 1950 that he and other scientists form a "citizens' committee" to debate arms-control proposals and appoint designated teams to represent US and Soviet views. Szilard wrote:

> When the Russians opposed the Baruch Plan, they did not tell us their real reasons for doing so, and what they told us of their reasons, they said in a language which is not intelligible to the American people. Our "Russian team," on the other hand, will not only tell us why they find the Baruch Plan unacceptable from the point of view of their "client," but they will tell us their reasons in a language which we can understand.[52]

Unfortunately, by this time better public understanding could do little to ease East-West rancor. Accused atom spies Julius and Ethel Rosenberg were arrested in June 1950, the same month North Korea invaded South Korea, ultimately drawing US troops into a war with Communist China's forces and raising the possibility that A-bombs might be used.[53] Weeks after the war began, Szilard drafted an open letter to US scientists, hoping again to arrange arms-control meetings with their Soviet counterparts.[54] Like so many of his proposals, this one laid out detailed suggestions for organization and meeting topics. And, like so many, it was disregarded and eventually dropped.

In September 1951, during a nuclear physics conference in Chicago, Szilard, *Bulletin* editor Eugene Rabinowitch, and English nuclear physicist Joseph Rotblat met one evening in a colleague's home and for the first time raised the question of arms-control discussions directly with Soviet scientists.[55] But when this overture yielded no Soviet response, Szilard truly believed a nuclear war inevitable and took two practical steps.

First, he made plans to create a school in Mexico, where his friends' children might survive the war. Second, in a fit of anger and frustration, he proposed a permanent institution to resolve those long-standing differences between the two superpowers that he feared would lead to war. His hasty and passionate draft, "You Do Not Want War with Russia?," is Szilard's half-sarcastic, half-desperate plea for help. As early as July 1948, Szilard had penned a measured letter asking intellectual leaders to join "an adequate movement aimed at the establishment of enduring peace." Then his goal was to clarify foreign policy issues before the coming presidential election. But in March 1952, with another election ahead, came this more fervent and demanding appeal.

"You Do Not Want War with Russia?" taunted his headline. "You will get it whether you want it or not—for it is in the making—unless you do something about it and do it quick." Szilard's badgering tone continued: "I say to you that there will be peace only if there is an overall settlement of all issues outstanding between America and Russia." The "problem," he said, "is not to write an agreement that Russia will sign but to write one which Russia will be eager to keep, not only for the next few years but ten years and twenty years hence." Rejecting traditional diplomacy, he argued that "to devise such an agreement requires imagination and resourcefulness rather than patience and firmness; it requires thought and perhaps some compassion rather than arms."[56]

Then Szilard's logic slipped loose, his reason outrunning common sense. Once "political issues" are settled, he argued, "there would no longer be a legitimate reason left for continued secrecy." He even wondered "why we should not grant immunity to spies of any nation and revoke the Espionage Act to permit our own citizens freely to cooperate with spies of any nation." Szilard was serious, logical, and wildly out of step with the anti-Communist paranoia then gripping his country.

To take "immediate action," Szilard proposed creating "a lobby for real peace in Washington" along with a bipartisan "political action committee" to influence the coming presidential election. As he would do ten years later with his Council for a Livable World, Szilard wanted to tithe participants to support these two activities.[57] It is not clear what fired Szilard to write this proposal with such passion and anger, but perhaps it was panic at the remarks of Donald A. Quarles, air force secretary, who had said on February 2, "I cannot believe that any atomic power would accept defeat while withholding its best weapons." Szilard clipped and saved the *New York Times* account of this speech and no doubt considered it proof that nuclear weapons might soon be used in Korea.

There is no record in Szilard's papers that he ever revised and sent this angry draft, but for us it reveals both his wild frustrations and his wily ingenuity. Fearful of both US and Soviet motives, Szilard realized that a peace agreement could succeed only by appealing to the interests and instincts of both partners. In short, it had to embody motives that would be self-enforcing. But his hopes to prevent a US-Soviet war and to slow the race for nuclear weapons withered just three days before the 1952 presidential election when the United States exploded its first H-bomb. Szilard asked the University of Chicago for a three-month unpaid leave in order to press even harder for arms control, but he could scarcely concentrate on any topic or project for long. He was clearly distraught, moving between Chicago, New York, and Princeton, later adding stops at Brandeis University and holiday visits to his wife, Trude, in Denver.

In his cold-war gloom, Szilard began drafting a retrospective analysis of the nuclear arms race and its possible solution, which he called "Meeting of the Minds." In August 1953 the Soviet Union tested its first hydrogen device, and in January 1954, Secretary of State John Foster Dulles declared that US policy for any Soviet adventurism might be—in the popular phrase—"massive retaliation" with nuclear weapons. Frightened by these chilling events, Szilard became more unfocused, convoluted, and imprecise with each draft of "Meeting." His thoughts about arms control were in turmoil. Fear now ruled his cherished reason.[58]

In September 1954, Szilard proposed another conference for US and Soviet scientists, but ill health and anxieties about his erratic biology career all but overwhelmed this project. By the spring of 1955, Szilard's personal and professional despair made him sarcastic about arms control. He told syndicated columnist Stewart Alsop what a bargain nuclear war would be; weapons were so abundant and powerful that the human race might be eliminated for only forty cents a head.[59]

Trying to link arms control and politics, Szilard urged Sen. Hubert Humphrey to use his Subcommittee on Disarmament for off-the-record conferences on "what kind and what degree of disarmament is desirable within the framework of what political settlement?" Szilard suggested convening Walter Lippmann, George Kennan, and other policy analysts out of government with representatives from the State Department and Harold Stassen's UN disarmament office. Szilard met with Humphrey, but nothing came of this idea.[60] Desperately, Szilard was trying to engage his country's brightest intellectual and political leaders in serious dialogue, in effect, creating another Bund. Again, no one cared to join him.

Ironically, the international forum that finally coupled Szilard's eclectic brainstorming with his fervent desire to curb the US-Soviet arms race evolved at just this time of despair: the Pugwash Conferences on Science and World Affairs. It began with a "manifesto" signed by philosopher Bertrand Russell and Einstein in 1955 in which they called on the world's scientists to "assemble in conference to appraise the perils" created by nuclear weapons.[61]

Industrialist Cyrus Eaton offered the scientists use of his estate in the Nova Scotia village of Pugwash. "At first, Russell thought it was a joke," recalled Rotblat, because in England " 'Captain Pugwash' was an indolent comic-strip character."[62] But when the first conference met in Pugwash, in July 1957, it embodied what Szilard had tried to achieve for more than a decade by uniting US and Soviet scientists for arms-control discussions. Szilard participated in the first meeting, and at a critical moment he helped direct the group's evolution. Ever the exuberant loner, Szilard also continued making fresh proposals for other US-Soviet meetings.[63]

Among scientists and arms-control activists, Szilard and Pugwash are forever paired. When accepting Russell's invitation to the first conference, Szilard suggested extending the meeting by several days, promised a paper on "how to live with the bomb" in order to achieve "stability in the atomic stalemate," and proposed the names of other American scientists.[64] A few days before the first conference, Szilard tried to convene some participants.

"Why do that?" asked Harrison Brown.

"So we can write the final report!" Szilard announced, and he arrived at Pugwash with a draft in his suit-coat pocket.[65] The Russians and Americans were nervous about the meeting, as this was the first of its kind to bring together atomic scientists from the two countries. But the Russians quickly warmed to Szilard's candor and wit. "They really loved Leo," recalled Ruth Adams, an editor at the *Bulletin* who attended the First Pugwash Conference and many to follow. "He never tried to disguise anything, and they appreciated that."[66]

The dominant US and Soviet delegations were joined by scientists from Australia, Austria, Canada, China, France, Great Britain, Japan, and Poland. The venue forced them all to relax, although most men still wore dark business suits day and night. Pugwash, a lobster-fishing village on the Northumberland Strait, was Eaton's birthplace. He proudly covered conference expenses and brought in three sleeping cars from his Chesapeake and Ohio Railway to accommodate guests; others slept in cottages or in family homes, as there was no hotel in this town of eight hundred. The twenty-five participants met around long folding tables in the village's largest chamber, a schoolroom in the brick Masonic temple. Between sessions they chatted informally over drinks at backyard cookouts and on a boat cruise in the strait.

Unlike most of his colleagues, Szilard was ebullient in all settings, taunting and quizzing about the arms race that many of them had helped to create. As the records noted politely, while delegates arrived the first day, "Szilard took the major share of the discussion." He also "formulated a number of specific questions which in his opinion should be the main concern of the conference," among them, would using tactical nuclear weapons help avert "an all-out atomic catastrophe," would President Eisenhower's "open skies" inspection policy enhance stability, and would "a potential danger for peace" be eliminated by uniting East and West Germany?[67] He even passed around drafts of his "final report" for comments.

In one session, the minutes note, Szilard "spoke at length about the needs and the prospects for a political settlement" to achieve stability. His flamboyant style amazed some colleagues and amused others, as he split the nuclear arms race into three phases:

1946-51	logical, insane, stable
1951-56	logical, insane, unstable
now	illogical, insane, unstable

The world became "unstable" in 1951 with the H-bomb, said Szilard, and now he feared that a peripheral political conflict—Berlin, perhaps, or a conventional war in the Mideast—might spark a nuclear war. To keep a nuclear conflict from escalating, Szilard proposed drafting in advance a "price list of cities, i.e., that the destruction of a city of one side would entitle that side to destroy an equivalent city of the other side but no more." He also proposed what would later be called "minimal deterrence" by urging the United States, the Soviet Union, and Britain to stop their nuclear tests and halt bomb manufacture, yet retain just enough weapons to deter each other.

During the conference, Szilard was asked to talk informally among his colleagues and to record what had actually been agreed on rather than what was said in speeches and papers. When the participants finally consented to issue a statement, however, Szilard was typically contrary and refused to sign it.[68]

Szilard relished his days at Pugwash, enjoying both the formal debates and informal chatter with his peers. His ironic humor and rigorous logic upset some of them but worked effectively with most, once they sensed how passionately he worried and thought about their common problems. "I am told that I puzzled some of my friends because I was not willing to go along with their statement that the most important step toward eliminating the danger of the bomb consists in stopping to test bombs," Szilard said at the close of the First Pugwash Conference. "Yet they refuse to draw the logical conclusion and to say the most important step toward eliminating the danger of the bomb is to explode every single one of the bombs in our stockpile as soon as possible in a test."[69]

Back in Chicago, Szilard proposed that the university sponsor a similar meeting of twenty scientists and pushed the idea in letters to Russians he had met at Pugwash: Dmitri Skobeltzyn, a science adviser at the United Nations, and Alexander Topchiev of the Soviet Academy of Sciences.[70] Inside the US government, too, Szilard's influence was reaching some official channels, at least indirectly. One route ran through the President's Science Advisory Committee (PSAC), created in 1957, where colleagues sympathetic to Szilard's ideas gained institutional access to the White House. "Szilard kept us interested in the subject" of arms control, recalled Hans Bethe, a PSAC member from its founding, because "many of us remembered what Szilard had said and we had a special subcommittee on arms control. I believe that Szilard had a lot of influence on all of us on that committee . . . and later on, that committee, in turn, sponsored the Arms Control and Disarmament Agency, a part of the government."[71]

When passing through London in December 1957, Szilard sat in on the Pugwash Continuing Committee meeting that was then deciding what to do beyond the first conference. On his own, and from recent conversations with German physicist Carl von Weizsäcker, Szilard had plenty of ideas, and although not a committee member, he challenged Chairman Russell's call to expand Pugwash into a scientists' movement with broad public appeal. Szilard argued for continuing small, private meetings lasting over several days, and he eventually succeeded, setting the pattern for later Pugwash conferences. And Russell soon founded the popular group he desired: the Campaign for Nuclear Disarmament, which organized mass demonstrations against nuclear weapons. "Szilard's presence that day made a difference," Rotblat recalled, "and Russell later conceded that this was the right decision."[72]

Following Szilard's suggestion, the Second Pugwash Conference was held for twelve days the following spring at a ski resort in Lac Beauport, near Quebec.[73] It focused on immediate political problems, seeking conclusions that might help governments to curb the nuclear arms race. Weizsäcker, who had worked on Germany's A-bomb during the war, was a participant. Others included Australian physicist Mark Oliphant, Topchiev and Skobeltzyn, chemists John Edsall and Linus Pauling, Manhattan Project physicist William Higinbotham, Rotblat, sociologist Morton Grodzins, aerial surveillance expert Richard Leghorn, Rabinowitch, and MIT engineer Jerome Wiesner (later President Kennedy's science adviser).

Absent from the meeting, although urged by Szilard to attend, was physicist Edward Teller, then an outspoken advocate for developing and testing "clean" nuclear weapons—with high neutron levels and little radioactive fallout.[74] Still, Szilard used Teller's name so often to personify scientists who were driving the arms race that he might as well have attended. Szilard ridiculed Teller's work in order to dramatize "how absurd the situation is into which the arms race has led us" by posing his own logical solutions to an illogical world. He told his Russian colleagues, "It will be in your interest to give us accurate maps of Russia" so that the United States might drop "clean" bombs on certain cities rather than "dirty" bombs.

Together with Leghorn, Szilard laid out more practical steps to achieve stability in the arms race, including "rules" that the nuclear powers would "pledge" to follow in any future war.[75] One rule would require that nuclear weapons only be used on the territory being defended; that is, if the Warsaw Pact invaded West Germany, the United States would only use them to repel invading forces in West Germany. "I am stressing the need for the internal consistency of the set of rules which must operate in both America and Russia if an agreement providing for disarmament is meant to be kept in force," Szilard argued, striving to impose his rational worldview on skeptical colleagues.[76]

Repeatedly, Szilard said their task was to learn "how to live with the bomb" and eventually how to use that stability to seek "political settlements" that might afford "real peace."[77] Reviving his idea from the First Pugwash Conference, Szilard proposed mutual inspection systems that would pay Americans to spy on America, Russians on Russia. To his colleagues this sounded serious, though farfetched. He also revived his "price list" for cities, now "divided according to size in ten categories" to make nuclear retaliation rational and predictable, with a month's warning "for an orderly evacuation." To his colleagues this sounded crazy.

To explain his point, Szilard discussed a hypothetical US invasion of Mexico after a Marxist takeover. Under his rules, "the Russian price might specify that if Mexico is invaded by American troops, Russia will give notice to two to four cities of the seventh category, and after four weeks, which is ample time for an orderly evacuation, Russia will destroy these cities, will demolish these cities, with the clean hydrogen bombs."

"Does this mean that after the Soviet Union destroys four of our cities, we can destroy four of theirs?" asked Grodzins.

"Yes, of the same category," said Szilard. Soviet chemist Topchiev interrupted, saying the two countries should "maybe pledge not to ruin cities at all."

"In that case," Szilard answered, "I cannot abolish war, because I cannot prevent the American government from invading Mexico."

"We are not going to organize Communist revolution in Mexico," Topchiev replied to a chorus of laughter.[78]

But delegates had trouble deciding whether to laugh or not when Szilard complained about "too much rigidity in thinking about to what use a bomb can be put" and suggested as a clever initiative that "our Russian friends" buy the city of Lyons from France and destroy it with three "clean" H-bombs.[79] Szilard also urged a "full exchange of information" between the United States and the Soviet Union to assure that bombs can be made "clean." And if the US government does not accede, he proposed—perhaps seriously—that they "enlist the help of the churches" and "organize prayer meetings" to ask God "that there should be a patriotic traitor among us who will inform the Russians on how to make clean bombs."[80]

For his part, Szilard expected his Pugwash colleagues to be skeptical. And they were. "I have some experience about what kind of reaction one gets if one proposes something unusual," he said, "like, for instance, setting up a chain reaction in the uranium atom." But when they stopped laughing at that crack and at his other quips, many scientists were still skeptical. Szilard seemed at once too logical and too whimsical.

Before the conference ended, Szilard, Leghorn, and Wiesner proposed that Topchiev convene a meeting that summer of Russian and American scientists, including some members of President Eisenhower's newly formed PSAC.[81] By June, Topchiev had agreed, and Szilard raised foundation money, arranged sponsorship by the American Academy of Arts and Sciences, and planned a September meeting in Moscow.[82] With that meeting in mind, Szilard began drafting a grand scheme for resolving the arms race, his "Pax Russo-Americana."[83]

Later versions of this manuscript served as the focus for Szilard's arms-control thinking during the rest of his life: in talks and proposals at Pugwash conferences, in letters to colleagues and newspaper editors, in a 1960 meeting with Khrushchev, and in satirical fiction as *The Voice of the Dolphins*. Szilard tried out these thoughts in May 1958 when, as MIT's Arthur D. Little Lecturer, he provoked academic and political guests at Endicott House in Dedham, Massachusetts, with the talk "Could a Stalemate Between the Atomic Striking Forces of America and Russia Be Made Truly Stable?" Unlike some arms-control activists, Szilard did not believe nuclear weapons could, or should, be eliminated; but they should be minimized and their use negated by political accord.

"When Russia and America can destroy each other to any desired degree, the overriding issue becomes the stability of the stalemate, and on this issue Russia's and America's interests coincide," Szilard argued.[84] When questioned about the value of achieving more stability in the current stalemate, he answered by mocking the premise on which US deterrence was based—Secretary Dulles's "massive retaliation." This, said Szilard, "would mean murder and suicide today. . . . Now such a threat is not believable, and therefore there is no deterrent in it whatsoever." His sarcasm may have miffed a few listeners when he said that "clean" bombs pose a greater threat than "dirty" ones because "you embarrass America more" by "making ten million people without shelter" than by killing them.[85]

Szilard's remarks clearly provoked a blustery exchange with Walt W. Rostow, then an MIT economic history professor and later a foreign policy adviser in the Kennedy administration. Rostow warned that the loss of China to the Communists in 1949 had a current parallel in Egypt, and "at this moment the Communist party in Egypt is saying exactly what the Communist party said about the [Chinese] Nationalists. . . ."

"We don't care a damn whether a country is democratic or not; what we care is to have a friendly government," Szilard complained. Citing the US-backed coup that ousted Premier Mossadegh in 1953, Szilard said: "We have subverted Iran very successfully, and we have a government which is friendly to us; but Iran is of course not a capitalist country, nor is it a democratic country. We are not propagating democracy. We want governments which are friendly to us. This is precisely what the Russians want."

"What makes a government friendly in the absence of a political conviction?" Rostow asked.

"A government is made friendly," said Szilard, giving no ground, "because we supply the arms for this government to remain in power."[86]

As founding chairman of the "Operating Committee on World Security Problems Raised by Nuclear Weapons" at the American Academy, Szilard schemed to use this panel to sponsor the Moscow meetings. Behind all his practical steps was Szilard's visceral fear of global nuclear war, a fear that intensified in the summer of 1958 as President Eisenhower sent US Marines into Lebanon to quell a rebellion and as Communist

China bombarded the Nationalist-held islands of Quemoy and Matsu off Formosa. "If a confrontation develops, I'm going to leave the country," he told Philip Morrison that summer.[87]

In September 1958, Szilard left the country, anyway, and met again with Topchiev and Skobeltzyn at the Third Pugwash Conference in Kitzbühel, a ski resort in the Austrian Tyrol. Still hoping for a Moscow meeting, the Russians promised their most knowledgeable and influential scientists. But plans for the meeting faltered when the State Department discouraged US participants from attending and the Rockefeller Foundation balked at covering the expenses. Then plans revived again in the spring of 1959, after Szilard added to his American Academy committee three scientists active in public policy issues, physicists Hans Bethe and Alvin Weinberg and environmental scientist Roger Revelle. At Revelle's urging, the US State Department finally favored the meeting.

For the Fourth Pugwash Conference in Baden, Austria, in the summer of 1959, Szilard simplified his "Pax Russo-Americana" paper into a talk about "How to Live with the Bomb." The Pugwash meetings by this time had become the leading forum for international discussion of the nuclear arms race. Issues such as an end to weapons tests and bans on specific technology were regular topics, and Pugwash veterans claim that their sessions highlighted ways to detect and measure nuclear explosions, laying the groundwork for the Treaty for a Partial Nuclear Test Ban of 1963 and the 1976 Threshold Test Ban that limited US and Soviet explosions to 150 kilotons. Pugwash discussions also developed and endorsed the "black box" approach to verifying underground tests with nearby monitoring devices, and some of the main provisions of the 1968 Non-proliferation of Nuclear Weapons Treaty were devised at a series of Pugwash meetings begun in 1958.

In his quest for "stability" in the arms race, Szilard complained that antiballistic missiles (ABMs) posed a special threat because they invited unlimited development of offensive nuclear weapons. "In particular," said physicist John Holdren, a Pugwash officer since the 1980s, "influential Soviet scientists, who initially supported ABMs as defensive and therefore inherently benign, were persuaded by their American counterparts in Pugwash that this view was dangerously incorrect." The consensus reached among Pugwash scientists led to the 1972 ABM Treaty.[88]

While Szilard assaulted his colleagues with rational schemes for saving the world that left them wondering whether he was joking, his humorous quips left them wondering if he were serious. At the Baden meeting,

Szilard offered toasts that poked fun at both the Americans' directness and the Russians' reserve. To A. P. Vinogradov, Szilard raised his glass and said, "While he may not always say what he thinks, he never says what he does not think. . . ."[89]

Szilard was just as jocular with the Soviet premier. In September 1959, a week before Nikita Khrushchev was to visit Washington, Szilard was in Geneva and sent the Russian leader "How to Live with the Bomb—and Survive." A copy would also go to the White House, Szilard wrote, "with the instruction that an oral report be rendered to President Eisenhower, if this appears to be warranted." Szilard even urged Khrushchev, in a cheeky aside, that "a similar procedure might, perhaps, be adopted by you also." He stressed the need to focus on "long-term implications" of current nuclear policies, and beyond prompting the Soviet premier on what to tell the US president, Szilard vowed to spend October in Washington "to discuss with certain members of the State Department the issues raised in the attached manuscript, which are semipolitical, as well as semitechnical." Szilard also promised Khrushchev that he would request discussions in Moscow with Soviet scientists.

Khrushchev never replied to this letter, although Szilard's blunt and teasing style appealed to the Soviet leader and they would later exchange many letters and meet privately. Szilard's feisty manner also appealed to some of his fellow scientists. One message that cheered Szilard came in a telegram referring to his decisive role in keeping the Pugwash movement low-key and private, "THE PUGWASH COMMITTEE MEETING IN LONDON SEND YOU BEST WISHES AND LOOK FORWARD TO DISAGREEING WITH YOU AT FUTURE PUGWASH CONFERENCES = BERTRAND RUSSELL."[90]

CHAPTER 25

Biology

 1946–1959

When a meeting of the Atomic Scientists of Chicago ended in a classroom at the University of Chicago's Social Science Building one March evening in 1947, Leo Szilard turned to a younger colleague and asked, "How would you like to join me in an adventure in biology?"

"I was very excited by the idea," Aaron Novick recalled years later. Szilard was a theoretical physicist who was then dabbling in the social sciences; Novick, a physical-organic chemist whose work during and after the war included nuclear and solid-state physics. "I knew very little about biology. But I knew a lot about Szilard. I admired his genius, I liked his enthusiasm for ideas, and I liked him personally." Novick's answer was quick and decisive: "I would be delighted!"

"You must think about it," Szilard cautioned.

"No need," said Novick. "I can tell you, I am certain." This was just the opportunity both men had been seeking. In 1946, when diverted from nuclear physics by his nemesis Gen. Leslie R. Groves, Szilard had secured a research professorship that linked biophysics and social issues and freed his restless mind for fresh discoveries. On his own, Novick was already intrigued with biology after reading *What Is Life?*, the provocative essay by physicist Erwin Schrödinger.[1]

Szilard and Novick had met on the Chicago campus during the war, at the Metallurgical Laboratory (Met Lab) of the Manhattan Project.[2] After the war, Novick had roomed with Enrico Fermi's colleague, physicist Herbert Anderson, in a carriage house behind the Quadrangle Club, the faculty club where Szilard lived. Szilard enjoyed brainstorming with younger associates and spent many late nights on a couch in that carriage house, talking and listening intently.[3] Novick had also joined Szilard in Washington, in the spring of 1946, to help lobby for civilian control of atomic energy, and through the Atomic Scientists of Chicago they worked with other Manhattan Project veterans for the atom's international control.

Biology appealed to both men because it offered them the chance to seek first principles in a field still lacking the kind of conceptual and theoretical breakthroughs that had revolutionized physics during the century's first decades.[4] As early as the 1930s there was a feeling that biology was ready for just such a dramatic theoretical advance, although this did not take place until about the time that Szilard and Novick decided to enter the field, with the rise of what is now called molecular biology.

While the impromptu offer in the classroom that night surprised Novick, just such an "adventure" had fascinated Szilard since his restless days in Berlin in the late 1920s. Szilard enjoyed reading *The Microbe Hunters*, Paul Henry de Kruif's 1926 saga about the discovery and early use of the microscope, and talked about it excitedly. He probably read *The Science of Life*, a visionary story about genetic engineering written in 1929 by H. G. Wells and others.[5] Szilard had met Wells in London that year and had tried (but failed) to contact his coauthor, biologist Julian Huxley, when promoting the idea of an intellectual policy group he called the Bund. Undoubtedly, Szilard also knew J. B. S. Haldane's inspiring 1924 essay *Daedalus, or Science and the Future*, about the human consequences of biological progress.

Szilard also knew about physicist Niels Bohr's imaginative approach to biology in the 1932 essay on "Light and Life," which urged applying a principle from quantum mechanics, "complementarity," to the understanding of biology.[6] By his own account, Szilard was "strongly tempted to go into biology" in the spring of 1933, when he landed in England as a refugee from Hitler's Germany. Szilard called on Archibald V. Hill, a biophysicist-turned-biologist who had won a 1922 Nobel Prize in his new field, and was advised to work as a physiology demonstrator at the University of London as a way to become a physiology instructor and, in the process, to learn biology.[7]

But biology had vanished from Szilard's mind by the fall of 1933, when he conceived and developed his concept of a nuclear "chain reaction." Now, fourteen years later, with the chain-reaction concept used to create the A-bomb he had dreaded, Szilard could balance his fervent arms-control efforts with renewed speculation about biology.

Szilard beamed with enthusiasm for his new field and inspired Novick and others he spoke with to share his fervor. He yearned to create a theoretical structure that would permit biologists to make intuitive leaps of faith to explain and unite what they knew empirically. Eclectic and acrobatic in his own thinking, Szilard hoped that the jumps and juxtapositions of his mind could reveal fundamental theories to explain life itself. In this way, he was still thinking as a physicist, ever eager to pose bold and imaginative hypotheses to explain basic phenomena.

There is a slim possibility that in his final months in Berlin, Szilard met German physicist Max Delbrück, who is credited as a later founder of molecular biology. At the time, Szilard was teaching a theoretical physics course with Schrödinger, who was also fascinated by the theoretical possibilities of biology. Delbrück returned from research with Bohr in Copenhagen in the fall of 1932 and began study under N. W. Timofeeff-Ressovsky, a Russian biologist working in Germany.[8]

After Delbrück immigrated to the United States in 1937, he took up genetic research at the California Institute of Technology (Cal Tech) and later, while at Vanderbilt University in 1943, joined Salvador Luria to demonstrate that bacteria adapt to new conditions—such as the presence of a virus—by Darwinian mechanisms, just as higher forms do. Virus-resistant mutants preexist in a population, they concluded, and are not induced by the selective agent (the virus) that is applied to isolate the mutants.[9] In 1943, Luria criticized bacteriology as the last stronghold of Lamarckism; and their demonstration of adaptation established bacteria as suitable objects for the study of genetic mechanisms so that principles applicable to all life could be discovered.[10]

Delbrück and Luria believed that they could better understand genetic mechanisms by studying one of nature's simplest creatures: the bacteriophage, or simply, phage. The phage is a virus that infects bacteria. Viruses reproduce in a living cell by using the cell's apparatus for reproduction, DNA or RNA. By definition, viruses contain either DNA (deoxyribonucleic acid: a large, stringlike molecule found in living cells that carries genetic information), which acts to redirect the bacterium's own biosynthetic systems to make more virus phage, or RNA (ribonucleic acid: single- or double-stranded molecules).

The final event in the infectious process is usually a breakdown of the bacterial wall (called lysis), which frees the newly reproduced phage particles. Some phage are tadpole shaped, with a head containing DNA within a wall of protein. The phage's hollow tail, also made of protein, can attach to bacteria and facilitate the transfer of DNA into them. Because of their simplicity, phages seemed ideal for studying how genetic material reproduces, mutates, and expresses genetic information.

In the summer of 1947, Szilard and Novick enrolled in a course on bacteriophage at the Cold Spring Harbor Laboratory on the north shore of Long Island, New York. Delbrück and Luria first organized this intensive course two summers before to recruit people to study phage and in three busy weeks participants were able to learn enough theory and technique to conduct their own phage research. "This isn't a course," Szilard huffed to his brother, Bela. "It's a church. Delbrück's phage church. And when you're in it, you've got to believe." Researchers who joined the "phage group" focused on the genetics of phage, anticipating that this study would help them to better understand the genetics of higher organisms. And since phages reproduce rapidly, as do the host cells, their reproduction provided a handy means for studying the evolution of many generations—here measured not in years or even months but in hours.

At work and at play the phage group became a close-knit alliance of cooperating researchers from very different backgrounds, and through the 1940s and 1950s it helped develop the new branch of science called molecular biology. But as one close observer noted, "With their fastidious rigor, their insistence on the simplest biological systems, their distrust of biochemists and earlier microbiologists, their self-conscious marking off of the group from others—their snobbery—the phage group has attracted attention even to a disproportionate degree."[11]

During the full days at the bay-front campus, students attended lectures, then paired off to learn laboratory techniques for growing and experimenting with phage. At night, participants relaxed around the grounds or attended impromptu talks and seminars. There were also occasional beer-and-pizza parties and square dances. Playful skits marked the course's graduation ceremony.

At Cold Spring Harbor everyone was informal, dressed for the summer heat and humidity in shorts or bathing trunks and T-shirts. Everyone, that is, except Szilard. Painfully modest all his life and, by age forty-nine, very portly as well, Szilard wore a three-piece cotton suit and tie to meals and to the labs, only doffing his coat to work over the lab benches.

"Leo, how can you stand it?" asked Frances Racker, Trude's sister and wife of biochemist Efraim Racker, when she saw Szilard.

"I am cooler than anybody else," Szilard insisted. "When you are dressed to sweat, you sweat. And when you sweat, you get cooler. So I am cooler than anybody else."

Around the laboratory, Szilard was freer about exposing his ignorance. In fact, he seemed to flaunt it. While many scientists dread making fools of themselves, Szilard enjoyed queries that were so simple even the experts squirmed. "How do you *know* this is a virus?" he asked at one lecture on replication in phage. "How do you *know* it replicates?" In this way he not only learned a new field in the most direct way but also challenged his colleagues to rethink their own knowledge and assumptions.[12]

Jacques Monod, of the Pasteur Institute in Paris, first saw Szilard at Cold Spring Harbor that summer, remembering him as a "short fat man" in the front row at a seminar. "He seemed to be sound asleep most of the time and, with his round face and potbelly, he certainly expressed little interest and no aggression. Yet, from time to time he would suddenly wake up, his eyes shining with intelligence and wit, to ask sharp, incisive, unexpected questions, which he would impatiently repeat when the answer did not immediately come straight and clear." When introduced, Szilard "started shooting question after question at me." Monod remembered. "Many of the questions seemed very unusual, startling, almost incongruous. I was not sure I understood them all, especially since he insisted on redefining the basic problems in his own terms, rather than mine."[13]

One June evening Szilard sat with Novick on the steps of a lab building and asked Monod about a finding in his doctoral dissertation: When bacteria were given a mix of glucose (a simple sugar) and lactose (milk sugar), they consumed the glucose completely before using the lactose. Monod realized that glucose blocks the consumption of lactose. But Szilard proposed, "Feed both glucose at a lower concentration with the lactose and the two sugars might be consumed simultaneously." He added that one might be able to achieve that result if the two sugars were fed to the bacteria in a continuous flow system, anticipating that as the glucose is consumed preferentially, its concentration would fall to a level that would allow simultaneous consumption of the lactose. This conversation led Szilard subsequently to conceive, design, and build a continuous-flow system he called a chemostat and apparently led Monod to build an equivalent system he called a biogen.[14]

Szilard's bench mate for the phage course was Philip Morrison from Cornell (who was later at the Massachusetts Institute of Technology), a physicist Szilard first knew at the Met Lab in 1942. Morrison's hand and concentration were much steadier, and although Szilard tried all the laboratory techniques, he was clumsy and impatient with procedures once the outcome was apparent. He soon realized—as he had in nuclear research— that thinking up experiments is far more fun than actually conducting them and, in the end, let Morrison perform most of their work. Once Szilard saw the point of an experiment and his mind began to wander, he sauntered outside to settle on the lawn and think.[15]

More eagerly than most, Szilard tried to extract universal laws from the welter of laboratory results. Like Schrödinger, Szilard was used to making hypotheses about the laws of nature to account for experimental results and contributed to molecular biology as a theoretical physicist. Szilard's transition was especially easy: He simply continued a habit of thought from childhood and boldly asked about anything that came to mind. "Usually Szilard listened to people only as long as they had something to say that interested him and made sense," Novick has said. "This meant that he sometimes turned away in the middle of a conversation. He did not waste time with unnecessary social niceties, which is why some people disapproved of him."[16]

In biology Szilard was "a cross-pollinator of ideas and an effective critic of others' work, an intellectual and ethical inspiration to younger scientists," the science writer Horace Freeland Judson noted.[17] The French molecular biologist François Jacob, who would share with Monod and André Lwoff a 1965 Nobel Prize for work inspired in part by Szilard, called him an "intellectual bumblebee" who carried ideas from place to place.[18] Crossing the Chicago campus with Novick one day, Szilard was stopped by a man who thanked him earnestly for a research idea he had suggested. "It worked very well," he said.

"Who was that, and what was the idea?" Novick wondered.

"I have no idea," said Szilard, shaking his head and smiling as they walked on.[19]

Szilard relished molecular biology because he could test ideas almost as soon as he conceived them. "While physics appears to be on the way out, biology does not seem to exist yet," he said soon after joining the field. "To me it seems quite probable that there must exist universal biological laws, just as there exist, for instance, physical laws [such] as the law of the conservation of energy or the second law of thermodynamics."[20] In physics, experiments often took months or years to design and vast sums of money to set up. More time was needed to analyze the results. But in biology, where physical changes in organisms occur within minutes or

Leo Szilard, with an unidentified girl and dog, photographed by Trude Weiss at a picnic in Wading River State Park, near Brookhaven, Long Island, June 1948
(Egon Weiss Collection)

From left: Max Delbrück, Aaron Novick, Leo Szilard, and James D. Watson at the Cold Spring Harbor Laboratory, June 1953 (Photograph by Norton Zinder. Courtesy of the James D. Watson Collection, Cold Spring Harbor Laboratory Archives)

Founders of the Emergency Committee of Atomic Scientists meeting at the Institute for Advanced Study in Princeton, November 1946. *Back row* (left to right): Victor F. Weisskopf, Leo Szilard, Hans A. Bethe, Thorfin R. Hogness, and Philip M. Morse. *Seated*: Harold C. Urey, Albert Einstein, and Selig Hecht. (AP/ Wide World Photos)

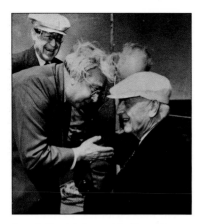

Leo Szilard jokes with financier
Cyrus Eaton on a cruise during
the first Pugwash Conference on
Science and World Affairs, off
Pugwash, Nova Scotia, July 1957
(Pugwash Conferences on Science
and World Affairs)

Leo Szilard sits "botching"—his
term for creative daydreaming—in
the Rocky Mountain National Park,
Colorado, in the 1950s. (Photograph
by Trude Szilard/Egon Weiss Collection)

Aaron Novick and Leo Szilard in the biology laboratory they created in
the basement of a synagogue near the University of Chicago (University of
Chicago)

Leo Szilard as sketched in the late 1950s by his longtime friend Eva Zeisel (Sketch by Eva Zeisel)

Leo Szilard on the Atlantic City boardwalk, March 1948 (Photograph by Gertrud Weiss/Egon Weiss Collection)

While attending a January 1957 biology conference at the Waldorf-Astoria Hotel, Leo Szilard and Jonas Salk discuss ideas for what would become the Salk Institute for Biological Studies. (Photograph by Karl Maramorosch)

From left: Leo Szilard with Matthew Meselson and Leslie Orgel at a biology conference in Boulder, Colorado, 1958 (Photograph by Gertrud Weiss Szilard/ Egon Weiss Collection)

Leo Szilard with his wife, Trude, at Memorial Hospital in New York City, 1960 (Courtesy of John Loengard)

Leo Szilard and Jerome Wiesner, at left, join in Soviet aircraft designer Andrei W. Tupolev's toast at a Pugwash Conference banquet in Moscow, November 1960. In the foreground are the wife of Soviet physicist E. K. Federov and Manhattan Project physicist William Higinbotham. (Egon Weiss Collection)

Leo Szilard with Inge and Giangiacomo Feltrinelli at Castello di Villadeati, near Asti, Italy, December 1960 (© Inge Schoenthal Feltrinelli)

Microbiologist Jacques Monod and Leo Szilard at a Cold Spring Harbor Laboratory biology conference in June 1961 (Photograph by Esther Bubley, © Jean Bubley)

Leo Szilard "lobbying from the lobby" at a desk in the Dupont Plaza Hotel, Washington, DC, 1961 (Photograph by Esther Bubley, © Jean Bubley)

Robert Grossman caricature of Leo Szilard for the cover of the *New York Times Book Review*, January 24, 1993. A bathtub dreamer, Szilard's thoughts are on the A-bomb and a dolphin, the subject of his popular 1961 political satire about arms control, *The Voice of the Dolphins*. (Courtesy of Robert Grossman)

Leo Szilard reading to a young girl. He enjoyed the company of children, praised their innate wisdom, and wrote animal stories to amuse them. (Egon Weiss Collection)

Harvard professor Henry Kissinger, Leo Szilard, and *Bulletin of the Atomic Scientists* editor Eugene Rabinowitch chat during a break at the fifth Pugwash Conference at Stowe, Vermont, September 1961. (Photograph by Trude Szilard/Leo Szilard Papers, Mandeville Department of Special Collections, University of California, San Diego, Library)

Eleanor Roosevelt and Leo Szilard at a 1961 seminar in Washington, DC (Photograph by Ike Vern)

Leo Szilard with Michael Straight, former publisher of the *New Republic*, at Straight's Green Spring Farm in Fairfax County, Virginia, May 1961 (Photograph by Esther Bubley, © Jean Bubley)

Francis Crick, Jonas Salk, and Leo Szilard in La Jolla, California, spring 1964 (Courtesy of the Salk Institute)

Leo Szilard's ashes are interred at Kerepesi Cemetery in Budapest on the 100th anniversary of his birth, February 11, 1998. (Barnabas Szabo, *Nepszabadsag*)

Szilard speaking at a Salk Institute seminar, February 1964 (Photograph by D. K. Miller, courtesy of the Salk Institute)

Leo Szilard's tombstone at Lake View Cemetery in Ithaca, New York (Photograph by William Lanouette)

hours, a single day's work could confirm or refute half a dozen hypotheses and inspire a dozen more for tomorrow.

The appeal of doing so many quick experiments drew Szilard and Novick back to Cold Spring Harbor later in the summer of 1947 and in five subsequent years.[21] On that first return, they made "genetic crosses" or matings between differently marked phages to see how the traits of the two parents assorted among the progeny, much as the Austrian botanist Gregor Mendel had done with sweet peas. But when Szilard and Novick tried to continue their work back in Chicago that fall, Novick was unable to qualify for a faculty appointment in biology and instead agreed to survive on whatever research grants they might muster, his income augmented by Szilard's out-of-pocket supplements. And since Szilard was not technically part of the biology department, he could not use its laboratories and equipment.

Impatient and clever, Szilard arranged for a laboratory all their own—in a synagogue at an abandoned Jewish orphanage owned by the university. This dingy brick structure, at Sixty-second Street and Drexel Avenue, stood several blocks south of the main campus, in the shadow of the rattling Sixty-third Street elevated-train line. Szilard and Novick took over a twenty-by-thirty-foot basement room cluttered with overhead pipes, and there created a laboratory. They scavenged for furniture discarded by other university labs, and at Sears, Roebuck bought kitchen cabinets, to serve as lab benches, and a dishwasher for their glassware. *Newsweek* called their setup "one of the weirdest of the radiobiology laboratories," but Szilard and Novick were grateful for a place to test their many ideas.[22]

Szilard's best ideas came to him as he soaked in his bathtub at the Quadrangle Club, free from all distraction. He emerged dripping wet each morning to scribble notes on a yellow legal pad, a routine that became famous in academic circles.[23] When British biologist Julian Huxley visited the University of Chicago in 1959, he happened to stay in a room at the club once occupied by Szilard. "He was tickled to use the tub where Szilard got his ideas," biophysicist John Platt later recalled. "Like Archimedes," Huxley said, and told the story over and over.[24]

Once dressed, Szilard walked downstairs to read the newspapers over a relaxed breakfast in the club's sunny dining room and at ten or so walked through campus, crossed the grassy Midway, and strolled in to greet Novick each day with the same question: "Any thoughts?" To Szilard, thoughts had power and reality all their own, and he was just as eager to hear where Novick's mind had wandered—awake and dreaming—as he was to spout his own fresh ideas.

On sunny days, Szilard dreamed up ideas by the tennis courts behind the Quadrangle Club, where he slumped in a chair, closed his eyes, and wondered about everything from understanding RNA to cracking the tax code. Some mild afternoons he sauntered a dozen blocks through the quiet Hyde Park neighborhood to a white frame house near Fiftieth Street and Kimbark Avenue, the home of law school professor Edward Levi (later chancellor of the university and US attorney general in the Ford administration). With few words, Szilard greeted Levi's wife, Kate, walked through the house, took a straight-backed wooden chair from the kitchen, and lugged it onto the small back lawn. There he sat under an elm. Sat and thought. And thought. Once in a while he scribbled a note. More often, he simply let his mind wander, opening and closing his eyes as he seized ideas and wrung from them every consequence and conclusion he could find. As the sunlight faded, Szilard returned the chair to the kitchen, thanked Mrs. Levi, and walked back to the club for dinner.[25]

One of the first experiments Szilard and Novick undertook was to clarify a difference of opinion between Delbrück and Luria, on the one hand, and geneticist Joshua Lederberg, on the other. Lederberg's experiments had led him to conclude that mating, or "genetic recombination," occurred in bacteria. Szilard and Novick designed what they considered to be a decisive test and sided with Lederberg. "I'll eat my hat if this isn't genetic recombination," Szilard wrote to Delbrück and Luria when describing their results. Luria agreed with Szilard, but Delbrück urged them to do more work. When they learned that Lederberg had already made an equivalent test, and discovered his results in a table listing several experiments, Szilard and Novick dropped this work and turned to a puzzling paradox that Delbrück had reported at the 1947 Cold Spring Harbor course.

Paradoxes fascinated Szilard because he considered them clues to defects in our understanding of the world, and this one led Szilard and Novick to discover a new phenomenon that came to be called phenotypic mixing. They found that if they infect bacteria with closely related viruses, such as the common T2 and T4 strains, some T2 viruses acquire the appearance (phenotype) of T4 viruses, but they remain genetically (in genotype) T2 and subsequently yield only T2 progeny. While in the anomalous stage, T2 viruses behave as if they were T4, in that they can infect bacteria normally resistant to T2, but their progeny cannot.[26]

Still intrigued with Monod's finding that bacteria choose which sugars they metabolize, Szilard speculated about this process in 1948 as he headed west for a vacation with Trude Weiss at the Stead Ranch in Estes Park, Colorado. Unlike most people, Szilard used his mountain vacations not to escape work but to pursue his thoughts even harder. "A mild anoxia" from the thin air, Szilard thought, made him dizzy with fresh ideas, which he caught like butterflies and scribbled on notepads wherever he happened to be.[27] "While he worked, he could not be disturbed at all," an acquaintance who met him in Estes Park recalled.[28]

From the Rockies, Szilard visited the Hopkins Marine Station at Pacific Grove, on California's Monterey Peninsula, where he joined Novick for a microbiology course given each summer by Stanford microbiologist Cornelius B. Van Niel. In letters to Trude in New York, Szilard raved about the "*splendid* sun" and the "good lectures." At first, botching went well as he sat on the rocks by the lab, watching sea lions splash as gulls yelped and circled the vacant canneries nearby.[29] Between the biology lectures, he scribbled drafts of a paper about international currency reform and proudly passed around manuscripts of two recently concocted satires: "The Diary of Dr. Davis," about negotiating with Stalin for nuclear disarmament; and "The Mark Gable Foundation," about ingenious ways to finance science—and to preserve people cryogenically.[30]

At Van Niel's lab that summer, Szilard made a very different discovery. "Unfortunately," he wrote Trude near the end of his stay, "I have very much fallen in love with the Pacific Ocean, and I do not want to go back to Chicago at all. I do not want to stay in Chicago much longer."[31] Szilard delayed his return by calling on friends at Cal Tech in Pasadena. In Los Angeles he boarded the El Capitan, a luxury express train for Chicago, but climbed off a few hours later at Santa Fe, New Mexico, and checked into Rancho Del Monte, a small hotel in the hills nearby. From there he took walks, read books about enzymes and biochemistry, and studied immunology and tissue culture.

Not always content when alone with his thoughts, Szilard interrupted botching to telephone his friend Edward Teller, nearby at Los Alamos with his family for the summer, but found he had already returned to Chicago. In a Santa Fe barbershop for a shave one morning, Szilard met by chance University of Chicago economist Rexford Tugwell, and this encounter led to a day together bouncing in a jeep through nearby canyons. With a balcony of his own in the adobe-style Rancho, Szilard daydreamed, wrote, and read in peace. But, ever the urbanite, he grew uneasy from "too few people for maximum privacy" and after a week boarded the train for Colorado.[32]

At the Stanley Hotel in Estes Park, Szilard met Novick and on the broad veranda explained the results of his thinking. All summer Szilard had wondered about growing a continuous culture of bacteria. Answering Monod's glucose-lactose question was one reason to devise such a contraption. Another was to save time. By the procedures then used, researchers who needed actively growing bacteria for phage experiments had to inoculate a culture and wait for two and a half hours before starting to work. Ideas for experiments came to Szilard so rapidly, so urgently, so fleetingly, that even half an hour's delay seemed insufferable.

"In great excitement," Novick later recalled, Szilard said that a continuous culture device might be made by starting with a vessel in which the bacteria could grow. They would be supplied with nutrient liquid that, when it flowed through the growth vessel, would wash out some of the bacteria. The trick, Szilard said, was simply to match the bacterial growth rate with the vessel's washout rate.[33]

Szilard had worked out the mathematics of the system, which showed that the size of the population of bacteria would be determined by the concentration of the controlling growth factor in the input medium, while the concentration of the controlling growth factor in the growth vessel would be determined by the washout rate and would be independent of the population size. Here was a way to keep a bacterial population growing indefinitely and at a rate set by the experimenter. Since the system would maintain a constant concentration of growth-limiting chemical, Szilard proposed that it be called the chemostat.

Back in their basement lab, Szilard and Novick tested the chemostat idea by building a simple apparatus with available flasks, tubes, and jars; and when it worked as they had hoped, they applied for money to develop their invention. But the National Institutes of Health (NIH)—on the advice of a consultant—concluded it wouldn't work and turned them down. "Later," Novick noted, "the NIH invited Szilard to apply again, but he declined."[34] The chemostat has become a standard tool for research in microbiology, physiology, and ecology, since it provides experimental conditions not otherwise available.

Szilard is remembered by his colleagues for the chemostat, but more importantly, for seeking the utterly practical—no matter how odd it might seem to others. Instead of warming the growth tube of the chemostat to 98.6 degrees Fahrenheit, Szilard thought it simpler to heat a whole room in his laboratory to that temperature. For the graduate students who conducted experiments wearing summer garb, "it was unbearable," recalled Stan Zahler, then studying phage with Novick.[35] But it worked.

To make his review of experimental records easier to read, Szilard devised a very utilitarian numbering system for his lab notebooks. It seemed odd, but always allowed him to view two consecutive pages. With conventional notebook numbering, you make the first right-hand page 1, then turn and number the first left-hand page 2. The second right-hand page is 3, the second left is 4, and so on. Using this method you cannot always see consecutive pages: 2 and 3 are visible side by side, but 1 and 2, or 3 and 4, are not. Szilard solved this problem by starting to number the *second* right-hand page 1 and the page to its left 2. When he turned the page again, he numbered the right 3, the left 4, and so on. Then, by holding vertical the page he was turning, Szilard could always view the notes of the last or the next page continuously. When using this system, he marked the pages with an "H," for Hebrew notation.[36]

Not all work at their Sixty-second Street lab was so practical, and at times it became cosmic—and comic. One afternoon, as Szilard, Novick, chemist Leslie Orgel, and biophysicist John Platt chatted over coffee, their talk drifted to the IQ of dinosaurs. Then someone wondered, "What's the IQ of God?"

"God must be very intelligent, because he's made all these intricate things, like DNA molecules and professors," Orgel said.

"Nonsense!" Novick interrupted. "Very stupid! In the first place, it has taken Him six billion years to do all this. And in the second place, He has done it by the clumsiest possible method, natural selection, just throwing away everything He couldn't use."

Grinning, Szilard cut in. "You forget," he said, "IQ stands for Intelligence Quotient. It's a *ratio* of mental age to geological age. And since God is both infinitely wise and infinitely old, His IQ is the ratio of two infinities, which can be a small finite number."

"Like that of a smart Hungarian!" Szilard's companions chanted.[37]

Having invented the chemostat, Szilard and Novick were eager to devise new ways to use it. One application was to study aging, since a bacterial population could be maintained for hundreds of generations and might serve as a model of a multicellular organism. At first, Szilard suspected that aging might be caused by accumulated deleterious mutations in a population of an organism's cells, but when tested, this idea proved wrong. They discovered, instead, that the "lifetime" of the chemostat population was too quickly limited by "evolutionary" changes to show them much about aging.

Still, Szilard and Novick were able to make accurate measures of the rate of mutation of a bacterium. They demonstrated that in the absence of a virus, mutants that are resistant to the virus would accumulate in the chemostat population as mutations occurred and could be plotted along a straight line—as long as they grew at the same rate as the whole population. The slope of the straight line should give a measure of the rate of mutation far more accurately, in fact, than could be obtained by other methods then available. But, to their surprise, when they studied the effect of growth rate and mutation rate, they discovered that even at different growth rates the mutation rate stayed constant—per hour rather than per generation. This paradoxical artifact was finally rationalized fifteen years later.[38]

They also saw that after about forty generations, the number of mutants resistant to a virus dropped sharply, then rose again at the same rate observed initially. This, they showed, resulted from an evolutionary change. The original population in the chemostat, along with its accumulated virus-resistant mutants, was replaced by the selection of a new bacterial population arising from a mutant that could grow more rapidly at the low concentration of the limiting growth factor. When this "faster" strain became established, virus-resistant mutants again accumulated in it as they had in the original population, and at the same mutation rate. In one study, followed for about 650 generations over six months, ten to twelve such evolutionary steps occurred. "Here," *Newsweek* reported about Szilard, "for the first time, as the bouncy, smiling physicist remarked, 'evolution has been made visible.' "[39]

Realizing that their chemostat afforded a sensitive means to measure rates of mutation, Szilard and Novick began experiments to test several common substances as possible mutagens. First they tried caffeine at levels found in a cup of coffee or tea, and these raised the mutation rate tenfold. As expected, X-rays and ultraviolet light were also highly mutagenic, even at low intensities. But in the course of their work, they also found substances that were antimutagenic—a completely new concept. Their presence eliminated totally the effects of caffeine and, for some mutations, even reduced so-called spontaneous mutation rates.

Szilard enjoyed caffeine and drank Coca-Cola all day long to sate his habit (and, like his bacteria, to satisfy his need for sugar). On his way into the Quadrangle Club for lunch one day, he paused to buy a Coke from a machine in the coatroom and asked if his guest would like one, too. "No, thank you, not before lunch," he said. Szilard swigged the Coke, and they walked upstairs to the dining room. On the way out, Szilard stopped at the machine again, bought a Coke, and offered one to his guest.

"But Szilard," he said, "you just got through telling me that Coca-Cola is full of caffeine. And that caffeine is a mutagen."

"That's okay," Szilard said. "I want to mutate into a native-born American so I can run for president."[40]

Szilard and Novick also studied the regulation of gene expression. They found that this process could be regulated by controlling the rate of formation of the protein coded by the gene but also that the chemical activity of the proteins themselves could be controlled by substances that the proteins made. This latter phenomenon, now called feedback inhibition, plays a critical role in the ways that cells control the formation of the many substances used in their metabolism and growth.

Szilard was so impatient for new findings and so eager for the questions these findings might pose that he only performed experiments when the result seemed to offer a surprise. He was simply too anxious for answers. New ideas crowded out the old, and happily so. "The most important property of a man's brain," Szilard told John Platt, "is the ability to forget things."[41]

Besides Szilard's constant urge to plunge into new pursuits without finishing work at hand, Novick said that "lack of time or bad luck" kept them from fully developing many ideas. But Monod thought that Szilard's creative nature itself kept him from performing decisive work. Had Szilard relentlessly pursued just a few of his ideas, Monod wrote, "his own specific achievements—written-up, formalized, and stamped—might have *appeared* greater, more definitely significant. Then however he would have been just as good, but no better, than many other highly distinguished scientists.

"Szilard was different," Monod concluded. "He knew that meaningful ideas are more important than any ego, and he lived according to these ethics." He was "a man to whom science was much more than a profession, or even an avocation" but "a mode of being."[42] Immunologist Melvin Cohn saw Szilard as a scientist more interested in discovering "how it might have worked than how it does work."

This mode of being also prompted Szilard to bend and break scientific conventions. His behavior fell outside the dichotomy that says science moves by evolution or by revolution. For Szilard, science advanced by subversion, by rigorous challenge to every discipline's most basic tenets, and by personal actions and reactions to ideas and events as they occurred. In Berlin, when he had seen Hermann Mark's modern X-ray equipment for studying fibers in 1923, Szilard decided they were better used to study the X-rays themselves. At Cold Spring Harbor, when Szilard couldn't keep bacteria at the right temperature and its gelatinous, agar-based medium solidified, he twisted the experiment into a study of agar and its properties. And, in Chicago, when impatient with the delays in peer-reviewed scientific journals, he bypassed them entirely by arranging regular meetings with the researchers he considered expert in particular topics.

In 1949 and 1950, Szilard organized the Midwest Phage, Marching, and Chowder Society, fortnightly brainstorming sessions at universities in Madison, Chicago, Urbana, Bloomington, and Saint Louis. A grant from the Rockefeller Foundation covered travel and meals, and Szilard used the encounters to question and challenge researchers at the forefront of molecular biology. Scientists described and discussed their latest results during these informal sessions, always under Szilard's feisty interrogation. Besides Szilard, Novick, and cosponsor Salvador Luria, the meetings included Alfred Hershey, Leonard Lerman, James Watson, Joshua and Esther Lederberg, Max Delbrück, Theodore Puck, Sol Spiegelman, Joe Bertani, Roger Stanier, Renato Dulbecco, and Bernard Davis.

Because Szilard relished the give-and-take of informal discussion, most of his ideas were carried away and tested by others or simply forgotten. But a few thoughts intrigued Szilard so strongly that he pondered them for years. One was his attempts to understand the aging process in all living things, and for this he developed the concept of "aging hits." In short, the number of chromosome defects that determine the natural length of life is set at birth.[43]

As with many of his ideas, Szilard merged fact and fiction in his brainstorming, and his fictional "Mark Gable Foundation," written in 1948, described how people age and how those with incurable illness might be preserved cryogenically and revived for corrective treatment decades or centuries later. In 1955, Szilard wrote "Process for Slowing the Aging of Man," in which he argued—this time, seriously—that life expectancy could be extended for persons with incurable diseases by alternating long states of low-temperature sleep with shorter periods of active living, since, when frozen, their body functions, including aging, are suspended.[44] In this way, Szilard argued, a forty-year-old man with an incurable disease and only five years to live might choose to sleep nine months a year and live with his family for three, sharing his children's development to adulthood.

In 1958, Szilard worked intensively on drafts of a paper on aging, proposing that aging hits determine our life span and that these occur randomly to deactivate chromosomes over time. "Thus, in its crudest form," Szilard explained, "the theory postulates that the age at death is uniquely determined by the genetic makeup of the individual . . . [and] the main reason why some adults live shorter lives and others live long is the difference in the number of faults they inherit."[45] The *New Scientist* magazine published an account of Szilard's aging paper, and *Newsweek* concluded from it that "females live longer because they receive a perfect 15,000 genes while men receive fewer . . . [and that] increased atmospheric radiation will make people of the future look older than they are."[46]

In biology, as in physics, Szilard continued his practice of taking out patents: for a "Process for Producing Microbial Metabolites," for the chemostat, for "Caffeine-Containing Products and Method of Their Preparation," and for cheese made with unsaturated fats—an early form of "lite" dairy products.[47] With Monod, Cohn, and Novick, Szilard developed a process for the industrial cultivation of microorganisms, based on the identical chemostat and biogen designs. During the 1950s, Monod used the proceeds of this patent to bribe border guards in order to secure the release of scientists from Hungary.[48]

While many of Szilard's ideas were dismissed as mind play, tried and disproved, or simply forgotten in the rush of his busy life, one did earn him lasting credit: negative feedback regulation of enzyme activity. When Szilard attended Monod's 1954 Jessup Lectures at Columbia University, he came away puzzled by what he heard about the induction in bacteria of the synthesis of betagalactosidase, an enzyme needed by bacteria for their consumption of lactose. The enzyme was only formed if lactose or an appropriate analogue was present in the bacterial growth medium.[49] When visiting Monod in Paris in 1957, Szilard urged him to test an idea proposed by New York University microbiologist Werner K. Maas: whether "induced enzyme formation" is under the control of a naturally occurring repressor, a molecule that somehow inhibits synthesis of the enzyme. In this view, the inducer (here, lactose) interferes with the repressor and allows synthesis of betagalactosidase.

In his lecture when receiving the Nobel Prize for research on this idea and in later remarks, Monod noted that it was Szilard who had kept him on the path to success by insisting that normally the switch controlling formation of the enzyme would be "on" except in the presence of a repressor, which turned it "off." Lactose induces the formation of betagalactosidase by inhibiting the repressor, Monod and his colleagues discovered, and Szilard had "decisively reconciled me with the idea (repulsive to me, until then) that enzyme induction reflected an antirepressive effect, rather than the reverse, as I tried, unduly, to stick to."[50]

"We all looked forward to Szilard's coming" to the Pasteur Institute, Cohn recalled. Right or wrong, Szilard's ideas were always novel and often exciting. "He was given an office in Monod's laboratory, and we all had to talk to him; everybody lined up for the chance." Szilard organized discussion groups and seminars and impressed his colleagues with the way he absorbed what they said, then days later put it together with what he had heard from others. "He followed the thread of the discussion and picked out from what was said the seminal ideas."[51]

Yet Szilard crushed bad ideas as eagerly as he cheered good ones, sometimes with dire results. In 1951 he convinced the Conservation Foundation to study advances in biological research that might be applied to human birth control and enlisted organic chemist William Doering from Columbia to help review current work. A grant from the foundation brought researchers to New York to explain their findings, and at one session a gynecologist from Boston proudly explained his discovery that hesperidin, a chemical in the rind of citrus fruit, was an effective birth-control agent in women. A charming man with a large practice, he had enlisted several patients to take part in his experiment, and to Szilard and Doering he reported proudly that none of the women studied had become pregnant while taking hesperidin. As proof that it worked, he said, when these women wanted to have children and stopped taking hesperidin, they became pregnant within a month or two.

"Leo jumped on that," recalled Doering, "and asked the doctor whether he had followed very carefully the length of each pregnancy—that is, the relation between the date when they said they wanted to become pregnant and the date of delivery. This went on and on, and it gradually dawned on the poor guy that because his patients were so attracted to him, rather than say, 'The hesperidin hasn't worked and I've become pregnant!' they came around instead with this reasonable story that they had to drop out of the experiment." A few weeks after Szilard's grilling, the doctor committed suicide.

"Leo was quite disturbed when he realized what he had done," Doering recalled. "Where he thought he was carrying out an objective, intellectually based conversation, he had been unaware of the emotional impact of what he was telling this man, namely, that he had been deceived—through the best of human intentions—by his patients."

Yet in other situations, Szilard's aggressive mind gave way to deep concern about the personal cares of his friends and relatives. He chatted by the hour with their children and told stories to them that always had some clever twist. Many times Szilard offered consoling advice about personal decisions, once taking a day to visit Doering's estranged wife in New Jersey just to be sure that their planned divorce was really the best action. Szilard also enjoyed brainstorming with young people, especially about their personal and career decisions.[52]

But a tension was always there between Szilard's fiery reason and his cool emotions. "Leo is almost frightening when he's on the trace of knowledge," Doering said later. "He literally pulls men's minds apart."[53] At conferences, Szilard's behavior was notorious. To some he seemed rigorous and logical; to others, obnoxious and lazy. The "Szilard index" became a new standard among biologists as a way to gauge each speaker's intellectual appeal. As John Platt recalled the procedure, whenever a dull speaker begins to talk, Szilard rises from his usual seat in the front row and marches majestically up the aisle—often through a slide-projector beam—to the door. Pausing there, he "stands with his hand on the knob for one sentence, nods as if to confirm his judgment, and departs." This index was part of Szilard's broader strategy for conference going: Don't be overwhelmed by all the papers and talks, good and bad. Instead, focus on just a few, but think about them intensely.

"There is a legend that I sleep in seminars but that I always wake up to ask a question that shows I have been listening and sleeping at the same time," Szilard told Platt over lunch at a 1958 biology conference in Boulder. "But it is only a legend—the truth is that I sleep; but when I wake up, I don't open my eyes until I am ready to ask the question."[54]

Outside formal meetings, Szilard enjoyed staging rump sessions: At Boulder, he bought two aluminum beach chairs, set them on the lawn or on the roof of the student union building, and sat there talking with one person at a time. He played mentor to a circle of younger scientists, whom he found more creative than colleagues his age, among them Leslie Orgel from Cambridge and Matthew Meselson from Cal Tech. Meselson later said that Szilard had changed his life with advice given at Boulder: Don't be driven by conscience to "finish up" dull projects, but plunge into work on your next appealing idea.[55]

"When I would tell Leo I was going to a seminar," Novick recalled, "he would later ask me if it was interesting. If I said yes, he would call the speaker and invite him for dinner, thereby getting a personal seminar and avoiding the chance of wasting time at a bad talk."[56]

Szilard's own freewheeling style, his zest for brainstorming, his need to push and probe and ponder ideas, set him apart from colleagues who toiled on well-focused research projects and doggedly saw them to conclusion. Scientists can cite few of Szilard's papers, for there are few, but will recount by the hour their exciting conversations with him. During his first visit to the United States, in 1954, Austrian zoologist Konrad Lorenz met Szilard at a party in the apartment of law professor Hans Zeisel, on Riverside Drive in New York. For half an hour, Szilard sat next to Lorenz in an armchair, staring morosely across the room, but then turned abruptly to say, "I was reading the other day . . ." and suggested an experiment in animal behavior. Lorenz looked at his wife in amazement. "That was the experiment I suggested to you last week," he said. For fifteen minutes Lorenz and Szilard began talking, soon in half sentences, neither waiting for the other to finish—or needing to. Zeisel's daughter, Jean, was so excited by the intellectual frenzy that she burst into tears.[57]

Recalling that episode, Lorenz remembered Szilard as "one of the most intelligent men I ever met." Later that night, he said, while riding in a taxi, Szilard asked him about Erich von Holst's "reafference principle," which explains sensations produced by the movements of a sensory organ, as when we detect after spinning around that the room is spinning in the opposite direction. "After moving only about two blocks, he had completely grasped the . . . principle and proved it by giving a striking example of its working which was entirely new to me."[58]

At a party in biologist Bernard Davis's New York apartment, as Monod, Jacob, Novick, and other distinguished researchers chatted in the living room, Szilard refused to join them. Instead, he took over a bedroom and, in turn, invited each guest in for a private chat, quizzing them on their latest work and findings, suggesting new experiments and novel interpretations of their research data, and reporting how that might relate to the work of others.[59] Participants found these interviews exhilarating, but also exhausting.

In New York another time, Szilard rushed into Maurice Fox's laboratory at the Rockefeller Institute, excited to report that "biological clocks"—cyclic activities in the behavior of living organisms—are not affected by temperature. "If there's no temperature coefficient, that means it's unlikely to be a chemical process. . . . Could it be an electrical circuit that is independent of temperature?" he wondered. Suspecting that an immutable law of biology might lie behind this fact, Szilard said, "If there's an undiscovered principle of physics, it seems likely that the biosphere will have employed it," and asked Fox to arrange a talk with some neurobiologists. Fox called a colleague, Theodore Shedlovsky, who offered to invite some electrophysiologists to lunch. And, he reported, Norbert Wiener, a pioneer in cybernetics, was around the institute and might also join them; at the time, Wiener was studying alpha rhythms in the brain, another cyclic phenomenon.

By lunchtime, more than twenty people had joined in, making a focused conversation impossible. "It was a disaster," Fox recalled. "It was clear from the beginning that Szilard wasn't interested in alpha rhythms of the brain and Wiener wasn't interested in biological clocks." The conversation quavered, wandered, and soon "degenerated into a discussion about the origin of life." Given another planet like Earth, would life emerge? Yes, many agreed, and their reasons for this dribbled out as Szilard's eyelids fluttered and he nodded off. Finally, someone said, "Given many planets like Earth, would man emerge on one of them?"

"What do you think, Leo?" Shedlovsky asked.

"No!" Szilard answered, his eyes popping open.

"But why not?" Shedlovsky asked.

"God wouldn't make the same mistake twice."[60]

For all Szilard's mental agility, few of his own ideas were ever pursued to theories or discoveries that are today recognized as his own work. Instead, he is remembered for the sincere and encompassing energy he brought to conversations and for the insightful questions and bold hypotheses that startled groups and stirred up ideas. In 1950–51, Szilard organized, with Richard L. Meier, a series of evening seminars in the Social Science Building at the University of Chicago, each time discussing a different problem that might merit research; one night it was world food supplies, another energy, a third, water redistribution from global climate change. Inspired at these discussions to study the "carrying capacity" of the earth, chemist Harrison Brown organized another seminar that Szilard often attended on the factors limiting the earth's human population and from this Brown wrote *The Challenge of Man's Future*.

These sessions aided Szilard's work in birth-control research for the Conservation Foundation, including, in 1951, his invention of an electric "birth-control clock," renamed a "fertility clock" to appease the Catholic church.[61] Designed for use at family-planning clinics in developing countries, the clock was to be reset monthly for each woman in order to sound alarms during her menstrual cycle when conception was most likely. Szilard had a model built at the University of Chicago, and Brown took it with him when he moved to Cal Tech. Brown even demonstrated the clock to the Ford Foundation, but found "little interest" there.[62] Szilard also devised a "fertility necklace," with colored beads that moved and locked for counting menstrual-cycle days. He wore this at some lectures on population control but never developed the idea further.[63] These gimmicks aside, Szilard is today credited with work on the study and preservation of sperm and with promoting research that led to the oral contraceptive at a time when few scientific or medical institutions would take the idea seriously.

Szilard's laboratory situation became unworkable in the fall of 1953, after Novick left for a year of research at the Pasteur Institute. Desperate for a more supportive situation, Szilard turned to Einstein, drafting for his friend and mentor a letter of recommendation to the new Albert Einstein College of Medicine in New York. When that attempt failed, Szilard admitted reluctantly, as he had when he first fled Hungary, "I would rather have roots than wings, but if I cannot have roots I shall use wings."[64]

He focused next on the University of Colorado Medical School in Denver, where Trude Weiss, his wife since 1951, taught public health. Szilard had craved a professorship there, in the new biophysics department created by his Chicago colleague Theodore T. Puck, but while Puck assured him that a place would be found, years passed with nothing more offered than a visiting professorship. "Puck is paying me for *not* being here," Szilard complained to Trude.[65] Instead, Szilard became a visiting professor at Brandeis University and a part-time consultant to Abe Spanel, the quixotic president of the International Latex Corporation in New York.

In a dispute unresolved for years, Szilard is both credited with proposing a novel method for the quantitative biological cloning of mammalian cells and chided as an interloper for doing so. The discovery—and dispute— occurred at Puck's department during the summer of 1954. Chatting over lunch with Philip Marcus, a former research assistant of Novick's and Szilard's at Chicago who was then Puck's student, Szilard learned that they were trying to increase the efficiency at which individual mammalian cells would grow into self-sustaining colonies (clones). These cloned cells would provide cell biologists with the powerful tools of genetic manipulation available to microbiologists and help usher in the era of somatic cell genetics.

Puck and Marcus sought to reduce the loss of diffusible nutrients by plating cells into microdrops of growth medium. If there were more than a hundred cells in a drop, Marcus said, growth occurred in virtually all of the drops. The efficiency of self-sustained growth fell off sharply as the initial number of cells in a drop was reduced; if there was only one cell, it had a 1 in 100 chance of developing into a full-size colony. How, Szilard wondered, might they maintain a high reproductive efficiency when cell numbers were small? He knew almost nothing about the field but found the problem fascinating.

According to Marcus, Szilard sat quietly for a few moments, then said, "Since cells grow with high efficiency when they have many neighbors, you should not let a single cell know it's alone." At first, Marcus thought Szilard's remark a joke—a flippant excursion into psychobiology. But what Szilard meant is that the cells should be grown in the same biochemical environment as that created by large numbers of cells. On a restaurant napkin, Szilard drew a mass culture, then sketched single cells on a glass plate over it.

"I told Puck about this idea, but he did not express an interest in it," Marcus later recalled. "But as a new graduate student I found the experiment too compelling." He had the department machinist make a small plastic platform to hold the glass slides on which the single cells were to grow while submersed in growth medium shared by a mass culture of cells. "I started the experiment on a Friday," he remembered, "and on Saturday I went back to examine the cells. Every place where I had put one cell there were two, and the day after—on Sunday—there were four." Marcus telephoned Szilard, and he and Trude visited the lab.

"We'll call Puck," Szilard said. Marcus did, but when Puck walked in, he looked at the dish, turned, and left in silence. "I didn't see him for a long time," maybe several days, Marcus recalled. "When I did, he said he had thought of a solution to prevent the viable cells of a mass culture from contaminating the single cells that grew on the glass slides above them."[66] While Szilard's idea was to separate the cells by a glass layer, Puck's solution, explained in later papers, was to X-ray the bottom cells, which kills their reproductive capacity but not their biochemical or metabolic activity. In May 1955, Puck and Marcus published their first paper about the biological cloning of mammalian cells and in it added a footnote thanking

Szilard, "who suggested a more advantageous geometrical arrangement" for placing the cell in the test dish.[67] Szilard later called this note his "punishment" for intruding in the work.[68]

After that, Szilard's relationship with Puck was strained, and in December 1955, Puck admitted that no permanent post would be offered in Denver. "With the greatest possible reluctance I have come to the conclusion that it is not possible for me personally to work with you scientifically," he wrote Szilard. "Your mind is so much more powerful than mine that I find it impossible when I am with you to resist the tremendous polarizing forces of your ideas and outlook." Puck feared his "own flow of ideas would slow up & productivity suffer if we were to become continuously associated working in the same place and the same general kind of field." Puck said, "There is no living scientist whose intellect I respect more. But your tremendous intellectual force is a strain on a limited person like myself."[69]

Instead, Puck suggested that Szilard continue to visit Colorado, and three or four other universities, as a "roving professor." That scheme collapsed because Szilard—in a typical fit of "independence"—refused to tell a National Science Foundation grants officer exactly how much time he might spend at each institution.

Szilard and Puck clashed again, in 1958, during a biology conference in Boulder. At the time, Szilard was drafting a paper about the aging process and asked if Puck was working on any related topics. When he heard no answer, Szilard asked Puck's assistant, Conant (Cody) Webb, for help locating reprints. But a few days later, Puck again accused Szilard of meddling with his assistants—and, possibly, infringing on his research. Now certain that he was not welcome at Denver, Szilard scouted for a way to flee his awkward situation in Chicago, where his hybrid faculty position continued without lab space.

※　※　※

When in Chicago, Szilard stayed at the Quadrangle Club and wandered about the campus calling on friends and associates. Physicist Samuel Allison often invited Szilard to solve rigorous mathematical problems whenever they met.[70] On his strolls, Szilard always looked in on chemist Nathan Sugarman and physicist Alexander Langsdorf—usually just in time for their coffee breaks. Szilard's only use of laboratory equipment at this time was to handle the beakers this group used to brew and drink coffee.

When Szilard could find no institution that honored his quirky blend of science and social policy, he decided to create one—at least in his mind. His brainstorming, and later behind-the-scenes plotting, led to the Salk Institute for Biological Studies, the intellectual refuge where he would spend his life's happiest, and last, days. As early as 1946, Szilard encouraged federal support for scientists' "leisure." As a way to assure their creativity, he asked that the proposed National Science Foundation pay researchers $12,000 a year for life, freeing them to pursue whatever ideas they fancied.[71] And in 1949 he advised the new Ford Foundation's directors on how to fund research on both scientific and political topics. So when Szilard devised his ideal research institution, it naturally merged many of his own projects and interests. His added twist was to tie the new institute to a famous scientist; this to give the place instant recognition and to aid fund-raising.

In 1953, Jonas Salk quickly became a medical celebrity when he announced development of a polio vaccine. Szilard first met Salk in October 1956, at a time when Salk himself was beginning to think about forming a research center in Pittsburgh and had signs of support from the March of Dimes antipolio foundation. They met again, early in January 1957, at New York's Waldorf-Astoria Hotel during a biology conference,[72] and two days after the conference ended, Szilard completed a memo he had been drafting with his Conservation Foundation colleague, chemist William Doering, on a proposal to create two interdependent research institutes.[73] One Szilard called the Research Institute for Fundamental Biology and Public Health, the other the Institute for Problem Studies. Together they would integrate science and social studies. Szilard sent the memo to Cass Canfield, a self-styled "urbane go-getter" with an abiding interest in birth-control research, who was head of Harper & Brothers publishers. Szilard's memo cited Salk as a scientist who had struck out from "the realm of pure science" to work on one of the "acute problems of our times"—and succeeded.

"Usually such diversions from pure research involve a great personal sacrifice, and those who engage in them must struggle against heavy odds," said Szilard, echoing his own lifelong frustrations. Some "recognized problems" that Szilard cited as worth pursuing were a biological method of birth control suitable to developing countries and the health risks from cigarette smoking. But, he warned, "unrecognized problems are of even greater importance." These occur in the realm of "political thought" and would not be solved by most social scientists because they care more for methods than results. "In the circumstances," he warned, betraying his own biases, "we may have to make a new start and to begin pretty much where Plato has left off." Szilard aspired to devise a new "form of democracy" suitable for developing countries and even suggested that the British Colonial Office might be employed to "field test" his institute's political solutions.[74]

A "confidential" appendix named researchers who might be invited to join the new institute, and among those who later joined the Salk Institute he named François Jacob, Renato Dulbecco, Melvin Cohn, and Edwin Lennox.[75] Like many of Szilard's proposals, this one dives too quickly into intricate schemes for hiring and organizing the staff and for financing day-to-day operations. But its appeal is twofold, as it both addresses current issues and suggests unknown areas for discovery. Besides, unlike almost any other institution at the time, this one matched Szilard's own unique exuberance.

Not content to enjoy his institute as a lovely idea, Szilard decided to press for its creation. He wooed Salk with a letter describing the Canfield memo, seeking his reaction to the dual research institutes, and asking him to consider becoming an Affiliate Member.[76] The next week, Szilard met Salk in Chicago to explain the proposal, and when Salk replied to Canfield in early February, saying that he supported this "good idea" but was too busy to join, Szilard pressed his case in a playful, teasing way.[77] You might attend for two reasons, Szilard wrote Salk. "First, as a 'duty' because these first few meetings may decide the shape of things to come. And second— more important—on grounds of the principle [of] 'pleasure before business.' " Szilard promised to argue for this pleasure principle when they met again, because, he wrote, "if we do not manage to get together an enjoyable group of Affiliate Members—at the very least—we have no right to be in this 'business' at all."[78] When Canfield saw a copy of this letter to Salk, he told Szilard, "You handled him perfectly. . ."[79]

In April, at a research conference in Gatlinburg, Tennessee, Szilard sat next to Salk on a bus ride in the Great Smoky Mountains and again urged him to join the institute.[80] Then, trying humor, Szilard wrote Salk proposing compulsory "insurance" against polio, just as Social Security is required for old age. This would be an inducement for vaccination, since parents escaped paying the premiums if their children were treated.[81]

When in Cambridge, England, that fall for a Pugwash conference on arms control, Szilard met with molecular biologists Seymour Benzer and Sydney Brenner, Dulbecco, and Cohn and at this meeting shaped a consensus on what an institute should be like. Among the glistening lab equipment and chemical smells of the Molecular Biology Center and in the musty splendor of the ancient college dining halls, Szilard proposed that they all try to enlist Salk to head the new institute. He was well known, Szilard argued, could testify well before Congress, and in this role the public could repay its debt for his work curing polio.[82]

In 1959, when Salk seemed determined to stay in Pittsburgh, Szilard dangled another lure: the offer of land—and affiliation with the University of California at San Diego (UCSD)—if he chose to locate in La Jolla.[83] "Frankly," Szilard warned, "I see no possibility of getting many first-class people to move to Pittsburgh."[84] And on a trip to Europe, Szilard wrote Salk twice more, urging him to explore the La Jolla proposal with James Watson and UCSD founder Roger Revelle.[85] Ultimately, this connection worked: The city of San Diego donated a twenty-seven-acre site on a cliff above the Pacific, near the new UCSD campus, and in 1961 the Salk Institute for Biological Studies was created.

A 1966 institute publication credited Szilard as "one of the moving spirits who helped to conceive the idea of the Salk Institute and to bring it into being,"[86] and by then it was prospering. But while Szilard was enticing Salk to join in his "dream" in 1957 and 1958, Szilard still lacked basic laboratory space and funding for his own scattered research. Still craving "roots," Szilard even considered the National Institutes of Health (NIH), in Bethesda, Maryland, just north of Washington, D.C. He went there in May 1958 to explain a research proposal and, while awaiting a decision, lingered as a consultant. Szilard liked the NIH's suburban campus and for a while roomed nearby at the institute's villas in Kensington and later stayed at the Kenwood Country Club.[87] Although Szilard feared water and always dressed formally in public, he enjoyed relaxing by tennis courts and swimming pools when he worked. By the villa pool one afternoon, Danish-born biochemist Herman Kalckar (an acquaintance from Berlin) recognized Szilard and stopped to chat.

"Would you like to swim?" Kalckar asked.

"No," Szilard replied, "I'd rather think."[88]

Because Szilard enjoyed lounging outdoors, he was able to find the "leisure" he considered necessary for serious work.[89] But he quickly came to hate the NIH's modern, air-conditioned offices. "I can't smell the grass," he complained, looking from his desk to the broad lawn—through windows he couldn't open. "Everything is sterile!" He asked colleagues with suburban houses to bring him grass they had mown, and this he stuffed into empty file drawers, giving his office a sweet country smell.[90]

In June 1958, Robert Livingston offered Szilard a $19,000-a-year full-time position at the NIH, "effective within a few weeks."[91] But when Livingston could not offer the 2,500-square-foot laboratory space Szilard wanted, he tried to arrange a joint appointment between the NIH and the Rockefeller Institute in New York.[92] Szilard reported on this proposal to Herbert Anderson, director of the Enrico Fermi Institute for Nuclear Studies in Chicago and still his boss. At this, Anderson became annoyed with Szilard's long absences from campus and replied sarcastically, "If I knew that all it would take to get you to lecture was to pay you, I'd be glad to double your salary."[93]

At the NIH, Szilard planned "memory and recall" as his general topic and in one note to Livingston proposed an "amusing experiment" that suggests the odd pursuits he may have had in mind. To test the link between memory and humor, Szilard wanted to anesthetize half a man's brain, then have him read jokes. "After the man is warmed up, ask him to *tell* the jokes which he has read," Szilard proposed. "Since the left half of the brain contains the speech center, he should have no difficulty in doing so. The question is, will the right half of the brain laugh about the joke that it hears?"[94]

At the NIH, Szilard became intrigued by dolphins, thanks to long talks with John C. Lilly, a visionary medical researcher who studied the animals' mental and linguistic development. Intrigued with intelligence, learning, and memory, Szilard questioned Lilly about how dolphins' brains function—often over lunches in the NIH cafeteria and dinners at Italian and Chinese restaurants around Washington's Dupont Circle. "Szilard was seeking the source and location of the memory function and was particularly interested in my work on the cortex," Lilly recalled.[95] In brain size and functions, Szilard learned, dolphins were remarkably close to humans. And, like humans, their brains controlled acts involuntary in most other mammals, such as deliberate breathing and penile erections. Two years later, Szilard would write Lilly and his discoveries into *The Voice of the Dolphins*, a popular political satire about arms control, scientific research, and other human foibles.

In the summer of 1958, before accepting the NIH appointment, Szilard left for a biology conference in Boulder, then attended a Pugwash conference in Kitzbuhel, Austria. He returned in September to learn that Livingston had reneged on his offer.[96] A friend advised Szilard that the job might have been canceled because some people perceived him "as tactless and even as somewhat intrusive in relationship to other people's work."[97] To Livingston, Szilard fumed that "at least part of the trouble comes from a fear that I might be a 'headache' from the administrative point of view because I may be expected to think, say, and occasionally do unusual things. . . ."[98] But worried about his financial security, Szilard wrote Livingston again, recounting how he needed the NIH appointment and offering to take two months' unpaid leave each year to trim his salary.[99] Szilard longed to gather a team of young and energetic scientists, he wrote, for "really imaginative experiments," not "trivial experiments, where you can virtually guarantee publishable results." But he could find no institution to welcome him.

During the next six months, Szilard published a paper "On the Nature of the Aging Process," spent two months at General Atomics in La Jolla consulting on advanced nuclear-reactor designs, wrote a proposal for the Los Alamos National Laboratory to take up molecular biology, worked through the American Academy of Arts and Sciences to plan a scientists' arms-control conference in Moscow, and again nudged Salk to site a research center at La Jolla.[100] In June 1959, Szilard applied to the NIH for a four-year research grant, then resumed his whirligig life with a three-month trip to Europe. He attended a Pugwash conference in Baden, visited scientists in Vienna and Zurich, and from Geneva wrote his first letter on arms control to Soviet Premier Nikita Khrushchev.[101]

A molecular biology conference in Copenhagen and a visit to the Karolinska Institute in Stockholm later in 1959 left Szilard wondering how to explain the way a microbial cell regulates the level of different enzymes by induction. An answer came to him while botching on a flight from Stockholm to London. He had to reconcile induction with antibody formation. Induction involves a small molecule turning on the synthesis of a specific enzyme, while antibody formation involves a large molecule turning on the formation of a specific protein (the antibody).

"The trouble is that in antibody formation you have to explain two phenomena simultaneously," he recalled.

One is that if you inject protein into a rabbit and then the same protein a month later, the rabbit responds with a much stronger antibody formation. It's called the secondary response. At the same time you have to explain a phenomenon which is called tolerance—when you inject large quantities of a foreign protein into a newborn rabbit, when that rabbit grows up, it cannot make antibodies against this specific protein. . . .

My explanation of the secondary response invokes a mechanism which is endowed with memory; the rabbit "remembers" that it was exposed to that protein before. I had a model for induced enzyme formation, but I had failed to see earlier that if I just changed a constant in it, made it bigger, the model became endowed with "memory."[102]

Here was more of Szilard's creative psychoscience, as when he endowed Maxwell's demon with a memory and urged Marcus to trick the single cells into thinking they were not alone. Here, again, was Szilard's mind over matter.

Returning to New York in October 1959, where Szilard had scheduled surgery for a prostate infection, he became so enchanted with his insights about antibody formation that he worked furiously to draft two related papers, turning for criticism and help to Maurice Fox at the Rockefeller Institute, to Howard Green and Baruj Benacerraf (then at New York University, both later at Harvard), and to Herbert Anker at the University of Chicago. "I didn't think I was seriously ill," Szilard recalled, and began fresh calculations for the papers.

Then, in November, the four-year NIH grant came through.[103] At last he had the financial—and institutional—security he long craved, the "roving professorship" that had eluded him for years. At last he could visit and brainstorm with his friend Novick at the University of Oregon, with George Beadle and Linus Pauling at Cal Tech, with George Klein at Karolinska, with Monod and his colleagues at the Pasteur Institute, with James Watson at Harvard, with the Lederbergs at Stanford, with Manhattan Project colleague Alvin Weinberg at Oak Ridge, and with other friends in New York, Philadelphia, Berkeley, and Cambridge, England.[104] At last, Szilard could have both roots and wings.

His joy was brief, however. The morning after Szilard finished his antibody manuscripts, he checked into the hospital and discovered that his problem was not the prostate. He had bladder cancer.[105]

CHAPTER 26

Beating Cancer

 1960

Room 812 at Memorial Hospital was seldom quiet. From the early calls by nurses and the clatter of breakfast dishes through doctors' visits in the late morning, to the parade of afternoon callers and the chitchat of surprise evening visitors, Leo Szilard's room buzzed with conversation. The telephone rang steadily. Papers rustled and flapped in their constant shifts from the cluttered bed to the low armchairs to the long window ledge as people arrived and places were cleared. Bright light beamed above Manhattan's skyline and through the broad windows, and with it came also the city's inescapable muffled roar. In calmer moments a Grundig stenograph whirred and clicked through bursts of Szilard's resolute dictation. Some visitors found it hard to believe that the plump, gray-haired man in the blue-and-white-striped robe at the center of this commotion was dying of cancer. Even late at night Szilard's room resonated with a busy cadence—his leonine snore.

This routine began almost as soon as Szilard transferred to the Sloan-Kettering Memorial Cancer Center on January 7, 1960. He moved from New York Hospital, where he was first diagnosed to have cancer, because doctors there had proposed removing his bladder by surgery. He balked at this and sent Trude scurrying to medical libraries for articles and books. He telephoned doctors and scientists he knew. He studied mortality statistics for bladder cancer that Trude had compiled. All this convinced him that the odds for survival were much better with radiation therapy and that Memorial was the place for such treatment. So, in his room there, Szilard planned to live out a gamble by trying to use his remaining time productively. "I'm not in distress," he told CBS newsman Howard K. Smith that spring. "It is true that I don't expect to live, but still I hope to be active for a few months, and perhaps for a year."[1]

In a part of the interview not broadcast, Szilard explained his thoughts about living and dying. "I always liked . . . the story of a man who had heart trouble and who went to see his doctor. And the doctor told him that he could live out his normal life expectancy if he were willing to go slow and to restrict his activities. The man thought this over—and finally, he told his doctor that he is not going to slow down. 'You see, Doctor,' he said, 'if worst comes to the worst, I'll be dead ten years longer.' This, I think, is a healthy attitude to take towards life and towards death."

Even as a child, Szilard recalled, he had tried to visualize what he would do if faced with early death and reached the same conclusion: Continue what you enjoy. But now, at age sixty-two, there was one difference. He had less interest in "short-range issues," such as the East-West standoff over Berlin, and much more interest in general problems, such as the bomb. "Now," he added, "that doesn't mean that if you want me to I couldn't give you a solution of the Berlin problem at the drop of a hat. . . ."[2]

Szilard was just as confident about directing his own treatment. He had his doctor authorize the nurses to tell the name and dose of all his medicines. He often took his own temperature and sometimes wouldn't tell the nurses what it was. He preferred to wash himself, with Trude's help, and to have her change his surgical dressings. And he awoke and slept when he pleased. Szilard had decided to remain active as long as he could by having two tumors removed endoscopically and then undergoing radiation treatment to his whole bladder.[3]

Two days after checking into Memorial, Szilard directed the first irradiation of his bladder. At the time, patients received lengthy but relatively low doses of radiation for several months. In many cases, this caused nausea and other uncomfortable complications. Szilard and his doctors decided, instead, on much higher doses almost daily, but for only a few weeks. The "patient tolerated treatment well," Szilard's chief physician, James J. Nickson, concluded, and after a few days of treatment, Szilard's only complaint was fatigue and diarrhea. An examination on January 25 revealed that around Szilard's cystoscopy the neoplasm, or uncontrolled tumor growth, could not be distinguished from inflamed bladder tissue.[4]

A recurrence of tumor growth would be detected in blood and urine samples, and when these were taken every few weeks, Leo and Trude waited anxiously for the results. Radiation therapy ended on February 13, and within a week Szilard's urine showed no signs of neoplasm. The news was also good for two samples taken in March. As the testing routine continued, the Szilards began to feel they were living month to month, sample to sample.[5]

Once word of Szilard's illness spread, letters and calls flooded in from friends and colleagues. It was as if he had survived to read—and savor—his own obituaries. His time in room 812 also gave Szilard his first continuous residence in years and made him the center of attention in an expanded world. His restless shuffle between Chicago, New York, and Denver and his forays around North America and Europe had ended. But from his hospital bed, Szilard soon realized, he attracted new attention for his quirky ideas about science and public policy. To *Life*'s science reporter Albert Rosenfeld, Szilard looked more like "a reclining, cherub-faced Roman emperor than a declining cancer victim."[6] Ironically, Szilard's gamble with death had yielded him a new and satisfying life.

For Trude, too, Szilard's year in the hospital gave fresh meaning to their nine-year marriage. This was the first time they shared a daily routine. Trude lived nearby, in Abby Aldrich Hall at the Rockefeller Institute, and as Leo's wife and doctor, she was at his bedside many hours each day, in new and fulfilling roles that his nomadic life had prevented. She met and ushered in visitors, arranged for stenographers, managed their social calendar, tidied his cluttered room, and opened his mail. She also watched Leo's symptoms and vital signs and dressed and changed the catheter tube and drainage bag he now wore.

"I don't think my doctors would agree that my case is incurable even though they may be willing to admit that they themselves cannot cure it," Szilard wrote to his former colleague Harold Urey the week he shifted hospitals. "In any case, I am not changed in any way, I am not much more of a sight than before, and there is no point in making too much fuss about dying." Confronting cancer, he said, involves "putting my affairs in order, and this I find very comforting."[7]

Yet there were also setbacks: recurrent fatigue and sudden temperature changes. The harder Szilard worked, the more vulnerable he felt—reminded anew of his gamble. He reread his will, listed for Trude his bank accounts (in New York, Frankfurt, and Zurich) and pensions (in the United States and Germany), and gave a book editor permission to revise a manuscript "in case I should take a turn for the worse or die."[8] Understandably, the friendly bustle in room 812 sometimes gave way to darker thoughts: to recurrent fears of a painful, lingering death. If his tumors spread, if his body failed him, then Szilard still wanted to be in control—to take his own life swiftly and painlessly. Characteristically, Szilard thought about the worst in a fresh way.

"How terrible it is," Szilard complained to chemist William Doering, "that you can't walk into a drugstore and buy something to kill yourself without pain." Friends like Doering soon realized Szilard had been studying poisons and suicide as well as radiation treatments. "Barbiturates are effective," Szilard reasoned, "but you have to take a lot of them and then be left alone for hours. Otherwise, someone can save you."

On a stroll to the eighth-floor solarium one day, while gazing at Manhattan's skyline and the East River, Szilard asked his friend Cody Webb about the effects of cyanide. Webb acted uneasy and Szilard dropped the subject, but later he learned from Doering that cyanide is so astringent it seems to choke you to death in an agonizing way. He asked Doering to mix a potion of cyanide and citric acid, to mask the poison's painful grip. Doering avoided this request as long as he could, but soon there came the long-distance phone calls to his office in New Haven. "Doering!" Szilard said brusquely, "I'm calling for the concoction you're going to make." Eventually, Doering delivered a small vial of cyanide, which Szilard would have to mix with orange or lemon juice.

But poison posed a further problem for Szilard: An autopsy would reveal the cause of death, negating life insurance payments to Trude. Szilard's thoughts sprang to a "suicide kit" that should be painless *and* leave no trace. He brainstormed on this device with a few close friends. Doering found the subject curious, and in its own way, healthy; Szilard was trying to make the end of his life easy for everyone concerned. But Benjamin Liebowitz was shocked, considered the idea ghastly, and refused to talk about it. Brother Bela, an eager coinventor since their childhood, vetted Leo's ideas conscientiously. They might even produce and patent something salable, the brothers agreed.

Research in the hospital library had convinced Szilard that asphyxiation would be the most difficult suicide to detect but would also be among the most painful. The pain, he learned, comes from the body's own survival mechanism. When you hold your breath for several seconds, your body absorbs oxygen and expels carbon dioxide into your lungs. Soon your lungs become uncomfortable, stinging as they swell with the carbon dioxide. To relieve this pain, you start panting naturally, both to exhale the carbon dioxide and inhale fresh oxygen. How, Szilard wondered, could he remove the carbon dioxide—to avoid the pain and the panting—yet still deplete the oxygen to lose consciousness? He decided on breathing in and out of an airtight plastic bag, with a filter to trap the carbon dioxide.

Szilard thought this simple device should be sold in drugstores. It had the added feature, he realized, as if drafting advertising copy, that a person once committed to suicide still had time to reconsider during the process—an impossibility with poisons or self-inflicted wounds. When asked to, Liebowitz refused to buy a plastic bag and plastic tubes or to be Szilard's accomplice, if necessary, by delivering and later removing the breathing device. But Szilard kept on scheming and devised another air bag that held cyanide vapors. These vapors could be released only by a second step, giving the person another chance to reconsider.[9]

For all the fear and foreboding, the year Szilard spent in the hospital was probably the happiest of his life. The radiation was completed within six weeks, and once fatigue from this passed, Szilard set a busy pace of writing, reading, dictating, and meeting. Before long, the hospital's operators were complaining that room 812 was tying up the switchboard. "For the moment I feel very well and even though I am confined to the hospital I am fully active and go out to dinner every day," Szilard reported that summer to Alexander V. Topchiev, head of the Soviet Pugwash group.[10]

To visitors, Szilard sometimes appeared frumpy in his broad-striped blue-and-white hospital robe, especially when he wore calf-length socks and garters as well. Callers remember that Szilard's bed and chairs were so strewn with books and manuscripts that he had to lead them to the solarium to find a place to sit. Szilard ambled in and out of his room impulsively and even took taxis to the St. Moritz Hotel, on Central Park South, where he rented a room to serve as his office. Rumpelmayer's, the delightful confectionery and pastry shop off the hotel lobby, was also a frequent stop. As his recovery progressed, Szilard and Trude went out to dinner often: to the homes of friends; to restaurants, including the Budapest, a lively Hungarian place on the East Side; and to the Czech National Club on Seventy-seventh Street, a neighborhood social center whose homey and inexpensive cafe served authentic mid-European dishes.[11]

Intermittently, Szilard tinkered in microbiology, but never on a single or sustained topic.[12] Public policy—and the stature he was gaining in it—was more fun. With his increasing celebrity, Szilard received offers from several publishers. He began dictating "Memoirs" and in several attempts managed to complete a few dozen pages.[13] Robert Livingston, his colleague and friend from the National Institutes of Health (NIH), tried to help by interviewing Szilard about his life and drafted some "Biographical Notes."[14] Szilard also signed a contract for a personal memoir with New American Library (NAL)[15] and discussed with Knopf a history of the A-bomb.[16]

"In the first draft I would go rather extensively into my childhood and even the childhood of my mother, but most of this will come out again, and only what is actually relevant to the history of the bomb will remain in the final draft," Szilard proposed.[17] For about a week in June, Szilard worked on an outline and rough draft for his "Memoirs,"[18] but current events and visitors distracted him constantly, and by the fall he had written very little. NAL later reclaimed its $5,000 advance when Szilard refused to finish a manuscript.[19]

It is intriguing that Szilard thought his mother should be covered so extensively and unfortunate for our knowledge of his life and personality that he did not dictate more about her. From what survives we know that Szilard revered his mother's moral influence and thought much of the instructive tales she told about her father's honesty and strict sense of ethics. In his hospital dictations Szilard also recalled from *The Tragedy of Man* a vision of the human race at the point of extinction and stressed anew the need for a narrow margin of hope to sustain valiant efforts— something he needed now for his own survival as well as for mankind's.

In all, Szilard's recollections and memoirs revealed very little emotion. He failed to mention his brother and sister, his cousins, his classmates, his infatuations and loves (and losses), or his long and deepening relationship with Trude. There is no instance of exuberance or despair, yet we know from his friends and colleagues that Szilard could be very animated, warm, brusque, bitter, and gentle. Despite this opportunity to bare his soul before a possibly early death, Szilard held back—not, as Trude would later say, because he "was much more interested in the next twenty-five years than in the past twenty-five" but because since adolescence he had worked to keep his emotion and reason separate, and in a lifetime of trying he had succeeded.[20] In the end, it was Szilard's reason that ran wild—exuberant reason in playful and profound excess.

With more time to study the newspapers and follow the details of statecraft, arms control was just the kind of "general problem" worthy of his final days. February's *Bulletin of the Atomic Scientists* featured Szilard's essay "How to Live with the Bomb—and Survive," in which he called for "a meeting of the minds" between Russia and America to maintain peace in the face of an unstable and changing strategic military balance.[21] In a long introduction, *Bulletin* editor Eugene Rabinowitch praised Szilard's "capacity to think years ahead of his contemporaries in a rapidly changing world" and argued that "this entitles him to attentive consideration, however bizarre some of the ideas expressed . . . may appear at first sight. . . ."[22] In this article, Szilard foresaw clearly the stalemate that eventually led to arms-reduction efforts in the late 1970s and 1980s.

He also restated his overly rational proposal to identify ten Russian and ten American cities, which in turn would be bombed—city for city—if a nuclear war started: his way to dramatize how and why such a war should never begin in the first place. The article gained him international attention. On March 7 the ABC program "Edward P. Morgan and the News" linked "How to Live with the Bomb—and Survive" with the call by Democratic presidential candidate John F. Kennedy to create an Arms Control Research Institute. Morgan praised Szilard's "bizarre but challenging plan" as an example of his "bold and imaginative thinking" on nuclear arms control.[23] *Newsweek* featured Szilard's article in a piece captioned "A room with a startling view."[24] This was echoed a few weeks later in the Soviet humor magazine *Krokodil,* which paraphrased the paired-cities scheme: "I Will Exchange: Detroit for Omsk? Philadelphia for Leningrad?"

Up to now international exchanges were widely used: among picture galleries, theatrical troupes, municipal delegations of cities . . . but whole cities!

However, the American scientist, Dr. Leo Szilard, proposes to exchange—it is true—not his own native city of Chicago, but Detroit and Philadelphia for Omsk and Leningrad.

Of course, every man in his right mind agrees that warming up of the international climate is not favorable to the "exchange" according to the prescription of the Chicago physicist. How much simpler would be the varying of this by exchanging . . . rooms within the borders of the same city. The magazine, "Newsweek," should help Leo Szilard by inserting the following advertisement:

> WILL EXCHANGE NEW YORK FOR NEW YORK
> ONE ROOM, NUMBER 812 (IN THE MEDICAL
> WING)—BRIGHT, WITH ALL COMFORTS—FOR
> A ROOM, NUMBER 6 (IN THE PSYCHIATRIC WING)

P.S. The thankful residents of Omsk, Detroit, Leningrad, and Philadelphia would assume payment for the ad.[25]

For the March *Bulletin,* Szilard wrote about the nuclear-test-ban debate: "To Stop or Not to Stop."[26] They should stop, he said, because testing allows both sides to develop more specialized weapons, such as those for tactical use on battlefields or for antiballistic missile defense—"a new kind of futile arms race." He proposed that scientists should be encouraged, with a tax-free $1 million reward from their governments, to report any secret weapons tests and be assured sanctuary afterward—an idea he had advocated for fifteen years.

His logic is impeccable, his proposals ingenious. But reading or hearing about this inspection scheme, people must have wondered if Szilard was serious. At the time, the *Herald Tribune's* syndicated columnist Marguerite Higgins described him as "an odd combination of scientific genius, eccentric . . . and politician interested in mobilizing forces to mould high policy."[27] And, in a way, Szilard himself had become part of the public policy debate as a celebrity who personified the terror that nuclear weapons posed and the hope needed to control them.

In March and April, Szilard was interviewed for several days about his views on arms control and disarmament by broadcasters Howard K. Smith and Edward R. Murrow—joined by such experts as Hans Bethe and Edward Teller, MIT engineer Jerome Wiesner, and former AEC commissioners Lewis Strauss and Thomas Murray. Murrow opened the first of two half-hour programs from the patients' balcony at Memorial Hospital; the Manhattan skyline at his back, he intoned that "the small world of Leo Szilard" is the story of "a man and a world on the danger list." Szilard's contribution to making the atomic bomb, Murrow said, "has left him with one driving purpose, and that is to try to help dismantle the era of terror he helped to create."[28]

In the most animated and best-remembered sequence, Szilard reminisced with Teller, the friend who had also become his chief adversary, over ways to curb the nuclear arms race. The two talked with obvious affection about their 1939 drive to Long Island to meet Einstein and draft a letter to FDR and about the Uranium Committee meeting that year where émigré scientists extracted the first federal money for chain-reaction research. But Szilard quickly became annoyed with Teller's declarations of patriotism and with his comparison of Szilard's crusade to make the A-bomb against Germany during World War II and Teller's own later efforts to create an H-bomb against Russia.

"I like to shock my audiences," Szilard said, and proposed that Teller and Klaus Fuchs, the Soviet atom spy, be depicted on a monument, shaking hands. "Because without you," Szilard added sarcastically, neither country would have H-bombs, and by Teller's logic both are "necessary if war is to be abolished." Still, Szilard kept open their friendship. "Come back and we'll talk . . . privately, without these microphones," he said. "It's much nicer."[29]

Teller and Szilard had a second, more celebrated clash that fall, on NBC-TV when they debated "Is Disarmament Possible and Desirable?" "We were in agreement that the danger [of nuclear war] was great," Szilard said, "but Teller meant this danger is great if the US government should listen to me, and I meant the danger was great if the US government should listen to him."

As their argument deepened, Szilard suggested, "I think, Teller, we should shake hands because maybe later on we don't. . . ." The audience laughed and applauded, but this sort of polite wordplay kept the two from sparring more aggressively over the main questions of the day: a moratorium on testing and schemes for actual nuclear arms reduction. The two points are often confused, Szilard said, because distrust of the Russians for cheating on underground tests is transferred to their possible cheating on arms reduction—a very different question. During one bitterly memorable exchange, when Teller accused Szilard of "irresponsible trustfulness" toward the Russians, he blamed Teller for his "irresponsible distrust."[30]

In addition to his many television appearances, Szilard's letters to newspapers appeared often. And besides Higgins, the nationally syndicated columnists Marquis Childs and Max Lerner wrote about his illness and his arms-control efforts.[31] The *New York Post* printed excerpts of the Szilard-Teller debates. *Newsweek* and *Life* magazines quoted Szilard. *U.S. News & World Report* published a three-and-one-half-page Szilard interview as part of a cover story entitled "Was A-Bomb on Japan a Mistake?" And *Harper's* magazine featured historian Alice K. Smith's warm and insightful profile "The Elusive Dr. Szilard" in its July issue.[32]

Awards gave Szilard added recognition and publicity during his busy year in the hospital, among them Humanist of the Year from the American Humanist Association and the New York Newspaper Guild's "Page One Award" for science. He received the annual Albert Einstein Gold Medal and Award from Lewis L. Strauss, and when Trude said that the roster of previous winners was impressive, Szilard replied: "Yes, and it is getting better and better."[33] A few days after Szilard received the Einstein award's $5,000 check, a call came from MIT president James R. Killian, who announced that Szilard would receive the Atoms for Peace Award and another, even more substantial, check. Understandably, the nurse on duty at the time noted in her log that Szilard was "quite cheerful this PM."[34]

In May, Szilard and Trude traveled to Washington, where he shared the 1959 Atoms for Peace Award with Eugene Wigner. At the same ceremony, Alvin Weinberg and Walter Zinn, two of Szilard's longtime associates, received the award for 1960. They were described as "the four men who, of all men living, have done most to originate and perfect the nuclear fission reaction," and Szilard was said to have been "untiring" in his "efforts to arouse men of all nations to the social and political implications of atomic energy." In his response, Szilard recalled that in 1945, as the war drew to its end, a young staff member on the Manhattan Project came into his office and complained that too much emphasis was placed on the bomb and not enough on the peacetime applications of atomic energy. " 'What particular peacetime applications do you have in mind?' I asked him, and he said, 'The driving of battleships.' "[35]

Emboldened by his celebrity, Szilard even dabbled in presidential politics but quickly discovered that his advice was not appreciated. He suggested that Sen. John F. Kennedy slip his chief Democratic opponent, Sen. Hubert H. Humphrey, a large donation to keep his faltering primary campaign alive, then, for maximum credit, leak this example of fair play to the press. Szilard also proposed setting up a postelection study program to instruct the new president in foreign policy issues. Both ideas were politely acknowledged and promptly ignored.[36]

At Memorial, which Szilard used more as a hotel than a hospital, urine samples taken in April and May still showed no return of the cancer, and his spirits continued to rise. In early June, Szilard left to speak at the Arden House Conference to Plan a Strategy for Peace at a site north of New York City. Szilard had come to the right place. He enjoyed the fresh air. He enjoyed the company; old friends who attended included Jerome Wiesner and Richard Leghorn from Pugwash meetings. And to an enthusiastic audience that night he spoke about how to "live with the bomb" as a way of "avoiding an all-out war."[37]

Szilard's fervor and humor impressed the conference participants, and he retired that night excited by this vision of a strategy to eliminate nuclear weapons. Too excited. As he awoke and stretched the next morning, Szilard suddenly felt vertigo. He turned pale. His heartbeat quickened. After breakfast, on the drive into the city, he felt "seasick" and vomited.[38] And, back in room 812, still dizzy even when lying in bed,

Szilard's complexion stayed pale. He began sweating. His pulse jumped again. A doctor on the ward that morning concluded that Szilard had suffered a "minor stroke."[39] An electrocardiogram gave Szilard sobering evidence that he should slow his pace. He had "myocardial disease and/ or extracardiac effect," Dr. Nickson discovered. The condition had developed during the busy spring—a reminder of Szilard's earlier heart attack in 1957.[40]

But Szilard's fatalism and his personal momentum drove him on to even bolder efforts: a personal appeal to Soviet Premier Nikita S. Khrushchev and a book showing how, by the late 1980s, the arms race might end. Two days after the stroke, on Sunday, June 5, the Soviet biologist Simeon E. Bresler stopped by for a chat, and their talk convinced Szilard that he and fellow scientists might be able to ease the tensions between the United States and the Soviet Union. Why not write directly to Khrushchev? Szilard wondered. After all, he had written to Stalin in 1947. Bresler said he would convey such a letter, and Szilard raised this possibility with Charles Bohlen, then a Soviet specialist at the State Department, sending him a draft letter that urged Khrushchev to back improved US-Soviet dialogue through Pugwash and other scientific meetings.

But Bohlen questioned a private citizen dealing with a foreign head of state and urged Szilard to address only Soviet scientists. Szilard replied by recounting his many efforts since 1945 "to arrange for informal discussions between politically knowledgeable American and Russian scientists." Two of these attempts collapsed because of US government criticism, Szilard noted, but after the Pugwash meeting at Baden in 1959, a new openness seemed evident from the Russians. Szilard promised to give Bohlen's suggestions "serious considerations"[41] and to await other State Department advice. But when no further word came after almost three weeks, Szilard sent his letter to Khrushchev.[42]

The same day, Monday, June 27, Szilard began dictating a satirical story he called "The Voice of the Dolphins," and both the Khrushchev letter and his satire were sparked by the same frustration. In May, Szilard had sent a copy of his proposal for arms-control negotiations, "Has the Time Come to Abrogate War?," to *Look* magazine, where editors accepted the article promptly but, to Szilard's annoyance, would not set a publication date. Next he sent the *Look* article to *Foreign Affairs*,[43] but the prestigious quarterly for the eastern foreign policy establishment rejected it in less than a week. "If they cannot take it straight, they will get it in fiction," Szilard vowed angrily.[44] As a youngster, he had enjoyed reading Bellamy's *Looking*

Backward and decided on the same device for a coldwar "history" written from the future. He took up the microphone on the Grundig stenograph and began a historical account of world events between 1960 and 1985, presenting his vision of how the nuclear arms race might be concluded.

"The Voice of the Dolphins" synthesized in fictional form the main points Szilard had made in his last three *Bulletin* articles. Looking out the window, crouching over the dictating machine, shuffling through copies of his articles, and scribbling notes, Szilard spun a tale peppered with details of his own life that took the world to the brink of nuclear annihilation, then safely back again. He recounted a revolution in Iran, a Chinese invasion of India, and a 1988 arms-control agreement between Russia and America—all events that later occurred. In Szilard's tour of the future, the world had solved its burgeoning food and population problems and straightened out the chesslike politics of Europe and Berlin, perhaps all due to the superhuman intelligence of dolphins swimming in a Vienna think tank. By humor, irony, ridicule, and wishful thinking the rational scientific and political saga unfolded. Szilard finished dictating a first draft of his story by evening.[45]

"The book is not about the intelligence of the dolphin but about the stupidity of man," Szilard would say later.[46] Still, dolphins played an important role. Szilard had first become interested in them in 1958 when he befriended marine biologist John C. Lilly at the NIH.[47] In the hospital two years later, Szilard read about a report Lilly had completed on the possibility that dolphins could imitate human speech. With that idea, Szilard had the "fiction" that would carry his serious message. He cited Lilly in the text[48] and linked his discoveries with the creation of a fictional US-Soviet Biological Research Institute in Vienna in 1963. The Vienna Institute, Szilard wrote, was inspired by Leningrad microbiologist Sergei Dressler, whose name, work, and travels resembled those of Simeon Bresler, Szilard's personal link to Khrushchev.

Although they were superhumanly brilliant, "on account of their submerged mode of life, the dolphins were ignorant of facts, and thus they had not been able to put their intelligence to good use in the past," Szilard's story explained. "Having learned the language of the dolphins and established communication with them, the staff of the institute began to teach them first mathematics, next chemistry and physics, and subsequently biology."

When the human researchers conducted experiments that the dolphins had suggested, these discoveries won Nobel Prizes for the next five years—

each time credited to the dolphins. Soon the dolphins devised a way to cultivate a fast-growing algae whose protein was nutritious, delicious, and a natural fertility depressant for women. Named "Amruss" (for America and Russia) the protein solved both world food and population problems while earning immense wealth for the institute.[49]

With its wealth, the institute sponsored a worldwide television program, free of commercial or political bias, that analyzed modern problems rationally and objectively. The program was called *The Voice of the Dolphins*. Quietly and in a very Szilardian way, the institute also bribed politicians around the world to behave responsibly, by paying huge sums to "retire" corrupt officeholders and by rewarding honest ones who made politically tough choices. While Szilard feared that he would not survive another year, his pensive and playful mind still worked to find new ways that might help humanity survive for at least another twenty-five.[50]

CHAPTER 27

Meeting Khrushchev

 1960

In the summer of 1960, at the time Leo Szilard was predicting in *The Voice of the Dolphins* how cooperation and peace might come about between the United States and the Soviet Union, a breakthrough in their relations was sorely needed. Soviet Premier Nikita Khrushchev had touted "peaceful coexistence" between the two superpowers during his visit to the United States in 1959, but in May 1960 he had abruptly canceled a Paris summit conference with President Eisenhower after an American U-2 high-altitude photo reconnaissance plane was shot down over Soviet territory. Khrushchev's visit to the United Nations in the fall gave the Soviet leader a platform to denounce UN and US policies; he banged his shoe on the table for emphasis. But if Khrushchev wanted to be conciliatory on nuclear arms control, then, in retrospect, Szilard personified American willingness to cooperate better than anyone else alive.

For his part, Szilard was known to the Russians because of the private views he shared at Pugwash meetings and his outspoken opinions, many made publicly this year at the expense of the Eisenhower administration. In May, Szilard's letter to the editor of the *New York Herald Tribune* about Eisenhower's "spy plane lie" expressed "indignation" at the US denials after the incident began. Szilard said he was "taken in by this cock-and-bull story and I resent being lied to by my own government." Another letter that month, in the *Washington Post,* proposed an exchange of engineers to monitor a nuclear-weapons-test moratorium. Publicity about Szilard's Atoms for Peace Award and the remarks he made at the ceremony further strengthened his stance as a scientist open to US Soviet dialogue. In June,

Szilard had been moved by Soviet biologist Simeon E. Bresler's visit to write to the *New York Times* that Russian disarmament assurances were not a hoax.[1] So when Szilard wrote to Khrushchev in late June, asking that US-Soviet dialogue might be improved through Pugwash and other scientific meetings, he brought to the exchange experience, authority, and conviction. And Khrushchev probably saw Szilard as a sympathetic—and prominent—contact within the American scientific community.[2]

But when Szilard received no reply to his June letter after waiting nearly two months, he wrote Khrushchev again in mid-August, this time inviting him to a visit at Memorial Hospital. As ever, Szilard preferred reaching the famous and powerful by trusted intermediaries,[3] so this letter was forwarded by Marshall MacDuffie, Khrushchev's friend from World War II, and was transmitted by Cyrus Eaton, sponsor of the Pugwash conferences. Khrushchev eventually replied to Szilard's first letter, saying he thought a scientists' meeting after the Moscow Pugwash conference that year was a "welcome" initiative, to which "Soviet scientists will respond. . . ."[4] So Szilard wrote Khrushchev again in mid-September, thanking him for his "very heartening reply" and stating that he hoped the US government would look upon informal discussions by scientists as "a more or less continuous process. . . ." And it wasn't necessary for Khrushchev to come to the hospital to visit, Szilard added, as "I feel very well . . . [and] can now be away from the hospital for days at a stretch."

Szilard first met Khrushchev at a lunch in his honor at the Biltmore Hotel in New York City on Monday, September 26, an event sponsored by Cyrus Eaton during the Soviet premier's visit to the UN General Assembly. Szilard met Khrushchev again on Tuesday evening, October 4, at a cocktail party. Szilard left disappointed that his conversation with Khrushchev had been so brief, so when the telephone in his hospital room rang the next morning, the message was a wonderful surprise: Premier Khrushchev can see you at 11:00 A.M. for fifteen minutes.

Excited, Szilard dressed and left Memorial Hospital immediately, hailed a taxi, and set off for the USSR mission to the United Nations. But when Szilard spotted a drugstore en route, he asked the driver to stop and wait. A few minutes later, he emerged holding a small paper bag and continued his ride. At the mission, Szilard was met by Mikhail A. Menshikov, the Soviet ambassador to the United States, and was then ushered to a salon. When Khrushchev entered, he and Szilard shook hands. Both men were cheery, rotund, and smiling as they sat down with Menshikov and a translator.[5]

"Sometime," Szilard told Khrushchev, "when time permits, I would like to have a leisurely conversation with you on the question of what the real issues are, the kind of thoughtful conversation we cannot have when we are in a hurry. Time is limited today, but before talking about a few serious matters, perhaps I can talk in a somewhat lighter vein."

Khrushchev agreed, and Szilard reached into his paper bag. He pulled out a Schick Injector razor and some extra blades and handed them to Khrushchev. "It's not expensive, but it is well made," Szilard said assuringly. "The blade must be changed after one or two weeks. If you like the razor, I will send you fresh blades from time to time. But this I can do, of course, only as long as there is no war."

"If there is a war," said Khrushchev, "I will stop shaving. Most other people will stop shaving, too."

On more "serious" matters, Szilard said he was upset to see that while in New York, Khrushchev stressed only his disagreements with the United States. What about the few points on which you agree?

"What points are those?" Khrushchev wondered.

"For instance," said Szilard, "you could say you are in agreement with Senator Kennedy on everything he is saying about Vice-President Nixon [his opponent in the presidential campaign] and add that you are in agreement with everything that Nixon is saying—about Kennedy."

Again trying to be "serious," Szilard produced a Russian translation of a letter he wrote for their possible meeting, along with the Russian texts of Khrushchev's letter to Szilard and a translated letter from William C. Foster to Szilard. Foster, then Nixon's adviser on arms control, wrote that he favored conferences among US and Soviet scientists. Szilard said that no matter who won in November, the new president would seek constructive solutions to US-Soviet problems. Khrushchev agreed.

Szilard then handed Khrushchev a seven-page memorandum, also in Russian. This was Szilard's "agenda" for their hoped-for leisurely discussion. Szilard looked at Menshikov for a signal about the time remaining, and Menshikov nodded to continue as Khrushchev read the memo.

"There is nothing in this paper to which I can object," Khrushchev said, and urged Szilard to talk about any topic. Sensing that his time with Khrushchev would soon end, Szilard spoke rapidly. He mentioned the need for verifiable nuclear arms reductions, and the one passage underlined in Szilard's memo caught Khrushchev's eye:

It should be possible . . . for Russia and America to create conditions in which Russia could be certain that secret violations of the agreement by America would be reported by American citizens to an international control commission, and America could be certain that secret violations of the agreement occurring on Russian territory would be reported by Soviet citizens to an international control commission.

The idea of an international arms-control authority, above national sovereignty, had appealed to Szilard since 1945. It was a leap of faith that both superpowers would have to make for serious limits on the nuclear arms race to occur. Surprisingly, Szilard later recalled, Khrushchev said that he "wholeheartedly accepted the [underlined] passage." If there were any doubt that we mean the same thing, Khrushchev said, he could sign it. Khrushchev also said he would do it right away, but Szilard glanced at Menshikov for the time.

"Why not just go on?" said Menshikov.

Let's consider the question of nuclear weapons testing, said Szilard, still talking rapidly. Some American strategists believed that their country might have to fight a war by using small atomic bombs against troops in combat, and these people press for continued nuclear weapons tests, Szilard said.

"Russia is not thinking in terms of using small atomic bombs against troops," Khrushchev interrupted, "because to prepare for this kind of warfare is too expensive and very complicated." More interesting to Khrushchev would be some statement from Americans about the future course of the arms race. Szilard urged Khrushchev to spell out disarmament proposals in a book, and Khrushchev invited Americans to produce such a book—and a draft agreement for the two nuclear superpowers. But, Szilard said, American citizens might not represent their government, while negotiators from the Soviet Union automatically reflected Kremlin policies. Nevertheless, Szilard concluded, he would try to organize material for such an "exchange of views." Again, Szilard looked at Menshikov, who nodded and said, "Why not just go on?"

To prevent accidental attacks, Szilard's memo stated, "either America or Russia ought to take the initiative at this time to arrange for the installation of telephone connections that would be readily available in case of an emergency." Khrushchev agreed this was a good idea and to underscore his point said, "It would have value, particularly if it becomes necessary to dispel quickly doubts that arise over some maneuver." Khrushchev said that just before he embarked on the yacht *Baltika* for his trip to New York, he was upset by reports about an American military maneuver; so upset, in fact, that he ordered "rocket readiness" for Soviet strategic forces—just in case. "And this readiness has still not been rescinded."

Szilard glanced at Menshikov again, who said, "Why not just go on?"

Two reasons can be cited for such a telephone link, Szilard continued. It could be useful in case of an emergency, and it would "dramatize the continued presence of a danger which will stay with us as long as the long-range rockets and bombs are retained." Khrushchev said he would be willing to have such a telephone installed if the US president were also willing. Szilard said he did not see how the president could object. "I find it difficult to get away from telephones," Khrushchev complained, "and even at the beach they mount a telephone for me. The only way to escape is to go into the water."[6]

To prepare for serious bilateral discussions, Szilard said, US and Soviet scientists should explore ways to negotiate a "first major step" to arms-limitation agreements. The Pugwash conferences offer one path to better understanding. But the ultimate goal is not just arms reduction but a broader trust with "safeguards against secret evasions" of an agreement. To pursue such a tentative agreement, Szilard wanted to keep in contact. Khrushchev, he said, could always reach him through the Soviet ambassador in Washington. "Who is the Soviet ambassador in Moscow?" Szilard asked.

"Topchiev will be able to arrange all the contacts that we might want," said the premier, mentioning the general secretary of the Soviet Academy of Sciences and the man Szilard knew through the Pugwash meetings. Indeed, that summer, Topchiev had invited Szilard to stay on in Moscow after the upcoming Pugwash meeting in order to hold informal discussions on US-Soviet arms-control initiatives. Now Szilard's Pugwash contacts, and his June letter to Khrushchev, seemed to be paying off.

Again, Szilard eyed Menshikov. Again came the encouraging "Why not just go on?" Would Khrushchev like to discuss Berlin? Szilard asked.

"Why not?" the premier answered.[7]

"Perhaps one could arrive at a solution of the Berlin problem without loss of prestige either for the East or for the West by proceeding as follows," Szilard said. "East Germany might offer to shift its capital from East Berlin to Dresden on condition that West Germany shifts its capital from Bonn to Munich. If that is done, then it would be possible to create two free cities: East Berlin and West Berlin, and there might be formed a confederation between East Berlin and West Berlin with a view of perhaps forming, at some later time, a similar confederation between East Germany and West Germany."

By Szilard's account, Khrushchev "appeared to get the point" but said he could not very well ask the East Germans to shift their capital. Yet on the broader question of living with a nuclear stalemate, the Soviet leader seemed more receptive. Szilard said he had just written "a little book," *The Voice of the Dolphins,* which explained the nature of the atomic stalemate and how it "may change in the course of the years to come and lead to a situation which may force disarmament on a reluctant world." But because most statesmen are too busy to read a book, Szilard said he had an excerpt that can be read in just over an hour. "A Russian translation will be prepared for you," Szilard said, prompting Khrushchev to say he would be Szilard's first reader in the Soviet Union.

In all, Szilard and Khrushchev spoke intensely for two hours and finally ended their meeting at 1:00 P.M. with some banter. First Szilard offered to show Khrushchev how to change the blades in his Injector razor and how to open it for cleaning.

Delighted with his gift, Khrushchev said he would like to give Szilard a present. How about a case of vodka?

"If I could, sir, I would like to have something better," Szilard replied.

"What do you have in mind?" asked Khrushchev.

"Borzhum," said Szilard.

Khrushchev grinned broadly. A few days earlier, during one of his long speeches at the United Nations, Khrushchev had sipped from a glass of mineral water and several times pointed to it. "*Borzhomi!*" he said. "Excellent Russian mineral water!" Now he told Szilard, "We have two kinds of mineral water in Russia. They are both excellent, and we shall send you samples of both."[8] Szilard and Khrushchev parted in good spirits.

Two days later, a large case—packed with two kinds of mineral water, canned food, caviar, and three smoked fish—arrived at Szilard's hospital room. A card conveyed Khrushchev's compliments and wishes for a speedy recovery. Szilard thanked the premier in a reply. "I am very grateful to [the translator] that he translated so many das and so few nyets. I am very grateful to you for having given me the opportunity to have such a conversation."[9]

There was no mention of this meeting in the American press, but in Russia the next morning, *Pravda,* the Communist party newspaper, carried this brief account on its front page:

RECEPTION OF AMERICAN SCIENTIST L. SZILARD

New York, 5 October (TASS). Today in the Mission of the Soviet Union to the United Nations N. S. Khrushchev received the prominent American scientist-physicist Leo Szilard and had with him a friendly meeting in which the Ambassador of the Soviet Union to the USA, M. A. Menshikov, participated.[10]

Clearly, Szilard and Khrushchev had enjoyed their encounter. But what had each gained? For Khrushchev it was an opportunity to open channels with Americans favorable to arms control at a time when relations with Washington were decidedly cool. Khrushchev had also favored Topchiev and his academy colleagues, who by their influential advice might be essential in gaining arms-control agreements within the Kremlin. Finally, in Szilard, Khrushchev now had a well-known ally among the American Pugwash group, an asset no matter who won the presidential election. For his part, Szilard hoped that with Khrushchev's support a constructive US-Soviet dialogue might flourish.

In Khrushchev, Szilard also seemed to have found a kindred spirit; like himself an establishment outsider in his own country, a bold and intuitive strategist who was personally committed to controlling nuclear arms, a risk taker who was not self-conscious and did what seemed right and true at the time.[11]

The day after meeting Khrushchev, Szilard wrote to Bohlen, then attending the General Assembly meeting in New York, and mentioned that "something came up yesterday which I feel I ought to discuss with you sometime." He also telephoned Bohlen and promised to call again, but there is no record in Szilard's papers that Bohlen showed any interest. Szilard wrote on October 13 to President Eisenhower, reporting the Khrushchev meeting and raising the "hot line" idea as something Khrushchev had approved. Typically, rather than writing directly, Szilard forwarded his letter through George Kistiakowsky, a science adviser to Eisenhower. This time Szilard's appeal to the White House gained some attention. A reply on November 10 from Secretary of State Christian Herter suggested that Bohlen and arms negotiator William Hitchcock would be available to meet with Szilard.[12]

Next Szilard wrote to presidential candidates Nixon and Kennedy, offering to report about the meeting with Khrushchev to each of them personally after the election—whether to the president-elect or the "Leader of the Opposition."[13] At the same time, Szilard offered accounts of the Khrushchev meeting to Kennedy foreign policy adviser and US representative Chester Bowles, New York governor Nelson Rockefeller, and Philip J. Farley, special assistant to the secretary of state for atomic energy and outer space.

Szilard strengthened his ties with Topchiev after Kennedy was elected president in November, when supporters of the Pugwash meetings decided to seek financial help from sponsors other than Cyrus Eaton. An heiress to the Gimbel department-store fortune agreed to give a luncheon at the Delmonico Hotel and invited C. D. Jackson, a publicist at Time-Life and President Eisenhower's senior speechwriter and "idea" man. Jackson was fiercely anti-Communist and used the occasion to berate Topchiev. The United States had nothing more to give in arms-control bargaining, Jackson insisted.

"In roulette we say, *je donne* to bet," Jackson instructed the Russian, who seemed bemused by the analogy.[14] At this, Jackson only became more insistent.

"We have no more *je donne!* No more *je donne!* No more *je donne!*"

Szilard, who attended the luncheon wearing his bathrobe over a shirt and tie, listened with chagrin at Jackson's tirade and finally interrupted. "Let me explain what Mr. Jackson is saying," Szilard told Topchiev. "You have got to do what we want, because we're crazy."[15]

The fourth annual meeting of the Pugwash Conferences on Science and World Affairs had been scheduled for September in Moscow, but American participants urged that the session be postponed until after the presidential election. By then Szilard felt well enough to attend and saw this trip—his first to the Soviet Union—as another chance to meet Khrushchev. He wrote Khrushchev on November 24, through Ambassador Menshikov in Washington, asking for an appointment. Ready to set the agenda for this meeting, as he had for their last, Szilard suggested that if they met, Khrushchev should first read—in the Russian translation, in Szilard's presence—from the *Voice of the Dolphins* excerpt. After that, Szilard suggested, they might discuss the US-Soviet nuclear stalemate.[16]

Despite a persistent fever, the day Szilard wrote Khrushchev he checked out of Memorial Hospital with Trude and rode by taxi to Idlewild (now Kennedy), New York's international airport, for an overnight flight to London. During the last leg of the trip, several members of the American Pugwash group said they were nervous about their first visit to Russia.

Talking among themselves, they wondered what treatment they might expect from their Soviet hosts and from the notorious secret police. Szilard tried to put the group at ease.

"You know," he said, "in a socialist country everything is better organized. That has bad points and good points.

"The bad point is if you go into a toilet the chances are you'll probably end up saying, 'Goddammit! There's no toilet paper in here!' The good point is that about three minutes later you'll hear a knock on the door. Someone will be bringing you toilet paper."[17]

When the Americans arrived at Moscow's airport, their Soviet hosts, acting very formal with foreign visitors, met the party and asked, "Who is the leader of your delegation?" The Americans had not thought to choose one and for some moments looked perplexed.

"We are *all* leaders," Szilard announced. The Americans laughed, but their Russian hosts did not like his answer. They still needed a leader and from then on considered Szilard head of the US delegation.[18] As a result, Szilard was the designated leader and was treated lavishly by his hosts. His suite at the Metropol Hotel had a high-ceilinged living room, a large bedroom, and a bathroom with a huge tub. He also had the prettiest interpreter and the biggest car.

In Szilard's suite that first evening he, Trude, *Bulletin* editor Ruth Adams, and several other Americans met for a party. They were all still nervous about the trip but also exhilarated by thoughts about the meeting that would begin in a few hours. They shared an optimism about being in the Soviet Union at all, and they saw the conference as a positive first step toward attaining something concrete in nuclear arms control. Toasts and laughter mingled with moments of serious discussion until well after midnight.[19]

For most participants, the Pugwash Conference that began the next morning seemed somber and boring. The gloom of Moscow's winter helped set the tone, for the sun did not edge over the snow-topped roofs until around 10:00 A.M. and set by midafternoon. Gone were the freewheeling summertime talks in small groups at remote seaside cottages or mountain resorts. Here participants filed into a long, bare-walled room for the opening session and sat on tightly spaced chairs at two narrow tables. The hosts at the Soviet Academy of Sciences seemed eager to demonstrate how rational and responsible they could be in their own capital city, but to some Americans they appeared pompous and inflexible.

At first, one Russian participant even seemed deranged. Every few minutes the famous aircraft designer Andrei N. Tupolev rose from his chair and paced the aisles, muttering *odin*, the Russian word for "one," over and over. Some Americans thought the old man crazy or senile, although he was then only in his early seventies. But a Russian colleague explained that Tupolev walked around every fifteen minutes as a longtime habit, one developed to preserve his sanity when in solitary confinement during a Stalinist purge. And *odin*, they said, was Tupolev's terse way of stating that all nuclear weapons should be reduced to one bomb, although he did not say which country should have it.[20]

Every move was carefully planned. After breakfast at the Metropol, delegates walked or were bused a dozen blocks to Dom Mir, or "House of Peace," where discussions began promptly at nine-thirty. At twelve-thirty, the group returned to the Metropol for lunch. Then it was back to Dom Mir for an afternoon session that lasted until five-thirty. Next came a reception, a dinner, or a visit to the ballet. The food was always rich and heavy, which Szilard loved but other Americans found unappetizing. The vodka and wine flowed. The Russian hosts were obviously trying to please their guests, but somehow their formality, the scrutiny by so many local officials, and the enlarged size of the gathering (in all seventy-five scientists from fifteen countries) worked against the creative discussions that had been a feature of earlier Pugwash meetings. As a result, too many things went unsaid, too many thoughts went unfinished, and many participants seemed weary and frustrated—a disappointing contrast to the high hopes they had borne to Moscow.

Historically, the meeting did lead to results, but as often happened, from actions taken behind the scenes. Two American participants would soon join the incoming Kennedy administration—Jerome Wiesner as science adviser and Walt W. Rostow as deputy special assistant to the president for national security affairs—and for them the week in Moscow offered a personal introduction to their Soviet counterparts. One day, when Wiesner tried to telephone Washington, he discovered there was no direct link; his calls had to be relayed through European and American cities. This gave the new hot-line proposal urgency and helped advance that idea within the Kennedy administration.

Informal discussions with Russian scientists also highlighted a striking imbalance in the deployment of US and Soviet nuclear forces. At the time, the United States was committed to expanding its submarine-launched nuclear warheads as part of a "triad" of weapons based on land, in the air, and at sea. Some of the Americans urged their Russian counterparts to consider doing the same thing—in the name of strategic stability—

and before long, submarine-launched missiles did become official Soviet policy.[21] Another day, Wiesner was able to begin arrangements for the release of US B-47 pilots then held by the Soviets.

For Szilard, an erratic participant at most conferences, being in Moscow was an opportunity to talk with people, preferably one-to-one, and especially a chance to meet again with Khrushchev. Szilard attended most formal receptions and dinners, sometimes exchanging cheery toasts and quips, but he participated in few discussion sessions.[22] He preferred to take his interpreter and driver around Moscow, looking up friends and acquaintances.

On one excursion Szilard visited the conference participant Nikolai N. Semenov, a physicist and physical chemist who had won the 1956 Nobel Prize in chemistry for research in chemical and kinetic chain reactions and the study of explosions. A tall, thin man with a trim mustache, Semenov met Szilard in the director's study at the Institute of Radiation Chemistry. There the two discussed radioactivity caused by pairing protons, and Szilard asked questions enthusiastically. Later, in Semenov's apartment in another wing of the institute building, he picked up a large cast-iron ashtray from the piano and handed it to Szilard. "This comes to you from your friend at MIT, Charles Coryell," Semenov said. "He sent it in advance of your arrival in Moscow." At first puzzled, Szilard studied the heavy object, then beamed a smile. The piece was adorned with a large dolphin. Before their dinner party ended that evening, Szilard presented a dolphin gift of his own: a Russian translation of excerpts from his book.[23]

From the Metropol Hotel, Szilard wrote Khrushchev, trying to arrange a meeting and offering to raise "a rather interesting possibility" that the Soviet Union might employ "to take the initiative and to take steps to stop the cold war. . . ."[24] Khrushchev had planned to meet the Pugwash delegates, but when he canceled because of illness, Szilard continued to try for a private get-together.

At the discussion sessions he did attend, Szilard made few comments—a contrast to his ebullient performances at earlier Pugwash conferences. With Rostow as chairman one afternoon Szilard read as his "paper" excerpts from *The Voice of the Dolphins*. He tried to be provocative, but the meeting's tone stayed serious, and he fell into making distinctions and quibbles.[25] Szilard's deep irony so confused a Chinese delegate that he took seriously the excerpts.

"I particularly dislike what he said about China," Chou Pei-yuan complained. "That such an article could have been presented as a scientific paper is to me a mockery of science, a mockery of the Pugwash Conference. Will Professor Szilard please excuse me for my frankness?" In particular, Szilard's passages about Sino-Indian relations "are extremely incorrect and are in reality slanders." Chou advised Szilard "to withdraw this book immediately from circulation for the sake of peace."[26] In his story Szilard had written, among other things, that between 1960 and 1985 China tried to extend communism to India, "but after ten years of Communist rule in India it began to dawn on the Chinese that the success of their own regime in China may have been to a large extent due to the civic virtues of the Chinese, which the Indians were totally lacking."[27]

Donald G. Brennan, an MIT mathematician and nuclear strategist, praised Szilard's "forceful perception" of arms-control problems at another session, referring especially to his understanding of the need for both flexibility and communication. Citing *The Voice of the Dolphins,* Brennan said,

> Although these writings are widely known among students of arms control, this perception has not been widely recognized because Szilard's papers have treated these functions in a rather implicit and allegorical manner, while explicitly discussing certain other matters (primarily inspection by public reporting and a bizarre form of a limited retaliation doctrine) that are largely unrelated to the basic problems of flexibility and communication.

The "inspection" scheme he probably had in mind involved international agencies paying scientists $1 million to report their country's secret nuclear tests. The "bizarre" retaliation was, no doubt, Szilard's idea for a "list" of cities to be traded in a nuclear war.[28]

If feelings were strained, and hopes frustrated, during this first Soviet Pugwash meeting, nevertheless, an air of camaraderie filled the main hall at the Soviet Academy of Sciences on the last night of the conference. Physicist William Higinbotham, a lively and politically active veteran of the Manhattan Project, brought his accordion to Moscow, and at the academy, Topchiev called him to the stage. Higinbotham warmed up the crowd with "I've Been Working on the Railroad" and "The Eyes of Texas Are Upon You." Soon a Russian joined in, pounding on a grand piano. Together they played the "Internationale." Higinbotham then "called" a few American square dances. Other dances followed, and the evening ended with the delegates swaying arm in arm and singing "Auld Lang Syne."[29]

But when his colleagues left for home on December 5, Szilard stayed on at the Metropol, still awaiting Khrushchev's reply about a meeting. In his suite Szilard tinkered with the *Dolphins* manuscript and other drafts.[30] A week passed, and a second, with still no word from Khrushchev. At the academy, Topchiev could only say the premier was ill. Finally, on December 20, Szilard wrote to Khrushchev with regrets that he was not well and that the two would likely not meet. Still, Szilard volunteered to stay on in Moscow for a few more days.

In this letter Szilard said he was trying to think up "some simple move" by the Soviet Union that might help to "bring the cold war to an end." In a scheme as odd as the Humphrey-campaign donation he had posed to Kennedy, Szilard now suggested that Khrushchev make a "bridal gift" to the incoming American president, who will face "a somewhat embarrassing problem" of an "outflow of gold" when he takes office in January. "What would happen," Szilard asked Khrushchev, "if you were to offer to him on behalf of the Soviet Union to loan the United States three to four billion dollars of gold for a period of three to four years?" Szilard said he had asked Rostow about this idea during the Pugwash meeting. "Professor Rostow thought that if this offer were to be made publicly, it would probably cause resentment," Szilard wrote. "But if it were made privately and if it were accepted by Kennedy, then this could be a very good thing. Naturally, Professor Rostow was not able to predict what Kennedy's reaction would be."[31]

Szilard lingered in Moscow a few days more and, while waiting, sent Khrushchev flowers. No reply ever came from Khrushchev about the "bridal gift" of gold, but the Soviet premier, speaking through Ambassador Menshikov in Washington, did thank Szilard for the flowers a few months later.

Just before the Western Christmas, Szilard and Trude left Moscow, flying through Kiev to Vienna. From there Szilard telephoned his boyhood friend Josef Litván. "Come to Vienna," Szilard said. "I will be here for a day or two and we can talk." Litván urged Szilard to visit Budapest, but he refused. He would never return to Hungary, Szilard said, because he feared the Fascists. They had hunted him as a young man in the wake of the Béla Kun regime in 1919. They had collaborated with the Nazis during World War II. Szilard insisted that he would never take the chance of being near them again.[32]

From Vienna, Szilard also telephoned Inge Feltrinelli, the wife of Italian publisher Giangiacomo Feltrinelli and a friend of Eva Zeisel's, whom he had met years before in New York.

"How is the weather in Milano?" asked Szilard.

"But Leo," she said, surprised and delighted to hear his voice, "I heard you were dying in New York City."

Soon the Szilards were on a plane to Italy. From Milan the four drove toward the Piedmont range, to the Feltrinellis' imposing Castello di Villadeati, near Asti. There Szilard, Trude, and their hosts walked the sunny terraces and snow-covered lawns, lounged in the Renaissance-style villa's imposing rooms, and joked and relaxed. Szilard recalled with Inge their New York friends. With Giangiacomo, Szilard talked about biogenetics, the Soviet Union, and the Italian Communist party, which Feltrinelli then worked in as an active member. The two men also chatted for hours about books. The Feltrinellis, who had first published Pasternak's *Doctor Zhivago* in the West, would publish the Italian edition of Szilard's *Dolphins* in 1962.[33]

In this hilltop retreat Szilard ended the fateful year 1960 enjoying as he had for months a life of comfort. His health improving, his hopes for the new American administration rising, his holiday hosts a delight to be with, his wife by his side, Szilard could later look back on this year as one of his happiest and most productive. How long, and where, he might live were for the moment questions of minor concern. In the photographs that survive from this visit to the villa, Szilard grins boyishly, as he rarely did in any other place or time.

CHAPTER 28

Is Washington a Market for Wisdom?

 1961–1963

He was still alive. That much Leo Szilard knew, and savored, as he returned to New York City from Europe in February 1961. He felt well, and tests at Memorial Hospital confirmed that his bladder cancer was still in remission a year after radiation treatment was completed. He could not know how long he might live or what to do with the indefinite time that was still his, but just being alive was itself a daily reward.

The Szilards moved into the Hotel Webster, an inexpensive but comfortable place on West Forty-fifth Street, and began calling on friends. The galley proofs for *The Voice of the Dolphins* had to be checked, but beyond that pleasant chore there was little for them to do now that Szilard's illness and his hospital room were no longer the center of so much attention. Still living from their suitcases, the Szilards reviewed the new medical-test data and wondered what they might try next. During her year's absence to attend her husband in New York, Trude's teaching post at the University of Colorado Medical School had been abolished. She had become unhappy with her work there even before Szilard's illness, so Trude had little desire to return to Denver.[1] Chicago held little appeal for Szilard and none for Trude. Where would they live, and what would each of them do next? They began looking around New York, a city they both enjoyed, a city now rich with concerned friends and acquaintances from their busy year spent at Memorial.

From Eva Zeisel, Szilard heard that her brother had a large apartment to rent on Central Park West. Szilard and Trude went to take a look. Perhaps, they thought, this could be their first "home" together, or, at least, a shared pied-à-terre. As they walked through the spacious apartment, Szilard immediately liked the large living room, with its wide French doors and dramatic entryway. "It matched his bourgeois tastes," Zeisel later reflected. Her nephews and nieces, and their teenage friends, were also around that day, filling the empty rooms with playful conversation and giving Szilard an idea.

"I could see myself seated here, greeting guests, talking with young people," Szilard thought aloud as they stood in the living room. But the apartment frightened Trude. "Too big," she said as she paced from room to room, shaking her head. 'Too much space. Too much to dust and keep clean." After the visit, Szilard and Zeisel stood on the front steps, chatting.

"This won't do for us," he said, looking disappointed. "Trude is looking for a place to live in. I'm looking for a place to die in."[2]

Later that February, Szilard was still fatalistic as he spoke on CBS-TV to interviewer Mike Wallace. "Right now," Szilard said, "I feel perfectly well, and I'm able to work. I don't think this will last forever, but for the time being I feel well." Is he optimistic? Wallace asked. "The doctors can't prophesy. This is a disease which can recur, and I don't think I should make ten years' plans," Szilard answered.

Asked if he "felt any guilt" about his part in beginning the nuclear arms race, Szilard said no, "because I was always aware of the dangers involved and I just chose the lesser of two evils. I thought we must build the bomb, because if we don't, the Germans will have it first . . . and force us to surrender." This, Szilard admitted, proved to be a false assumption. "But you cannot do better in life than to try to find out what the situation is and then act on that basis. I never blame myself for having guessed wrong."

Wallace recalled Einstein's comment about Szilard, made in 1930 when considering the idea of the Bund. Szilard, he said, "may be inclined to exaggerate the significance of reason in human affairs."

"Well, that's probably true," replied Szilard. "But I think that reason is our only hope. So when I exaggerate the significance of reason, I am just hoping. . . ." He did not believe in a personal God, Szilard went on to say, "but in a sense, I am a religious man" because "I think that life has a meaning. . . . I will say that life has a meaning if there are things which are worth dying for." Szilard said he would "spend four to six weeks in Washington" to assess the month-old Kennedy administration. "I will discover there," he declared, "if there is a market for wisdom."[3]

On the day after Wallace's interview, February 28, the Szilards rented a small room at the Dupont Plaza. Billed as "Washington's newest hotel," this modern glass-and-white-brick structure stood among town houses and shops at the north side of busy Dupont Circle. The neighborhood was studded with turn-of-the-century mansions, some now turned to clubs, others to restaurants. At the time, before social dropouts and drugs arrived and the neighborhood's boutiques moved to Georgetown, Dupont Circle was Washington's most cosmopolitan area. Diplomats from Embassy Row came to the Riggs Bank on one corner to attend to their international transactions. Art galleries, flower shops, and delicatessens crowded the tree-lined streets.

The Szilards' hotel was inexpensive but convenient, a five-minute taxi ride to the White House, the State Department, and most foreign embassies. Their room was modern and modest, but with a pleasant view through treetops to the irregular town-house roofs. And once the bags and papers were spread about, the room seemed crowded—a clutter that was tolerable for only a short stay. But Szilard intended to be there for only a few weeks and within hours of arriving set to work telephoning and writing his friends in the White House, seeking appointments, offering advice.

"Szilard here," he announced to begin each call, often surprising people who thought he still lay dying in Memorial Hospital. "I spent a month in Moscow," Szilard announced in notes to Walt W. Rostow, an assistant to Kennedy for national security, and to Jerome Wiesner, the president's science adviser. He asked Wiesner for "advice on how to go about" finding that "market for wisdom." "Yes, who's got it?" Wiesner asked in reply. They planned dinner to discuss events since their time together in Moscow and to assess the new administration's plans for arms control.

Wiesner and other White House recruits from academia had discovered that a handful of people seemed to be shaping foreign policy. "It was a $300 billion family business," he recalled with a chuckle years later. "Eight or nine of us kept guard on all the business of the world." Szilard's access to this family at first seemed easy. Access *was* easy, but Szilard had no talent—or patience—for the bureaucratic scramble that led and followed decisions. "Being inside the White House, we often got caught up in day-to-day details," Wiesner realized. "Szilard's contribution was to look at the world at large and motivate people like us to think about it."[4]

During the 1960 presidential campaign, Kennedy had complained that fewer than a hundred federal employees worked full-time on disarma-

IS WASHINGTON A MARKET FOR WISDOM?

ment, "the most glaring omission in the field of national security and world peace of the last eight years." President Eisenhower had created a Disarmament Administration within the State Department in September 1960, and after the election, Kennedy named John J. McCloy as his disarmament adviser. A lawyer long involved in government, McCloy had served as an aide to Secretary of War Stimson at Potsdam and had helped draft the 1946 Acheson-Lilienthal Report proposing international control of atomic energy. Like Stimson, he believed in the need to control the bomb. McCloy had later served as president of the World Bank and as US military governor and high commissioner for Germany, so he brought stature and authority to his new appointment. In June, McCloy recommended creating an autonomous Arms Control and Disarmament Agency (ACDA), an idea Kennedy adopted and sent to Congress. The new agency was enacted in September, and McCloy was named its first board chairman. William C. Foster, a Republican who had advised Vice-President Nixon, became the ACDA's first director.

After several direct approaches, Szilard realized that if he could influence McCloy and Foster at all, the tactics must be subtle: The two men, like his friend Teller, deeply mistrusted the Russians. When Szilard wanted to have "discussions" about disarmament, the brief notes he received from McCloy during the spring and summer questioned his faith in the Russians' willingness to negotiate a test-ban treaty and challenged Szilard's criticism of US policies. After six months of polite but firm exchanges, Szilard finally lost interest.[5]

Foster also questioned Szilard's faith in direct negotiations and saw his role as "a one-man State Department" with Khrushchev both irksome and threatening. Mention of Szilard's name usually led Foster to mutter to aides about the Logan Act, the federal law barring private citizens from conducting diplomatic relations.[6]

Still, Szilard persevered. He wrote and met with Wiesner, with McGeorge Bundy, Kennedy's national security adviser, and repeatedly with Bundy's assistant, MIT economist Carl Kaysen. He called on Rostow. He met with Chester Bowles, the retired Connecticut governor and congressman now under secretary at the State Department. Tempting Bowles's political instincts, Szilard predicted that disarmament might not take twenty-five years to achieve—as outlined in *The Voice of the Dolphins*—but could come about within eight, potentially the maximum term of the new administration. Yet the month he had just spent in Moscow, Szilard reported, convinced him that the Soviets would not "jump" at the first offer America made for a nuclear test-ban treaty; it was not their chief concern at the time.[7]

Szilard met Lewis L. Strauss, the former Atomic Energy Commission (AEC) chairman, for lunch at the Hay-Adams Hotel. He dined at the University Club with Glenn T. Seaborg, the current AEC chairman and a principal adviser on US-Soviet negotiations for a test-ban treaty. Always the bumblebee of news and gossip and ideas Szilard told Seaborg that the respected French nuclear official Bertrand Goldschmidt had recently said he thought the United States should improve its representation on the UN's International Atomic Energy Agency. The next morning, Szilard sent Seaborg two names to consider: former AEC commissioner Thomas Murray and Farrington Daniels, Szilard's former boss at the Met Lab whom he had recently seen in Washington.[8] Szilard also wrote to Soviet specialist Charles Bohlen at the State Department, who said he would be delighted to get together.

Szilard's life at the Dupont Plaza was ideal for peddling his "wisdom" around the capital city, and the neighborhood itself seemed to match his offbeat style. He could walk out the back door of his hotel to meet people at Pierre, a French restaurant in an imposing Italian Renaissance villa at Connecticut Avenue and Q Street. He could sit in one of Washington's first sidewalk cafés. He could stroll up Connecticut to another Italianate favorite, the Golden Parrot Restaurant at Twentieth and R streets. Edwardian panels, a stained-glass bay window, fancy stone fireplaces, coffered ceilings, and a grand piano in the lounge all reminded Szilard of bourgeois times past, of Budapest cafés and London hotels.

The food around Dupont Circle, and nearby at the Cosmos Club, the scientists' association in a French-style mansion, was ordinary. But so were Szilard's tastes. His appetite was easily sated when he ate at the bars and grills that lined Connecticut Avenue just below Dupont Circle. A favorite, the Rathskeller, served European food southern style.[9] Szilard began most days in his hotel's modest restaurant, heaping orange marmalade by the spoonful onto toast or bypassing the toast entirely and simply scooping it onto his plate before eating it. Coffee he mixed with half a cup of sugar and cream. Snacks in his room throughout the day included East European favorites such as pâté or herring in sour cream spread on dark bread. A utilitarian about eating, Szilard seldom bothered with serving plates; he simply scooped the food from jars and cans.

The appearance of *The Voice of the Dolphins* in April added to Szilard's celebrity among the Washingtonians he was there to influence, and he made a point of sending most of them copies. By the spring Szilard had outgrown his eighth-floor hotel room, its dresser and telephone table now covered with files and books. Besides this clutter, a card table Trude used to hold medical equipment and Szilard's bladder appliances made movement about the room awkward. Szilard's solution was not to rent a separate office but to spend more time working in the Dupont Plaza's lobby. At first content to sit on a modern sofa, his papers spread about him, he eventually commandeered a desk for guests by the front window. There he sat most days, with a view of the Circle and the town houses along New Hampshire Avenue, thumbing through his pocket diary, greeting friends, editing manuscripts, and meeting people for appointments. A reluctant bellhop soon strung a telephone to this desk, and before long stenographers sat by, taking dictation. "I can work very happily in this lobby," he told a *Life* magazine reporter. "I have never owned a house, and don't feel the need of owning one."[10]

Trude made sure that a book rack near the lobby newsstand was always stocked with copies *of The Voice of the Dolphins,* and Leo made sure that no one missed it. After lunch at the hotel restaurant one day, Szilard passed the rack with Thomas I. Emerson, a Yale law professor first known when he had worked to draft the McMahon bill in 1945. "He stopped and picked up a copy and passed it on to me," Emerson recalled. "Somehow it was clear that I was to pay for the book. I looked at it casually and put it back on the shelf. The next day I came back and bought a copy."[11]

Szilard was encouraged enough by his many meetings and telephone calls to stay on in Washington beyond the four to six weeks he' first thought it would take to find his "market for wisdom." Gradually, over the spring and summer, he became increasingly involved as events and policies grew more troubling. On Monday, April 17, the Central Intelligence Agency launched an armed invasion of the Bay of Pigs on Cuba's south shore, at first backing and then betraying a ragtag army of anti-Castro refugees. Szilard's first angry response came that day in a cable to a friend: "They ain't in the market."[12]

His second response was a letter to the editor, comparing this US-backed landing with Russia's 1956 invasion of Hungary and declaring that "two wrongs don't make a right." His protest letters appeared in the *Washington Post,* the *New York Herald Tribune,* and the *St. Louis Post-Dispatch,*[11,] but Szilard yearned to do more.

A week after the Cuban landing, Szilard was elected to the National Academy of Sciences (NAS), and at once he thought of a way to use this group for a third response. "For the second time in my life I find myself drafting a petition to the President," he explained in a memorandum about the Bay of Pigs that he mailed to more than half the NAS's members.[14] The United States, he said, "transgressed" the UN Charter, and the president "is entitled to know whether or not the policies of his administration offend our moral sensibilities. . . ."

In the petition he mailed, Szilard said the invasion had "created the impression that henceforth the United States may intervene with her own troops in civil wars in order to prevent the establishment, or stabilization, of governments which look to the Soviet Union or China, rather than to America, for economic assistance and military protection." He urged the president "to adopt a policy with respect to our obligations under the United Nations Charter which is in conformity with the moral and legal standards of behavior that we are demanding of others. . . ."

Szilard also sent a copy of this petition package to the president, not directly but through Wiesner.[15] And in a more playful manner he complained to Kennedy's national security adviser, McGeorge Bundy. "Szilard could be pixieish . . . in a way that did not include gravitas," said Marcus G. Raskin, at the time a consultant on defense policy to the National Security Council. Szilard had Raskin present Bundy with a memorandum that asked him to "tell me your side of the Bay of Pigs expedition in Cuba, so that in the event that there's an international war-crimes trial you will have on record with me your defense." Szilard's playful memo "just sent Mac right up the wall," Raskin later recalled.[16]

By his June 5 deadline, Szilard received 129 replies to the 364 packages mailed out, including 56 signed petitions. (Among the signers were friends and colleagues, such as Edward Condon, Max Delbrück, Salvador Luria, geneticist Hermann J. Muller, Harlow Shapley, Cornelius B. Van Niel, and Victor Weisskopf.) In all, fourteen academy members opposed either the idea of the petition or its content, sixteen abstained, sixteen said they approved of the petition but would take other actions themselves, and twenty-seven members gave no reason for not signing.

In a letter to Kennedy the next day, Szilard claimed that "about one in six of those to whom I wrote responded by signing the petition. . . ." But he hesitated to draw conclusions, instead urging Kennedy to consider that "there is probably no group in the population whose membership would be as reluctant to sign a petition" as this one. "It would be my guess that most of them have never signed a petition in their life." In his usual oblique way, Szilard did not send the petition results directly to Kennedy but passed them through Edward R. Murrow, the newly appointed director of the US Information Agency. Szilard turned to Murrow "because what we think about ourselves is even more important than what others may think about us . . . ."[17]

The inconclusive tallies no doubt tempered Szilard's zeal when he sent his petition to the White House. But his oddly equivocal tone may also have come from the way some fellow scientists had censured both his message and his methods. Physicist Llewellyn H. Thomas at the IBM Watson Research Laboratory at Columbia University wrote the president directly, with a copy to Szilard. "I regard the statements made in this petition as typical special pleading," Thomas said. Szilard "is against sin and frames his statements so that those who disagree with him would seem to be for sin. Further, the actions which he urges on you seem to be phrased vaguely and with tendentious question-begging epithets."[18]

Szilard must have felt more troubled reading the reply from physicist James Franck, a friend he had known and admired both in Berlin and at the Met Lab. Szilard had also served on the Manhattan Project committee that Franck chaired in 1945 that urged the A-bomb not be used on Japanese cities. Then Franck had been grateful for the moral and political direction Szilard had given their work. Now he questioned Szilard's whole enterprise.

> 3309 Avon Road
> Durham, North Carolina
> May 21, 1961

Dear Szilard,

I am more than sorry that I can not sign your petition. In spite of the fact that I agree with you that our Cuban adventure meant all that you express in your letter, there are three reasons why I refuse to sign your petition.

One is that I am certain that Kennedy himself knows that what he did in Cuba was not only an error in judgment, but also a violation of the democratic principles to which we adhere.

Two, I am absolutely against it that scientists as a class believe that their scientific reputation is a proof that they are also experts in political reasoning. As citizens, we have just as much right as others to tell our opinions and, if we judge ourselves as a relatively intelligent group, we can also hope our opinions should carry some weight. However, here you ask only scientists who are members of the National Academy to join your appeal. I believe that there are a great number of people, for instance in humanities, who would be at least as competent as we are as a group. You can argue with some right that you do not deny that, but everyone should stir up his own group in times in which basic political decisions seem to go haywire. However, we scientists had in the last years more [than] enough opportunity and duty to say our opinions about matters directly connected with our science. We endanger our influence in these particular questions if we speak up as a group in matters not directly connected with our profession.

> Three, right now, I believe that it would be quite necessary that we as scientists not insist on a foolproof mutual control in the discussions with Russia about the bann [sic] of atomic tests. All of us know that such a control is not possible, and, if it were, it would be a snooping around which we would dislike just as much as would Russia. . . .

He closed with "a few personal words" about *Dolphins,* saying "my wife and I did not know whether we should laugh or cry. Your satire and your style are wonderful. . . . If you can, drop me a few lines to tell me why you are in Washington and how you are. With greetings and all good wishes. Yours as ever, James Franck."[19]

This candid advice from an esteemed friend must have shaken Szilard, for after he received it, he ceased all efforts to publicize or use his academy petition. He also abandoned the use of petitions as a way to influence public policy. Besides, Szilard may have finally realized, petitions were for outsiders, which he had become. How, he must have worried, could he become a Washington insider?

Franck was not the only person wondering why Szilard was in Washington. Soon after Szilard's letters to newspapers about Cuba appeared, Byron R. White, deputy attorney general, telephoned FBI director J. Edgar Hoover to request background information on Szilard.[20] When the bureau could provide no new material and only reported that Szilard associated with known "liberals" and supported "internationalization of the atomic program," White apparently dropped the matter.

For his part, Szilard was so eager to advise his government that while circulating a petition critical of the president and assailing White House aides, he also sought to meet with Kennedy in person, hoping to advise him on the forthcoming "summit" meeting with Premier Khrushchev in Vienna.[21] Typically, Szilard was oblique, addressing Kennedy's appointments secretary, Kenneth O'Donnell, but also calling Harris Wofford, a former colleague of Chester Bowles's who was then a special assistant to the president. "He is one of the finest men in our midst," Wofford advised O'Donnell, "with a roving imaginative mind, and an angle on Khrushchev and arms negotiations which the President might appreciate."[22] Szilard also alerted Bowles in the State Department, who replied, "I hope it will be possible for you to see the President before he leaves for Europe. . . ."[23] But no meeting was arranged. Szilard wanted to urge Kennedy to be jocular and flexible with Khrushchev in their first meeting, in effect to behave just as he had during his own 1960 meeting in New York. The same advice to be informal came to Kennedy en route to the summit, from veteran diplomat W. Averell Harriman. But when Kennedy met Khrushchev in Vienna, their first encounter was testy. Bold statements led to bellicose threats. "It will be a cold winter," Kennedy told Khrushchev as they parted.[24]

From his desk in the Dupont Plaza's lobby, Szilard continued to dispense advice to his government. Anticipating a Kennedy-Khrushchev showdown over Berlin, Szilard drafted several memos proposing a political settlement. One appeared in the May 1961 *Bulletin of the Atomic Scientists* and was reprinted in the *Congressional Record.*

The first Pugwash meeting to be held in the United States was scheduled for that September at a ski resort in Stowe, Vermont, and during the summer Szilard drafted enough memos to make a book-length manuscript on "How to Secure the Peace in a Disarmed World." This was his step-by-step plan to eliminate nuclear weapons and restrict conventional weapons to regional peacekeeping armies. Routinely, now, Szilard sent his memos to William C. Foster and Adrian Fisher at the ACDA, to Wiesner at the White House, and to Prof. Henry Kissinger at Harvard, then an administration consultant.[25]

Memos on other topics also fluttered from Szilard's lobby desk that spring and summer. He proposed creating a private think tank, similar to the military-oriented RAND Corporation, to operate in the field of arms control and general disarmament. He asked financier and Pugwash sponsor Cyrus Eaton to consider funding a joint US-Soviet research center in molecular biology. He offered President Kennedy advice on responding to civil unrest in Brazil.[26]

No doubt with his Bund in mind, Szilard proposed creating a "National Society of Fellows" that would provide systematic advice to the US government. These ten to twenty scientists and scholars would address current issues full-time. Besides probing topics that make headlines, these fellows would draw attention to "unrecognized problems" and to "problems which have been recognized but neglected." For the fellows' agenda, Szilard cited topics he had fancied for years: new forms of democracy suitable for developing countries; fertility and population controls; novel ways to arbitrate international conflicts.[27] More speculative still were Szilard's proposals in the summer of 1961 to study ways to use expanded leisure time and, perhaps, even find a biological means to eliminate the need for sleep.[28]

All the while, Berlin continued to haunt Szilard's calculations. One chapter of the paper he drafted for the Pugwash meeting at Stowe, "Political Settlement in Europe," repeated his scheme to declare Berlin a "free city" by moving the two German capitals. Within hours of completing this chapter, his speculation gained a new significance. After nightfall on Saturday, August 12, East German border guards sealed the crossings between the divided city and began building the Berlin Wall—first with barbed wire, soon after with concrete blocks and mortar. Tensions were already high between Khrushchev and Kennedy. The actions that night in Berlin had ominous strategic overtones, and Szilard, who personified and internalized the fate of the world, must have been shocked.

Henry Kissinger was also preparing to attend the Pugwash meeting at this time, and when he read Szilard's papers on a political settlement in Europe, he wrote: "I agree with the philosophy contained in your draft proposal on disarmament and have been working with a group at the Disarmament Administration [ACDA] in order to improve our position." The two had dined that summer. Now Kissinger suggested that they meet for lunch.[29]

But while Kissinger said he liked Szilard's approach to solving the Berlin crisis and would discuss it in the White House, to his former Harvard colleague McGeorge Bundy, Kissinger was more skeptical about Szilard's ideas, especially those on Berlin. Forwarding Szilard's Berlin paper to Bundy, Kissinger wrote that Szilard "claims that he knows his proposal is negotiable with the Soviets because of his conversation with Khrushchev last year." To Bundy, Kissinger was "not very enthusiastic" about Szilard's approach but admitted that if the United States was moving to recognize East Germany, then his proposals do "have the advantage of a certain ingenuity." Kissinger added that Szilard "will be glad to do the negotiating" over Berlin, a thought that infuriated Bundy. "He should *not* be encouraged to feel he has any role as a US negotiator," Bundy urged.[30]

Szilard was a more successful strategist when working through Wiesner. During the Pugwash meeting in Moscow, the Soviet and Chinese scientists were at times openly hostile to each other, reflecting their governments' growing conflicts. Szilard urged that a Chinese participant be invited to the Stowe meeting as a way to begin a dialogue fractured in Moscow. But the US State Department refused an entry visa, in part because Washington did not recognize the Communist regime in Beijing. Szilard urged Wiesner to press the issue, and the latter had President Kennedy request a visa from the reluctant Dean Rusk, secretary of state. A visa was issued, but for other reasons no Chinese delegate attended.[31]

At the time, Szilard felt ebullient and cocky, again enjoying his celebrity. A three-page "Close-Up" on Szilard entitled "I'm Looking for a Market for Wisdom" appeared in *Life* magazine the first week in September. He was pictured strolling around Washington "badgering his friends in government to buy his brand of political wisdom." Readers saw him working away at his lobby desk and holding forth on disarmament while seated on the broad lawn at the Virginia farm of the *New Republic*'s publisher, Michael Straight. This "disputatious, free-spirited man" described his own role in the capital by saying, "The most important step in getting a job done . . . is the recognition of the problem. Once I recognize a problem I usually can think of someone who can work it out better than I could." A page of Szilardisms included this thought on credit and fame:

In life you must often choose between getting a job done or getting credit for it. In science, the important thing is not the ideas you have but the decision which ones you choose to pursue. If you have an idea and are not going to do anything with it, why spoil someone else's fun by publishing it?

On nuclear disarmament he was quoted as saying, "It is not necessary to succeed in order to persevere. As long as there is a margin of hope, however narrow, we have no choice but to base all our actions on that margin."[32]

When this *Life* profile appeared, Szilard, and the world, watched as that margin of hope narrowed. In mid-August, President Kennedy had responded to the Berlin Wall by sending a 1,500-man battle group to West Berlin to strengthen US forces stationed in the divided city. On September 1 the Russians unexpectedly broke a nuclear-testing moratorium they had honored for nearly three years; in the next two months they would explode fifty weapons, including the largest ever recorded by either superpower, a 56-megaton blast. A few days after the Soviets resumed testing, President Kennedy emphasized the national fallout-shelter program he had first announced in May. All public buildings would be surveyed, and suitable sites would be marked with yellow and black signs. Food and water would be stockpiled. Alarm systems would be improved to "make it possible to sound attack warning on buzzers right in your homes and places of business," *Life* reported in a special cover story.[33]

At Stowe during the second week of September, the weather was warm and sunny, but the normally friendly Pugwash meeting was decidedly cool. "Disarmament and World Security," the conference theme, had everyone on edge. The Russians, who had to be coaxed at the last minute to attend at all, were obliged to defend their government's nuclear tests, while the Americans had to explain their own civil-defense plans. Compounding tensions during the conference, the United States resumed its nuclear weapons tests. The special issue of *Life* magazine that was passed around was headlined "How You Can SURVIVE FALLOUT" and promised that "97 out of 100 people can be saved." Included were detailed plans for building home shelters, among them "A $700 Prefabricated Job to Put Up in Four Hours."[34]

Scientists at the Pugwash conference were cautious with one another during the opening sessions, unsure how to begin or sustain helpful dialogue. Indeed, for the first few days the only people who seemed pleased to be in Stowe were the Russian delegates' wives; on a shopping spree in local stores, they returned each night hauling to the hotel bags of goods they could not buy at home. Determined to lift his colleagues' gloom, Szilard moved among the delegates in the autumn sunlight, chatting and cajoling. At one point during the formal discussions, he walked over to sit among the Russians, and from there proposed a topic all might agree on: that the nuclear testing moratorium should be resumed.

"Does the American delegation agree?" asked the chairman.

"Yes, we do" came the answer.

"Does the Soviet delegation agree?" There was a nervous silence. A long pause.

"*Da!*" It was Szilard's voice. Everyone laughed. Tensions eased. And with the laughter serious and friendly conversation at last began.[35] During that Pugwash meeting came the first focused discussions about a permanent ban on atmospheric nuclear tests, an initiative that would result two years later in the US-Soviet Limited Test Ban Treaty.[36]

Passing through Boston at the time of Pugwash, Szilard also turned his irrepressible mind to lighter affairs. Over dinner at the Hong Kong

Restaurant in Harvard Square, he mused to stepnephew John Schrecker about how to attract readers to *The Voice of the Dolphins*. "Make posters," Szilard suggested, "and put them in the window of the Harvard Coop," the university store. When Schrecker suggested this was too obvious and perhaps a little crass, Szilard replied: "Let's run an ad. In the *Harvard Crimson*." A popular advertisement for a newspaper at this time declared, "In Philadelphia, nearly everybody reads the *Bulletin*." Szilard found a scrap of paper in a pocket and scribbled copy for his own ad:

> In Philadelphia almost everybody reads "The Voice of the Dolphins," five stories of political and social satire by Leo Szilard. (In second printing Simon & Schuster; paperback $1) on sale in the Harvard Coop. If you do not buy it to-day you will forget it.[37]

Back in Washington by mid-September, Szilard resumed his personal crusade for arms control by writing to Khrushchev. Within the Kennedy administration, Szilard reported, there had begun—"appearances to the contrary—some serious thinking about general disarmament." Szilard also reported that he had "so far not given up hope that a constructive approach will be made with respect to the problem of Germany," and to pursue this, he planned to remain in Washington "for the time being." This letter, with a package of Schick Injector razor blades, Szilard forwarded to Khrushchev through Pugwash participant Alexander Topchiev.[38] Just after dispatching this letter, Szilard drafted another, to McGeorge Bundy and Carl Kaysen at the National Security Council, offering to return to Moscow and discuss with Khrushchev a political settlement for Europe. But Szilard had doubts and never mailed the second letter, perhaps already aware of Bundy's anger at his efforts as a mediator.[39]

Another letter to Khrushchev followed in early October, this one sent via Ambassador Menshikov. Szilard repeated the "free city" plan for Berlin he had posed at their first meeting a year before. He urged Khrushchev to agree that the long-term goal for both Russia and America should be "to have Europe as stable as possible." And he said, if Khrushchev "would be willing to think through with me the implications of a political settlement . . . I would be glad to fly to Moscow, at a time convenient to you." But the only reply Szilard received, through the Soviet embassy, was Khrushchev's thanks for the "good wishes" and the blades.

Szilard's Berlin solution was featured in the *New Republic* in October, yet neither side took his pragmatic steps seriously. Pondering how his voice might be heard, Szilard pushed anew for the National Society of Fellows, gaining fresh encouragement from Joseph L. Rauh, Jr., a Washington lawyer active in liberal politics, and from Pentagon lawyer Adam Yarmolinsky.[40] But private foundations he approached were cool to the project, and it languished.

"In the Arms Control and Disarmament Agency Leo was viewed by the bureaucrats with whom I worked as a well-meaning eccentric who understood nothing of politics," noted Richard Barnet, the deputy director in the Office of Political Research. "But his genius was in understanding precisely the unreality of bureaucratic realism."[41] Szilard resisted their "realism," just as he resisted joining the scholastic debates that characterized many academic arms-control studies. He was after a simpler, more straightforward truth. It was quick and clever in reaction to slow-paced procedure. It was cool and ironic in reaction to the terrifying times.

And yet Szilard gradually accepted another lesson about Washington, from his friend Herbert York. A physicist who was then leaving his job as the Pentagon's director for defense research and engineering to become chancellor at the University of California in San Diego, York advised Szilard that "people in government pay more attention to the advice they invite than to the advice they are offered."[42] In the popular phrase, government officials expect scientists to be "on tap, not on top," and those on tap were chosen to be there.

As Szilard's frustrations with Washington grew, Trude's seemed to fade. She was delighted to be there, and sensing that their temporary visit was becoming permanent, she found appealing work in her field as a clinical associate professor in preventive medicine at the Georgetown University School of Medicine and a consultant in the health statistics branch of the Pan American Health Organization. These part-time appointments allowed her to share in her husband's busy life and yet be free to travel with him—a necessity because of his bladder and heart conditions. The tube Szilard wore, which allowed urine to collect in a plastic bag strapped to his leg, she had to check for infection every few days. If the bag itself was not emptied after a few hours, it made a sloshing noise as Szilard walked. Rather than trying to conceal this, he called attention to it as a way to dramatize his precarious health and, some friends suspected, to be enjoyably shocking.

In print, too, Szilard could not resist shocking readers as a way to teach a graver lesson. In 1960, when his hopeful scheme for disarming the world gained no serious attention, he added dolphins and wrote it as fiction.

Now that his search to stabilize the nuclear arms race was being ignored, Szilard felt it was time for some more fiction. With Russia and America testing larger and larger weapons almost daily in the fall of 1961, Szilard needed to find a way to dramatize the folly of the nuclear arms race. His city-for-city proposal the year before had made sense, at least to him, and he had revived it at Stowe. But there was no guarantee that world leaders would act rationally during a nuclear confrontation. (The Cuban Missile Crisis and its lucky resolution were still a year away.) Perhaps, he thought, fiction would make them face the facts.

In his latest letter to Khrushchev, Szilard had argued for "conversations" between Moscow and Washington on "how to keep to a minimum the amount of destruction—if a war should break out which neither of the two nations wanted." It is important, he wrote, "that it should be possible even if one or two cities were destroyed through an unauthorized attack" to "get things under control" promptly.[43] One way to avoid intentional attacks, Szilard now thought, would be to assure swift and certain retaliation if nuclear weapons were ever used. To gain that assurance, he mused, why not simply mine a few US and Soviet cities with H-bombs?

Preposterous? Of course, but no zanier, thought Szilard, than the whole precarious and wasteful nuclear standoff itself. Szilard wrote "The Mined Cities" to describe a rational and inexpensive way to maintain a reliable strategic stalemate while at the same time dismantling most nuclear weapons. Set in the year 1980, Szilard's piece has a patient regaining consciousness after being preserved cryogenically for eighteen years— awaiting a cure for some illness. He is in Denver, one of fifteen American cities in which a team of Russians in an underground fortress maintains an H-bomb. The same number of Russian cities has American-manned fortresses with bombs. All other nuclear weapons have been destroyed, and with the savings from scrapping huge nuclear arsenals, the government can afford to pay each mined city's family $3,000 a year to compensate for their anxiety. Szilard's dialogue explains the mined-cities concept.

A [the patient]: Is Denver in any greater danger, now that it is mined, than it was before, when there was the air base nearby?
B [the doctor]: No, of course not. . . .

They are still a military target, but now they will be warned before a nuclear explosion and given time to evacuate. Simple ground rules have been devised so that the teams, all family men, were paired with teams in another city: The Russians under Denver came from Kiev; the Americans under Kiev came from Denver. Each team will detonate a bomb only if its own home city has been destroyed. The plan eliminates the concern that a third country's atomic attack on Russia or America could set off World War III and a nuclear holocaust.

This dialogue, at turns both playful and profound, also revealed Szilard's own barbs and peeves.

A: Who thought up these mined cities?

B: Szilard had proposed it in an article . . . in 1961, but the idea may not have been original with him. His proposal was presented in the form of fiction and it was not taken seriously.

A: If he meant his proposal seriously, why didn't he publish it in serious form?

B: He may have tried and found that no magazine would print it in a serious form. . . .

In fact, before the piece appeared in the *Bulletin of the Atomic Scientists* in December 1961, the *Saturday Evening Post* and other magazines had rejected it.

"In spite of its fictional form," the *Bulletin* editors said in a preface, "the article is technically correct. The form permits the author to be more enlightening by being more entertaining." *Newsweek's* article "The Bomb Under Denver" called Szilard's "mined cities" idea "a product of an intelligent imagination and a search for the bold act that might eliminate the danger of thermonuclear war. Short of complete disarmament, such a system might work, Szilard writes. . . ."[44]

But for all the satisfaction he may have enjoyed by proposing and publicizing his "mined cities" scheme, Szilard had begun to realize his limits both as a humorist and as an "outsider" in Washington—a serious and self-important city that squanders laughter and lives by cliques. Being a jocular outsider was Szilard's advantage when he first arrived, but by the fall, after eight months of "lobbying" from the lobby of the Dupont Plaza, he had learned enough about how the capital works to realize that dispensing wisdom was not enough.

Over a dinner at the Raskins', Szilard heard his friend James R. Newman, a brilliant mathematician he had collaborated with on the McMahon bill, complain about his own career. Then Newman turned on Szilard and his dubious position.

"Szilard," he said, "you're not known beyond Dupont Circle."

"That's why I never leave there," Szilard replied.[45]

Szilard's peculiar status in Washington was dramatized again, more painfully, when he learned that a Soviet delegation—including some friends from the Pugwash conferences—would soon arrive. Szilard enlisted his nephew, Andy Silard, to drive the forty miles out to the new Dulles Airport in rural Virginia. Patiently they waited by the exit from customs, but when Szilard's friends came into view and he approached to offer a welcoming handshake, they were whisked past by State Department and security officials. During the long drive back to the city Szilard sat fuming with anger, not saying a word.

CHAPTER 29

Seeking a More Livable World

 1961–1963

When at Brandeis University to receive an honorary degree in October 1961, Leo Szilard voiced his fears about the nuclear arms race at a dinner for trustees and fellows.

"Suppose you are right?" someone asked. "What can we do?"

"I had to admit," Szilard later recalled, "that this was a legitimate question and that I had no answer."[1] Somehow, he realized, thinking up clever ideas for arms control is not enough. You also need the power— political, legal, financial—to enforce your views.

Szilard mused about this problem back in Washington and in the Virginia countryside, where he attended a "Strategy for Peace" conference at Airlie House. His interlude at the rustic conference center offered a welcome time to think. To daydream. And from these musings came an answer to the questioner at Brandeis and one of Szilard's most successful legacies—a political-action committee for arms control.

But his years in Washington brought Szilard fresh heartaches as well. Attempts to influence President Kennedy and Premier Khrushchev were frustrating failures, and his response to the Cuban Missile Crisis harmed his stature in the arms-control community. Szilard's "sweet voice of reason" never became shrill despite his many setbacks, although to those who still listened it was beginning to sound more and more desperate.

Typically, Szilard joined few discussion groups at the Airlie conference; he preferred to buttonhole people in the halls or to wander in the fresh autumn air—through the formal garden, around the duck pond, or along the farm roads that swung through rolling pastures. When living in England, Szilard had felt "in Oxford but not of Oxford" because he had no affiliation with a college, the heart of the university's social and intellectual system. In Washington, too, Szilard had no affiliation. He was still a professor of biophysics from the University of Chicago. Not a consultant to the Arms Control and Disarmament Agency (ACDA). Not a member of the President's Science Advisory Committee (PSAC). Not a fellow at a local think tank or a consultant to a congressional committee. As he had discovered during his first months in the capital, Washington is a city where what you do is often less important than where you do it.

At Airlie, Szilard thought about creating "a sort of lobby" that "would not only speak with the voice of reason" but also "deliver the votes." And as he paced around the grounds that fall weekend, ideas spun together. "Neither reason nor votes alone mean very much, but the combination," he thought, could be "unbeatable."[2] He may have recalled his 1952 "lobby for real peace in Washington" idea, his brief hope of influencing that year's presidential elections. This time his idea would become reality.

How, Szilard wondered, could an informed minority use its unity and its money effectively? A rational cost-benefit calculation led him to target the US Senate. At some point, he hoped, the United States and Soviet Union would sign treaties, first to ban nuclear tests, eventually to limit nuclear weapons. Treaties must be ratified with the "advice and consent" of the US Senate. Here Szilard combined the calculus for law and democracy. Each state has two senators, regardless of size or population. It is much cheaper to be elected in a less populous western state, so a western senator's treaty vote comes at a bargain. By raising relatively small amounts of money, Szilard calculated, he hoped over time to help elect western senators who shared his own arms-control views. Senators serve for six years, so the "investment" need not be made as often as in the biennial House elections. Then, when arms-control treaties come before the Senate in a few years, votes for ratification would be there. His was a clever scheme, working at the margin of power to widen the "margin of hope."

Szilard called aside Roger Fisher, a Harvard Law School professor who specialized in international negotiations, and sat him on the stone steps by the side of Airlie's large yellow farmhouse. Fisher listened, liked the peace-lobby idea, and invited Szilard to describe the plan at a law school forum.[3]

As soon as Szilard returned to Washington, he began drafting a speech and a proposal.[4] At the time, President John F. Kennedy's administration was proving to be a disappointment. First came the Bay of Pigs invasion, then the stormy Vienna summit with Khrushchev and overreaction to the Berlin Wall, and now a defiant campaign for fallout shelters. Behind these events was a graver fear, for conversations with White House advisers convinced Szilard that the administration had shifted from a "no-first-use" policy for nuclear weapons to a "counterforce" strategy that picks military targets for a preemptive strike.

The red-brick Georgian-style Lowell Lecture Hall at Harvard was packed with students and faculty on Friday afternoon, November 17, 1961, when Fisher led Szilard to the stage. His talk was called "Are We on the Road to War?," but Szilard began in a droll, almost teasing way, his words clustered and clipped by his Hungarian-German accent. He was really there, Szilard said, "to invite those of you who are adventurous to participate in an experiment that might show that I am all wrong."[5]

The title question he answered right away: "We are headed towards an all-out war" with Russia, and "our chances of getting through the next ten years without war are slim." Just as quickly came his solution. It is conceivable, Szilard said, that "a rebellious minority" might "take effective political action" to change the American government's attitudes about Russia and the arms race. A new organization might combine a thoughtful 10 percent of the voters with the "sweet voice of reason" and "substantial political contributions." That combination, he said, might become "the most powerful lobby that ever hit Washington."

This lobby's "political objectives" echoed Szilard's own, among them: Renounce a first-strike policy against Russia; avoid "meaningless battles in the cold war" that might escalate to nuclear confrontation; and improve East-West relations. Private discussions should begin to explore "how to secure the peace in a disarmed world," he said, how to bring about "an orderly and livable world" in Southeast Asia and Africa by devising new forms of democracy for these regions, and how to develop effective and acceptable birth-control methods for poor countries.

If a "sizable minority" were to back these objectives, Szilard said, he might "go further" and propose a new Council for Abolishing War, complete with scientists and scholars on the board and members who would pledge "2 percent of their total income" to support candidates who share these policies. The council's elitist rationality resembled Szilard's Bund from thirty years before, but with one important difference: This time he was trying to influence not just public officials but elective politics.

For the "experiment," Szilard urged the audience to ask how many other people might join such a lobby. Perhaps, he surmised, 25,000 students "would go all-out in support of this movement," and if each brought ten members, it would "attain 250,000 members within twelve months. This would represent about $25 million a year in political contributions. . . ."

The hall erupted with applause. "For those who heard him then," remembered historian Barton J. Bernstein, Szilard's "words were an inspiring call for action in a country where some liberals were losing heart with the Kennedy administration. . . ." They "were not offended by the implicit elitism of this arrangement," nor were most troubled that it "seemed politically daring and risky" for the time.

To some, however, Szilard's idea seemed dangerous. A navy official promptly warned US intelligence that the proposed council was "subversive and Communist inspired," and the FBI increased its watch on Szilard. He scoffed when asked later if he feared that the group might be infiltrated, saying, "In order to manipulate me, the others would have to be much brighter, and I don't think there are many of that kind."[6] Szilard even welcomed being hauled before the House Un-American Activities Committee for leading his new "movement." "This would be a very nice spectacle," he said, "and I have no doubt at all who would come out on top." Besides being more clever than congressmen, Szilard declared, scientists should lead his movement because—unlike politicians—they have "integrity and purity."

"Why not mothers in the same context?" asked a voice from the audience.

"The trouble with mothers," Szilard admitted over a rumble of laughter, "is there are so many of them I would not know how to choose."[7]

Szilard was hardly charismatic, and the "movement" he advocated depended more on logic and reason than on emotion. But when he met with students in Holmes Hall the next day, Szilard was eager to answer their questions, patient and giving with his time and ideas.[8] They talked for hours, until Szilard had to leave for the airport. He delivered the same speech that night at Swarthmore College, again to enthusiastic response.

A capacity audience of students and faculty from Western Reserve University packed into Severance Hall in Cleveland on a late November evening as Szilard opened his second "Road to War" speaking tour.[9] And a lively crowd in the University of Chicago's Mandel Hall gave Szilard an affectionate homecoming welcome. This talk gained Szilard his first national press for the proposed council, including local and national talk-show appearances, stories and a cheering editorial in Chicago's papers, and coverage by the *New York Times,* the *Washington Post,* and the wire services. ABC's national six o'clock television news ended a description of Szilard's efforts with the comment "We wish him good luck."[10] *Commonweal* magazine praised Szilard's "rare combination of idealism and hard practicality." *Newsweek* quoted the talk in a feature on human nature and war: "If you observe how we, as a nation, respond to Russia's actions and how they respond to our responses, you can see a pattern of behavior emerging which, more likely than not, will lead to war within a period of, say, ten years." FBI agents also monitored Szilard's movements, adding fresh press clips to his swelling file.[11]

Now Szilard had to decide how his council might actually operate; he enjoyed devising institutions but had little patience for running them. Help came at the right time when Allan Forbes, Jr., a filmmaker and anthropologist in Cambridge, Massachusetts, volunteered his services. Forbes said he had "been on the fringes of the peace movement in various ways for several years and have just recently backed out, more or less in disgust, due to the futility of the standard approach. . . ." Szilard's idea for the council offered the hope he and other activists needed.[12]

Publicist Harold Oram, who had known Szilard since their work together on the Emergency Committee of Atomic Scientists in 1946, joined him and molecular biologist Maurice Fox for dinner in Washington to brainstorm about fund-raising. Oram mentioned a meeting planned by several philanthropists in New York, and a few days after Fox attended and described the council, two checks arrived at the Dupont Plaza; donations to Szilard from Sallie and Eleanor Bingham, wealthy sisters from the renowned Louisville publishing family. Their $7,000 was the council's first "seed money."[13] As the 1961 Christmas holidays approached, Szilard received more good news from peace-movement supporters Ralph and Jo Pomerance, who sent a check to finance a speaking tour and promised more help later.[14]

But mindful of a recent kidney infection, Szilard resisted making plans. "Even though I have no trouble and no symptoms at present," he wrote in late December, "I have made it a rule not to make any long-term commitments except at one month's notice at most."[15] In *Pageant* magazine, he was quoted as saying: "Death is part of life. If it didn't exist, one would have to invent it. There is nothing alarming in thinking that after your death you'll be in the same state as you were before birth." For the article, Szilard even composed his own epitaph, "He did his best."[16]

In January 1962, Szilard flew west with his "Road to War" speech, this time renaming his "experiment" the "Council for a Livable World." Capacity crowds greeted him at the University of California in Berkeley, at Stanford University, at Reed College, and at the University of Oregon.[17] He also addressed a SANE (National Committee for a Sane Nuclear Policy) rally at Santa Monica Auditorium, near Los Angeles.[18] Throughout this hectic schedule Szilard's wife, Trude, was with him, introduced at most stops as "my doctor."[19] Also along, but less conspicuous, was a team of FBI agents.

The speeches generated a radio interview on WBAI in New York, broadcast nationwide by the Pacifica network, and a feature in the *Christian Science Monitor*.[20] But good press alone could not make the council a success, and those able to help were frustrated by Szilard's impetuous style. Enrico Fermi's widow, Laura, and Szilard's Manhattan Project colleague Samuel Allison wrote from Chicago to complain that "a small group of faculty, students and interested citizens" had met a few times to discuss ways to implement the Mandel Hall speech but could do no more "unless we have definite information on the evolution of your ideas." Without an "immediate" statement the interest Szilard's speech had aroused there "will wane fast," they warned, and urged him to appoint scientists to the proposed board quickly.[21]

But Szilard did almost nothing for months, still unsure what reactions his experiment might produce. Some responses were personal. Psychologist Margaret Brenman Gibson wrote to praise his "Road to War" speech and described reactions to it by peace activists in western Massachusetts. "My young son, aged 8, hearing me on the phone for hours asked, 'Mommy, how much is 2 per cent of a quarter? I could give some of my allowance.' When told that would be half-a-cent a week, he replied: 'I'd even give three cents a week if that would help.' "[22] Some responses were professional. In February, theater and film director Arthur Penn invited Szilard to his Manhattan home to meet friends in the arts, including comics Mike Nichols and Elaine May. They and other entertainers volunteered to host fund-raising parties in New York and Beverly Hills for the council.[23] "Certainly I'm with you for 2 per cent of my income whenever you ask," actress Anne Bancroft wrote Szilard. "And whether you ask or not you have many of my prayers."[24]

While in New York, Szilard also spoke at Sarah Lawrence College in suburban Bronxville, the last stop on the nine-campus tour. By late February about four hundred people had pledged 2 percent of their incomes, far from the twenty-five thousand Szilard had thought necessary to begin the council's political work. Still, Szilard loved to devise detailed plans for his "institutional inventions," as friend Edward Shils called them, and in a memorandum in February 1962 defined interconnected roles for the council's fellows, a panel of political advisers, a board of directors, and the associates.[25]

But for all the layers of authority that Szilard devised, it was the new council's board of directors that became his principal interest and the driving force for his Washington activities. Its founding meeting, in June 1962, chose Szilard and Yale chemist William Doering as cochairmen. Members included Forbes, Fox, and Penn; Ruth Adams, an editor at the *Bulletin;* University of Chicago sociologist Morton Grodzins; physicist Bernard T. Feld; James Patton, president of the Farmer's Union; photographer Charles Pratt, Jr.; and attorney Daniel M. Singer, counsel for the Federation of American Scientists, who would serve as secretary and treasurer. Adams later asked Szilard how he had assembled this remarkable group, and he admitted to but one criterion: "Your sense of humor."[6]

Jennifer Robbins, the council's first secretary, also had a sense of humor but was hired for quite another reason. When she appeared for an interview, Szilard sat her on a couch in the Dupont Plaza lobby and began to pace around.

"Do you speak German?" he asked.

"No, sir," she said.

"Good," said Szilard. "I want to be able to speak to Trude without being understood." After testing her note-taking skills, he hired Robbins and led her to a tiny room on the seventh floor, a converted maid's closet so narrow that Szilard had to shut a desk drawer to sidle past a couch. On the couch were piled letters and checks, many postmarked and dated months before. By mid-August the couch was clear, and a filing system was in place.[27]

As office manager, Szilard hired Ruth Pinkson, once a secretary to Henry Wallace. "I have three kids at home," she said when interviewed, "and I have to have a nine-to-five job, five days a week."

"No problem! No problem!" Szilard assured her. But before long, he began dictating letters only in the late afternoon and expected her to work into the evening.

"I'm sorry," she said, "but I have to go home to my children."

"You can't go home!" Szilard declared. "If you don't do this, we won't have a world!" Believing in the cause, Pinkson found a way to work late.[28]

Szilard could be blind to his supporters' needs as well. Groups he addressed that summer—in Greenwich, Connecticut, and in Stockbridge and Woods Hole, Massachusetts—found his speeches disconcerting. Forbes blamed this on Szilard's "total lack of tact," although that seems to have changed little over the years.[29] More likely, Szilard's message changed, from inspiring followers to importuning for money. He wearied his listeners with details about how cleverly the boards, the council, the lobby, and the panel would interact. And he sounded too vague about the criteria for picking candidates the council would support.

"He was very intemperate and impatient with the processes of American democracy," said editor and publisher Michael Straight, a friend since the 1940s. "He was much more interested in moving step-by-step ahead and in working out, in his own marvelous ways, ways of circumventing Congress and circumventing anybody else in order to strike a deal with the Russians." He wanted to short-circuit the political system, to get right to those with power. Questioned after the Woods Hole speech about why he chose to focus on the Senate, Szilard said: "Because there are fewer of them, and they're easier to buy."

In the audience, Straight groaned at that answer. "Leo, you can't say those things in public," Straight said after the talk.

"Why not?" asked Szilard. "It's true!"[30]

Some who saw Szilard thought he might be weakened or distracted by his cancer, but he shrugged off fatigue and occasional infections. "I feel fine," he told a *Time* magazine reporter in the spring of 1962. "But I don't want to mislead people into thinking I am cured, because I do not know if I am. There is no telling how long I will be well." To *Newsweek*, Szilard said, "Even now, I have no five-year program in mind." He was "grateful" for his recovery but hardly on top of the world, the magazine reported. "It's a rather heavy world," he said. "I feel that the world, in fact, is on top of me. . . ."[31]

Despite its tentative start, the council collected enough money to play a role in the 1962 congressional elections. Its first candidate was George McGovern, a former US representative from South Dakota who had failed in 1960 to unseat the state's incumbent senator, Republican Karl Mundt, and was then working on food programs in the White House.[32] (Szilard urged another White House aide, Harris Wofford, to quit his job and join the council's staff. At the time, Wofford declined, but in 1991 he was appointed US Senator from Pennsylvania, then reelected with support from the Council for a Livable World.) Szilard's move into practical politics upset at least one old friend, physicist Edward Teller. "Szilard, I have a little difficulty to agree with you when you are imaginative," Teller chided during a nationally televised debate on arms control. "When you are becoming a political realist, that is the time when I have real difficulty to agree with you completely."[33]

In April 1962 the United States resumed above-ground nuclear-weapons tests in response to Russia's break of a moratorium six months before. Then, in June, Defense Secretary Robert McNamara officially declared U. S. policy on the "counterforce" use of nuclear weapons against military targets, not cities. The Russians denounced him as a madman who was out to create "rules for the holocaust."[34] Szilard was also appalled.

"Did you read what McNamara said?" Szilard asked Forbes in an anxious telephone call. "In effect, he said America is prepared for a first strike!"[35] That made dialogue with the Russians more critical than ever, Szilard thought, and he spent the summer of 1962 thinking about ways to recruit and use a few scientists and scholars for yet another scheme—the "Angels Project."

These musings came together during a Pugwash conference in August 1962 in the paneled common rooms and mossy courtyards of Gonville & Caius College in Cambridge, England. There a Russian scientist—whom Szilard would only identify as "R"—encouraged him to open more focused talks. Szilard decided to enlist government policymakers and consultants who would be "on the side of the angels" by being disposed to forgo temporary diplomatic advantages "for the sake of attaining an agreement with the Soviet Union that would stop the arms race." The Angels would work out policy options that might then be weighed by their governments. Among those Szilard first considered for this task were physicist Freeman Dyson and Jerome Wiesner. But while many friends encouraged Szilard, physicist Hans Bethe declined an invitation, saying there were now enough other channels for US-Soviet dialogue.[36]

At the Soviet embassy, Szilard explained his Angels Project to Ambassador Anatoly Dobrynin. He seemed intrigued but raised a question that cut to the heart of Szilard's plans and his role in Washington. In the United States, many individuals outside government advise on policy at the highest levels, Dobrynin said, but in Russia there are none. How, he asked, can you arrange comparable participants?[37] Ignoring this complaint, Szilard drafted a detailed proposal for Khrushchev.[38]

Szilard's enthusiasm for the Angels Project grew from his conviction that if reasonable men could just sit down together and discuss their fate rationally, a solution would be found. He felt that President Kennedy was "an amateur, wet behind the ears . . ." and no match for the Russians, "intellectuals by comparison who are respected for their training." And he feared that Russia and America might allow a minor cold-war conflict to escalate into a global nuclear confrontation. America's fall elections had politicized just such a case: Cuba, Russia's ally just off America's shores. After Soviet defensive, short-range missiles had been discovered in Cuba, Senate Republicans backed a resolution in October warning Cuba against subversive activities in the hemisphere.

On October 13, Kennedy was shown U-2 reconnaissance-plane photos of offensive, intermediate-range missiles being installed southwest of Havana, and in secret his administration planned to challenge the Soviets directly. US air and sea units were deployed around Cuba on Saturday, October 20, and on Monday the twenty-second, Kennedy briefed congressional leaders at the White House, then prepared to address the nation on television.

"Come right down, Forbes," Szilard said in a phone call to the council's vice-president. "Kennedy is to speak about Cuba. I fear he will overreact."[39] Forbes flew in from Boston that evening and joined Szilard, Trude, and national security analyst Marcus Raskin in the Szilards' cluttered bedroom. On television Kennedy sounded serious and stern as he announced that US intelligence had discovered Soviet offensive, intermediate-range missiles in Cuba despite the Kremlin's promise not to station them in this hemisphere. It would be US policy "to regard any nuclear missile launched from Cuba against any nation in the Western Hemisphere as an attack by the Soviet Union on the United States requiring a full retaliatory response upon the Soviet Union," Kennedy said.[40]

Szilard stared at the television screen, his face pale with fear, his eyes wide with panic. Kennedy *was* overreacting!

"What do you do in a situation like this?" someone asked.

"A blockade is an act of war," said Szilard.[41] "An act of war." About a dozen young people came up to Szilard's room after Kennedy's speech, seeking advice and reassurance.

"What can be done?" one asked.

"Nothing," Szilard answered. "It is hopeless." In a moody and rambling talk he confessed that he had failed to control the weapon he helped create. Now, soon, it might destroy the world. He was too old, Szilard said. He only hoped that young people would pick up the broken pieces.[42]

Szilard slept fitfully that Monday night and the next morning appeared pallid and weary as he moved about the hotel nervously—telephoning, hanging up and staring, pacing the lobby and halls. With help from policy analysts Arthur Waskow and Raskin, Szilard stood at a pay phone in the Dupont Plaza and placed a call to the pope, hoping he would intervene with Kennedy. But the call never went through. Szilard walked to the bank to withdraw his money.

"Pack the bags," he said to Trude when he returned.

"Where are we going?" she asked.

"I don't care."

"What should we take?" she asked.

"Everything," said Szilard. "Give up the room."

Trude suggested they go to Canada, but Szilard shook his head. "Maybe Mexico," Szilard said. But she worried about sanitation there and Leo's recurrent infections.

"Geneva," Szilard finally said. "Let's go to Geneva. You always liked Geneva." Trude telephoned Lisbeth Bamberger, who had become a "gal Friday" for the council. When she arrived, Szilard said, "Come with us! You have to have the courage to leave when you recognize the time has come." Bamberger telephoned airlines and travel agents for Trude and helped the Szilards pack, but finally decided to stay.[43]

It was Szilard's habit to pay whenever he invited someone to a meal, but at lunch with Raskin that frantic Tuesday, October 23, he said, "Today we go Dutch." Szilard said he needed "all the cash on hand I can get." He advised Raskin to always be prepared in three ways: keep your passport in order, keep plenty of cash, and always keep a bag packed. (One bag Szilard always kept was his "Big Bomb Suitcase," with essential family, academic, and patent records, in case he had to flee a nuclear war.) "I will take Erika with me if you want," Szilard said, referring to Raskin's three-year-old daughter, whom he enjoyed entertaining on Sunday walks. Raskin admired Szilard and began to wonder what to do. Was this advice right? Should he flee, too?[44]

The Szilards littered their room with new suitcases and packed until after midnight. Before dawn the next morning they stuffed fourteen bags into a taxi to National Airport and from there flew to New York. That Wednesday, October 24, as about twenty-five Soviet-bloc vessels steamed toward Cuba, the US Navy closed its blockade around the island. Then the Soviets recalled some ships.

But as the White House redefined its quarantine procedures, a new problem arose in Washington, this time over the US Navy's own conduct. The White House planners now wanted to be sure that the blockade did not humiliate the Russians. Otherwise, "Khrushchev might react in a nuclear spasm. . . ."[45] Just the irrational blunder Szilard had feared.

All day Szilard telephoned friends, urging them to leave, asking their opinions. He called geneticist Joshua Lederberg and in an anxious voice asked him what he thought he would do. He also called Richard Garwin, an H-bomb designer on the President's Science Advisory Committee, and the two physicists met for lunch at a small East Side restaurant. Even in panic, Szilard strained to be rational about the few facts they knew. "Leo maintained that the Soviets would not put offensive nuclear missiles in Cuba because they had nothing to gain," Garwin recalled. "And I said they would because they had nothing to lose." As it turned out, "we were both wrong," Garwin reflected years later. "They did put them in, which made Leo wrong, and they did have much to lose, which made me wrong."[46] Szilard and Garwin were not the only nuclear scientists frightened by the blockade. At Harvard, President Eisenhower's science adviser, chemist George Kistiakowsky, dismissed his classes early, saying, "War is at hand."

Ironically, Szilard was supposed to attend a nuclear strategy conference in Cambridge that Wednesday. He missed it and a meeting for fellows of the new Salk Institute the next day at the Gotham Hotel on Fifth Avenue.[47] On Thursday morning, he and Trude were in a taxi driving toward Idlewild. Fog over Western Europe diverted their flight from Geneva, and as their plane touched down in Rome, Trude said, "Now you can see the pope." Szilard decided to try but, telephoning from his hotel, learned that Pope John XXIII was involved each day in the Ecumenical Council (Vatican II) that he had opened two weeks before.

The next morning, Saturday, October 27, the Szilards flew on to Geneva and from the airport took a brief taxi ride to the sprawling campus of the European Center for Nuclear Research (CERN). The center's director at the time was their friend physicist Victor Weisskopf. Szilard knocked at Weisskopf's office door, then stepped in.

"Leo," said Weisskopf, looking up from his desk in surprise. "What are you doing here?"

"I'm the first refugee from America, there's persecution of liberals, JFK is a sick man, there will be nuclear war in a few days." Szilard said he had fifteen pieces of luggage at the airport and wondered if he might have use of a desk.

Weisskopf laughed nervously, then offered Szilard an office at CERN and told him the latest he had heard by radio about the crisis.[48] Some Soviet vessels had been boarded; others passed without inspection. Unknown to the public that day, President Kennedy had dispatched his brother Robert, the US attorney general, on a secret visit to Ambassador Dobrynin at the Soviet embassy. Behind the confrontation was a three-step compromise. Soviet missiles would be withdrawn from Cuba. In return, the United States would not invade the island and would remove its obsolete, intermediate-range Jupiter missiles from Turkey.[49]

Szilard was unaware of how frightened and ultimately cautious Kennedy had become as the crisis progressed. As far as he knew, it was Khrushchev who had been the statesman. Szilard situated himself and Trude at the old-fashioned Hotel Bernina, on Place de Cornavin opposite Geneva's railway station, and set about trying to revive his Angels Project—a scheme he thought was now needed more than ever. Martin Kaplan, director general of the World Health Organization (WHO) and a friend from Pugwash conferences, offered the Szilards offices at the Palais des Nations, the imposing headquarters for several UN agencies. This placed Szilard conveniently between the Soviet mission along the north shore of Lake Leman and his hotel downtown. At the Bernina, Szilard wandered in and out of the lobby with papers under his arm and gray hair flying about. "He was here but not here," recalled the concierge, Elias Blatter. "Appearing, disappearing. . . ."[50]

At about that time, Robert Livingston reported to Szilard that Defense Secretary McNamara thought the Cuban Missile Crisis might have improved opportunities for the Angels Project,[51] and an invitation came from Khrushchev to proceed and to visit Moscow. "I like this proposal," Khrushchev said, now eager for ways to ease US-Soviet tensions.[52] Szilard replied immediately, sending the names of possible American participants. These included Fisher, George Rathjens of MIT, Henry Kissinger and Harvey Brooks of Harvard, Lew Henkin of Columbia Law School, Dyson, James Fisk of Bell Telephone Labs, and Don Ling of Morristown, New Jersey.[53]

The Angels Project, Szilard thought, could be the first positive step between Russia and America after the wrenching experience of the missile crisis.[54] But as Szilard happily made plans to leave for Moscow on November 26, Livingston telephoned from Washington.[55] Some "misunderstandings" with the Kennedy administration would have to be cleared up first. Livingston had told Raskin about Szilard's plan to visit Khrushchev, and Raskin had told McGeorge Bundy, who fumed. Szilard would *not* negotiate with the Soviet premier! Bundy insisted.[56]

Dejected by this news, Szilard unpacked and again wrote to Khrushchev, citing a misunderstanding.[57] When Szilard read in the papers that Khrushchev was about to visit French president Charles de Gaulle, he hopped on a train for Paris. There he called his biology colleagues François Jacob and Jacques Monod at the Pasteur Institute, seeking help to penetrate the Foreign Ministry. But when this approach failed, Szilard returned to Geneva and wrote Khrushchev again, this time enclosing a rectangular pocket watch whose case pulled apart to reveal the time.

"I wanted to give you a souvenir that would give you pleasure," Szilard said. "It's not a formal gift. The watch is very convenient; I have the same sort myself." The action of opening the watch also winds it. "It's hard for us older people to wear wristwatches," Szilard said, "because they can interfere with the circulation. I hope you will use it."[58] Then, musing about Bundy's complaint, Szilard returned to Washington, now somewhat chagrined about his leaving.

❋   ❋   ❋

"Welcome back, Forbes," Szilard joked when the two men met in the Dupont Plaza lobby in mid-December. Forbes had been "disgusted" with Szilard when he left for Geneva, and the council's office manager, Ruth Pinkson, "felt so let down" she had wanted to quit. "He had preached to me, 'We are saving the world!,' and then he left us," she recalled. But Forbes and Pinkson stayed on and moved the office to the Dupont Circle Building, then a low-rent haven for public-interest groups.

Back in Washington, Szilard seemed depressed and showed little interest in the council.[59] But he was amused to learn that during the missile crisis a homemade fallout shelter built in California by Manhattan Project chemist Willard Libby had been damaged in a brushfire. A former US atomic energy commissioner and a Nobel laureate, Libby opposed arms-control negotiations, prompting Szilard to remark that the shelter fire proved not only "that God exists, but that He has a sense of humor."[60]

At 5:00 P.M. on December 31, Szilard appeared at the West Gate of the White House for an appointment with Bundy. In their meeting Szilard tried to stress that the Angels Project was not meant to displace official negotiations, only to precede them by identifying promising topics. This made no impression on Bundy. "Maybe," he reflected years later, "because it was New Year's Eve."[61] But when Carl Kaysen, Bundy's assistant, showed more interest in the project, Szilard drafted two memoranda for the participants.[62] Still, Szilard could raise little foundation money to fund the first meeting, nor could he be sure that the administration would allow his best candidates to participate.[63]

In mid-February 1963, Szilard met again with Bundy and this time annoyed him by offering gratuitous advice about how the administration should handle "domestic political pressure" then building "on the issue of Cuba."[64] To get around Bundy, Szilard proposed staging a "hearing" for expert "witnesses" from in and out of government, with the transcript sent to the president.[65] Nothing came of this idea. When the council failed to attract the $25 million he had first hoped for, Szilard quipped that that much was not really needed; you can be just as effective with congressmen by threatening to withhold donations as you can by promising to make them.[66]

In fact, the council raised more than $79,000 for the 1962 Senate elections, giving more than $20,000 to McGovern and the rest to Joseph Clark of Pennsylvania, Wayne Morse of Oregon, J. William Fulbright of Arkansas, and Frank Church of Idaho. The donation to McGovern, one-fifth of his campaign expenditure, was credited with his slim 597-vote recount victory. Just the political leverage Szilard had had in mind.

"I'd have lost without the council," McGovern later said.[67] (When the council celebrated its thirtieth anniversary in 1992, it had helped to elect eighty senators who favor arms control, and it continues to be one of the most effective political-action committees in Washington.)

At the council, Forbes had defended Szilard's flight to Geneva by explaining it as a quasi-official trip connected somehow with arms control and a Moscow visit. Szilard claimed it was the only rational response at the time, although sometimes he sounded defensive. "If I were to stay in Washington until the bombs begin to fall and were to perish in the disorders that would ensue, I would consider myself on my deathbed, not a hero but a fool," he declared to his friend psychiatrist René A. Spitz.[68] Szilard—still behaving as an interloper who would not learn local customs and conventions, still scheming to speak directly to those in power—seemed impatient among scholars and scientists concerned with the government and its policies.

In 1963, when his friend Marcus Raskin cofounded the Institute for Policy Studies, a progressive think tank in a town house at Nineteenth Street and Florida Avenue, near Dupont Circle, Szilard asked to become a fellow and was given a small office. One day social scientist Harold Orlans invited Szilard to lunch at the more respectable Brookings Institution to give staff members an opportunity to learn about the council. But Szilard seemed oblivious to his listeners' knowledge of Congress, Orlans recalled, and delivered a monologue past the customary 2:00 P.M. quitting time. Szilard also misbehaved by asking for a second helping of the crab dish served for lunch, something never before done by a Brookings guest. "His flight to Switzerland during the missile crisis hurt his reputation," Orlans noted, "both as a sound analyst and as a leader of the antinuclear cause."[69]

In his analysis at the time, Szilard coined two useful phrases: "saturation parity," to define the minimal deterrence that might stabilize the US-Soviet nuclear arms race; and "sting of the bee," to describe what minimal deterrence a country would need and how it might use it. The irony of the sting, like that of a nuclear retaliation, is that the unavoidable outcome for the bee is death. Once they reached a nuclear stalemate, Szilard had argued since the 1950s, the superpowers would finally realize they were safer scrapping arms than racing to build them—a condition that finally occurred in the late 1980s, about when *The Voice of the Dolphins* had predicted.[70]

"Preparations for the Angels Project have been moving rather slowly in Washington," Szilard complained to Wiesner at the White House, "perhaps because I was too disheartened to push them with vigor." He asked Roger Fisher to assume responsibility for enlisting the American participants.[71] Instead, biology recaptured Szilard's fancy, and he admitted to Jonas Salk that Washington was "not the best place" to work and that he needed to "reexamine" how long he might stay.[72] If Trude could find suitable work in La Jolla, Szilard said, then he would like to join the new Salk Institute—his first long-term commitment to research in more than a decade.[73] Szilard accepted Salk's offer of a nonresident fellowship the same day the letter arrived and requested the option to become a resident fellow.[74]

In June 1963, Szilard left Washington for a Cold Spring Harbor biology symposium, then flew to Geneva for an informal WHO meeting on research and a "Discussion on Molecular Biology" at CERN, a meeting that laid plans to create the European Molecular Biology Organization (EMBO). Szilard proposed that grants be made to EMBO by both the US and Soviet academies of science, a typically freewheeling idea that was perfectly logical but politically zany.[75]

While he was busy with biology in Geneva, Szilard's Angels Project finally gained cautious support from President Kennedy, provided it would not be considered an official US effort.[76] From his office at the Palais, in July, Szilard resumed his contacts with Khrushchev. But now signals began to cross, just as the White House had feared. At the time, a delegation led by W. Averell Harriman was in Moscow negotiating a US-Soviet nuclear-test-ban treaty. Szilard told Khrushchev that Carl Kaysen, Bundy's aide and a member of the Harriman team, "is fully familiar with all aspects of the 'Angels Project' . . ." and would be able to judge its usefulness. Americans who were then prepared to participate, Szilard reported, included Fisher, physicists Marvin Goldberger and Murray Gell-Mann of Cal Tech, and Steven Muller of Cornell. The next day, at the Soviet mission, Szilard cabled Kaysen in Moscow, asking him to answer any questions that Khrushchev might ask about the Angels. Kaysen was furious and complained that such a role was inappropriate.[77]

When Khrushchev realized the problems Szilard had created, he suggested that because of "certain difficulties" in organizing the Angels Project, such a meeting should best take place as part of the next Pugwash conference in Dubrovnik, Yugoslavia, that September, and not independently. "The Chairman of the Council of Ministers of the USSR, N. S. Khrushchev, conveys to you and to your wife his best wishes for good health and success in your noble work in defense of peace." Szilard's cable to Fisher was terser: "Negative response cancels Angels meeting. . . ."[78]

※　　※　　※

In August 1963, US and Soviet negotiators concluded a ban on nuclear tests in the atmosphere, in space, and under the seas, and at month's end the Moscow-Washington hot line—actually a teletype link—clattered into service. The US Senate ratified the Test Ban Treaty in September, leaving many arms-control activists delighted. But Szilard thought that euphoria over the limited test ban obscured his more ambitious search for comprehensive arms reductions, and in a hotel room in The Hague he penned a draft of "A Farewell to Arms Control." This restatement of his minimal-deterrence writings had a plaintive title meant to suggest that he had little more left to give.[79] The "outsider" was still outside.

At the Pugwash meeting in Dubrovnik that September, Szilard joined Joseph Rotblat and Henry Kissinger in a working group on "Denuclearized Zones" but said almost nothing during the discussions.[80] Instead of dominating debate, as he had at other conferences, Szilard was content to sit in the sun and daydream, not about arms control but about aging, sex selection, birth control, and—between naps—ways to eliminate the human need for sleep.[81]

Even before the Test Ban Treaty was ratified, Szilard had concluded that while it "seems that some of the key people in Washington are now beginning to listen to me . . . time is running out on the Kennedy administration. . . ." Unless goals were set by the end of the first term in 1964, Szilard reasoned, the administration could do little by the end of Kennedy's second term. At the same time, Szilard reflected on his hectic pace and perilous health. "While I would, of course, like to live as long as possible," he wrote, "I also want to accomplish as much as possible while I am alive. So far I have not given up a single working day on account of my heart condition. . . ."[82]

That pace continued on his way home from Dubrovnik in September; Szilard stopped in Rome, checked into the Hotel Regina, and drafted a memorandum that put his Angels Project in the hands of a real expert: the pope. He suggested that the Vatican take a lead in arms control by convening US and Soviet scientists. Szilard left a package at the Vatican containing memos and reprints about his career and proposals and waited two days for a reply, passing the time by riding a horse-drawn carriage around St. Peter's Basilica, reading in the city's lush flower gardens, and sitting in cafés.[83] But when no call came, Szilard returned to Geneva, spent ten days in London dispensing arms-control wisdom "in the shadow of the shadow Labor cabinet . . . ." and returned to Washington.[84]

He knew by then that his effectiveness there was at an end. And having just retired, at age sixty-five, from the University of Chicago, he also knew that the Salk Institute offered his most promising outlet for creative work. Nonetheless, Szilard kept tinkering. He helped organize seminars for senators and their staffs, explaining the interplay between the technical and political requirements for arms-control agreements. And he tried to help Kennedy ease tensions over Berlin by suggesting in a letter that incidents between US and Soviet troops over the right of access to the divided city might be avoided if each side agreed to comply with the border guards' requests before they were made. He offered to send a similar letter to Khrushchev.

Answering for the President, Bundy's reply was curt. He thanked Szilard for his "imaginative note." "Yours is a characteristically original suggestion, but I doubt if it would be useful for us to conduct our relations with Chairman Khrushchev through you."[85] It was the last contact Szilard would have with the Kennedy administration. The day after Bundy wrote that reply, Kennedy was assassinated and Vice-President Lyndon B. Johnson was sworn in as the thirty-sixth president of the United States.

Fitfully, Szilard weighed his prospects in Washington with plans for the Salk Institute. In December he redrafted "Sting of the Bee," his thoughts about minimal deterrence, trying to define a simple and logical halt to the nuclear arms race. Szilard had first called his article "A Farewell to Arms Control" but a few days before Christmas made the title even more insistent: "Let Us Put Up, or Shut Up."[86]

CHAPTER 30

La Jolla: Personal Peace

 1964

La Valencia, a pink stucco, Spanish-style hotel with clay-tiled roofs, crowned the hillside at La Jolla, California, in 1964, giving the peninsula and the lazy seaside village a fairy-tale quality. From its terraces and windows the Pacific's waves could be seen crashing on the rocks by the green park below and heard thundering into the sandstone cove nearby. Leo and Trude Szilard checked into this quaintly comfortable place in mid-January, refugees from a harsh Washington winter. "We stayed in the best suite at La Valencia," she later recalled, "and in two weeks I was sold."[1]

Leo Szilard had been "sold" on La Jolla for years, ever since visiting this Pacific outpost in the mid-1950s as a consultant to General Atomics (GA). At the time the Szilards arrived this sunny January, the town's main scientific center was the Scripps Institute of Oceanography along the beach to the north. Above the Scripps pier, on the crest of a steep hill, stood a eucalyptus grove where former army barracks were being patched and painted into the new campus for the University of California at San Diego (UCSD). And on the sandy cliffs up the coast from Scripps and the UCSD spread the site for the Salk Institute for Biological Studies, then a tree-shaded construction site and a cluster of trailers tied together by wooden decks, the first tentative offices and laboratories.

Szilard had helped found the institute—his long-sought fusion of biological and social science—and anticipated a return to basic research and brainstorming after three years' lobbying for arms control in Washington. The locale was striking, although oddly bucolic for an urbanite like

Szilard. And even in this dream world, Szilard could feel restless during these final months of his life, at peace with himself but still at odds with the elusive mysteries of nature.

In the months before this visit, Szilard had been concerned about Trude's reaction to La Jolla, then a placid village at the north end of San Diego. He knew that she enjoyed her public health work in Washington and would miss their eclectic social life in the capital. He did not know what new and compelling interests she might find in this idyllic coastal setting. Szilard had mentioned in several recent letters to friends that he "will not live forever" and worried that if they moved to La Jolla, Trude should have a reason to remain there on her own. Her delight with the tropical climate and with the sociable community of scientists gathering at the Salk Institute soon put his mind at ease.

That spring, Szilard told Margaret Brenman Gibson, a board member at the Council for a Livable World (CLW), that he had decided to live in La Jolla because at his age it offered "a foretaste of paradise." Indeed, so at ease did Szilard seem during his sunny January visit that he and Trude found themselves looking around for not only an apartment to rent but also a house they might buy—the first time in his life that such stability seemed appealing.[2] Yet mentally Szilard was as restless as ever. In his suitcase came a draft of "Sting of the Bee and Saturation Parity" for revision and review by his colleagues. Its ominous last sentence captured Szilard's anxieties about the nuclear arms race as well as any he had ever written: "To make progress is not enough, for if the progress is not fast enough, something is going to overtake us."[3]

With time, life in La Jolla enticed Szilard from the worries of arms control to the wonders of science. He poked his nose into the makeshift laboratories where his institute colleagues worked. He wandered about the concrete slabs rising nearby—the structures taking shape in architect Louis Kahn's sweeping courtyard and reflecting-pool design for the institute. Almost every day he sat in a deck chair on the wooden patio, scribbling in his biology notebooks and squinting into the sun's reflection on the blue sea far below.

At first, the Szilards were intrigued by a beachfront cabana on La Jolla Shores, at the Beach and Tennis Club across the bay from the cove and the village center. But before they returned to Washington in early February, the couple settled, instead, on a quiet cottage behind the Del Charro Motel.[4] Situated on Torrey Pines Road, the winding, two-lane route that tied the village to the new science and research establishments growing on the cliffs above, this was an ideal location for both work and play. A stable behind the motel had been converted to a two-room apartment, and from the large windows the Szilards viewed palms and flowers that framed the lawn. The arched ceiling, with wood beams exposed, gave the place an informal air. Yet there were maids and room service, along with a small dining room in the motel. This small but elegant place seemed just right for Trude to live in, for Leo to think in.

Back in Washington that February, Trude packed their belongings at the Dupont Plaza and prepared for the move, saying farewell to friends and colleagues at Georgetown University and the Pan American Health Organization. She would later call her interlude in Washington "the happiest time of my life."[5] Szilard dashed about town to conclude his arms-control lobbying. In a February note to McGeorge Bundy, who had remained as national security adviser in the Johnson administration, Szilard sent a copy of "Sting of the Bee." "In the unlikely case that you should find time these days to read this paper, I should appreciate the opportunity to answer any questions which you might have before February 20," Szilard wrote. "On that date I am leaving for the West Coast, where I intend to stay 'permanently.'"[6]

Over lunch at the garish Genghis Khan Restaurant with Herbert (Pete) Scoville, the former technical director of the Central Intelligence Agency who was then working at the Arms Control and Disarmament Agency, Szilard talked about ways to verify arms agreements.[7] He tried to meet Defense Secretary Robert McNamara in order to explain the Angels Project, and when this failed, Szilard finally abandoned the whole idea.[8] On Capitol Hill, he paid farewell calls to senators helped by the council, including Fulbright, Church, Clark, and McGovern.

In his cluttered room at the Dupont Plaza and in his small office at the Institute for Policy Studies nearby, Szilard polished his "Sting of the Bee" manuscript and drafted a statement criticizing a multilateral nuclear force.[9] He also wrote to magazine editors, seeking popular outlets for his ideas, but only the *Bulletin of the Atomic Scientists* would agree to publish his manuscripts.

While he was in Washington, *Holiday* magazine featured a long and flattering profile, "Leo Szilard: The Conscience of a Scientist," written by Tristram Coffin and illustrated with dramatically posed photographs by Arnold Newman. The article cited science historian Alice K. Smith's conclusion that Szilard was one among five men in the past century who have done the most to change our times; the others were Lincoln, Gandhi,

Hitler, and Churchill. "The creative scientist has much in common with the artist and the poet," the article quoted Szilard as saying. "Logical thinking and an analytical ability are necessary attributes to a scientist, but they are far from sufficient for creative work. Those insights in science which have led to a breakthrough were not logically derived from preexisting knowledge; the creative processes on which the progress of science is based operate on the level of the subconscious."

What excites him now, Szilard said, is biology. "The mysteries of biology are no less deep than the mysteries of physics were one or two generations ago, and the tools are available to solve them provided only that we believe they can be solved."[10]

But while anticipating the pleasures of science, Szilard could not stop worrying about the perils of the nuclear arms race or yet escape his role in it. After the move to La Jolla in late February and just before he began dictating his ambitious theory on memory and recall, Szilard still mused about the strategic nuclear situation. Both the United States and the Soviet Union were testing large nuclear warheads. The "missile gap," a Kennedy campaign device in 1960, was actually prompting a Soviet buildup; at that year's end the USSR had fewer than 100 intercontinental ballistic missiles (ICBMs) and submarine-launched ballistic missiles (SLBMs), while the United States had more than 600 ICBMs and 160 SLBMs and continued building them rapidly. Understandably, when the United States proposed, in January 1964, that both superpowers stop producing fissionable materials for nuclear weapons, the plea was promptly rejected.

"I am rather pessimistic about getting arms control in the predictable future," Szilard wrote to his friend the German physicist Carl von Weizsäcker, "but this does not mean that we ought to give up fighting for it; on the contrary, I believe the time has come to fight for it in earnest."[11] But how would he fight? Szilard revealed a growing pessimism two weeks later when declining an invitation to be commencement speaker at the Albert Einstein College of Medicine. To Dean Marcus Kogel, Szilard complained about the irrelevance of his work on arms control. At first, he wrote, he thought his latest paper on "Sting of the Bee" might have offered a theme for an address. But, he continued,

> The events of the last few weeks have brought home to me rather forcefully, however, that to a very great extent our policies are determined by expediency and nothing but expediency. I rather doubt that the rational considerations upon which my article is based can prevail if they are not supported by moral

considerations—the kind of moral considerations which are the very antithesis of considerations of pure expediency. It seems to me that the best thing I could do for the time being is to keep silent, and silent I propose to keep.[12]

In early March, with the frustrations of arms control seemingly put aside, Szilard focused on the mysteries of the human mind as he began dictating "The Molecular Basis of Long-Term Memory."[13] But within minutes of sitting down in his office, a meandering thought led him astray, and he dictated, instead, a memorandum on antibody formations. His speculation about memory was interrupted again the next day when Szilard described the induction of the sugar betagalactosidase and dictated a second paper on antibody formations.[14] It wasn't until the third day, after a lively lunch with some economists at La Valencia, that Szilard rode back to the Salk Institute, sat down again, dictated a memo about antibody allotypes, and at last returned to his memory and recall paper.[15]

Even in this peaceful setting, his mind raced in urgent and unpredictable ways. When finally thinking again about memory, Szilard began in his own subconscious freely associating, testing, and twisting ideas. He focused on the topic for hours at a time, almost every day, and a few days later he boasted to his friend Eugene Wigner that "this paper is a result of my *not* thinking about the problem for three years, and I am quite pleased with it."[16]

Most days at the institute, Szilard sat on the open veranda that connected the temporary laboratory buildings, staring at the Pacific and botching in his playfully persistent way. Salk remembers him in a deck chair, blissfully "sitting in the sun, thinking and churning, as in a cocoon, weaving ideas."[17]

If the Salk Institute runs short of funding, Szilard walked in to tell immunologist Melvin Cohn one sunny day, the fellows should simply invest in a huge ship. Anchored offshore, beyond the US territorial limit, it would feature gambling and prostitutes, and the profits could be used to finance important research. Another time, Szilard walked in to tell Cohn that rules should be devised to limit useless and irrelevant scientific papers. Under Szilard's scheme, "when you get your Ph.D., you are given 100 slips for publications. You can use one for each article you write. No more. That way you will decide which ideas are important enough to publish." Szilard also thought about the human body; not just how it worked, but how it *ought* to work.

"You know, Cohn," Szilard interrupted one day after a time lounging on the veranda, "humans should be born with platinum teeth." Then he toddled off to think about how that much platinum might be concentrated by nature in the earth's evolutionary scheme. "Often Leo was interested in how it might have worked rather than how it works," Cohn recalled.

At first, this brainstorming routine amused Cohn, until he realized that when daydreaming, Szilard sometimes stopped breathing; he nodded off and became very still, then suddenly heaved up and gasped for the breath he had lost. At institute seminars, too, Szilard sometimes appeared to doze, then sprang to life just when a critical point seemed apt. Rather than a sign of distress, this behavior was taken as Szilard's style of focused thinking. And although Szilard did sometimes nod off after heavy lunches, no speaker could be sure when he might be interrupted with an incisive question or comment.

At one neurobiology seminar, when a boring and pretentious speaker droned on, Szilard put the whole enterprise into an amusing perspective. The speaker was so confident about his interpretation of how the brain works that at one point he turned to the front row and stared at Francis Crick, a fellow of the institute and 1962 Nobel laureate (with James D. Watson and Maurice H. F. Wilkins).

"Francis," he asked, " what do you think of that idea?" "Oh," Crick said, embarrassed for the speaker but not wanting to hurt his feelings, "Seymour [Benzer] here didn't understand it, and I was spending so much time explaining it to him that I missed some of your points."

"Well, Seymour," the speaker persisted, "what do *you* think of it?"

"I'm still thinking about it," Benzer answered. "I'm not quite sure."

"Dr. Szilard," the speaker asked next, "what do you think of my idea?"

With no pause, Szilard answered, "I think it explains how *your* brain works."

At other seminars, Szilard applied his well-known "index" to guest speakers, listening for a few minutes to decide if the ideas were original and, if they were not, rising from the front row and striding out the door. Szilard could be more polite, but just as insistent, with his institute colleagues. "Mel," he once scolded Cohn in a strong voice, "you listen to me now, because when you understand this you'll need it."[18] He characterized academic pettiness by saying to Salk about his successful ideas for developing a vaccine, "Jonas, they'll never forgive you for having been right." In another conversation with Salk, Szilard declared his "three stages of truth," demonstrating how well he understood his colleagues, if not his own habits of thought. Confront scientists with a new idea, Szilard said, and most will say, "It's not true!" Next, they'll say, "If true, it's not very important." Finally, they'll say, "We knew it all along!"[19]

Cohn and Szilard were both interested in how human immune systems work and for a while focused on betagalactosidase as a possible triggering mechanism. This inquiry led nowhere, but from it Szilard became the first theorist to press for a unified concept of the immune system. Many of Szilard's other ideas in biology were wrong, Cohn believes, because he applied a physicist's search for elegance and parsimony to a process that is evolutionary, at some stages needlessly complex, and forever changing.

"He read very little but used people as his information sources," Cohn recalled. "He had a way of putting order into confusion. When considering a problem he could think of thirty fields at once," relating one to another, and not only analogously but with working details from each. After listening intently, Szilard tried to synthesize what he had just heard: to clarify apparent contradictions, to pose new questions, to discover paradox, and to make his informant restate the topic in Szilard's own terms. It is better to be clear and wrong than to be right and confused, Szilard believed. By consuming other people's information, research results, and hypotheses, all the while trying to find ways to order and unify the confusion, Szilard raised his "sweet voice of reason" amid a chorus of dissonant creativity.

When Cohn said he liked *The Voice of the Dolphins,* Szilard was at first pleased, then pained.

"Why aren't you saying it's great literature?"

"Because it's *not* great literature," Cohn replied. Consider a story by Kafka, by Sartre, by Faulkner, he continued. "They have complexity and overtone. They require an emotional response." Szilard couldn't see that.

"I'm emotionally moved by extraordinary reasoning," he declared.[20]

Szilard generalized so aggressively because he tried to explain almost everything—from moral values to paintings to music to immune systems—as quantitative statements about universal truth. Szilard judged one painting as better than another not because it is prettier or technically more complex, or somehow inspired but simply because it "reduces the entropy of the universe" more efficiently, he once explained to Cohn. And when speculating about how the natural universe works, Szilard would not talk about making his guesses correct but about making them clear and concise. Experience appealed to Szilard not for its individual excitements but for the chance that a multitude of events might somehow reveal a common pattern or purpose or law.

In his own engrossing way, Szilard consumed foods as voraciously as he took in thoughts—in a direct, eclectic, and unaffected manner and often without restraint. Some mornings he poked his smiling face into Cohn's lab to say, "Let's go eat!" Szilard seldom cared where they went or what they ate. The important thing about eating, he believed, was not to feel hungry; so the more often you ate, the better you should feel. On one occasion, however, his lifelong passion for sweets led to a distressing problem. The Szilards came to the Cohns' house for dinner and began the evening by sitting on the patio. Szilard was under a fruit tree, and when Cohn told him the ripe fruit were sapota mangos, he plucked and ate one. "Delicious," he announced, and nibbled on another. Ignoring the hors d'oeuvres, Szilard plucked another sapota. And another. Suddenly, he groaned deeply and seemed to faint. He felt weak and looked ill; so distressed, in fact, that he had to be rushed to Scripps Hospital to have his stomach pumped.[21]

His unquenchable appetite for work got Szilard into medical trouble of another kind a few weeks later. He seemed impatient and energetic once he was able to concentrate on the memory paper. He dictated and rephrased, edited and rewrote his ideas with new force. When stumped in this thinking, Szilard turned to another institute fellow, mathematician and historian Jacob Bronowski, for help with a statistical problem: how to count two overlapping neurons in the brain that are connected through a synapse.[22] The two men enjoyed statistics and each other's company. They met every morning for a week to consider the problem, at first over coffee in the Bronowskis' seaside garden, then in their offices. They filled pages with their calculations, Bronowski carefully jotting in his leatherbound diary, Szilard scribbling in a cheap composition notebook. And they punctuated this intense work with abrupt belly laughs, usually over the many East European stories they shared: about morality in small villages or smart-alecky boys and wise rabbis.[23]

Seeking help elsewhere, Szilard also telephoned the mathematicians Mark Kac and Leo Goodman, but before they could reply, Bronowski produced an answer. His statistical problem with the neurons solved, Szilard worked briskly to complete the first part of his memory paper and sent it off to the *Proceedings of the National Academy of Sciences*. But the effort left him so weary that he had to cease all work and rest at his cottage for several days.[24]

The paper that Szilard completed for the *Proceedings* described a "hypothetical biological process on which the capability of the central nervous system to record and to recall a sensory experience might conceivably be based."[25] This problem, his colleague Aaron Novick later noticed, was in a sense the same one he had begun his career with as a physicist in the early 1920s. Then trying to exorcise Maxwell's demon, Szilard had asked "whether the brain violated the laws of thermodynamics." With memory and recall, Novick said, Szilard "sought to understand how the brain could 'learn.' "[26] Over breakfast in the Del Charro dining room one day, Szilard met Nicholas Kurti, a visiting physicist from Oxford whom he had known since the 1930s, and explained how his theory on human memory was going. Then he leaned forward. "Kurti, it will not be right," Szilard whispered in a conspiratorial way. "But it will be impossible to prove it wrong!"[27]

Szilard's "Memory and Recall" paper is more significant for what it attempted than for what it achieved, although many questions he raised still remain unanswered. The theory it offered provided an effective chemical basis for recording and tracing the many parts that constitute a single memory in different brain cells. Bronowski later explained to Salk that the paper "made it possible to understand how a memory can be reinforced by repetition and also how it can be extinguished (that is, forgotten) if it is not reinforced." Bronowski concluded that "Szilard's theory was bold in general conception and yet searching in its detail: for example, it took account of the constant fluctuations in the level of chemical and electrical activity in the brain which no other theory had tried to accommodate."[28]

As Szilard became more comfortable at the Salk Institute and in his cottage at the Del Charro, he consciously embraced other disciplines, perhaps hoping with each encounter to spot fresh, unifying details. He organized Sunday brunches for friends and associates, which the motel staff catered. Tables and chairs were arranged around the lawn for jolly, freewheeling conversations with economists, medical doctors, local artists, political scientists, and educators. On one visit to La Jolla that spring, physicist Freeman Dyson joined the group, and a few colleagues from the Salk Institute became regulars, among them Cohn and the Bronowskis.[29]

Most of Szilard's ventures out-of-doors were for fresh ideas rather than fresh air, and while many other institute colleagues enjoyed Southern California's natural life-style and the nonurban setting, efforts to move him beyond his habitual physical lassitude invariably failed. For example, when Cohn suggested a drive northeast into the Anza-Borrego Desert for a hike among the rare spring wildflowers, the Szilards agreed. But on the Sunday morning when Cohn and his wife set out in their white convertible, along with the Szilards and visiting biologist Rita Levi-Montalcini, Leo was ill at ease: He wore his habitual dark suit and tie and slumped in the backseat, seeming to notice almost nothing. Throughout the day, he was oblivious to the dramatic terrain, and while the others climbed and scampered among colored rocks and flowering bushes, with Trude eagerly photographing the desert's fresh light and color, Leo sat in the car, staring ahead, botching.

"Like an Indian guru, Leo had an inner world that was totally sufficient for him," Cohn observed later. "You had to interact with his world to force him to consider your world outside." Trude returned with several stunning photographs of the dramatic flowers and landscape, so perhaps in this way Szilard eventually did see the desert's bloom in springtime. But for all the hours he reclined in Cohn's convertible, he might just as well have been seated at his desk or slumped in his deck chair.[30]

On April 1, Szilard realized his longtime dream of becoming a resident fellow at the institute. With a yearly salary of $25,000, another $10,000 for a secretary and travel expenses, and the promise of more money for researchers and laboratory equipment, the institute offered him financial security and intellectual challenge, which he had rarely enjoyed together.[31]

"An appointment for life," he liked to boast to old friends. "For life." Those words sounded wonderful. But now that he had this ideal position, Szilard still needed to fret. In a letter that spring to Abram Spanel, a friend and collaborator who had long worried over Szilard's retirement problems, he praised La Jolla as "a wonderful place" to live and boasted about his "life" appointment. All that will be fine, Szilard added, "except, of course, if I should outlive the institute."[32]

At the institute, Szilard soon turned to a second topic that had long intrigued him, the aging process. He began to lay plans to host a conference on the subject and, working with chemist Leslie Orgel, drew up topics and a guest list. Over the years, Szilard had both challenged and annoyed his colleagues (and other experts) by confronting them with basic and seemingly simple questions, which they often could not answer. His approach to biological aging took the same quizzical form, as a few of the forty-eight questions he posed reveal.

Can aging be defined? What is the evidence that it is a general phenomenon throughout an organism rather than a phenomenon involving only key molecules or cells? How do genes affect the rate of aging? Do all organisms and tissues now growing age? Do complex molecules change their properties with time? Are hormonal differences with age causes or effects? Are individual cells capable of living indefinitely? Do individual cells age more rapidly under radiation? For each topic, Szilard also asked direct questions about experiments. How could they be designed? Conducted? Evaluated?[33]

This approach suited Szilard, and the new institute, because it explored scientific questions that might not be solved, indeed might not be seen by some researchers as legitimate questions. Most scientific research, Szilard believed, focused instead on the many questions that we know, ones that have a reasonable chance of being answered with obvious solutions.

During the last week in April, Julius Tabin came to La Jolla and stopped in at the Del Charro. A physicist who had first met Szilard during the Manhattan Project at Chicago, Tabin later specialized in patent law and worked for General Atomics at the time Szilard was a consultant there in the 1950s. They met for dinner, and as the two men drove into town to Szilard's favorite French restaurant, they talked about many things— Szilard's fondness for his new life in La Jolla, his work at the institute, and Tabin's work as a patent lawyer in Chicago. Only later, as their leisurely meal progressed, did Szilard reveal he had a reason for their meeting.

After years of neglect and rejection, Szilard said, he still craved some compensation from the federal government for his early pioneering work in atomic energy. "He made it quite clear that he was not interested in seeking such compensation for himself," Tabin recalled about their conversation that night. Rather, "he wished to provide for [Trude] in the event that anything should happen to him." Over dinner, Szilard recounted vividly the pressure that General Groves had used in their 1943 meeting to gain his signature on a patent agreement, adding the fact that a notation was later typed on the assignment stating that the signature was obtained without coercion. Tabin found that "highly unusual" and an indication that Szilard probably was pressured to sign away his patent rights. Szilard said that he would be satisfied to receive the Atomic Energy Commission's Enrico Fermi Award, a $50,000 tax-free prize for distinguished achievements in atomic science. In 1962 the award had gone to Teller; in 1963, to Oppenheimer.[34] Now, Szilard thought, it was his turn. With that, he said, he would make no further claims.

Next Szilard revealed that during his last visit to Washington he had discussed with sympathetic senators the possibility of introducing special legislation to compensate him for the early inventions. He said he preferred not to file for compensation; this would only create lengthy legal proceedings and press coverage that might imperil his chances for the Fermi Award. Above all, Tabin remembered, Szilard "felt that time was running out" and that he should not delay in taking some action. At Tabin's request, Szilard agreed to gather up all his documents bearing on the atomic patents.[35]

Perhaps it was his whole life's forward momentum, its restless surges and pursuits. Perhaps it was mere habit. Perhaps a deeper unrest he could never explain. But for some reason, Szilard had been at the Salk Institute for only about three months—and officially a resident fellow for about three weeks—before he entertained thoughts about leaving, at least for a while. "There was something driving Leo," Salk recalled later. "He was not conscious of it, but he kept on going. Moving about. Searching to find something in the wind."[36]

In January 1964, Szilard had declined an invitation to spend the fall semester as regents lecturer at the University of California in Santa Barbara. But in April, when invited again, Szilard held on to the letter and puzzled over it. His friend Robert Hutchins, the former president and later chancellor of the University of Chicago, was then heading the Center for the Study of Democratic Institutions in Santa Barbara and for more than three years had written exuberant invitations to Szilard to visit. This strong tie may have added to the lectureship's appeal, for on May 27, Szilard asked Bronowski, the Salk Institute's deputy director, about whether he might accept the regents lectureship that fall.[37]

A few days earlier, Szilard had told Trude about other travel plans. Before writing the second and third parts to his "Memory and Recall" paper, Szilard said, he needed to appraise the latest research on neurobiology. This he could do by spending two or three weeks at MIT, "to learn about their experiments and perhaps to start some new ones."[38]

Still, Szilard was ambivalent about leaving La Jolla, and when invited to participate in Cornell University's centenary celebration, he replied with mixed feelings. He wrote on May 13:

> As a rule, I have in the past declined such invitations because, while they sound very attractive at the time when you receive them, when the time comes to fulfill the obligation, it usually turns out to interfere with some other urgent activity. Because I moved to La Jolla, which is somewhat isolated, upon receipt of your letter I did some soul-searching to see whether I ought to change my attitude towards invitations of this particular kind. I came to the conclusion that I should not.[39]

La Jolla, Szilard reported to researcher and CLW supporter Betty Goetz Lall in May, "is a wonderful place and very good for my work in science, but it puts me out of circulation regarding the field of arms control." Yet Szilard could not escape what had for years been his life's governing purpose.[40] In March the *Boston Herald* published an attack on the CLW by conservative columnist Holmes Alexander. Calling the council "one of these unilateral disarmament groups," he denounced "The Liveable (with Communism) Worlders" and named the candidates receiving their money. A week later, in the *Herald* and several other papers, Alexander attacked again. His "Cold War Comes into Wyoming" compared Sen. Gale McGee's acceptance of council money with other candidates' support by Robert Welch's ultraconservative John Birch Society.[41]

On March 10, H. Ashton Crosby, the council's executive director, mailed Szilard the two *Boston Herald* clips, adding: "I am not worried about the impact on the Council but I am worried about what these articles may do to our Senators." Crosby reported that he and CLW counsel John Silard (Bela's son) were planning to draft a statement "from somebody high up in the Administration that they support the Council's efforts and know that it is a responsible organization."[42] By May, after several letters were exchanged between Alexander and Crosby, the council became the subject of a Senate debate that would continue for months.[43] Szilard finally became annoyed enough to join the dispute when North Dakota's Quentin Burdick charged in a speech on the Senate floor that the council had advocated that the United States "should unilaterally disarm." Writing to Burdick on May 28, Szilard challenged him to show where such a notion was ever stated in council documents. All Szilard had ever favored, he insisted, was "multilateral disarmament, carried out step by step with proper guarantees."[44]

Szilard dictated but did not sign that letter. There is no record of how he spent the next day, a Friday, but apparently he worked from the cottage to plan a research program for his memory and recall studies. By telephone he invited chemist Anna Beck to be one of his research assistants. Laboratory space in the new institute buildings would be available within a few months.[45] He envisioned several years' work, pursuing clinically the ideas he was still devising in theory. On that Friday night, CLW board member Matthew Meselson joined Trude and Leo. After a meal, they all chatted and enjoyed the Bob Hope show on television, then Leo went off to bed "feeling fine."[46]

Trude awoke before dawn the next morning, Saturday, May 30. The alarm clock glowing in the darkness read 3:55. The room was quiet. Too quiet, she thought. Leo wasn't snoring or breathing heavily. Then Trude touched him and discovered that he was still. She tried to arouse him. No response. Terrified, she began artificial respiration, then pulmonary resuscitation, her medical skills overtaking collapsing emotions. Still no response. She now realized there was no life to revive. Leo had died in his sleep, taken by a massive heart attack.[47]

Trude telephoned La Jolla's emergency number, then their friend Robert Livingston, then Jonas Salk. The San Diego County coroner sent a man to the cottage, who certified Szilard's death and said that an autopsy was required. Salk telephoned institute biologist Edwin Lennox and Szilard's new urologist in La Jolla, James Whisenand. A call came from Trude to her sister, Frances, in Mount Vernon, New York. "The most terrible thing in the world has happened," she whispered, breaking the sad news. Frances called Bela, who lived nearby. In La Jolla, Livingston telephoned the Associated Press, and at 9:54 A.M., Pacific daylight time, bells clanged on AP tickers in newsrooms around the country to signal an important story: "BULLETIN, SAN DIEGO, CALIF., MAY 30 (AP)—DR. LEO SZILARD, A TOP US NUCLEAR PHYSICIST AND ONE OF THE MEN CREDITED WITH KEY WORK IN DEVELOPING THE ATOMIC BOMB, DIED TODAY AT HIS LA JOLLA HOME. HE WAS 66.

"HIS DEATH WAS ATTRIBUTED TO A HEART ATTACK."[48]

Within an hour, a complete story moved on the wires, crediting Szilard as the scientist "who helped the United States become an atomic power, and then later campaigned for peace. . . ." Correctly the AP account mentioned Szilard's role in the 1939 Einstein letter to Roosevelt, his code-sign of the first nuclear reactor with Fermi, his 1947 letter to Stalin, and the 1960 Atoms for Peace Award. Incorrectly, it noted that his name is pronounced "ZilARD, not SILard."[49]

By that time, Szilard's body had been moved to the San Diego County Morgue, where Whisenand and a coroner's pathologist prepared for the autopsy. The task wasn't easy, Whisenand recalled, "with Jonas Salk and Ed Lennox breathing down our necks" and with the widow determined to have detailed information about their subject's bladder. They found "no gross evidence of any tumor or abnormality in the urinary tract," Whisenand later reported to James Whitmore at Memorial Hospital in New York.[50] At Trude's request, both the bladder and urinary tract were removed during the autopsy and sent to Memorial Hospital, where a detailed biopsy later confirmed that the cancer had been completely eradicated.[51]

But what they found in Szilard's heart confirmed that a life of indulgent eating and little exercise had, at last, proved fatal. "He had died so quickly" from a coronary thrombosis, Whisenand reported, that there had been no time for any secondary effects or changes in his heart. Serious coronary arteriosclerosis and scars from his 1957 heart attack were apparent. Salk later reassured Trude that her attempts to resuscitate Leo could not have succeeded. She had done all that was possible, but by then nothing could have saved him.

Wire-service dispatches later that day praised Szilard as a "tireless campaigner for peace" and the *New York Times,* in a page 1 obituary the next morning, called him "one of the great physicists of the century" and detailed the many achievements in his long and varied career. His demeanor, the paper reported, "was that of a volatile owl." In London, the more stately *Times* called Szilard "a leading American physicist who played an important part in the wartime development of nuclear energy. . . . He was not of Fermi's stature—few are—but he had qualities of quick imagination combined with persistance *[sic]* which gave him a position of importance during the whole of the early formative period of American work on nuclear energy. He has left his mark on history as well as physics."[52]

Rose Szilard Detre was just leaving Denver for Budapest when she learned about her brother's death and decided she could best represent the family by continuing her trip. She bore the sad news to her friends and relatives. Alice Danos, Szilard's high school sweetheart, told Rose she would miss Leo, although they had not spoken for nearly three decades. "Just knowing all those years that he was alive and up to his surprises" made her feel good, she said.[53] In Princeton, Eugene Wigner heard the news on his radio and sat down, tearfully, to pen a note to Trude, praising "the friendship of a lifetime" with "the most imaginative man I ever knew." Chemist John Polanyi recalled Szilard's "greatness and sweetness." Albert Rosenfeld, *Life's* science editor, considered him "one of the great and good human beings of our time."[54] And Edward Teller, Szilard's personal friend and professional critic, gave a characteristically ironic twist to his response. "I cannot but think of that legendary, restless figure, Dr. Faust, who in Goethe's tragedy dies at the very moment when at last he declares he is content."[55]

Szilard had told Trude he wanted no funeral service, but in her grief she needed to mark his loss somehow. Her sister, Frances Racker, and Bela flew in from New York and joined Trude for a private farewell on Monday morning, June 1. In a small, plain room at the La Jolla Mortuary, the three sat on a bench by a plain wooden coffin. Trude placed a single rose atop the long pine cover. There was no music. There were no tributes. Just a few quiet words whispered among the three. Then the coffin was taken to the Cypress View Mausoleum in San Diego, where, at 1:00 P.M., Szilard's remains were cremated.[56] Trude could not decide what to do with Leo's ashes and left them in a copper urn at the mausoleum.[57]

Jonas Salk and the other resident fellows also defied Szilard's wish and held a memorial service at the unfinished institute site, on Saturday, June 13, at 3:00 P.M. Participants clustered in rows of folding metal chairs in a surrealistic setting, among the concrete slabs and steel scaffolds of the new institute complex, overlooking the Pacific. A string quartet played Haydn. Ruth Adams, Szilard's longtime friend from the *Bulletin* and the Pugwash conferences, gave a warm and friendly eulogy. Salk puzzled over Szilard's creative spirit. "Some minds convert detail to principle quickly, while others move ponderously," he said, "defending all the way what earlier was believed to be true. Szilard wanted merely to know the facts, which he then soon assembled into new forms of thought. This capacity to perceive the essence of things and to see and formulate fundamental principles was the nature of his wisdom."

For Salk, the mental processes that animated Szilard also revealed a natural evolution to higher states of consciousness, an idea he pursued earnestly in later years. "There is much we need to learn about what it was that made him feel at peace and at one with himself here," Salk said. At the time the institute began, Salk saw Szilard as a pillar on which the "humanistic" approach to science would be built; biological research, he knew, would advance on its own, as it always had. So with Szilard's death, Salk feared—correctly, it turned out—that his institute's special mission would be hard to achieve. "Leo cared not to carry the torch but simply to light it; and when there were not others to carry, he did so himself. . . .

"Of the torches he lighted, many have been carried far beyond his own visions." Szilard's one driving motive, Salk concluded, was "toward relief of human suffering. His interest in peace and in problems of disease were all of one piece. As we have said, he was a humanist with a powerful intellect, but he was also a man with a warm heart."[58]

In his own eulogy, Lennox marveled at Szilard's heroic and ingenious recovery from cancer, wondered why he had not been as successful with the fatal heart attack, and could only conclude: "God never would have got Leo if he had been awake!" After the service, Szilard's friends chuckled again when they heard about his idea for scattering his ashes. "Put them into brightly colored balloons," Szilard had instructed Orgel, "and release them over the ocean. That way, at least it will delight all the children."[59]

Epilogue

"What would Leo think?" his many friends and colleagues ask today as they ponder the state of the world since his death. With a knowing and mysterious air, Leo Szilard was so farsighted and so certain about where this planet was headed that he gave us all a glimpse into the future, at least for a while. "I still miss his unusual intelligence and wit," Szilard's friend James D. Watson said more than two decades after his death. "He was irreplaceable in the truest sense."

Szilard's predictions in 1960 regarding the course of the nuclear arms race seemed right on schedule in 1988, for the United States and Soviet Union at last recognized that the stalemate they had reached with their arsenals left no alternative but to reduce weapons stockpiles drastically and to turn to domestic problems. Yet for all his fascination with the politics of Europe, Szilard nowhere in his writings foresaw the rapid and revolutionary sweep of democracy that followed the cold war's thaw.

In death, as in life, Szilard remained something of a mystery. His wife, Trude, lived on in La Jolla after his death, collecting and editing his papers from the jumble of bags and suitcases in which they were stored, producing the three volumes published by the MIT Press. When she died of cancer in 1981 (from the heavy smoking that Leo had long warned was unhealthy), she was buried at her mother's plot in Ithaca, New York. The tombstone there listed Trude as "the wife of Leo Szilard," with the years and places of birth and death for both. But while Trude's remains were actually in Ithaca, Leo's were not. They stayed in San Diego, in a setting he would probably detest. After his remains were cremated in 1964, the ashes were placed in an annex to the Cypress View Mortuary in San Diego, in a cigar-box-sized plastic container that was stacked on a rack also used to store corpses before their cremation. Trude never could decide what to do with the ashes, so they remained there, in what Cypress View called "permanent storage."

In 1986, Bela Silard decided that his brother's ashes should at least be on public view and arranged for them to be moved into the Cypress View Mausoleum, a garish marble extravaganza complete with doleful Muzak, inspirational religious paintings, and gauche art reproductions. The ashes were placed in a smaller, Styrofoam box and slipped into a marked niche in the main section of the sprawling and gaudy mausoleum, in the "Columbarium" in the "Court of Apostles," his resting place until 1998. With keen eyesight or a pair of binoculars you could just make out LEO SZILARD 1898–1964 on the four-by-eight-inch brass plate on the niche, a few inches below the fourteen-foot-high ceiling.

In his instructions to Trude, Szilard said there was no need to retain his ashes, as there is no difference between them and anyone else's "except perhaps slight variations in calcium content, which are of no interest." As "an innovation," he suggested that she may wish to spread the ashes using balloons because "it is more pleasing for people to look up rather than to look down," and in the Columbarium, at least, visitors could "look up," although not in the way Szilard had intended.[1]

As to the other earthly reminders of Szilard, before his death he did not receive the AEC's Enrico Fermi Award or any further compensation for his chain-reaction patents. Nor did he ever receive a Nobel Prize, as so many of his colleagues and collaborators did. He was nominated once for his work in physics, however, and was mentioned by others in their Nobel lectures both for biology and for peace. Szilard's determination in pressing a theory to explain human immune systems was praised by biologist Jacques Monod, who pursued the idea that won him and others the 1965 prize for physiology or medicine. And, in 1985, the Soviet doctor Evgeni Chazov, when accepting the Peace Prize for the International Physicians for the Prevention of Nuclear War, said that Szilard's science-fiction view of uranium had dramatized the stupidity of the nuclear arms race and that his social conscience had driven him to undo the peril his scientific discoveries had created.[2]

The American Physical Society established a Leo Szilard Award for science and public policy, and when Soviet physicist Andrei Sakharov received it in 1983, he said that Szilard's devotion to public service "sprang from his innate, acute feeling of personal responsibility for the fate of mankind on our planet, and for the possible consequences of science's great victories."[3] The American physicist John H. Gibbons, director of the congressional Office of Technology Assessment (OTA), said when he received the award in 1991 that "Szilard should be the Patron Saint of OTA!" because of his many efforts to "clarify what the real issues were" just as the "Voice of the Dolphins" program did. "OTA's mission is akin to Szilard's fervent hopes for better communication between science and society."

In the same spirit, Szilard would have been pleased with *Science* magazine's early assessment of the Salk Institute in 1972, when it praised an "Elite Pursuit of Biology with a Conscience"—just the qualities he had sought when devising the Bund half a century earlier.[4] On the other hand, Szilard would not have been so pleased with the way the institute's dual science-and-society approach has since fared. Szilard personified that fusion, and with his death the social aspects of science lost out to the practical. "Leo was the balancing influence between myself and others," Salk later reflected. "I insisted that our research be broad, while others wanted it to be very narrow, very focused."[5] After Szilard's death in 1964, Jacob Bronowski inherited the leadership role for the social sciences but did little with it, and after his death in 1974, the institute focused increasingly on applied research.[6] By the 1980s it was the center of a developing biotechnology industry in San Diego County, and in 1990 *Science* concluded that while Salk "wanted an ivory-tower sanctuary," he "got a high-powered research lab" instead.[7]

Salk acknowledged that Szilard was "more interested in getting something done than in who got credit for it,"[8] as he did by inspiring other scientists to unite research with real-world problems. One disciple who followed Szilard's example is Richard Garwin, a physicist who seizes on the policy implications of science and technology, as he did in the 1970s with the controversy over the effect that supersonic transport planes would have on the atmosphere and in the 1980s with his critical analysis of "Star Wars" defense systems. Another disciple, biochemist Matthew Meselson, not only worked actively in the Council for a Livable World but also applied research on chemical weapons in the 1980s to demonstrate that "yellow rain" falling in Southeast Asia was not chemical defoliant, as the Reagan administration had claimed, but bee feces. And Harrison Brown, an editor of the *Bulletin of the Atomic Scientists,* integrated biology and social science into his studies of the earth's "carrying capacity" decades before global environmental issues became prominent.[9]

In applying science himself, both by research and in creative fiction, Szilard was ahead of his time when he advocated and advanced fertility and birth-control studies and when he dealt with the medical opportunities and psychological costs of preserving disease victims cryogenically. He experienced the terrors that lead many thoughtful people to choose euthanasia over a long and painful death, and as this issue receives increased attention today, he would no doubt be trying to market the kind of "suicide kit" he first devised in 1960.

Szilard's faith in the peaceful benefits of atomic energy has certainly been rewarded in the development of medical technology, although his hope that nuclear power would help developing countries to prosper has proven impractical. Overstated, too, was Szilard's faith in his breeder reactor, which has proven to be a dangerous and costly electricity producer in every country that has tried to build one. Szilard never anticipated the need to deal with nuclear waste and would surely be miffed by the public debate that today surrounds this issue, although, no doubt, he would also have a clever solution, or two or three.

For a scientist of his stature, there is still little public recognition for Szilard's life: No professorship or university chair or institute bears his name. But as he wished in a moment of whimsy, Szilard is memorialized in at least one place along with Copernicus, Galileo, Newton, and Einstein: There is a crater named for him on the moon.

Szilard's legacy is best captured in his mode of thinking, a playful but persistent attack on the world and all the woes and wonders in it. With that feisty spirit in mind, it is still useful to do as his friends do: to pause, to smile, and to ask: "What would Leo think?"

Chronology of
Leo Szilard's Life

1898 February 11	Born Leo Spitz in Budapest, 28 (now 50) Bajza Utca.
1900 October 4	Family changed its name to Szilard and later moved to 33 Varosligeti Fasor.
1908 September	Attended technical eight-year high school in Budapest (to June 1916).
1916 September	Entered Budapest Technical University to study engineering.
1917 fall	Was drafted into the Austro-Hungarian army.
1918 fall	Returned to civilian life and engineering studies.
1919 spring	Began political activities during the Béla Kun government's four-month Communist rule in Hungary, organizing with his brother the Hungarian Association of Socialist Students.
1919 December	Escaped from Hungary to study engineering at the Technical Institute (Technische Hochschule) in Berlin.
1920 fall	Transferred to the physics department of the University of Berlin, where he met physicists Max von Laue, James Franck, and Albert Einstein.
1922 August	Received a Ph.D. in physics under von Laue and began teaching as his first assistant in 1925.
1922 fall	Applied the principle of entropy to information, the basis of modern "information theory."

1927 December	Filed the first of more than thirty patents with Einstein for an electromagnetic pump, which became the basis of cooling systems in "breeder" nuclear reactors in the 1950s and 1960s.
1929 January 5	Patented the concept of the cyclotron.
1929 fall	Met Gertrud (Trude) Weiss, his future wife.
1931 December 25	First visited New York, working on research in theoretical physics at New York University.
1933 March 30	Left Germany for Vienna and then London, where he worked to settle academic refugees.
1933 September/ October	Developed the idea of a nuclear "chain reaction" and the concept of a "critical mass" to create it.
1934 March	First patented the chain-reaction concept.
1934 summer	Conducted atomic research at St. Bartholomew's Hospital, London; invented the Szilard-Chalmers effect for isotope separation.
1935 winter/spring	Visited New York, then began research at Oxford.
1935 spring	Began extensive efforts to encourage scientists to self-censor research to keep atomic developments secret from Germany, his first attempts at nuclear arms control.
1938 January 2	Landed in New York; conducted research at the University of Illinois, Rochester, and Columbia.
1939 spring	Worked on atomic research at Columbia with Enrico Fermi, Walter Zinn, and others.
1939 summer	Collaborated with Fermi to design the first nuclear reactor; urged censorship of atomic developments by US, French, and British scientists.
1939 July	Told Einstein about chain reactions and the weapons potential of his equation $E = mc^2$. Proposed and drafted a letter for Einstein's signature to Pres. Franklin D. Roosevelt warning of atomic weapons—an alert that led to creation of the Manhattan Project to develop A-bombs.
1942 February 1	Moved to Chicago with other Columbia scientists, becoming chief physicist of the Manhattan Project's Metallurgical Laboratory at the University of Chicago.
1942 December 2	With Fermi put into operation the world's first chain-reaction atomic "pile" (reactor) of their design.

1943 January	Prepared a memo on the first of three designs for a "breeder" reactor (a name he coined) to create plutonium for fuel and A-bombs.
1943 March 29	Became a US citizen.
1944 August 10	Proposed postwar arrangements for national and international control of atomic energy (to curb what he predicted would be a US-Soviet arms race) almost one year before the first A-bomb was tested.
1945 March 15	With another Einstein letter sought an appointment with President Roosevelt to present scientists' views about wartime and postwar use of A-bombs. FDR died before their meeting.
1945 May–June	Helped write the Franck Committee report urging a demonstration of A-bombs before use on Japanese cities.
1945 July 1	Organized a scientists' petition against dropping A-bombs on Japan, an effort hampered by his superiors (first bomb tested July 16; bombs dropped on Japan on August 6 and 9).
1945 October	Led scientists' protest against the May-Johnson bill, which would keep atomic energy under military control, and testified before Congress (October 18 and December 10) in favor of the successful McMahon bill, which would create a civilian Atomic Energy Commission.
1945 fall	Helped organize the Federation of American Scientists, a group active in arms-control issues, and the *Bulletin of the Atomic Scientists,* a public forum for scientific and arms-control issues.
1945 December	Proposed the first of several initiatives to convene US and Soviet scientists for arms-control discussions.
1947 spring	Began an extensive public-speaking tour, mostly at universities, proposing political solutions to a US-Soviet atomic arms race.
1948 summer	Invented, with Aaron Novick, the chemostat to test bacteria in a steady state. In satirical writing and by informal organizations opposed anti-Communist pressures on US campuses.
1951 October 13	Married Gertrud (Trude) Weiss in New York City.

1955 May 18	Received, together with Enrico Fermi, a joint US patent on the nuclear reactor.
1957 January	Proposed that Dr. Jonas Salk, developer of a polio vaccine, found and lead a new research center combining science and social issues which later became the Salk Institute for Biological Studies in La Jolla, California.
1957 spring	Helped plan the First Pugwash Conference, involving US and Soviet scientists and policymakers in discussions on peace and disarmament (held in July at Pugwash, Nova Scotia).
1957 December	Suffered a mild heart attack when in Paris.
1959 fall	Diagnosed as having bladder cancer. Hospitalized in New York City. Directed his own radiation treatment. During the next year wrote and edited science fiction and political satires that were published in 1961 as *The Voice of the Dolphins*.
1960 May 18	Received the US Atoms for Peace Award.
1960 fall	Met with Soviet Premier Nikita Khrushchev in New York, gaining his assent to a Moscow-Washington hot line.
1961 February	Moved to Washington to promote arms control.
1962 June	Organized the first political action committee for arms-control and disarmament issues, the Council for a Livable World.
1962 October	Fled to Geneva during the Cuban Missile Crisis and from there tried to reach Khrushchev to further US-Soviet dialogues.
1963 December	Wrote on "minimal deterrence" as a concept to guide arms-control negotiations.
1964 February	Moved to La Jolla to work at the Salk Institute, becoming a resident fellow on April 1.
1964 spring	Began a concerted research program and wrote on "Memory and Recall."
1964 May 30	Died of a heart attack in his sleep in La Jolla.

Acknowledgments

Researching and writing this biography has been an eight-year effort to capture Leo Szilard's elusive spirit. In this quest many people have helped me to retrace Szilard's life and works in Europe and North America. For their assistance and encouragement I am deeply thankful. Among those whose aid I gratefully acknowledge are the following:

Lynda C. Claassen, Lillian Gutierrez, Geoffrey Wexler, Kim Palmer, Steve Coy, Jackie Dooley, Renee Robinson, and Evelyn Sander at the Leo Szilard Papers, Mandeville Special Collections Department, University of California, San Diego; Sylvia Bailey and June Gittings at the Salk Institute for Biological Studies; Marjorie Cirolante and Eddie Reese at the National Archives; Cooper Graham at the Library of Congress; Joshua Lederberg and Ann Quatela at Rockefeller University; Darwin H. Stapleton, Madeleine Tierney, and Lee Hiltzik at the Rockefeller Archive Center; Paul Marks and Evelyn Saulpaugh at the Memorial Sloan-Kettering Cancer Center; Melanie Marhefka, James Yntema, and Daniel Meyer at the Joseph Regenstein (University of Chicago) Library, Special Collections Division; Jeri Nunn, John Verso, and Ronald Grele at the Columbia University Oral History Research Office; Louis Brown at the Department of Terrestrial Magnetism, Carnegie Institution of Washington; Judy Goodstein at the California Institute of Technology Library; Nancy Bressler at the Albert Einstein Duplicate Archive in Princeton University; John Stachel, David Cassidy, Robert Schulmann, and Ann Lehar at the Albert Einstein Papers Archive at Boston University; Helen W. Samuels and Kathy Marquis at the MIT Manuscript Archive; Dale Mayer, Mildred Mather, and Shirley Sondergard at the Herbert Hoover Presidential Library; Barbara Anderson, Megan Desnoyers, and Michael Desmond at the John Fitzgerald Kennedy Library; Susan Elter at the Franklin D. Roosevelt Library; Paul Forman, Roger Sherman, and Andrew Szanton at the Smithsonian Institution; Glenn Stout and Wendy Marcus at the Boston Public Library; Jerome

501

Grossman, Rosalie Anders, Julie Cohen, and Michael Litz at the Council for a Livable World; Larry Arbiter at the University of Chicago, News and Information Bureau; Len Ackland, Lisa Grayson, and Ruth Grodzins at the *Bulletin of the Atomic Scientists;* Spencer Weart, Jean Hrichus, Julie Martin, Elisabeth Elkind, Douglas Egan, and Ann Kottner at the American Institute of Physics; Jennifer Belton and Cathy Wall at the *Washington Post* library; Ted Slate at the *Newsweek* library; Mary Ellen Adamo at the *Forbes* magazine collection; and Nati Krivatsy at the Folger Shakespeare Library.

In England, Colin Harris at the Bodleian Library, Oxford; Michael Brock and David Smith at Nuffield College Library, Oxford; Ramila Chauhan, Fiona MacColl, and Jill Breen at the British Library of Political and Economic Science, London School of Economics; Sally Grover and Nina Cohen at the Library of the Royal Society; Natalia Macher and Edith Salt at the London office of the Pugwash Conferences on Science and World Affairs; and Liz Fraser at the Society for the Protection of Science and Learning, (formerly AAC). My special thanks to Esther Simpson, Szilard's friend and AAC colleague.

In Budapest, Istvan Lang, Annamaria Furst, George Litván, and Attila Pók at the Hungarian Academy of Sciences; Laszlo Nagy, Emil Horn, and Robert Szabo at the Museum of Contemporary History; Istvan Kovacs and Gábor Pallo at the Technical University; Ferenc Szabadváry at the Museum for Science and Technology; Eva Litván at the Kiscelli Museum; Tibor Sandor at the Ervin Szabó Municipal Library; and Eva Kaszas at the Lukács Archive.

In Geneva, Roswitha Rahmy at the Archives of the European Center for Nuclear Research (CERN) and Yves Felt, John Krige, and Dominique Pestre at the Study Team for CERN History.

Enid C. B. Schoettle, Joyce Nixon, Kathryn Mitchell, and Laurice H. Sarraf at the Ford Foundation; Ruth Adams, Rachel Williams, and George Hogenson at the John D. and Catherine T. MacArthur Foundation; Arthur L. Singer, Jr., at the Alfred P. Sloan Foundation; and Clifton Mitchell at the Brookings Institution.

Sharon Mylask, Tom Power, and Frank Tauss at Idea Tech; Eugene Racanelli and Tina Oliver at Kinko's Capitol Hill.

Egon and Renée Weiss and Efraim and Frances Racker for their encouragement and their sharing of numerous items from the Gertrud Weiss Szilard estate; Gar Alperovitz, Brian Balogh, Anna Bettelheim, John F. Bresette, John C. Culver, Warren Donnelly, Robert Doyle, Elie Feuerwerker, Floyd Galler, John H. Gibbons, Vitalii Goldanskii, Margaret Gowing, Gail Griffith, Paul Hendrickson, Gregg Herken, James P. Hume, Marvin Kalb, William Kincade, Richard Leghorn, Howard J. Lewis, Patricia Lindop, William Martin, Henry Myers, Ayub Ommaya, Barbara and Owen O'Neill, Humphry Osmond, Nancy Palmer, Bill Perkins, Paul Ress, Jean Richards, William Scott, Martin Sherwin, David Shoaf, John and Janet Silard, Thomas Simons, Ralph G. H. Siu, Alice K. and Cyril Smith, Margaret Spanel, Istvan Szemenyei, Joseph Tarantolo, Valentine and Lia Telegdi, Kosta Tsipis,

Francis Wagner, Horst and Maria Luise Wagner, George Weil, David Wiener, and Ruth Wuest for their helpful comments and suggestions during my research.

Josef Ernst and Gábor Pallo for determined research and translation. Priscilla Johnson McMillan and Stanley Goldberg for their critical reviews of the manuscript; Dan Grossman, Carol Gruber, James Hershberg, Ira Kaminow, Ralph W. Moss, Sharyl Patton, Paul Pavlovich, Marc Trachtenberg, Frank von Hippel, and Helen Weiss for their suggestions of information and sources; Barton Bernstein, Albert and Susan Cantril, and Robert E. Hunter for their helpful brainstorming; and Pnina Abir-Am, Charles Fenyvesi, Maurice Fox, G. Allen Greb, Howard Green, Helen Hawkins, Karl Maramorosch, Philip Marcus, Aaron Novick, John Platt, Thomas Powers, Theodore Puck, Frances Racker, and John Yakaitis for reviewing and correcting sections of the manuscript.

Claudia Andrews, David Austin, Brian Daley, William and Helen Hawkins, Paul Keegan, Hinrich and Ursula Lehmann-Grube, Thomas and Irene Litz, Robert and Joany Mosher, Michael and Claude Sullivan, Tony and Mary Alice Wolf, and my brothers John, Joseph, Peter, and Robert and their families for their hospitality during my research travels. And my Kennedy School, Wilson Center, and GAO collegues for enduring all my Leo Szilard stories.

The Ford Foundation, the John D. and Catherine T. MacArthur Foundation, the Alfred P. Sloan Foundation, Rockefeller University, the National Endowment for the Arts, the National Endowment for the Humanities, the Council for a Livable World Education Fund, and the Hoover Presidential Library Association for financial support during my research and writing.

I am grateful to Gene Dannen for providing copies of his correspondence with C. H. Collie and for helpful endnote corrections; to Alice K. Smith for providing copies of her interviews with Manhattan Project scientists; and to James Hume for providing copies of his files on Szilard's patent applications and negotiations with the Manhattan Project; and to Thelonious Sphere Monk, whose music—so sprightly and so loose—created a perfect atmosphere for editing.

I am very grateful to my agent, F. Joseph Spieler; and to my editor, Robert Stewart, as well as Roberta Corcoran, Carol Cook, Theresa Czajkowska, and Mark LaFlaur at Scribners, for their helpful criticisms and steady encouragement.

I am especially grateful to Bela Silard for his patient and diligent work at recollecting and reconstructing so many telling details of his brother's life and to his wife, Elizabeth, for her amusing perspectives on the Szilard family. Bela, in turn, is thankful to Ruth Henry and Eileen Reilly for their assistance.

I am delighted that Frederick Reuss and Robin Moody led me to Skyhorse Publishing, and at Skyhorse I am most grateful to Jay Cassell and Kelsie Besaw for helping me add so much to this new edition.

And I am deeply grateful to my wife, JoAnne, and my daughters, Nicole and Kathryn, for making Leo Szilard's life—and my struggle with it—a part of their own.

Although aided and sustained by all these generous people, I alone am responsible for this book's errors and omissions.

Notes

The richest source of documentary information about Leo Szilard is his collected personal papers (LSP), which were assembled in La Jolla, California, after his death by his widow, Gertrud (Trude) Weiss Szilard, and donated to the University of California at San Diego by her brother and heir, Egon Weiss. Where possible, my citations include the box and folder numbers as (box #/folder #).

The second-richest source is the memory of Bela Silard, Leo Szilard's brother. In countless personal and telephone interviews, Bela described, corrected, and revised anecdotes and facts about his family and his life with Leo. All of Bela Silard's written contributions are cited as the Bela Silard Papers (BSP).

In addition, Egon Weiss inherited several documents found in Trude's apartment at the time of her death in 1981, including health and financial records for herself and Leo. For my purposes, the most important documents from this collection are the more than three hundred letters from Leo to Trude, which I had translated from the German for use in this volume. Incidental letters between Trude and Leo's mother are also in this collection, which I designate as part of the Egon Weiss Papers (EWP).

All books and articles listed in the Selected Bibliography are cited only by author in the Notes, together with appropriate page numbers.

INTERVIEWS

The people I have interviewed for this book are listed below. Where an interview was the source of specific information, the date of the interview appears in the endnote.

Pnina G. Abir-Am, James L. Adams, Ruth Adams, Harold Agnew, Edoardo Amaldi, Herbert L. Anderson, James Arnold, Pierre Auger, David Baltimore, Sidney W. Barnes, Etienne Bauer, Anna Beck, Baruj Benacerraf, Charles H. Bennett, Barton J. Bernstein, Hans Bethe, Anna F. Bettelheim, Lazislas Bihaly,

Elias Blatter, Walter Blum, Sydney Brenner, John F. Bresette, Rita Bronowski, Harvey Brooks, Harrison Brown, McGeorge Bundy, Lydia Cassin, Melvin Cohn, Arthur Cooke, Norman Cousins, Edward C. Creutz, Francis Crick, Alice Eppinger Danos, Bernard D. Davis, Kingsley Davis, Jean-François Delassus, Manny Delbrück, Roland Detre, William Doering, Paul Doty, Renato Dulbecco, Freeman Dyson, John T. Edsall, Walter M. Elsasser, Ugo Fano, Bernard T. Feld, Inge Feltrinelli, Richard P. Feynman, Roger Fisher, Allan Forbes, Maurice Fox, Lawrence Freedman, Milton Friedman, David H. Frisch, Richard Garwin, Vitalii I. Goldanskii, Stanley Goldberg, Marvin Goldberger, Mildred Goldberger, Maurice Goldhaber, Bertrand Goldschmidt, Margaret Gowing, Gerda Gray, Howard Green, Clifford Grobstein, Ruth Grodzins, Carol Gruber, Jules Gueron, David H. Gurinsky, Morton Hamermesh, Charles Hartshorne, Gertrude S. Hausmann, David Hawkins, Frances Hawkins, Helen Hawkins, Mariana Heller, William A. Higinbotham, Dorothy C. Hodgkin, Rollin D. Hotchkiss, Patrick Hogan, James P. Hume, David R. Inglis, François Jacob, Gerald Johnson, Herman M. Kalckar, Martin M. Kaplan, Carl Kaysen, Esther Scheiber Kelerman, William Kincade, Albert Kornfeld (Albi Korodi), Nicholas Kurti, Ralph Landauer, Ralph E. Lapp, Peter Lax, Joshua Lederberg, Richard Leghorn, Leonard Lerman, Max Lerner, Edward Levi, Milton Levinson, Elsbeth Liebowitz, John C. Lilly, Patricia Lindop, Herman Lisco, Eva Litván, George Litván, Robert Livingston, Franklin Long, Antonio deLozado, Salvador Luria, Oloe Maalóe, Renata Maas, Werner K. Maas, Norman Macrae, Norene Mann, Philip I. Marcus, Hermann Mark, M. A. Markov, Paul Marks, John Marshall, Jr., Samuel Marx, George McGovern, James Warren McKie, Richard L. Meier, Horst Melcher, Matthew Meselson, Philip Morrison, Phylis Morrison, Aaron Novick, Harold Oram, Leslie E. Orgel, Harold Orlans, Humphry Osmond, Harry Palevsky, Gábor Pallo, Rudolf Peierls, Max Perutz, Gerard Piel, Maria Piers, Ruth Pinkson, John R. Platt, Attila Pók, Ralph Pomerance, Theodore T. Puck, Isidor I. Rabi, Efraim Racker, Frances Racker, Marcus G. Raskin, George Rathjens, Joseph L. Rauh, Jr., Roger Revelle, Alexander Rich, Jennifer L. Robbins, Walter Orr Roberts, Maurice Rosenblatt, Joseph Rotblat, David Rudolph, Jack Ruina, Abram L. Sachar, Robert Sachs, Jonas Salk, Rose Scheiber, Thomas C. Schelling, Lisbeth Bamberger Schorr, John Schrecker, Frederick Seitz, Martin J. Sherwin, William A. Shurcliff, Elizabeth Silard, Janet Silard, John Silard, Lady Charlotte Simon, Esther Simpson, John A. Simpson, Daniel M. Singer, Alice K. Smith, Cyril Smith, Margaret R. Spanel, Michael Straight, Roger Stuewer, Ferenc Szabadváry, Istvan Szemenyei, Julius Tabin, Theodore B. Taylor, Lia Telegdi, Valentine Telegdi, Alfred Tissieres, Kosta Tsipis, Frieda Urey, Frank von Hippel, Spencer R. Weart, Conant Webb, George L. Weil, Irwin Weil, Alvin M. Weinberg, Nella Fermi Weiner, Egon A. Weiss, Renée Weiss, Victor Weisskopf, John A. Wheeler, Rolf Wideröe, Jerome B. Wiesner, Eugene P. Wigner, Robert R. Wilson, Leona Steiner Wolf, Naomi Liebowitz Wood, Ramsay Wood, Christopher Wright, Herbert F. York, Stan Zahler, Eva Zeisel, Hans Zeisel, Walter Zinn.

ABBREVIATIONS

AAC Academic Assistance Council Records, reorganized as the Society for the Protection of Science and Learning (SPSL). The society's papers are in the Bodleian Library, Oxford. Now the Council for Assisting Refugee Academics.

AEP Albert Einstein Papers. About forty-three thousand documents in the Einstein Archive are now housed at the Hebrew University in Jerusalem. Duplicate archives are maintained at the Seeley G. Mudd Manuscript Library, Princeton University, Princeton, New Jersey (AEP/P) and at the Albert Einstein Papers project at Boston University, Boston, Massachusetts (AEP/B). Most Einstein documents are identified by an index number, which I cite with the date whenever possible.

AIP Niels Bohr Library, American Institute of Physics, College Park, Maryland.

AKS Alice K. Smith. Her book about the postwar scientists' movement to control nuclear weapons, A *Peril and a Hope,* appeared in editions by the University of Chicago Press (AKS/UC) and the Massachusetts Institute of Technology Press (AKS/MIT). All other AKS citations are to papers in her possession in Cambridge, Massachusetts.

BAS *Bulletin of the Atomic Scientists* Files, JRL.

BFP Bernard T. Feld Papers, MIA.

BSP Bela Silard Papers, in Lanouette/Szilard Papers, MSS 659, LSP.

CIT California Institute of Technology Archives, Pasadena, California.

CRN CERN (European Center for Nuclear Research) Archive, Geneva.

DTM Archives of the Department of Terrestrial Magnetism, Carnegie Institution of Washington. Now the Carnegie Institution for Science.

EFP Enrico Fermi Papers, JRL.

ERP Eugene Rabinowitch Papers, JRL.

EWP Egon Weiss Papers, Carlsbad, California.

FDR Franklin D. Roosevelt Library, Hyde Park, New York.

FLP Frederick Alexander Lindemann (Cherwell) Papers, Nuffield College, Oxford.

FSP Francis Simon Papers, Library of the Royal Society, London.

GSP Gertrud Weiss Szilard Papers, identified as the "Gertrud Szilard Materials" in boxes 84 to 92 and 107 of the LSP.

GSS Gertrud Weiss Szilard Scrapbook. Five numbered scrapbooks containing newspaper and magazine clips about Leo Szilard and political developments surrounding the use and control of nuclear weapons. These are in the Leo Szilard Papers (LSP).

JFK John Fitzgerald Kennedy Library, Boston, Massachusetts.

JGC Jerome Grossman Collection, Boston Public Library.

JRL Joseph Regenstein Library, University of Chicago.

JSP Jonas Salk Papers, MSS 1, Mandeville Department of Special Collections, Central University Library, University of California, San Diego.

LCM Manuscript Division, Library of Congress.

LLS Lewis L. Strauss Papers, Herbert Hoover Presidential Library, West Branch, Iowa.

LSP Leo Szilard Papers, MSS 32, Mandeville Department of Special Collections, Central University Library, University of California, San Diego. Number citations are to the box and folder. (E.g., LSP 2/3 is to the Leo Szilard Papers, box 2, folder 3.) The Szilard Papers have been reorganized twice in the course of my research, each time with no references kept to the documents' previous locations. Where I was able to trace the documents to their current (and, I hope, final) locations, I have cited box and folder numbers. For all other documents, I have simply cited the Leo Szilard Papers (LSP).

MED Manhattan Engineer District Records, National Archives.

MIA Institute Archives, Massachusetts Institute of Technology, Cambridge, Massachusetts.

MIT *The Collected Works of Leo Szilard* is published in three volumes. For brief citation I have listed Bernard T. Feld and Gertrud Weiss Szilard, eds, *The Collected Works of Leo Szilard: Scientific Papers* (Cambridge, Mass.: MIT Press, 1972) as MIT Vol. I; Spencer Weart and Gertrud Weiss Szilard, eds., *Leo Szilard: His Version of the Facts* (Cambridge, Mass.: MIT Press, 1978) as MIT Vol. II, and Helen Hawkins, G. Allen Greb, and Gertrud Weiss Szilard, eds., *Toward a Livable World* (Cambridge, Mass.: MIT Press, 1987) as MIT Vol. III.

MPP Michael Polanyi Papers, JRL.

OSR Office of Scientific Research and Development (OSRD) Records, National Archives.

PCF Files of the Pugwash Conferences on Science and World Affairs, London Office.

RAC Rockefeller Archive Center, Pocantico Hills, North Tarrytown, New York.

RMH Robert Maynard Hutchins Papers, JRL.

S-1 S-1(uranium bomb project), National Archives.

SIA Salk Institute Archive, the Salk Institute for Biological Studies, La Jolla, California.

SMR Leo Szilard Medical Records, Memorial Sloan-Kettering Cancer Center, New York (copies obtained by permission of Egon Weiss), in LSP.

VNP John von Neumann Papers, LCM.

WBP William Beveridge Papers. British Library of Political and Economic Science, London School of Economics and Political Science.

CHAPTER 1

1. Most episodes in this chapter were first drafted or dictated by Bela Silard, who also provided anecdotes and details in many interviews and letters.

Two other essential sources are the manuscripts "Recollections" by Louis Szilard, 1953, translated by Bela Silard (LSP 2/6); and "Memoirs" by Tekla Szilard, 1939, translated by Anna Bettelheim (BSP).

Spitz/Szilard family documents are in 1 and 106 LSP. Other details are from John Lukacs, *Budapest 1900* (New York: Weidenfeld & Nicolson, 1988). For a photograph of the Fasor near the Szilard home and for other views of the Garden District, see the last few pages of *Budapest Anno* (Budapest: Corvina, 1984).

2. Realizing that Leo loved his felt hat and all it stood for, his parents bought him a new one. According to Bela, the replacement was black, which displeased Leo, but had an even wider brim that seemed to make up for the color change.

3. MIT Vol. II, p. 3. Szilard dictated this episode in May 1960.

4. "Recollections" (LSP 2/6).

5. By a second link between the Spitz and Klopstock families, the youngest Klopstock brother, Adolf, married the second-youngest Spitz sister, Gizella.

6. "Memoirs," p. 185 (BSP).

7. Ibid., p. 186.

8. "Recollections," p. 143 (LSP 2/6).

9. "Memoirs," p. 188 (BSP).

10. McCagg, p. 186.

11. "Recollections," pp. 145–46 (LSP 2/6).

12. Ibid., p. 146.

13. "Memoirs," p. 190 (BSP).

14. "Leo Szilard—Biographical Notes," p. 2. From a February 21, 1960, interview by Robert Livingston (LSP 2/9).

15. Notes by Bela Silard in "Recollections" (LSP 2/6).

16. "Recollections," pp. 154–56 (LSP 2/6).

CHAPTER 2

1. Most episodes in this chapter were first drafted or dictated by Bela Silard, who also provided anecdotes and details in many interviews and letters.

Two other essential sources are the manuscripts "Recollections" by Louis Szilard, 1953, translated by Bela Silard (LSP 2/6); and "Memoirs" by Tekla Szilard, 1939, translated by Anna Bettelheim (BSP).

2. For more details on the Vidor Villa see *"Vidor Zsigmond dr. ur fasori villája," Magyar Pályázatok IV*, 1906 (No. 2, pp. 25–31); and Dr. Peter Buza, "Lustjo—Vidor," *Budapest a fovaros folyoirata*, Vol. 20 (March 1982, pp 18–20).

3. Louis Kossuth became finance minister of Hungary's independent, anti-Austrian government after the March 1948 revolution, and when the Austrians counterattacked, he led a government of national defense from

Debrecen. In 1849, Hungary's Parliament declared the country an independent republic, with Kossuth as its first president.

4 "Leo Szilard—Biographical Notes" by Robert Livingston, February 21, 1960 (LSP 2/9).

5. Louis Szilard was a founder of the Petrofi Lodge. See Joseph Palatinus, A *szabadkomuvesseg bunei* (Budapest, 1939). My thanks to George Litván for this information. George Litvan to the author, August, 18, 1987.

6. MIT Vol. II, p. 3.

7. *New York Post,* November 24, 1945.

8. R. B. Turnai to Szilard, October 4, 1953 (LSP 18/31).

9. Rose Scheiber interview, July 16, 1987.

10. "Memoirs," p. 193.

11. Ibid., pp. 193–94.

CHAPTER 3

1. Most episodes in this chapter were first drafted or dictated by Bela Silard, who also provided anecdotes and details in many interviews and letters.

   Two other essential sources are the manuscripts "Recollections" by Louis Szilard, 1953, translated by Bela Silard (LSP 2/6); and "Memoirs" by Tekla Szilard, 1939, translated by Anna Bettelheim (BSP).

   Szilard's electricity text was by Gyözö Zemplén, once a professor at Budapest's Technical University.

2. A Leclanché battery had one carbon and one zinc electrode suspended in a jar of acid.

3. MIT Vol. II, p. 4; *Ertesitoje,* Budapest Hatodik Keruleti Allami Forealiskola 19-ik evi (BSP).

4. MIT Vol. II, p. 4.

5. The only adolescent friendship that survived after Szilard left Hungary was with Albert Kornfeld (Albi Korodi). The two first met in Budapest when they received science prizes in 1916, and later studied together in Berlin.

6. Alice Eppinger Danos interview, July 16, 1987.

7. The Eppinger family lived at 28 Fasor.

8. Alice Danos interview, June 10, 1986.

9. Emil Freund owned the Köbánya Brewery, in the Budapest suburb of that name southeast of the Garden District.

10. "Recollections," p. 175 (LSP 2/6).

11. MIT Vol. II, p. 5.

12. Ibid., pp. 5–6. See also Szilard's October 30, 1961, draft dictation, p. 10, for "Are We on the Road to War?" (LSP).

13. Documents relating to Szilard's military service (LSP 1/9 and 1/10).

14. Report Card and other academic records (LSP 1/9).

15. MIT Vol. II, p. 5. The "fission," or splitting, of the uranium atom became known to the scientific community in January 1939 and, in large part because of Szilard's inventiveness, led to the successful creation of a nuclear chain reaction by December 1942.
16. Copy dated January 3, 1920 (LSP 1/12).
17. MIT Vol. II, p. 5.
18. Information on Szilard's engineering studies comes from his transcripts (LSP 1/2 and 1/11). See also Szabadváry, pp. 187–90.
19. Before the war, high school graduates could start officers' training directly after graduation and serve for one year. In wartime only the title survived; Szilard's tour of duty was unlimited.
20. Leo Szilard Biographical Table (LSP 2/9).
21. MIT Vol. II, p. 6.
22. Alice Danos interview, June 10, 1986.
23. The camp was located at the entrance to the Ziller Valley, between the towns of Wörgl and Jenbach, a few miles east of Innsbruck.
24. MIT Vol. II, p. 6.
25. Undated postcard, translated by Roland Detre, Rose Szilard's husband.
26. MIT Vol. II, p. 7.
27. Ibid., pp. 6–7. See also September 26, 1918, "Kopfzettel" (LSP 1/10).
28. MIT Vol. II, p. 8.
29. Szilard ended his military career as a titular ordnance cadet in Reserve Unit No. 4, a rank conferred on October 1, 1918 (LSP 1/10).
30. November 27, 1928, Certificate, translated by Alexander Somlyo (LSP 1/18).
31. Szilard's 1960 dictated "Memoirs" (LSP 40/10).
32. "Memoirs," p. 199 (BSP).
33. Attila Pók, "Jászi as Organizational Leader of a Reform Movement," p. 9. My thanks to the author for sharing this manuscript.
34. MIT Vol. II, p. 14. Baptismal Register, Reform Church, District VI–VII Budapest (LSP 1/11).
35. "Memoirs," pp. 200–201 (BSP).
36. PIA 638 fond-l/1920-IV-13-2096, Institute of the History of the Party. Attila Pók to the author, August 8, 1986. A detective completed the report on February 15, 1920, absolving the Szilard brothers from further suspicion.
37. "One should never undertake anything at all," by George Klein. Klein to the author, October 28, 1986. Translation from the Swedish by Margareta Feller, pp. 3–4.

CHAPTER 4

1. MIT Vol. II, p. 8.
2. Der Polizeipräsident in Berlin, September 28, 1931 (LSP 1/21).

3. MIT Vol. II, p. 8.

4. Berliner Adressbuch 1920, Fiche 7, p. 652, col. 3, Frankel. Lady Charlotte Simon interview, May 27, 1986.

5. Polizeiliche Abmeldung, February 1, 1920. Each time Szilard moved, he dutifully reported old and new addresses to the police and tucked the receipts in his suitcase (LSP 1/24).

6. PIA 638 fond-1/1920-IV-13-2096, Institute of the History of the Party. Attila Pók to the author, August 8, 1986.

7. James Franck, who reported this anecdote to Michael Polanyi, thought well of Szilard. The incident is reported in Polanyi to Franck, May 18, 1961 (MPP). See also Polanyi's oral-history interview, February 15, 1962, p. 8 (AIP).

8. *"Abgangszeugnis"* September 28, 1926 (LSP 1/12).

9. Eugene Wigner interview, October 12, 1984.

10. Wigner/Szanton "Recollections" ms., pp. 132ff. Andre Gabor to Bela Silard, March 13, 1984 (BSP).

11. MIT Vol. II, p. 9.

12. Ibid.

13. MIT Vol. II, pp. 10–11.

14. Ehrenberg, "Maxwell's Demon." pp. 103–10.

15. James Clerk Maxwell, *Theory of Heat* (London: Longmans, Grun and Co., 1871) p. 328.

16. Carl Eckart, MIT Vol. I, p. 31.

17. N. Katherine Hayles, "The Information Perspective in Literature and Science: Chaos or Order?" September 23, 1985, Woodrow Wilson International Center for Scholars.

18. "Leo Szilard Biographical Notes." RG 6876, p. 1 (RAC). See also LSP 2/9.

19. Published as *"Uber die Ausdehnung der phänomenologischen Thermodynamik auf die Schwankungserscheinungen"* in *Zeitschrift für Physik,* 32, pp. 753–88 (1925). Translated in MIT Vol. I, pp. 70–102. My thanks to Gene Dannon for this information.

20. Szilard also carried this idea further than the mathematicians of the 1920s by showing that under certain conditions the only probability distribution with a *single scalar sufficient statistic* is the distribution devised in the nineteenth century by Yale mathematician Josiah Willard Gibbs. And, in creating his own application of sufficiency, Szilard anticipated by decades the simultaneous and independent papers of G. Darmois, B. O. Keepman, and E. J. G. Pitman. My thanks to Paul Penniman for research on this point.

21. Maurice Fox interview, May 24, 1988.

22. Published in 1929 as *"Uber die Entropieverminderung in einem thermodynamischen System bei Eingriffen intelligenter Wesen,"* in *Zeitschrift für Physik,*

53, pp. 840–56 (1929). MIT Vol. I. German text pp. 103–19; English, pp. 120–29.

According to Carl Eckart, Szilard said he wrote this paper about six months after finishing his thesis. See MIT Vol. I, p. 32.

23. MIT Vol. I, p. 121.
24. Ehrenberg, p. 109. For a different translation, see MIT Vol. I, p. 125.
25. "Leo Szilard's Influence on Physics," Essay dated June 13, 1964 (JSP 400/3).
26. Translation from the von Laue and Planck review of Szilard's *Habilitations-schrift* by Maria Luise Wagner. Original in *Gutachten zum Habilitationsge-such Dr. L. Szilard's"* dated December 11 and 22, 1926, by von Laue and Planck (LSP 1/3). *Zeitschrift für Physik* 53, p. 840 (1929). See MIT Vol. I, pp. 103ff.
27. See Eckart in MIT Vol. I, p. 32; Ehrenberg, p. 109. On February 21, 1952, Karl Darrow, secretary of the American Physical Society (APS), wrote to Szilard inviting him to speak at a symposium on "Entropy and Informa-tion" for the May 1–2 APS meeting in Washington. Szilard replied on the twenty-seventh: "All I could do is to present—as an introduction to the topic—the original considerations which I published in 1927. I have not done any further work in the field since I wrote this one paper" (LSP 4/10).
28. Lazislas Bihaly interview, May 25, 1986.
29. Eckart, June 13, 1964, p. 3 (JSP 400/3). Philip Marcus interview, December 3, 1986.
30. Speech on "Education" at Brandeis University, October 23, 1953 (LSP 42/20).
31. Eckart essay, June 13, 1964, p. 1 (JSP 400/3).

CHAPTER 5

1. Alice (Eppinger) Danos interview, September 4, 1986.
2. Alice Danos interview, June 10, 1986.
3. Alice Danos interview, September 4, 1986.
4. Alice Danos remembers that this decisive visit occurred in August 1924, but it may have been in March 1925, another time that Szilard returned to Budapest. Alice Danos interviews, June 10 and September 4, 1986.
5. By this Szilard meant that, unlike the queen or drone, he had no sex life; he only worked. Alice Danos interviews, June 10 and September 4, 1986, and July 16, 1987.
6. For details on this and later moves, see Polizeiliche Abmeldung certificates (LSP 1/24).
7. Leona Steiner Wolf to Bela Silard, 1988.
8. The widow of Jacob A. Philipsborn is listed in the Berliner Adressbuch for 1925, Fiche 53.

9. Gerda Philipsborn to Albert Einstein, February 3, 1929; from 59 Brixton Hill, Norton House, London (AEP/B 46 094).
10. Szilard to Albert Einstein, March 22, 1930 (AEP/B 35 586 1 + 2, 35 587 1–3).
11. Szilard to Albert Einstein, April 1, 1930 (AEP/B 35 588).
12. Gerda Philipsborn, April 11, 1932. Polizeiliche Abmeldung (LSP 1/24).
13. Szilard to Eugene Wigner, October 8, 1932, address care of Philipsborn, Berlin W. 30, Motzstrasse 58 (LSP 21/4).
14. Eugene Wigner to Michael Polanyi, probably 1933, on United States Lines stationery (MPP). Bela Silard translation. The German science historian Horst Melcher suggested in an October 20, 1987, interview that Gerda Philipsborn's London address was that of her brother. Gerda Philipsborn is buried at Jamia Millia Islamia (National Islamic University) near Delhi, where a daycare center is named in her honor. The university was founded by Dr. Zakir Hussein, a noted educator who served as India's president from 1967 until his death in 1969. He first befriended Ms. Philipsborn in Berlin in the 1920s.
15. *Livret d'Etudiant* (University de Lausanne) delivré à Mlle. Gertrud Weiss; Exmatricule 28 Juin 1929 (EWP).
16. Details of Trude's move to Vevey and Berlin are contained in a letter by Frances Racker, her sister, to Bela Silard, February 6, 1985 (BSP). Additional details are contained in Trude's student book for Berlin University (EWP).
17. Trude said years later that she had first met Szilard through the Polanyi family. See Gertrud Szilard to Freeman Dyson, September 10, 1980 (EWP). The dialogue with Claire Bauroff is based on a recollection by Trude Szilard, as told to Maurice Fox. Fox interview, November 12, 1985.

    It is possible that Trude met Szilard through the Polanyis but turned to her Aunt Claire for advice on the translation fee, as I have reported here. It is also possible that Professor Schrecker or Claire Bauroff knew Szilard and made the first introduction.
18. Maurice Fox interview, November 12, 1985.
19. Eugene Wigner oral-history interview, November 21, 1963, p. 8 (AIP).
20. "Leo Szilard 1898–1964: A Biographical Memoir by Eugene P. Wigner," *Biographical Memoirs,* Vol. XL (New York: Columbia University Press, for the National Academy of Sciences of the United States, 1969).
21. Eugene Wigner interview, October 12, 1984.
22. Wigner, Biographical Memoir.
23. Eugene Wigner oral-history interview, November 21, 1963, p. 8 (AIP). Eugene Wigner interview with Andrew Szanton, March 31, 1987.

24. Wigner, Biographical Memoir; Wigner to the author, January 16, 1984; Wigner interview with Alice K. Smith. March 15, 1960: Wigner interview. October 12, 1984.

25. Eugene Wigner interview with Andrew Szanton, March 31, 1987. Wigner, *Symmetries and Reflections* (Bloomington: Indiana University Press, 1967), p. 259.

26. Eugene Wigner oral-history interview, November 30, 1966, p. 3 (AIP).

27. Eugene Wigner oral history, November 21, 1963, p. 10 (AIP). Another Hungarian friend who attended Einstein's seminar was Albert Kornfeld (Albi Korodi).

28. Marschak to Michael Polanyi, July 8, 1973 (MPP). My thanks to William Scott, Polanyi's biographer, for this citation and to economist Ira Kaminow for his historical analysis of this anecdote.

29. Hermann Mark interviews, October 28 and November 25, 1985.

30. Victor Weisskopf interview, December 16, 1985.

31. Hermann Mark interview, November 25, 1985.

32. Eva (Striker) Zeisel interview, August 19, 1985.

33. Suzannah Lessard, "The Present Moment," a profile of Eva Zeisel, *The New Yorker*, April 13, 1987, p. 40.

34. Eva Zeisel interview, August 19, 1985. Hans Zeisel interview, February 25, 1987. Victor Weisskopf interview, December 16, 1985. *Darkness at Noon* was based on Eva's own years in a Soviet prison, accused of plotting to kill Stalin and only released through the contacts of her uncles, Michael and Karl Polanyi.

CHAPTER 6

1. Eugene Wigner interviews, October 12, 1984, and March 5, 1986.

2. Wigner/Szanton "Recollections" ms, Chap. 5. This refers to interview notes prepared for *The Recollections of Eugene P. Wigner as Told to Andrew Szanton* (New York: Plenum Publishing Corporation, 1992).

3. Theodore von Laue to the author, November 18, 1985.

4. Stanley Goldberg, "Albert Einstein and the Creative Act: The Case of Special Relativity," Chap. 9 in *Springs of Scientific Creativity*, eds. Aris, Davis, and Stuewer (Minneapolis: University of Minnesota Press, 1983), p. 237.

5. Hermann Mark interview, October 28, 1985. Albert Kornfeld interview, June 10, 1986. Lazislas Bihaly interview, May 25, 1986. In a 1940 interview with the Federal Bureau of Investigation, Einstein recalled that in the early 1920s he saw Szilard almost every day. See November 8, 1940, FBI Report, No. 77-153 EMR (LSP 95/9).

6. Esther Salaman, "A Talk with Einstein," *The Listener*, September 8, 1955. Clark, 1972, p. 37.

7. Wigner/Szanton "Recollections" ms, Chap. 9.
8. Named for American physicist Arthur Holly Compton. For making this discovery in 1923, Compton shared the 1927 Nobel Prize in physics with C. T. R. Wilson. Szilard would later work with Compton at the University of Chicago's Metallurgical Laboratory, part of the US Manhattan Project to build the first atomic bombs.
9. Hermann Mark to Bela Silard, June 26, 1984. Mark interview with Alice K. Smith, April 13, 1960.
10. Szilard's "Book" manuscript, beginning "Apology (in lieu of a foreword)" (LSP 40/4). See also Bronowski, *The Ascent of Man* (Boston: Little, Brown and Company, 1973), p. 254, and Szilard to Einstein, January 18, 1924 (AEP).
11. Lazislas Bihaly interview, May 25, 1986.
12. MIT Vol. II, p. 12.
13. *The New York Times,* April 25, 1929. Clark, 1972, p. 37.
14. Clark, pp. 37–38.
15. See, for example, Einstein's September 15, 1928, letter to Szilard about reading a letter by Spinoza (AEP/P Box 46/33–271).
16. For details of the Szilard-Einstein patent applications, see MIT Vol. I, pp. 540–42.
17. Einstein to H. N. Brailsford, April 24, 1930. C. Eichhorn translation (LSP 7/27).
18. Szilard to Einstein, June 30, 1931 (LSP 7/27).
19. See Szilard to Wigner, October 8, 1932 (LSP 21/4).
20. Szilard to Einstein, July 23, 1931 (AEP/P Box 49/35-600-1 and 2).
21. Bela Silard translation (AEP/B 21 441-1). Founded in 1930, the Institute for Advanced Study was then headed by its first director, Abraham Flexner. Rudolph Ladenburg, a physicist whom Polanyi and Szilard knew in Berlin, was then teaching at Princeton.
22. Einstein to Polanyi, October 19, 1932 (AEP/B 21 442).
23. Szilard moved into Harnack House on October 23, 1932 as Polanyi's guest, living in the Müller Zimmer for visiting scholars. *"Wohngaste des Harnack-Hauses in Oktober 1932,"* in the series *"KWG Generalverwaltung 2513 Harnack-Haus Gastelisten 23.7.1930–1938,"* is in the Archive of the Max Planck Institute in Berlin-Dahlem.
24. See Szilard to Einstein, November 10, 19, and 24, 1931 (AEP).

CHAPTER 7

1. Polizeiliche Abmeldung, September 12, 1922. The address, between Mark-graftenstrasse and Jerusalemerstrasse, is now occupied by the Axel Springer building, in a neighborhood once bisected by the Berlin Wall. For records of this and other moves, see (LSP 1/24).

2. Polanyi to Professor Lorenz, October 16, 1922 (MPP), Bela Silard translation. My thanks to William Scott for sharing this letter. The Professor Stern mentioned is undoubtedly Otto Stern, recipient of the 1943 Nobel Prize in physics.

3. "Leo Szilard's Influence on Physics" by Carl Eckart, June 13, 1964 (JSP 400/3). Von Laue, Szilard, and Schrödinger would teach "Discussion of New Work in Theoretical Physics" at the Kaiser Wilhelm Institute for Physical Chemistry in the winter semester, 1930–31. Universität Berlin, *Vorlesungsverzeichnis Wintersemester 1930–31*, p. 55. Von Neumann's seminars begin in 1927.

4. Schrödinger to Donnan, August 26, 1933 (LSP 17/12).

5. Occupation began on January 11, 1923.

6. Mark to Bela Silard, June 26, 1984. Mark interviews, October 28 and November 25, 1985. Mark interview with Alice K. Smith, April 13, 1960. Mark's undated "Data on Meetings with Leo Szilard," written for the author in 1985.

7. See MIT Vol. I, p. 697.

8. Technische Hochschule Berlin issued "Matrikel Nr. 32770" on May 30, 1923, for acceptance, valid until March 31, 1927 (LSP 1/17).

9. Szilard to Einstein, November 1, 1926 (AEP). Szilard to Ortvay, File K785/538, Hungarian Academy of Sciences library, Budapest.

10. "A Simple Attempt to Find a Selective Effect in the Scattering of Roentgen Rays," *Zeitschrift für Physik*, 33, p. 688 (1925); and "The Polarization of Roentgen Rays by Reflection from Crystals," *Zeitschrift für Physik*, 35, p. 743 (1926).

11. Roland Detre interview, October 1, 1986.

12. Wigner/Szanton, "Recollections" ms. notes.

13. Eva Zeisel interview, August 19, 1985.

14. Dennis Gabor to Alfred Rosenfeld, May 30, 1960 (LSP).

15. "Discoveries and Their Uses," a 1978 talk by Dorothy C. Hodgkin, p. 4. Hodgkin places this conversation in 1928; Gabor, in a 1960 letter to Rosenfeld, places it in 1927.

    "Dennis Gabor," p. 133, in T. E. Allibone's *Biographical Memoirs of Fellows of the Royal Society* 26 (1980): 107–47. Gabor to Alfred Rosenfeld, May 30, 1960. Gabor to Szilard, July 13, 1961 (LSP 8/23).

    Andre Gabor to Bela Silard, March 13, 1984. T. E. Allibone to Bela Silard, March 14, 1984.

    "Dennis Gabor: A Biographical Memorial Lecture," by T. E. Allibone. *Israel Journal of Technology*, Vol. 18, 1980, pp. 201–208, see especially pp. 202–203. Oration by Professor Le Poole at Delft, 1971; quoted in Dennis Gabor, p. 133.

16. MIT Vol. I, pp. 527 and 707.

17. Dennis Gabor to Alfred Rosenfeld, May 30, 1960 (LSP).
18. Roland Detre's "Family Recollections," written for the author in September 1986, pp. 3–4. When Leo heard that Bela planned to marry his friend Elizabeth, he was more tolerant. "I am opposed to marriage," he said, "but in your case I make an exception."
19. Roland Detre interview, October 1, 1986.
20. MIT Vol. II, pp. 23ff; Szilard's letter to Peter Odegard, January 10, 1949, with a draft for the advisory committee members of the new Ford Foundation, included his thoughts on alternative forms of democratic government and recalled his idea for the Bund (LSP 8/14, 14/23, 34/15, and 68/3).
21. Enclosure in Szilard to Odegard, January 10, 1949 (LSP 14/23).
22. Szilard to Michael Polanyi (MPP 2/5).
23. MIT Vol. II, p. 25.
24. *"FWU zu Berlin Verzeichnis der Vorlesungen, Wintersemester* 1929–30," p. 52.
25. MIT Vol. II, p. 13. Szilard may have recalled meetings for the Young Plan. Germany signed on August 1929, and the plan was ratified in March 1930. The plan established a new schedule that gave the Reich more favorable annual terms but prolonged payments until 1988.
26. MIT Vol. II, p. 24.
27. Ibid., p. 25.
28. Ibid., p. 29.
29. Ibid., p. 26.
30. Ibid., p. 28.
31. Ibid.
32. Ibid., fn. 9. Szilard to Odegard, January 10, 1949, enclosure (LSP 14/23).
33. Szilard to Einstein, March 22, 1930 (AEP 21 434).
34. Brailsford to Einstein, March 31, 1930 (AEP 45 650).
35. Einstein to Brailsford, April 24, 1930, C. Eichorn translation (LSP 7/27).
36. Universität Berlin, *Vorlesungsverzeichnis,* 1930–31, pp. 5 and 55.
37. See Szilard's curriculum vitae, stamped October 9, 1933 by the Academic Assistance Council (AAC SPSL Sect. 4, Drawer 9). In March 1930, Szilard and Rupp sent an article on *"Beeinflussung 'polarisierter' Elektronenstrahlen durch Magnetfelder"* to the science magazine *Die Naturwissenschaften.*
38. Universität Berlin, *Vorlesungsverzeichnis,* 1931, p. 54.
39. See MIT Vol. II, pp. 16–17.
40. MIT Vol. I, p. 528. See Szilard's *Curriculum Vitae* of June 18, 1932, for details on his research. Rolf Wideröe interview, July 7, 1987.
41. MIT Vol. I, pp. 528 and 722. See also the 1953 note by W. B. Mann, *The Cyclotron* (New York: John Wiley & Sons, 1953), p. 93, about Szilard's ideas on frequency modulation and phase stability.
42. MIT Vol. I, pp. 147–48.
43. Szilard to Niels Bohr, March 26, 1936 (LSP 4/34). MIT Vol. II, pp. 44–45.

CHAPTER 8

1. See draft in Szilard to Einstein, June 30, 1931 (AEP 35 598). Szilard received the visa on July 6, 1931. Szilard's erratic movements can be traced in his tattered passports and alien registration documents (LSP 1/6).
2. L. P. Eisenhart (chairman of the department of mathematics and dean of faculty at Princeton University) to Szilard, September 24, 1931 (LSP 15/24).
3. Einstein to the American consul general, October 24, 1931. Bela Silard translation (LSP 7/27).
4. Universität Berlin, *Vorlesungsverzeichnis*, 1931–32.
5. Szilard was assigned to Room 30, Berth C, Tourist Class (LSP 82/1).
6. *Brandeis University Bulletin*, February 1954, pp. 6–7. As Szilard added during the interview: "Today [1953–54], of course, it is not too difficult to think of things that will make it disappear. . . ." Things he had helped create, such as the A-bomb.
7. Chase Safe Deposit Co. receipt dated January 20, 1932 (LSP 1/23).
8. Eugene Wigner interview, October 29, 1987. Serge A. Korff, "The Physics Department of University College during the Three Decades from 1940 to 1970," p. 8, and his September 24, 1970, manuscript (New York University Archives). Szilard CV stamped October 9, 1933, by Academic Assistance Council (AAC SPSL Sect. 4, Drawer 9). Statement attached to immigration form N-400 (LSP 1/25, 2/9). The Kenmore Hall was at 145 East Twenty-third Street.
9. Szilard to Lady Murray, April 24, 1934 (LSP 12/14–15). MIT Vol. II, pp. 36–37. In a similar protest, which Szilard proposed two years later, the boycott was to take effect if eight-tenths of the Nobel laureates who were approached agreed to sign.
10. Universität Berlin, *Vorlesungsverzeichnis*, 1932, p. 58. The only record of Szilard's Washington visit is his receipt for a three-dollar money order, purchased at a post office on Pennsylvania Avenue, perhaps to pay the fee for his US citizenship application or for a patent. Receipt for April 29, 1932 (LSP).
11. MIT Vol. II, p. 13. Universität Berlin, *Vorlesungsverzeichnis*, 1932–33, p. 5.
12. MIT Vol. II, pp. 16–17.
13. *Nature* 131, 421, 457 (1933). Stent and Calendar, pp. 24–25. See also Victor Weisskopf's review of *Redirecting Science* by Finn Aaserud in *Science*, Vol. 251 (February 8, 1991), pp. 684–85.
14. Delbrück Oral History 1978, p. 42 (CIT). Manny Delbrück interview, February 9, 1987.
15. Szilard's papers reveal no evidence of a visit to Copenhagen, although he did visit Stockholm in 1928 (according to Hermann Mark and stamps in his passport) and again in the 1950s.

16. See Szilard's Brandeis University speech of December 8, 1954, p. 2 (LSP 42/25).
17. MIT Vol. II, p. 16.
18. Szilard to Wigner, October 8, 1932, Bela Silard translation, with revisions by Maria Luise Wagner (LSP 21/4).
19. Polanyi to Einstein, October 13, 1932, Bela Silard translation (AEP 21 441-1). The Institute for Advanced Study was then headed by its first director, Abraham Flexner.
    See also Einstein to Polanyi, October 19, 1932 (AEP 21 442). My thanks to William Scott for calling our attention to this letter. Scott to Bela Silard, May 7, 1986 (BSP).
20. Wigner to Polanyi, October 18, 1932, Bela Silard translation (MPP).

CHAPTER 9

1. MIT Vol. II, p. 13.
2. "Reminiscences," p. 96. Polanyi to Lapworth, January 13, 1933, Bela Silard translation (MPP 2/11).
3. "Memoirs," dictated 1960 (LSP 40/10). MIT Vol. II, p. 13.
4. Rose Scheiber interviews, September 4, 1986, and July 16, 1987.
5. Alice Danos interviews, June 10 and September 4, 1986.
6. The February 3, 1933, letter arrived at 58 Motzstrasse, home of Gerda Philipsborn and her mother.
7. After World War II, Harnack House became the officers' club for the US occupation forces in Berlin.
8. Joachim Fest, Hitler (New York: Random House, 1975), pp. 396–98.
9. MIT Vol. II, pp. 13–14. "Reminiscences," p. 96.
10. Meyer, p. 21; Shirer, pp. 193–94.
11. The New York Times, March 28, 1933, pp. 1 and 10, and March 29, p. 10.
12. This incident appears to be a temporary measure taken at the beginning of the anti-Jewish boycott. Other non-Aryan travelers left and entered Germany for several years.
13. MIT Vol. II, p. 14.
14. Hermann Mark to Bela Silard, June and July 1984; interview, October 28, 1985. Mark memo, "Data on Meetings with Leo Szilard," p. 3. (undated but received at a November 25, 1985, interview).
15. "Leo Szilard: The Man Behind the Bomb, A Postscript with Gertrud Weiss Szilard," interview by Helen Hawkins. KPBS-TV, San Diego (LSP 108). Esther Simpson to the author, January 14, 1993.
16. Szilard to Einstein, April 1933, Bela Silard translation (AEP/P 21 443).
17. According to Marschak, both Schlesinger and Kuhnwald committed suicide five years later, when Hitler arrived in Vienna. Marschak to Edward Shils, October 4, 1964 (LSP 92/4).

18. "Leo Szilard: A Memoir" by Edward Shils. *Encounter,* December 1964, pp. 35–41.
19. Jacob Marschak to Edward Shils, October 4, 1964 (LSP 92/4).
20. "Reminiscences," pp. 97–98. Szilard wasted little time, setting off by train for London a few days later and arriving there on April 21. While en route to London, the summer semester began at Berlin University, and its catalog listed Szilard as teaching two courses: a "Discussion of New Work in Atomic Physics" with Lise Meitner and a "Discussion of New Work in Theoretical Physics" with Erwin Schrödinger. Universität Berlin, *Vorlesungs-verzeichnis 1933,* p. 60.
21. Undated Hungarian letter "No. 3," Bela Silard translation (LSP 1/23).
22. *Lost London* (New York: Weathervane Books, 1971), p. 199. Architectural critic Hermione Hobhouse called the structure's style "riotously Plateresque." The Imperial was known for three features: ornate Turkish baths for "ladies and gentlemen" decorated with glazed Doulton tile; a Winter Garden where a string orchestra played; and, overlooking the square, a huge clock with noisy chimes. Just above the clock's face was the bust of a stern-looking man, and with each chime of the hour his red metal tongue stuck out—a scornful gesture by the architect toward the neighbors in the Georgian row houses who had protested the flamboyant hotel's construction.
23. Szilard to Beveridge, April 22, 1933 (AAC SPSL Sect. 3, Drawer 8).
24. Szilard to Einstein, April 23, 1933 (AEP/P 115 and LSP 7/27).
25. Einstein to Szilard, April 26, 1933. Clark, 1972, p. 575 (LSP 7/27).
26. Szilard to Beveridge. MIT Vol. II, p. 31.
27. Szilard to Einstein, April 26, 1933 (AEP 21 446). As a scientist, Weizmann was well connected with academics in many countries; as a British subject, he helped influence the government's administration of Palestine. In 1948 he would become the first president of the Republic of Israel.
28. Szilard to Beveridge, May 4, 1933 (LSP 4/30).
29. MIT Vol. II, p. 14.
30. Szilard to Beveridge, April 27, 1933 (AAC SPSL Sect. 3, Drawer 8).
31. "Preliminary Note on Academic Assistance Council with Appendices A.B.C.D." (WBP Bev IXa 46).
32. Esther Simpson interview, October 3, 1985 and (AAC SPSL Sect. 3, Drawer 8).
33. Szilard to Dr. D[onnan], May 7, 1933. MIT Vol. II, pp. 32–34.
34. Hans Bethe interview, November 21, 1985.
35. MIT Vol. II, pp. 32–33.
36. Clark, 1972, p. 575.
37. Szilard to Liebowitz, n.d. (LSP 12/4).
38. Szilard to Beveridge, May 23, 1933 (AAC SPSL Sect. 3, Drawer 8).
39. Esther Simpson interview, May 25, 1986. Szilard to Beveridge, May 23, 1933 (AAC SPSL Sect. 3, Drawer 8).

40. Academic Assistance Council files; Szilard to an unknown addressee, August 11, 1933 (LSP 66/10). MIT Vol. II, p. 35.
41. The Notgemeinschaft Deutscher Wissenschaftlicher im Auslande was headed by physicist Max Born.
42. Szilard to Gibson, July 19, 1933. July 22, 1933, letter to Szilard (AAC SPSL Sect. 3, Drawer 8).
43. L. W. Jones memoranda to Warren Weaver, June 28 and July 11 and 27, 1933. RAC, 200D, NYU, Szilard (RAC).
44. August 11, 1933, letter to unknown addressee (LSP 66/10). MIT Vol. II, p. 36.
45. "Memoirs," dictated 1960, p. 6 (LSP 40/10).
46. August 11, 1933, letter to unknown addressee (LSP 66/10). MIT Vol. II, p. 35.
47. Midland Bank to Szilard, June 9, 1933 (LSP 75/8). MIT Vol. II, p. 17.
48. MIT Vol. II, pp. 34–36.
49. Wigner to Polanyi, June 25, 1933, Bela Silard translation (MPP 2/12).
50. Ibid. Then visiting Budapest, Wigner heard from Bela that Leo wanted to return to scientific work. This news, Wigner wrote Polanyi, "would please me very much if Szilard's wish is serious. However, I am afraid that at one time about two or three months ago [he] did not yet know what he wanted to do." Wigner to Polanyi, June 30, 1933, Bela Silard translation (MPP 2/12).
51. Donnan to Einstein, August 14, 1933 (AEP/P 453-1). Von Laue and Schrödinger to Donnan, July 21, 1933 (AAC SPSL Sect. 3, Drawer 8).
52. Max Volmer to Donnan, August 2, 1933 (AAC SPSL Sect. 4, Drawer 9, L. Szilard), (LSP 20/2).

Fritz G. Houtermans, a Dutch-born physicist Szilard knew from Berlin, reported to the Oxford physicist Frederick A. Lindemann that von Laue, Schrödinger, and Einstein all wished to help Szilard find a position. At the time, Lindemann advised the Imperial Chemical Industries trust for research grants. Houtermans was staying at the Royal Hotel, just off Russell Square, and may have met with Szilard during his visit. Lindemann could do nothing then, but promised, "If any possibilities arise I will get in touch with Dr. Szilard." Lindemann to Houtermans, July 24, 1933 (AAC SPSL Sect. 4, Drawer 9), (FLP D. 91).

Meanwhile, Donnan wrote to Einstein, explaining Szilard's situation and closing: "I need scarcely say that anything written by you would carry the greatest weight." Donnan to Einstein, July 25, 1933 (AEP 21-453-1).
53. Szilard to unknown addressee, August 11, 1933 (LSP 66/10). MIT Vol. II, pp. 34–36.
54. Szilard to Wigner, August 17, 1933 (MPP 2/12).
55. Szilard to Einstein, August 14, 1933, Josef Ernst translation (AEP/B 21 452).
56. Einstein to Donnan, August 16, 1933, Josef Ernst translation (AEP/P 21 454 and AAC SPSL Sect. 4, Drawer 9).
57. Ehrenfest to Donnan, August 22, 1933 (LSP 32/7).

58. Schrödinger to Donnan, August 26, 1933 (AAC SPSL Sect. 3, Drawer 8 and Sect. 4, Drawer 9).
59. Szilard to unknown addressee, August 11, 1933 (LSP 66/10). MIT Vol. II, p. 35.
60. MIT Vol. II, p. 19.
61. Ibid., p. 16.
62. "Memoirs," dictated 1960, p. 6 (LSP 40/10).

CHAPTER 10

1. The third kind, "gamma" radiation, carries no electrical charge and is the most penetrating of the three; while alpha particles can be stopped by a sheet of paper and beta by a thin sheet of metal, gamma particles can only be blocked by a dense material such as lead.
2. Clark, *The Scientific Breakthrough*, p. 152.
3. "Memoirs," dictated 1960, p. 7 (LSP 40/10). Chadwick's letter to *Nature* appeared February 27, 1932, while Szilard was in New York. Szilard returned to Germany in May and probably read or heard about the neutron that summer.
4. Szilard's recollections of the "moonshine" quote come from "Memoirs," dictated 1960, pp. 7–8 (LSP 40/10) and from "The Sensitive Minority Among Men of Science," a speech delivered on December 7, 1954 (LSP 42/25). MIT Vol. II, pp. 17–18. See also *Nature* 132: 432–33, September 16, 1933.

   His "chain reaction" idea occurred either the day after Rutherford's speech or within a few weeks, depending on which of Szilard's versions is true. It seems unlikely that he would be out walking the day he awoke with a bad cold, and other evidence also suggests at least a few days' delay in his contemplations. "Memoirs," dictated in 1960, p. 7 (LSP 40/10).

   Szilard never said where or when he stopped for that red light. Southampton Row is six blocks long and has more than a dozen crossings with traffic lights. Szilard remembered the incident several different ways. In "Creative Intelligence and Society: The Case of Atomic Research, the Background in Fundamental Science," a public lecture delivered on July 31, 1946, Szilard said his realization came in October 1933, "a week or two" after Rutherford's speech; in "The Sensitive Minority . . ." he remembered that it came the afternoon he read about Rutherford's speech—or September 12, 1933.
5. Szilard interview taped in New York, May 1960 (LSP 101/1–3). MIT Vol. II, p. 17.
6. Notes, "Rough Draft, Outline for Book," June 1960 (LSP 30/9). MIT Vol. II, p. 17.
7. *Nature*, September 16, 1933, p. 433.
8. "Creative Intelligence and Society. . ." MIT Vol. I, p. 183.
9. Max Lerner, "Life of a Man," *New York Post* magazine, "Page Four," March 4, 1960, p. 34; Max Lemer interview, January 23, 1986.
10. The address was on the south side of the street, in a block since demolished. See October 18, 1933, letter from the AAC's assistant secretary to Szilard (AAC SPSL Sect. 3, Drawer 8).
11. When free from thoughts about the atom, Szilard continued his errands for the AAC and the Jewish Board. On October 29, 1933, Szilard joined a panel

of "academicians" to address a meeting of Jewish organizations. Speakers recounted the waves of dismissals from German universities to an audience that included the round-faced Wall Street investment banker Lewis L. Strauss, there to represent several Jewish groups from the United States. As Strauss remembered the gathering, "the impotence of the conference could never have been deduced from its trappings, which were those of an important international conference with committees and subcommittees, agendas, and *rapporteurs*." Strauss and Szilard did not meet then, but would be drawn together five years later in two significant ways: first to explore medical and commercial uses for the atom, then to fret that Germany might already be making atomic bombs. But in 1933 the only person in the conference hall with such fears was Szilard himself. Strauss, p. 108.

12. Lady Charlotte Simon interview, May 27, 1986.

13. MIT Vol. II, p. 17.

14. "Reminiscences," p. 101 (LSP 40/10).

15. See "Atomic Transmutation," *Nature*, September 16, 1933, p. 433. Glasstone notes in *Sourcebook on Atomic Energy* (fn., p. 417) that Szilard's "general ideas were correct" for a chain reaction (beryllium^9 absorbs a neutron and gives off two to become beryllium^8 and then splits into two helium^4 atoms) but that it could not work in beryllium because the neutrons liberated had relatively low energies.

16. "Memoirs," dictated in 1960, p. 8 (LSP 40/10). Glasstone, pp. 293–94. MIT Vol. II, pp. 17–18.

17. MIT Vol. II, p. 17.

18. Ibid., p. 18.

19. The director was C. C. Paterson. Szilard to Polanyi, December 11, 1933 (MPP 2/13).

20. Esther Simpson to the author, October 16, 1985. December 13, 1933, statement from Midland Bank, Russell Square Branch (LSP 75/8).

CHAPTER 11

1. Wigner to Polanyi, January 12, 1934 (MPP 2/14).

2. Wigner to Mrs. Michael Polanyi, October 7, 1933 (MPP 2/13). University of Wisconsin physicist Gregory Breit was making "unparalleled efforts" to help Szilard at NYU but confronted "a not very honorable feeling of nationalism" about supporting foreign scholars, Wigner reported. Wigner to Michael Polanyi, November 6, 1933, Bela Silard translation (MPP 2/13).

3. Szilard to Polanyi, December 11, 1933 (MPP 2/13).

4. Strauss, p. 164. Toni Stolper to Gertrud Weiss Szilard, September 13, 1968; copy in "Album" on AE-FDR letter (EWP).

5. In January 1934, Szilard reported to Polanyi "working out methods for the production of fast electrons" and meeting with Sir Hugo Hirst, the founder

of General Electric (U.K.). By mid-February he applied for a patent on an accelerator to produce "fast charged particles," such as electrons or protons, but warned Hirst that there was no "immediate important application for fast electrons." In fact, Szilard's patent posed two concepts that would become standard in cyclotrons in years to come: frequency modulation and phase stability.

Although Szilard urged Hirst to have GE follow the "probably very fast development" in accelerators, his own thoughts were on the neutron chain-reaction concept.

6. *The New York Times,* January 29, 1933, Sect. 4, p. 1. Szilard had sent information to the AAC for a job in Bangalore and a few weeks later wrote to an acquaintance about "the Delhi vacancy." Szilard reported having direct information "from friends in Delhi," perhaps Gerda Philipsborn, with whom he had lived in Berlin. A few weeks later Szilard would report, "I have written to Delhi offering to visit Delhi if they should decide that they would rather have me in preference to other candidates, and that I could decide on the spot. The University of Delhi is not a full university but will possibly be developed into something of the sort in the future. I could probably become chairman of the department (which practically does not exist). There are no professorships, only readerships, and the salaries are very poor."

Szilard to P. Gent, January 9, 1934, "Papers R *[sic]* Szilard sent to me for Bangalore application" and Note in the AAC files dated December 12, 1933 (AAC SPSL Sect. 4, Drawer 9, L. Szilard). Szilard to Michael Polanyi, January 29, 1934 (MPP 2/14).

7. "Memoirs," dictated 1960, p. 9 (LSP 40/10).

8. MIT Vol. I, pp. 529 and 615. The third element to sustain a nuclear chain reaction, plutonium, is produced by bombarding uranium[238] with neutrons. This would not be realized theoretically until 1940, and the element itself would not be discovered until 1941.

9. Szilard to Hirst, March 17, 1934. MIT Vol. II, p. 38.

10. MIT Vol. I, p. 722. Szilard to A. Benjamin, Electrical & Musical Industries, Ltd., December 31, 1934; A. Benjamin to Szilard, February 12, 1934, with technical quibbles about the "Microbook" patent (LSP 7/30).

11. Einstein to Thomas B. Appleget, March 14, 1934. RG 1.1, series 200D, box 153, folder 1881 (RAC). See also Weaver to Einstein, March 20, 1934 (AEP/P Box 27, 21 455).

12. Szilard to Fritz Lange, March 21, 1934 (LSP 11/20).

13. Maurice Goldhaber interview, December 4, 1985.

14. MIT Vol. II, pp. 36–38.

15. Murray to Szilard, April 27, 1934 (LSP 12/15); Szilard to Garnett, May 9, 1934 (LSP 8/23).

16. Eugene Wigner Memorandum, April 16, 1941: "Ideas Before the Discovery of Nuclear Fission" (LSP 65/23).
17. Szilard to Oliphant, May 20, 1934 (LSP 14/26).
18. This visit is mentioned in a letter by Walter Adams to Rutherford. See also Rutherford to Adams, May 30, 1934 (AAC SPSL, Sect. 4, Drawer 9, L. Szilard).
19. MIT Vol. II, p. 46. Rutherford recounted this meeting to Harvard spectroscopist Kenneth Bainbridge, whom he met in the hall. See "Orchestrating the Test" in Wilson, ed., *All in Our Time* p. 203. See also Blumberg and Owens, p. 86.
20. Szilard to Rutherford, June 7, 1934 (LSP 16/28).
21. Szilard to Felix Heim, July 19, 1934 (LSP 9/11).
22. Railing to Szilard, April 26, 1934, and Szilard to Railing, July 20, 1934 (LSP 8/28).
23. Paterson to Szilard, July 27, 1934 (LSP 8/28).
24. MIT Vol. II, pp. 37–38. Szilard to Railing, July 20, 1934 (LSP 8/28).
25. St. Cuilson and Paterson letters to Szilard, August 9, 1934, and Szilard to Paterson, August 14, 1934 (LSP 8/28).
26. At Szilard's urging, Adams also asked Thomson if it might be worth approaching manufacturers "because of the possibility of its eventually being of importance for practical application." Szilard to Walter Adams, July 23, 1934 (LSP 3/28). Adams to Thomson, July 24, 1934 (AAC SPSL Sect. 4, Drawer 9, L. Szilard).
27. G. P. Thomson to Walter Adams, July 26, 1934 (AAC SPSL, Sect. 4, Drawer 9).
28. Szilard to Simon, July 31, 1934 (FSP FS/14/3/S).
29. Szilard to Professor Singer, June 16, 1935 (LSP 18/3).
30. Szilard to Wigner, August 7, 1934 (LSP 21/4).
31. MIT Vol. I, p. 139.
32. Szilard to Hopwood, August 28, 1934 (LSP 9/40). See also MIT Vol. I, p. 143.
    To an international conference on nuclear physics in London that September, Szilard explained his experiment this way: "By irradiating 25 gm. of beryllium with the penetrating radiation from 150 mgm. radium and exposing 100 c.c. ethyl iodine to the radiation excited in the beryllium we could induce radioactivity in iodine, and separate chemically the radio-iodine from the ethyl iodine in the form of a silver iodide precipitate." *International Conference on Physics, London,* 1934, *Papers and Discussions in Two Volumes* (Cambridge: The Physical Society, 1935). Vol. I, *Nuclear Physics,* pp. 88–89.
33. Szilard and Chalmers, "Chemical Separation of the Radioactive Element from Its Bombarded Isotope in the Fermi Effect." *Nature* 134, p. 462 (September 22, 1934). MIT Vol. I, pp. 143–44.
34. *International Conference,* pp. 88–89.

35. MIT Vol. II, p. 20.
36. Szilard to Tristram Coffin, July 3, 1963 (LSP 6/22).
37. Szilard telegram to Polanyi, September 2, 1934 (MPP 14/9). Szilard to Polanyi, October 3, 1934, Josef Ernst translation (MPP unidentified correspondence 3/1).
38. MIT Vol. II, p. 40.
39. *Nature* 134, p. 880 (December 8, 1934). The work was completed on November 26, 1934. MIT Vol. I., pp. 140 and 148.
40. MIT Vol. I., p. 140.
41. "Atom Energy Hope Is Spiked by Einstein. Effort at Loosing Vast Force Is Called Fruitless." *Pittsburgh Post-Gazette* headline, Section 2, p. 1, December 29, 1934, reporting Einstein's visit to the city.
42. MIT Vol. I, p. 140.
43. *Nature* 135, p. 98 (January 19, 1935). MIT Vol. I, p. 140. Ugo Fano interview, March 17, 1987.
44. Nicholas Kurti interview, May 26, 1986.
45. See L. L. Whyte to Szilard, January 29, 1935 (LSP 21/2).
46. Nicholas Kurti interview, May 26, 1986.
47. Szilard to Lindemann, March 4, 1935 (LSP 12/7) and March 30, 1935 (FLP D.23 1).
48. Wigner to Polanyi (MPP undated letters).
49. Szilard to C. K. Ogden, June 4, 1935 (LSP 14/23).
50. Isidor I. Rabi interview, August 19, 1985.
51. Dirac to Szilard, May 7, 1935 (LSP 7/15). "Book 1960" files (EWP).
52. Anna Kapitza interview (with Weiss and Patton), September 1988.
53. See Smithsonian Institution Negative No. 80-9568, Science Service of the Niels Bohr Library (AIP).
54. M. L. Oliphant, P. Harteck, and E. Rutherford, "Transmutation Effects Observed with Heavy Hydrogen," *Nature* 133 (March 17, 1934), p. 413.
55. Hans Bethe interview, November 21, 1985. Bethe to the author, December 31, 1987.
56. FBI report, October 23, 1942 (LSP 95/9). Szilard to Professor Singer, June 16, 1935 (LSP 18/3).
57. The work with Chalmers led Szilard to conclude, mistakenly, that a "double" neutron, one with a mass of 2, might trigger such a process; in fact, he was observing an "isomer" of indium (having the same mass and atomic number) in a distorted radioactive state.
58. Szilard to Lindemann, June 3, 1935. MIT Vol. II, pp. 42–43.
59. June 6 draft for June 16, 1935, Szilard to Charles Singer (LSP 18/3).
60. Szilard to Charles Singer, June 16, 1935 (LSP 18/3). Dorothea Singer to Alice K. Smith, September 1, 1960. My thanks to Alice K. Smith for sharing this correspondence.

61. Szilard to Charles Singer, June 16, 1935 (LSP 18/3).
62. Esther Simpson interview, May 25, 1986.
63. Jean dePeyer Einersen to the author, December 16, 1987, and January 18, 1988.
64. Jean dePeyer Einersen to the author, January 18, 1988.
65. Szilard to Einersen, August 5, 1960. Quoted in her letter to the author on March 18, 1988.
66. DePeyer Einersen to Szilard, July 30, August 13, and September 12, 1960; his reply August 10, 1960 (LSP 7/12). Esther Simpson interview, May 25, 1986.
67. MIT Vol. II, p. 19.
68. MIT Vol. I, p. 186.
69. Walter Elsasser to the author, January 1, 1988.
70. Walter Elsasser interviews, November 12, 1985, and February 20, 1987.
71. J. Coombes to Claremont Haynes & Co., October 8, 1935 (EWP). Director of Navy Contracts to Szilard, March 25, 1936 (LSP).
72. Nicholas Kurti interview, May 26, 1986.
73. "Absorption of Residual Neutrons," *Nature* 136, p. 950 (December 14, 1935). MIT Vol. I., pp. 140, 150–52. Rutherford to Szilard, December 17, 1935 (LSP 16/28).
74. Maurice Goldhaber in MIT Vol. I, p. 141.
75. Szilard to Breit and Wigner, January 12, 1936 (LSP 5/6). Niels Bohr to Szilard, February 4, 1936 (LSP 4/34).
76. Niels Bohr, "Neutron Capture and Nuclear Constitution," lecture to the Royal Danish Academy of Sciences, reprinted in *Nature* 137, p. 348 (February 29, 1936).
77. Cockcroft to Szilard, January 22, 1936 (LSP 6/21).
78. Szilard to Fermi, March 13, 1936 (LSP 8/6).
79. Lindemann to Gen. Leslie Groves, July 12, 1945, with "Top Secret" attachment (MED 201, Szilard and Cherwell files).
    At this time, Szilard had new reason to fear that nuclear weapons might be made. In his recent Nobel Prize lecture, Frédéric Joliot-Curie had predicted: "If we look back and take a glance at the progress achieved in ever-increasing measure in science, then we are justified in supposing that the investigators who are able to build up or break down elements at choice, will also learn how to realize transformations of an explosive character, veritable chemical chain reactions." Frédéric and Irène Joliot-Curie, *Oeuvres Scientifiques Complètes* (Paris: Presses Universitaires de France, 1961), p. 552; Otto Hahn translation in *New Atoms* (New York: Elsevier, 1950), p. 29.
80. Szilard interview with Connie Lien, *Glamour,* January 5, 1962 (transcript, p. 2, in LSP 40/26).

81. Szilard to Fermi, March 13, 1936 (LSP 8/6).
82. Goldhaber to Szilard, March 18, 1936. MIT Vol. II, p. 44.
83. Navy patent correspondence (LSP 4/2).
84. MIT Vol. II, pp. 44–45.
85. Szilard to Gertrud Weiss, March 26, 1936, Bela Silard translation (LSP 20/9).
86. Szilard left England for Calais on April 14, 1936, and entered Switzerland from Delle on April 16 (LSP 1/6). Gertrud Weiss Registration Certificate #598560 (EWP).
87. Szilard to Cockcroft, May 27, 1936 (LSP 6/21).
88. Szilard to Segrè, March 30, 1936 (LSP 17/18).
89. Szilard to Rutherford, May 27, 1936. MIT Vol. II, pp. 45–46.

CHAPTER 12

1. Nicholas Kurti interview, May 26, 1986.
2. Germany signed a treaty guaranteeing Austrian neutrality on July 11, 1936.
3. Attachment to October 11, 1943, court settlement; see especially pp. 4 and 8 (LSP 3/31-2).
4. Szilard to Arno Brasch, January 16, 1937 (LSP 5/4).
5. Szilard to Trude, March 31, 1937. Josef Ernst translation for this and all other personal letters between Leo and Trude (EWP).
6. Szilard to Trude, April 4, 1937 (EWP).
7. Hoover/Kreithe Navy report, 1942 (LSP).
8. Szilard to Trude, April 7, 1937 (EWP).
9. Szilard to Trude, April 13, 1937 (EWP).
10. Szilard to Trude, April 20, 1937 (EWP).
11. 1/4 Stmt, to N-400 (LSP 1/25, 2/9). FBI report, October 6, 1942 (LSP 95/9). See also April 20, 1937, Szilard to Trude (EWP).
12. Szilard to Brasch, July 3, 1937 (LSP 5/4).
13. Lindemann to Szilard, July 30, 1937 (FLP D.237).
14. Szilard to Trude, August 27, 1937 (EWP).
15. Szilard to Trude, September 3, 1937 (EWP).
16. Szilard to Trude, September 1 and 6, 1937 (EWP).
17. Roland Detre interview, October 1, 1986. Szilard to Trude, September 1, 1937 (EWP).
18. See Szilard to Lindemann, September 15, 1937 (FLP D.237).
19. Goldhaber in MIT Vol. I, p. 141. Maurice Goldhaber interview, December 4, 1985.
20. Joliot-Curie to Szilard, November 26, 1937 (LSP 10/23).
21. Szilard to Joliot-Curie, December 24, 1937 (LSP 10/23).
22. ICI to Szilard, December 31, 1937 (LSP). MIT Vol. II, p. 21.
23. Isidor I. Rabi interview, August 19, 1985.

24. Sent by Karl Polanyi on a Thursday from London, Bela Silard translation (MPP undated correspondence).
25. Gertrud Weiss Szilard's "Identification and Personnel Data for Employment of United States Citizen," September 8, 1961 (LSP). Frances Racker interview, November 22, 1985.
26. Szilard to Francis Simon, July 22, 1938, Josef Ernst translation (LSP 18/1).
27. Strauss correspondence in Szilard, Leo (LLS). See also Szilard's 1/4 Stmt, to N-400 (LSP 1/25, 2/9).
28. Szilard to Lindemann, March 11, 1938 (FLP D.237).
29. Pfau, pp. 52–53.
30. Szilard to Francis Simon, July 22, 1938, Josef Ernst translation (LSP 18/1).
31. Strauss, p. 165. "Memorandum concerning my recollection of the conversations which I had with Mr. Strauss during 1938," February 16, 1949 (LSP 41/11). Strauss remembers this as "toward the end of 1937" (p. 164), but Szilard was not in the United States then. See also Pfau, p. 53.
32. As he had done in London, Szilard turned to the handiest radiation source, a hospital, and in April arranged to have Dr. M. Lenz at the Montefiore Hospital for Chronic Diseases irradiate six cigars with 100 kilovolts for ten minutes on a side. "I hope," Lenz wrote when returning the cigars to Szilard, "that your friend finds the taste unchanged." Lenz to Szilard, April 15, 1938 (LSP). Cited in Rhodes, p. 239. In May, a Havana tobacconist wrote Szilard that his irradiated cigars had no difference in taste from untreated samples. Correspondent to Szilard, May 7 or 27, 1938 (LSP).
33. Elsbeth Liebowitz interview, October 2, 1986.
34. Szilard recommended irradiating the trichina worms, not to kill them but to sterilize them, thus preventing them from laying eggs. This strategy proved to be effective years later: Pork irradiation became a commercial process in many countries by the 1970s and was approved by the US Food and Drug Administration in 1986.
35. Szilard's US-U.K. arrangement was initialed and dated May 6, 1938, by Lindemann. Chapman to Szilard, May 17, 1938; Lindemann to Szilard, June 6, 1938 (LSP 12/7).
36. Segrè oral-history interview, p. 31 (AIP). Rhodes, p. 240.
37. Szilard to Francis Simon, July 22, 1938 (LSP 18/1).
38. Lindemann to Szilard, July 30, 1938 (LSP 12/7).
39. Simon to Szilard, August 23, 1938, Bela Silard translation (LSP).
40. Szilard to Simon, September 9, 1938 (FSP FS/14/3/5).
41. Maurice Goldhaber interview, December 4, 1985.
42. MIT Vol. I, p. 155ff.
43. Szilard to Trude, October 2, 1938 (EWP).
44. A night letter to Lindemann late in October announced: "HAVING ON ACCOUNT OF INTERNATIONAL SITUATION WITH GREAT REGRET

POSTPONED MY SAILING FOR AN INDEFINITE PERIOD STOP
WOULD BE VERY GRATEFUL IF YOU COULD CONSIDER ABSENCE
AS LEAVE WITHOUT PAY STOP WRITING STOP PLEASE COMMU-
NICATE MY SINCERELY FELT GOOD WISHES TO ALL IN THESE
DAYS OF GRAVE DECISIONS SZILARD." MIT Vol. II, pp. 21 and 48.

45. Sidney Barnes interviews, November 8 and 11, 1985.
46. Victor Weisskopf interview, December 16, 1985.
47. Sidney Barnes interviews, November 8 and 11, 1985.
48. Szilard to Bohr, November 11, 1938 (LSP 4/34).
49. Invitation to Dunning's January 16, 1939, seminar (LSP 6/26).

CHAPTER 13

1. MIT Vol. II, p. 60.
2. James Arnold interview, February 11, 1987.
3. MIT Vol. II, pp. 50–52.
4. Niels Bohr had learned the news from Lise Meitner, the Austrian physicist who worked at the Kaiser Wilhelm Institute for Chemistry in Berlin but was then on holiday in Sweden, having just fled Nazi racial restrictions. She had learned of the experiment in a letter from Strassmann and with her nephew Otto Frisch immediately recognized—and named—the process of nuclear "fission."
5. Herbert L. Anderson, "The Legacy of Fermi and Szilard." *Bulletin of the Atomic Scientists,* September 1974, pp. 56–61, and October 1974, pp. 40–47.
   See also Lawrence Badash, Elizabeth Hodes, and Adolph Tiddens, "Nuclear Fission: Reaction to the Discovery," IGCC Research Paper No. 1, p. 16 (hereafter referred to as IGCC #1).
6. Wells described atomic warfare in his 1914 book *The World Set Free,* which Szilard first read in Berlin in 1932.
7. MIT Vol. II, p. 53.
8. Szilard telegram to Trude, January 23, 1939 (EWP).
9. IGCC #1, p. 16.
10. MIT Vol. II, p. 60.
11. IGCC #1, pp. 22 and 47. Edward Teller to the author, September 16, 1988.
12. IGCC#1, pp. 23 and 47.
13. *An Early Time—Edward Teller.* Los Alamos Scientific Laboratory Film.
14. MIT Vol. II, p. 54. At the moment, Szilard and Teller were among the few physicists who had looked beyond the stunning scientific news of fission to consider its use as a weapon. Another who saw its consequences was J. Robert Oppenheimer, a brash and brilliant thirty-five-year-old theoretical physicist at Berkeley. At first, when Luis Alvarez told him about uranium fission, Oppenheimer "proved" with stunning logic that it was impossible, posing several arguments about the technical barriers that must be over-

come. But a few hours later, once Oppenheimer had seen the pulses on a neutron monitor, he began speculating—about power-production plants and about bombs. Within days Oppenheimer guessed that a "ten cm cube of uranium deuteride . . . might very well blow itself to hell." IGCC #1, pp. 22 and 25.

15. MIT Vol. II, p. 54.
16. Bernard T. Feld interview, January 13, 1987.
17. See, for example, Szilard's "Book" ms "Apology (in lieu of a foreword)," pp. 4–5 (LSP 40/4).
18. MIT Vol. II, p. 61.
19. Ibid., p. 69–70, note 24.
20. Ibid., p. 55.
21. Ibid., pp. 55 and 63.
22. A high-energy "fast" neutron moves at about one-thirtieth the speed of light, or 10 million meters a second. "Slow" neutrons move at about 2,200 meters a second.
23. I. Miller to S. E. Krewer, February 21, 1939, and S. E. Krewer to Gertrud Szilard, March 22, 1978 (LSP 10/26).
24. IGCC #1, pp. 23 and 47. *Newsweek,* February 6, 1939.
25. IGCC #1, pp. 23 and 47. "Release of Atomic Energy Disclosed . . . ." Science Service, January 28, 1939.
26. "Is World on Brink of Releasing Atomic Power?" Science Service, January 30, 1939.
27. "Releasing of Atomic Energy . . . " Science Service, January 30, 1939. My thanks to Ralph Lapp for this source.
28. Szilard had proposed a similar organization to Fermi in March 1936. Szilard to Strauss, February 13, 1939. MIT Vol. II, p. 63.
29. MIT Vol. II, pp. 63–65.
30. Receipt from "US Appraiser Stores, 201 Varick St., 7th Subs, West Houston Street, New York" (LSP).
31. Walter Zinn to the author, September 12, 1987.
32. William L. Laurence, *Dawn over Zero* (New York: Knopf, 1946).
33. Szilard Memoranda February 24 and March 3, 1939 (LSP 41/3).
34. Remarks at a banquet for the eightieth anniversary of *The Nation,* December 3, 1945. Reprinted in MIT Vol. II, p. 55.
35. Teller, *The Legacy of Hiroshima,* p. 10.
36. Herbert L. Anderson, "The Legacy of Fermi and Szilard," p. 61. The paper "Neutron Production and Absorption in Uranium" by Anderson, Fermi, and Szilard was published August 1, 1939, in the *Physical Review,* Vol. 56, pp. 284–86.
37. Szilard cable to Strauss, March 4, 1939 (LLS Szilard, Leo 1936–39, Box 109).

38. Strauss, p. 174.
39. Szilard's US Patent Application No. 10,500 was filed March 11, 1935; others were filed in England on March 12, May 9, June 14 and 28, July 4, and September 20 and 25, 1934 (LSP 35/40 and 36/1–11).
40. Strauss appointment book (LLS Box 19).
41. Szilard Memorandum to Compton, November 12, 1942 (MED 201).
42. Wigner interview in "The Man Behind the Bomb," Japanese Television documentary, 1979.
43. Others attending the meeting were Wigner, George Placzek, Leon Rosenfeld, and John A. Wheeler. See Wheeler's "Discovery of Fission," *Physics Today,* November 1967, pp. 49–52. IGCC #1, pp. 25 and 48. Stuewer, p. 282. Wheeler interview, November 18, 1985. Barton Bernstein introduction, MIT Vol. Ill, p. xxix.
44. H. von Halban, F. Joliot-Curie, and L. Kowarski published "Liberation of Neutrons in the Nuclear Explosion of Uranium," in *Nature,* pp. 470–71. The issue was dated March 18, 1939, but was available in England a few days earlier. IGCC #1, p. 45. "Uranium Atoms Split by Neutrons . . ." Science Service, March 16, 1939.

    Had Joliot ignored Szilard's February plea to withhold publication? Szilard thought so, but in Paris the publication had a different meaning, at least to the three authors. "Joliot disregarded Szilard's letter for a very simple reason," coauthor Lew Kowarski recalled years later. "Szilard, being a Hungarian, wrote with this exquisite courtesy," saying: "The only purpose of this letter is to prepare you for the remote possibility of a cable, which we might send you in some weeks. We hope that this letter is an unnecessary precaution." A cable was to follow if neutrons were released when uranium fissioned. "There was no cable, and Joliot disregarded the letter. It's that simple." Lew Kowarski oral-history interview, p. 79 (AIP).
45. When Szilard returned to Columbia and discovered that Fermi had argued the case for suppressing publication effectively, he was impressed with his fairness and sense of honor—an impression that carried much weight whenever the two later disagreed. Szilard "Book" ms "Apology (in lieu of a foreword)," p. 5 (LSP 40/4).
46. Quote recounted in Ralph Lapp interview, April 6, 1987. For background on the self-censorship debate, see Spencer Weart, "Scientists with a Secret," *Physics Today,* No. 29, February 1976, pp. 23–30.
47. *Newsweek,* March 27, 1939, p. 32.
48. See, e.g., Wigner to Dirac, March 30, 1939. MIT Vol. II, pp. 71–72.
49. Szilard memorandum to Weisskopf, March 31, 1939 (LSP 20/25).
50. Lew Kowarski oral-history interview, p. 80 (AIP). The Nazis did note the halt in publication. See "Farm Hall" transcript FH 4 for August 6–7, 1945, pp. 16–17 (MED, RG 77, Box 163).

51. Cables in MIT Vol. II, pp. 73–74 (LSP 10/23).
52. Tekla Szilard to Leo Szilard, April 1, 1939, Bela Silard translation (BSP).
53. Sherwin, p. 25. For details of Szilard's continued efforts to suppress neutron research, see Blackett to Weisskopf, April 8, 1939, MIT Vol. II, p. 74; Goldhaber to Szilard, April 12, 1939 (LSP 8/35); Tuve to Breit, April 12, 1939, DTM. See also IGCC #1, pp. 25–26 and 48. Isidor I. Rabi interview, August 19, 1985.
54. *Physical Review,* April 15, 1939, pp. 797–98 and 799–800, respectively. IGCC #1, pp. 45–46.
55. Wigner to Szilard, April 17, 1939 (LSP 21/4).
56. The article by von Halban, Joliot-Curie, and Kowarski appeared in *Nature* on April 22. See also Szilard to Joliot, July 5, 1939. MIT Vol. II, p. 80.
57. For the following description of worldwide reaction to Joliot's publications I am indebted to Badash, Hodes, and Tiddens, whose study "Nuclear Fission: Reaction to the Discovery in 1939" (IGCC #1) documents scientific activities and publications comprehensively.
58. IGCC #1, p. 28.
59. Wyden, p. 32.
60. IGCC #1, p. 27.
61. Herbert York, p. 29. IGCC #1, p. 28.
62. Lew Kowarski oral-history interview, p. 87 (AIP).
63. *The New York Times,* April 20, 1939, p. 39. *Washington Post,* April 29, 1939, p. 30.
64. *The New York Times,* May 5, 1939, p. 25. IGCC #1, p. 29.

CHAPTER 14

1. Today's nuclear power plants use the system that Szilard, Fermi, and Anderson tried—with water as both a moderator and a coolant. But now the water's absorption of neutrons is compensated by enriching the fissionable uranium^{235} isotope in the fuel from 0.7 to 3 percent.
2. MIT Vol. II, p. 81.
3. Fermi to Szilard, June 26, 1939 (LSP 8/6). See also "Neutron Production and Absorption in Uranium" by Anderson, Fermi, and Szilard, *Physical Review,* Vol. 56 (August 1, 1939), pp. 284–86.
4. Fermi to Szilard, July 1, 1939 (LSP 8/6).
5. Szilard to Fermi, July 3, 1939. MIT Vol. I, pp. 193–94.
6. Szilard to Strauss, July 3, 1939. MIT Vol. II, p. 88–89.
7. Szilard to Fermi, July 5, 1939. MIT Vol. I, pp. 194–95.
8. C. P. Brower to Szilard, July 6, 1939 (LSP 14/4).
9. Szilard to Fermi, July 8, 1939. MIT Vol. I, pp. 195–96.
10. MIT Vol. II, p. 82.
11. Fermi to Anderson, July 18, 1939 (LSP 8/6).

12. Szilard to Richards, July 9, 1939 (LSP 16/14).
13. G. J. Gardner to Szilard, July 21, 1939 (LSP 14/4).
14. Szilard to Trude, July 10, 1939 (EWP).
15. MIT Vol. II, pp. 82 and 89–90. Hewlett and Anderson, p. 16.
16. Szilard to Wigner, February 1, 1956. MIT Vol. I, pp. 190–92.
17. Szilard to Trude, July 14, 1939 (EWP).
18. IGCC #1, pp. 27 and 48–49.
19. Szilard to Trude, July 10, 1939 (EWP).
20. Jungk, pp. 83–84.
21. Clark, 1972, pp. 668–72.
22. Szilard notes for interview, April 18, 1955. MIT Vol. II, p. 83.
23. Sources for this account are many and varied, but the version here represents the latest assessment of the evidence.

    For the date, which has long been in dispute, I find a proposed visit mentioned in a newly discovered July 10, 1939, letter by Szilard to Trude Weiss.

    For accounts and details of the drive and the visit I rely on these sources:

    1. Szilard's April 18, 1955, notes for a radio interview the day Einstein died (AEP 39 488 1–3).

    2. Various drafts of Szilard's recollections, including a "Memorandum about the Einstein Letter," August 19, 1955 (LSP).

    3. Szilard's printed recollections (MIT Vol. II) with notes by Gertrud Weiss Szilard on the handwritten drafts then in her possession.

    4. Jungk's *Brighter Than a Thousand Suns,* which Szilard later praised for its accuracy.

    5. Ralph Lapp's August 2, 1964, article for the *New York Times Magazine,* and my interview with Lapp on April 6, 1987.

    6. Mrs. Elwood Martz's October 12, 1964, letter to Lapp about the Einstein cottage, where she then lived.

    7. Eugene Wigner interviews, October 12, 1984, and March 5, 1986.

    8. Interview with Robert McAll, the cottage's current owner, August 10, 1992.
24. MIT Vol. II, p. 84.
25. The meeting date is estimated from Szilard's reference to "a new 'Strauss' " in a July 14, 1939, letter to Trude Weiss (EWP).
26. Sachs said in 1959 that he had drafted a version of the Einstein letter, although he did not take credit when testifying about the initiative in 1945. See his testimony on S.R. 179 in 1945 and his September 29, 1959, letter to AEC historians Hewlett and Anderson. See also Sachs to Szilard, July 26, 1939, Sachs Papers, Container 79 Szilard, Dr. Leo (FDR).
27. Szilard to Einstein, July 19, 1939. MIT Vol. II, pp. 90–91.
28. Einstein to Sam Marx at M-G-M, July 19, 1946 (AEP 57 162).

29. MIT Vol. II, p. 84.
30. Janet Coatesworth to Gertrud Szilard, August 29, 1964 (LSP 85/5).
31. Szilard to Einstein, August 2, 1939. MIT Vol. II, pp. 92–93.
32. Szilard to Trude, August 3, 1939 (EWP).
33. Bela Silard translation of Ten Commandments draft, marked on the back "Hist. Box Hist.-I bbs-49," included with an October 30, 1940, version, · which is mimeographed (LSP 34/20).

    According to his wife, Szilard first began to draft the commandments when he was in Europe and always thought they were not translatable from the German. This is the first date on which all Ten Commandments were typed. On a version dated October 30, 1940, Szilard corrected the grammar of the second commandment to "should be directed toward."

    As late as 1960, when approached by historian Alice K. Smith, Szilard still insisted that his commandments were untranslatable and should be read only in German. Upon his death in 1964, his widow asked Jacob Bronowski, a colleague of Szilard's at the Salk Institute for Biological Studies, to translate the commandments (LSP 89/9). Bronowski's translation and the German text *"Zehn Gebote von Leo Szilard"* appear in MIT Vol. II, before the contents page.
34. Undated Einstein note to Szilard, Bela Silard translation (BSP). Reference is made to this note, then in the possession of Gertrud Szilard, in the introductory text to Document 56, MIT Vol. II, p. 96.
35. Bela Silard translation. MIT Vol. II, p. 96.
36. Jungk, p. 86. The long version, which was sent to President Roosevelt, has been on display at the presidential library in Hyde Park, New York, since 1945. The short version, which Szilard retained, passed to his widow, then to her brother, Egon Weiss, who sold it for $50,000. A seller offered the letter at Christie's in New York for $60,000 to $80,000. The letter was purchased at auction by publisher Malcolm Forbes on December 19, 1986, for $220,000, and was then on display in the lobby of the Forbes Building on Fifth Avenue in New York City. Also displayed was a handwritten note by Einstein to Szilard transmitting the signed letters. See *The New York Times,* December 19, 1986, p. C28, and December 20, 1986, p. 15; *Washington Post,* December 20, 1986, pp. C1 and C11.
37. "Memorandum," August 15, 1939. MIT Vol. I, pp. 201–3.
38. Rhodes, pp. 311–12.
39. Einstein to Lindbergh, Lindbergh papers, Yale University Library.
40. Szilard to Einstein, September 27, 1939. MIT Vol. II, p. 100.
41. Wigner to Szilard, September 26, 1939. MIT Vol. II, pp. 103–104.
42. See, for example, Szilard to Charles E. Chapin Co., October 3, 1939; Smith Bolton (United States Graphite Co.) to Szilard, November 10, 1939; and Szilard to Bolton, November 15, 1939 (LSP 6/11, 19/21).

43. Moore, p. 268. Hewlett and Anderson, pp. 17–19. Jungk, pp. 109–11. Sachs testimony before Senate Special Committee on Atomic Energy, November 27, 1945.

44. Rhodes, pp. 313–15.

45. Roosevelt replied to Einstein on October 19, thanking him for the letter and materials and explaining that he had created a special interdepartmental committee. "I am glad to say that Dr. Sachs will cooperate and work with this Committee and I feel this is the most practical and effective method of dealing with the subject." MIT Vol. II, p. 96. FDR Secretary's File, Sachs (FDR).

46. R. B. Roberts and J. B. H. Kuper, "Uranium and Atomic Power," *Journal of Applied Physics,* Vol. 10, No. 9; Hewlett and Anderson, p. 20.

47. Eugene Wigner interview, October 12, 1984. Sachs to Wigner, October 17, 1939, Eugene Wigner Papers. Hewlett and Anderson, p. 20.

48. Teller, *Energy from Heaven and Earth,* p. 144.

49. Eugene Wigner interview, October 12, 1984.

50. Szilard to Pegram, October 21, 1939. MIT Vol. II, pp. 109–10.

51. Szilard Memorandum to Lyman J. Briggs, "The Possibility of a Large-Scale Experiment in the Immediate Future." October 26, 1939. MIT Vol. I, pp. 204–206.

52. The dinner was held on November 7, 1939. Szilard to Sachs, November 5, 1939. MIT Vol. II, p. 112–13.

53. Szilard to Benjamin Liebowitz, December 4, 1939. MIT Vol. II, pp. 113–14. According to Liebowitz, Szilard later repaid the loan.

CHAPTER 15

1. MIT Vol. II, p. 115; (LSP 41/1).

2. MIT Vol. II, p. 115.

3. *The New York Times,* February 1, 1940, p. 1. Cf. H. von Halban, Jr., F. Joliot, L. Kowarski, F. Perrin, *Journal de Physique et le Radium,* ser. 7, 10: 428–29 (1939). MIT Vol. II, p. 115, note 1.

4. MIT Vol. II, pp. 115–16.

5. MIT Vol. I, pp. 207–56.

6. MacPherson, pp. 141–43. Rhodes, pp. 331–32.

7. Einstein to Sachs, March 7, 1940 (AEP 39 475 1–3 and 39 476 1–2). Sachs to Roosevelt, March 15, 1940; Roosevelt to Sachs, April 5, 1940. MIT Vol. II, pp. 120–22.

8. Szilard to Sachs, April 22, 1940. MIT Vol. II, pp. 123–25.

9. Szilard to Dr. Bourse, April 19, 1940 (LSP 4/20).

10. Rhodes, p. 327. In April 1940, something was being considered potentially in Japan as Takeo Yashuda, director of the Aviation Technology Research

Institute, ordered a subordinate to begin work on a report about nuclear fission for the Imperial Japanese Army.

11. Szilard memorandum to Pegram and Fermi, April 7, 1940 (LSP 15/8).
12. MIT Vol. II, pp. 116–17. Hans Bethe interviews, November 21, 1985, and March 3, 1986.
13. Szilard's "Book" ms "Apology (in lieu of a foreword)," pp. 4–5 (LSP 40/4).
14. Fermi talk on "Physics at Columbia University," January 30, 1954. Ms in Fermi file, University of Chicago, News and Information Office. Reprinted in *Physics Today*, November 1955, pp. 12–16. George Weil interview, April 7, 1983.
15. *Time*, August 20, 1945, Atomic Age reprints. *Reviews of Modern Physics*, January 1940, Vol. 12, pp. 1–29.
16. Turner to Szilard, May 27, 1940. MIT Vol. II, pp. 126–27. Clinch River Breeder Reactor Program history in *Perspectives*, 1968, p. 121.
17. "Creative Intelligence and Society: The Case of Atomic Research, the Background in Fundamental Science," MIT Vol. I, p. 178ff., Turner quote, p. 188. Herbert Anderson interview, November 4, 1982.
18. Szilard to Turner, May 30, 1940. MIT Vol. II, pp. 127–29; (LSP 19/7); Turner to Szilard, June 1, 1940. MIT Vol. II, pp. 132–33.
19. Breit to Szilard, June 5, 1940. MIT Vol. II, p. 133. Briggs to Szilard, June 7, 1940 (LSP 5/9).
20. Rhodes, p. 338.
21. See Szilard memo to Compton, November 12, 1942 (LSP 6/30). Szilard memo to Harold Urey, MIT Vol. II, pp. 117 and 129–31.
22. The meeting was held on June 15, 1940. See *Columbia Alumni News*, September 1945, pp. 5–6.
23. Szilard to Fermi, June 19, 1940 (LSP 8/6).
24. Szilard to Fermi, July 4, 1940. Vol. II, pp. 133–35.
25. MIT Vol. II, p. 135, n17.
26. George Weil interview, April 7, 1983.
27. MIT Vol. II, p. 143. Hans Bethe interview, March 3, 1986.
28. Szilard to Lawrence, July 12 and August 6, 1940 (LSP 11/27).
29. Bertrand Goldschmidt interview, July 9, 1987.
30. Lt. Col. S. V. Constant report, August 13, 1940 (LSP 95/7).
31. FBI Reports, Serial 01042016, in FOIA CO 2.12-C (1001) of January 15, 1981, by Gannett to Gertrud Weiss Szilard (LSP 95/1).
    Hoover, who later made a point of reporting about all Szilard's suspicious actions, now seemed under pressure from the military to produce a favorable account. On October 9 he cabled the New York FBI office: "MR. SZELARD SPECIAL INQUIRY, NATIONAL DEFENSE PROGRAM. COMPLETE INVESTIGATION AT ONCE AND SUBMIT REPORT NO LATER THAN FOURTEENTH INSTANT HOOVER."

Six days later he cabled his agents to "COMPLETE INVESTIGATION AND SUBMIT REPORT AT ONCE HOOVER" (LSP 95/9).

Another telegram query about the "Szelard" investigation followed a day later. By October 24 the FBI reported from Washington on Szilard's immigration status and past movements.

FBI report, October 24, 1940; File No. 77-2571 (LSP 95/9).

Four days later, FBI agents reported from New York City about Szilard's activities in and around Columbia. FBI report, October 28, 1940; File No. 62-6878 VA (LSP 95/9).

And on the day Szilard was appointed to the NDRC at Columbia, FBI agents called on Einstein in Princeton to inquire about Szilard's loyalty. FBI report, November 8, 1940, p. 3; File No. 77-153 EMR (LSP 95/9).

32. Elizabeth Silard interview, March 15, 1987.

33. "Reminiscences," p. 122.

34. Undated application due October 15 and marked 1940 but not completed on cover sheet (LSP 81/2). Stephen L. Schlesinger (secretary to the Guggenheim foundation) to the author, March 3, 1988.

35. Feld, pp. 24–25.

36. Ladenburg to Briggs, April 14, 1941 (MED).

37. John Marshall, Jr., interview, February 2, 1988. Szilard wrote "Control of the Chain Reaction" on July 19, 1941 (LSP 23/11).

Szilard thought of cooling with bismuth, a grayish-white metal used to fuse alloys. His design would carry off the heat that fission generated by circulating liquid bismuth or a bismuth-lead solution through the system. Here Szilard's Berlin work with Einstein proved helpful; the bismuth coolant would be circulated by the kind of electromagnetic pump they had designed a decade earlier for refrigerators. December 9, 1940, patent description, MIT Vol. I, p. 724.

38. *PIC*, July 22, 1941, pp. 6–8.

39. MIT Vol. II, p. 144. See also "Proposed Conversation with Bush, February 28, 1944," Part IV, in Vol. II, pp. 176–79.

40. "Suggestions for a Search for Element 94 in Nature." Report A-45, September 26, 1941. Cited in MIT Vol. I, p. 376.

"Memorandum Raising the Question Whether the Action of Explosive Chain-reacting Bodies Can Be Based on an 'Expulsion' Method," written by Szilard October 21, 1941. Report A-56 (Columbia University). Cited in MIT Vol. I, p. 376.

Draft of "Memorandum on the Contribution of Fast Neutrons to the Chain Reaction in a Uranium-Carbon System," October 28, 1941 (LSP 41/5).

"Preliminary Report on Fission Caused by Fission Neutrons," November 14, 1941, by Szilard with John Marshall, Jr. MIT Vol. I, pp. 266–75.

"Preliminary Report on the Capture of Neutrons by Uranium in the Energy Region of Photo Neutrons from Radium-Beryllium Sources," December 5, 1941, by Szilard and John Marshall, Jr. MIT Vol. I, pp. 276–79. "The Capture of Neutrons by Uranium in the Energy Region of Photo Neutrons from Radium-Beryllium Sources," December 6, 1941 (LSP 30/18).

41. Barton J. Bernstein, "Leo Szilard: Giving Peace a Chance in the Nuclear Age," *Physics Today*, September 1987, p. 40; based on Alice K. Smith interview with Eugene Wigner, March 15, 1960. My thanks to Alice K. Smith for sharing this information.

In fact, no paper I could find was authored by Feld, Marshall, and Szilard, although to Teller and others this still makes a fine story. For Teller's account, see Teller, Edward with Judith Shoolery. *Memoirs: A Twentieth-Century Journey in Science and Politics.* Cambridge, MA: Perseus Publishing, 2001, p. 111, n 3.

42. Bush to Roosevelt, November 27, 1941 (MED).
43. Herbert L. Anderson, *Bulletin of the Atomic Scientists*, September 1974, p. 42. Hewlett and Anderson, p. 49. Rhodes, pp. 388–89.
44. Roosevelt to Bush, January 19, 1942. Hewlett and Anderson, p. 49.
45. MIT Vol. I, pp. 262–65 and 280–87.
46. See site-selection details in Szilard to Fermi, December 31, 1941 (LSP).

CHAPTER 16

1. Hewlett and Anderson, pp. 54–55.
2. Szilard "Memorandum for Professor A. H. Compton," January 19, 1942; Szilard "Progress Report to Dr. Doan and Dr. Hilberry on the Search for a Site in the New York Area," January 20, 1942 (EWP).
3. Physicist Isidor I. Rabi, who talked with the participants after the war, said that had Pegram been more assertive, he could have convinced Compton to keep the uranium work at Columbia. Rabi interview, August 19, 1985.

As Anderson reported to Szilard on January 21, his seven favored sites were (1) the Pegassus Club polo field in Bergen County, New Jersey; (2 and 3) the Fokker Hangar or the Goodyear blimp hangar at Bendix Field in Bendix, New Jersey; (4) a Curtiss-Wright hangar in Valley Stream, Long Island; (5) a concrete-and-steel industrial building in Clifton, New Jersey; (6) a golf course in Yonkers, New York; and (7) the Armonk airfield hangars in Armonk, New York. Anderson memorandum to Szilard, January 21, 1942 (LSP 4/12).

Szilard memorandum to Anderson, Fermi, Feld, Marshall, Teller, Zinn, Alvarez, Compton, and Lawrence, January 26, 1942 (EWP). Szilard Rough Draft to Compton, January 27, 1942 (LSP 6/30). It is possible the Compton letter was not sent.

4. Civil Service Commission investigation, July 6–9, 1959, p. 2 (LSP 96/2).

5. FBI report, October 13, 1954, p. 2 (LSP). MIT Vol. II, pp. 146–47.

6. Szilard to Trude, May 10, 1942 (EWP).

7. Navy secretary's consent, March 24, 1942. Signed J. R. Cannon (LSP). Compton to Conant, May 16, 1942 (MED).

8. Compton to Conant, May 16, 1942. RG 227, Bush-Conant, folder 217, Szilard, Leo (MED).

9. Szilard to Bush, May 26, 1942. MIT Vol II, pp. 151–52.

10. S-1, RG 227, Bush-Conant, folder 217, Szilard, Leo (MED).

11. Compton to Bush, June 1, 1942 (MED).

12. Szilard to Compton, June 1, 1942 (MED).

13. See MIT Vol. II, p. 193, n10.

14. Spencer R. Weart, "The Road . . ." p. C8-317. Eugene Wigner interview, October 12, 1984.

15. Bush to Conant, June 3, 1942, S-1, RG 227, folder 217, Szilard, Leo (MED). Gannett to Gertrud Weiss Szilard, January 15, 1981 (LSP). Undated pink routing slip to Bush and Conant. S-l, RG 227, Bush-Conant, folder 217, Szilard, Leo (MED).

16. Briggs to Bush, June 9, 1942 (LSP).

17. Enrico Fermi, "Physics at Columbia," January 30, 1954. *Physics Today*, November 1955, pp. 12–16.

18. MIT Vol. I, pp. 332–39 and 346–50.

19. Szilard Memorandum for Allison, Compton, Fermi, Wigner, Doan, and Hilberry, July 1, 1942 (EWP).

20. "A Magnetic Pump for Liquid Bismuth" by Szilard and Feld, July 14, 1942. MIT Vol. I, pp. 351–58.

21. Szilard to Trude, back of letter dated July 8, 1942 (EWP).

22. Szilard to Trude, July 8 and 28, 1942 (EWP).

23. After several visits, Brush company officials told Szilard there was "a sporting chance" uranium metal could be made by the same process used to make beryllium metal, but Szilard's efforts at the Met Lab to buy more uranium for experiments failed, and the idea was dropped. See Szilard to Compton, July 31, 1942; Szilard telegram to Dr. C. B. Sawyer of Brush Beryllium Company, July 29, 1942; Szilard memo to Compton, July 30, 1942 (LSP). Creutz to the author, September 27, 1986, and October 22, 1986. David Gurinsky to the author, November 21, 1986.

24. Glenn Seaborg, *History of Met Lab Section C-l, April 1942 to April 1943* (Berkeley, Calif: Lawrence Berkeley Laboratory, 1977), pp. 192ff.

25. Szilard to Trude, August 9, 1942 (EWP).

26. Szilard September 19, 1942, Memorandum (copy in LSP, original in EWP).

27. "What Is Wrong with Us?," September 21, 1942 (LSP). See also MIT Vol. II, pp. 153–60.

28. Groves, pp. 39–41.

29. Memorandum from Frances Henderson to Don Bermingham *(Time)* on "Groves v. the Scientists," March 8, 1946 (LSP). See also "The Elusive Dr. Szilard," *Harper's,* September 1960. Szilard University of Chicago Faculty Questionnaire, 1958/9.
30. Charles D. Coryell, oral-history interview, pp. 88–89. Columbia Oral History Research Office.
31. Groueff, pp. 35–36.
32. Compton to Szilard, October 7, 1942; Szilard to Compton, memo on "Your note concerning engineering of bismuth cooled plant, October 7, 1942," October 9, 1942 (EWP).

   DuPont engineer Crawford H. Greenwalt at first meeting thought Szilard "a queer fish—but I believe honest." Greenwalt Diary, January 3, 1943, p. 19, Hagley Museum and Library, Wilmington, Delaware.
33. Szilard to Trude, October 11, 1942 (EWP). Report CE-301 for month ending October 15, 1942, cited in MIT Vol. I, p. 376.
34. Samuel K. Allison interview with Alice K. Smith, April 12, 1960.
35. John Marshall, Jr., interview, February 2, 1988.
36. Szilard to Trude, Friday. Found between October 19 and November 2, 1942 (EWP).
37. MIT Vol. II, p. 149.
38. Szilard Aide Memoir, October 30, 1942 (EWP).
39. October 26, 1942, cable (MED 201 Szilard, Leo).
40. Szilard Aide Memoir, October 30, 1942 (EWP).
41. Compton to Groves, October 28, 1942 (MED 201 Szilard, Leo).
42. Secret draft, War Department Office of the Chief of Engineers, Washington, October 28, 1942 (MED 201 Szilard, Leo).
43. Groves reminiscences, October 17, 1963. See Gruber, pp. 80 and 86.
44. Compton to Szilard, October 30, 1942 (MED 201, Szilard, Leo).
45. Szilard to Trude, October, undated (EWP).
46. Szilard to Trude, November 2, 1942 (EWP).
47. Szilard to Compton, November 4, 1942 (LSP).
48. Szilard to Trude, November 5 and 16, 1942 (EWP).
49. Szilard to Compton, November 4, 1942 (LSP). Compton to Groves, November 13, 1942 (MED 201 Szilard, Leo).
50. Herbert L. Anderson, "The Legacy of Fermi and Szilard," *Bulletin of the Atomic Scientists,* September 1974, p. 43.
51. Szilard's "Uranium Aggregates for Power Unit," November 23, 1942. A memo to Wigner on bismuth cooling was also sent the same day. See Report CP-357, cited in Vol. I, p. 376.
52. Anderson, "The Legacy of Fermi and Szilard . . ." See also Len Ackland, "Dawn of the Atomic Age," *Chicago Tribune Magazine,* November 28, 1982, p. 23.
53. Charles Hartshorne to the author, January 10, 1986.

54. Humphry Osmond to Margaret Spanel, November 13, 1986. Osmond to the author, March 8, 1987.
55. Compton, p. 139.
56. Anderson, "The Legacy of Fermi and Szilard . . ." pp. 44–45.
57. "The First Reactor," USAEC Division of Technical Information. LC 50-60514. 1960.
58. *CBS Reports* interview, April 2, 1960, transcript p. 47 (LSP). See Vol. II, p. 146, attribution to Mike Wallace interview, WNTA-TV, February 27, 1961. See also *See It Now,* November 1952, with Edward R. Murrow.

CHAPTER 17

1. Szilard to Compton, December 3, 1942 (LSP).
2. Szilard to Samuel K. Allison, December 4, 1942 (LSP).
3. Pierre Auger interview, July 3, 1987. Szilard enjoyed this story and retold it in a television interview the day Einstein died.
4. Szilard to Trude, December 26 and 31, 1942 (EWP).
5. Szilard to Marshall MacDuffie, June 13, 1956, pp. 1–3 (LSP).
6. Szilard to Compton, January 8, 1943. "Memorandum on the Production of 94 and the Production of Power by Means of the Fast Neutron Reaction." MIT Vol. I., p. 175.
7. Weinberg 1982 interviews for William Lanouette, "Dream Machine," *Atlantic Monthly,* April 1983. Weinberg interview, December 17, 1985, for this biography. "Liquid Metal Cooled *Fast* Neutron Breeders," by Szilard, UC-LS-60. March 6, 1945. This memorandum cites Szilard's presentations at breeder discussions on April 16 and 28, 1944. MIT Vol. I, pp. 369–75.
   Although they scarcely produce any extra fuel, breeders that have operated in the United States and France were cooled by electromagnetic pumps of the kind that Szilard and Einstein designed in the 1920s.
8. Szilard to Trude, January 15, 1943 (EWP).
9. Szilard to Compton, January 2, 1943 (MED).
10. MIT Vol. II, p; 149. Hans Bethe interview, November 21, 1985.
11. On February 22, 1943, T. O. Jones wrote to Maj. John Lansdale mentioning that Szilard's "mail was being watched" (MED 77, 201).
12. Gruber, p. 80, n 25. T. O. Jones to Maj. John Lansdale, February 22, 1943 (MED RG 77, 201). Szilard to Trude, February 8, 1944 (EWP).
13. Szilard to Trude, March 6, 1943 (EWP).
14. FBI report, November 30, 1954 (LSP). (FLP D.237). See also Confidential June 24, 1943, Memo to Officer in Charge, pp. 1 and 56 (LSP).
15. July 12, 1945, letter and memo to Groves (FLP D.237 and D.35).
16. Groves to Calvert, June 12, 1943 (MED 201 LS).
17. The account of Szilard's June 20 and 22, 1943, visit to Washington comes from FBI memorandum, June 24, 1943 (LSP).

18. Confidential June 24, 1943, Memo to Officer in Charge, p. 5 (LSP).
19. Lavender to Szilard, August 23, 1943 (MED 201 Szilard, Leo).
20. Samuel K. Allison interview with Alice K. Smith, April 12, 1960.
21. Connie Fay to Szilard, January 2, 1962 (LSP).
22. John Marshall, Jr., interview, February 2, 1988.
23. Szilard to Trude, July 5, 1943 (EWP).
24. Szilard to Compton, July 20, 1943. Compton to Szilard, August 4, 1943 (MED).
25. Compton to Szilard, August 4, 1943 (LSP).
26. Szilard to Trude, August 4, 1943 (EWP).
27. James Hume interview, February 14, 1986. See also Hume to Szilard, January 27, 1944, and enclosures. I am grateful to the late James Hume for sharing his Szilard files with me. Hume interview, October 3, 1986.
28. Szilard to Compton, August 7, 1943 (LSP).
29. Szilard to Trude, August 16, 1943 (EWP).
30. Szilard to Trude, August 19, 1943, letter number 3 (EWP).
31. Friday (September 24) letter by Szilard to Trude, found between letters dated September 22 and 26, 1943 (EWP).
32. Groves to Szilard, October 8, 1943 (LSP and MED RG 77, 201LS). Szilard cable to Lavender, October 8, 1943 (LSP and MED 201 Adam, Mr. Isbert; Szilard, Leo).
33. Lavender memo, October 12, 1943 (MED 201 Szilard, Leo).
34. Four-hour meeting on November 21, 1943, noted in Hume to Szilard, January 27, 1944. James Hume interview, October 3, 1986.
35. James Hume interview, October 3, 1986.
36. The psychoanalyst Szilard had in mind was Taseau Bennet. Szilard to Trude, October 31, 1943 (EWP).
37. Szilard to Lavender, November 24, 1943 (MED).
38. James Hume interview, October 3, 1986. Metcalf "Memorandum of a conference held at the Chicago Area Office, US Engineers, on 3 December 1943," December 27, 1943. MUC.PA-418(MED).
39. Szilard to Compton, "Report about conversation with General Groves, Friday, December 3rd," December 11, 1943 (MED).
40. Szilard to Hume, December 3, 1943, and Metcalf Memo, December 27, 1943. MUC.PA-418 (MED). See also December 23, 1946, FBI report, p. 5 (LSP).
41. Szilard to Bush, December 13, 1943 (MED S-l, RG 227, Bush-Conant, folder 217 Szilard, Leo).
42. Bush to Conant, December 18, 1943; reply on December 23 (MED).
43. (Bush-Conant, 277 MED).
44. See December 3, 1943, account in Szilard to Moses, January 17, 1944 (LSP).
45. Frederick Seitz to the author, May 26, 1987.

46. Davis, p. 163. See also Rhodes, p. 451.
47. Szilard notes for "Proposed Conversation with Bush," February 28, 1943, Part IV, p. 5. MIT Vol. II, p. 178.
48. MIT Vol. II, pp. 161–63.
49. Conant to Bush and Bush to Conant, both March 9, 1944 (MED).
50. See MIT Vol. I, pp. 295ff. and 376ff.
51. Szilard to Trude, April 20, 1944 (EWP).
52. Szilard to Trude, May 20, 1944 (EWP).
53. See Szilard to Trude, March 24 and July 31, 1944 (EWP).
54. Szilard to Trude, November 11, 1944 (EWP).
55. Memorandum, August 10, 1944. MIT Vol. II, pp. 189–92. Manuscript copy (EWP).
56. Charles D. Coryell, oral-history interview, pp. 87–88. Columbia Oral History Research Office.
57. Crick, pp. 13–14.
58. Szilard to Bush, October 14, 1944, with an October 17 memo attached by Compton (LSP).
59. Jeffries to Compton, November 18, 1944. AKS/MIT, pp. 18 and 23.
60. Szilard draft speech on US-Russian relations, October 30, 1953 (LSP).

CHAPTER 18

1. Memorandum March 6, 1944 (LSP). "Reminiscences," pp. 122–23.
2. MIT Vol. II, pp. 196–204.
3. Ibid., p. 181.
4. The meeting opened with Stimson addressing questions put to FDR in a March 3 memorandum by James F. Byrnes, his director of war mobilization, seeking results of the $2 billion secret program. Byrnes had admitted he knew little about it, and Stimson listed the eminent scientists engaged in the project, including four Nobel laureates. Hewlett and Anderson, p. 340.
5. Hewlett and Anderson, p. 340; cf. Titus, "Collective Conscience," p. 7; Stimson in *Harper's*, February 1947, pp. 97–107.
6. Bohr to Roosevelt, March 24, 1945 (FDR).
7. MIT Vol. II, pp. 205–207.
8. LP Disc Recording, "Leo Szilard: Inventor of the Atomic Bomb." George Garabedian Production, 1984 (LSP 107). MIT Vol. II, p. 180. Hewlett and Anderson, pp. 203 and 681 n43. Conant to Bush, July 31, 1943, and Bush to Conant, September 22, 1944 (MED/OSR).
9. *U.S. News & World Report*, August 15, 1960, pp. 68–71.
10. MIT Vol. II, p. 182.
11. Rhodes, p. 636. "Reminiscences," p. 124–25. MIT Vol. II, pp. 182–83.
12. AEC Document 286 in Sherwin, p. 169.
13. Hewlett and Anderson, p. 345.

14. MIT Vol. II, p. 184. Szilard later reversed his view. "I saw his point at that time, and in retrospect I see even more clearly that it would not have served any useful purpose to keep the bomb secret, waiting for the government to understand the problem and to formulate a policy; for the government will not formulate a policy unless it is under pressure to do so, and if the bomb had been kept secret there would have been no pressure for the government to do anything in this direction." "Reminiscences," p. 127.

15. MIT Vol. II, p. 184. "Reminiscences," p. 128.

16. Hewlett and Anderson, p. 355. Byrnes, p. 284.

17. Hewlett and Anderson, p. 690, n38. Szilard's S-14 dictation, May 22, 1956, transcript p. 21 (LSP 107). Walter Bartky interview with Alice K. Smith, July 8, 1957.

18. Groves diary, May 30, 1945 (MED 77).

19. Szilard LP Disc Side D (LSP 107). MIT Vol. II, p. 185. "Reminiscences," pp. 128–29.

20. MIT Vol. II, p. 185.

21. Hewlett and Anderson, p. 356. "Notes of the Interim Committee Meeting, May 31, 1945." Appendix L of Sherwin, pp. 295ff.

22. Sherwin, Appendix, pp. 295–304. (Byrnes is cited on p. 301.) As the meeting's secretary recorded: "Mr. Byrnes expressed the view, which was generally agreed to by all present, that the most desirable program would be to push ahead as fast as possible in production and research to make certain that we stay ahead and at the same time make every effort to better our political relations with Russia."

23. This is based on notes of the meeting secretary and is not a direct quote from Groves.

24. AKS/MIT, p. 41.

25. Szilard to Norman Hilberry, June 6, 1945 (LSP).

26. Feis, p. 50 n. AKS/MIT, p. 42. Other members of the Franck Committee were Thorfin R. Hogness, Donald Hughes, C. J. Nickson, Eugene Rabinowitch, Glenn T. Seaborg, and Met Lab director Joyce Stearns.

27. Feis, p. 51.

28. Compton to Nichols, June 4, 1945, MUC-AC-1306/7 (MED 201 Szilard).

29. Ibid.

30. Barton J. Bernstein introduction to MIT Vol. III. Stimson in Harper's, 1947. Teller, "Seven Hours . . ." p. 191. Sherwin, Appendix M.

31. Groves to Compton, June 29, 1945 (MED 201 Szilard, Leo).

32. Groves to Cherwell, July 4, 1945. (FLP D.35 and 49.3).

33. "Reminiscences," p. 130.

34. Szilard to Waldo Cohen, July 4, 1945 (LSP).

35. Edward Teller, "Seven Hours . . ." p. 191.

36. (MED 201 Szilard, Leo.) Teller claims he dated the letter July 2 but actually wrote it days later. A more likely explanation is that he knew about the petition in advance and drafted his reply on the second, even before receiving a copy.
37. Teller to Szilard, July 2, 1945. MIT Vol. II, pp. 208–209.
38. Teller, "Seven Hours . . ." p. 191.
39. Ralph Lapp interview, April 2, 1987. MIT Vol. II, pp. 212–13. See also Edward Creutz to Szilard, July 13, 1945 (MED 201 E. Creutz).
40. Lieutenant Parish to Groves, July 9, with a copy of Szilard's July 4 letter to Teller, and Teller's reply (MED). Ralph Lapp interview, April 2, 1987.
41. MIT Vol. II, p. 213.
42. The results of Daniels's poll, in votes and percentage, are as follows:

      (1) Use the weapons in the manner that is from the military point of view most effective in bringing about prompt Japanese surrender at minimum human cost to our armed forces. (23, 15%)

      (2) Give a military demonstration in Japan, to be followed by a renewed opportunity for surrender before full use of the weapons is employed. (69, 46%)

      (3) Give an experimental demonstration in this country, with representatives of Japan present; followed by a new opportunity for surrender before full use of the weapons is employed. (39, 26%)

      (4) Withhold military use of the weapons, but make public experimental demonstration of their effectiveness. (16, 11%)

      (5) Maintain as secret as possible all developments of our new weapons, and refrain from using them in this war. (3, 2%)

      (MED DCN-55442, p. 9; see also Young to Szilard, January 31, 1961, LSP).
43. "Reminiscences," p. 130.
44. Szilard dictation, May 22, 1956 (LSP 100). Claude C. Pierce, Jr., to Groves, November 26, 1947. Papers of Leslie R. Groves (MED 200, Entry 2, Correspondence 1941–70, Box 7, Folder "P"). My thanks to Stanley Goldberg for this source.
45. MIT Vol. II, pp. 211–12.
46. Szilard to Creutz, July 10, 1945, and Creutz to Szilard, July 13, 1945 (MED 201 E. Creutz).
47. Szilard to Compton (MED NDN-55430).
48. Compton to Nichols, July 24, 1945 (MED). For all versions of the petition, see National Archives, Manhattan Engineer District, Record Group 77, Harrison-Bundy Files, Box 153, Folder 76.
49. Groves directive (MED 5E).
50. Fletcher Knebel to Szilard, October 25, 1961 (LSP). Knebel and Bailey, *Look,* August 13, 1960, pp. 19–23. Szilard "Reminiscences," p. 132, n57.

51. FBI Secret Report, p. 007 (LSP). Creutz to the author, October 22, 1986. George Weil interview, April 7, 1983.
52. Harry S Truman, *Year of Decisions* (Garden City, N.Y.: Doubleday & Co., 1955), p. 390ff.
53. FBI "Secret Report" (LSP).
54. By year's end, more than 130,000 had died, and latent effects claimed another 70,000.
55. Sayen, pp. 151 and 316, from an interview with Helen Dukas, November 9, 1977. Lapp, "The Einstein Letter That Started It All," *The New York Times Magazine*, August 2, 1964, p. 54. Clark, 1972, p. 708.
56. Groves, pp. 333–36; Farm Hall transcripts (MED RG 77, Box 163).
57. Transcript for "The Mike Wallace Interview," February 27, 1961 (LSP 40/22).
58. Leigh Fenly, "The Agony of the Bomb, and Ecstasy of Life with Leo Szilard," *San Diego Union*, November 19, 1978, pp. D-1 and D-8.
59. (Copy in LSP, original in EWP.) Ironically, a War Department certificate dated August 6 and signed by Stimson soon came to Szilard. It read: "This is to Certify that Leo Szilard University of Chicago has participated in work essential to the production of the Atomic Bomb, thereby contributing to the successful conclusion of World War II. This certificate is awarded in appreciation of effective service"(EWP).
60. Fenly, "The Agony . . ." p. D-8.
61. Robert M. Hutchins, oral-history interview, November 21, 1967. Columbia Oral History Research Office, p. 81. Szilard to Hutchins, August 8, 1945 (LSP).
62. Szilard's May 22, 1956, dictation, transcript, p. 26 (LSP 107).
63. On August 13, Szilard drafted another letter to the White House urging President Truman to halt further atomic bombing (LSP).
64. Hartshorne to the author, January 10, 1986. Hartshorne interview, April 10, 1986. Szilard's May 22, 1956, dictation, p. 26 (LSP 107).
65. Langsdorf 1983 Questionnaire to L. Badesh, IGCC #1, p. 36. Martyl interview (Weiss and Patton), Leo Szilard Biography Project.
66. For details on Szilard's efforts to release the petition, see Capt. James S. Murray to Szilard, August 27 and 28, 1945; Szilard to the editors of *Science*, August 18, 1945; Szilard to Hutchins, August 28, 1945 (LSP). Szilard, "Reminiscences," p. 133. Files on the petition are in LSP 40/15, 73/15, 73/16, and 89/6.

　　In rejecting Szilard's request, the army ruled that "[E]very paragraph . . . either contains some information or implies 'inside' information . . . which implies that internal dissention [sic] and fundamental differences in point of view disrupted the development and fruition of the District's work—an implication which you as well as I know is not founded on sober fact and which, if released at this time, might well cause 'injury to the interest or prestige of the nation or governmental activity.' " Murray to Szilard, August 27, 1945. For Szilard's response to this ruling, see Szilard to Hutchins, August 28, 1945 (LSP and MED 201 Szilard, Leo).

67. Szilard to Hutchins, August 29, 1945 (LSP). "Reminiscences," p. 133, n60.
68. FBI "Secret Report," p. 12 (LSP).
69. "Memorandum," First Version, Rough Draft, August 14, 1945, with September 13 and 17 revisions (LSP). See also Lapp, p. 115.
70. FBI "Secret Report," pp. 13, 53, and 241 (LSP). Hartshorne to the author, January 10, 1986.
71. FBI report, December 23, 1946, p. 246 (LSP).
72. See, for example, papers prepared for "Round Table: Hiroshima and the End of World War II." The Society for Historians of American Foreign Relations, Seventeenth Annual Conference, June 19, 1991, Washington, D.C.
73. David Dietz, "Era of Atomic Energy: Man's Control of Weather Seen Possible in Future," *New York World Telegram* (Scripps-Howard), August 17, 1945. (GSS No. 1, p. 48).

CHAPTER 19

1. AKS/MIT, p. 89. This is a paraphrase from the *Chicago Tribune,* September 2, 1945.
2. AKS/MIT, pp. 89–91.
3. AKS/MIT, p. 91, and BAS 33/41.
4. Historian Alice K. Smith cites Szilard's memo as the only evidence that the atomic scientists were concerned with domestic control of the atom; their dominant issue at the time was international control. Here Szilard anticipated the struggle ahead by his intense ratiocination with concrete details. AKS/MIT, p. 92.
5. McMahon proposal, September 6, 1945. Franck proposal made September 9, 1945.
6. Hewlett and Anderson, p. 423.
7. Stimson to Truman, September 11, 1945; Congressional Quarterly Service, p. 242.
8. Papers of the Atomic Scientists of Chicago (BAS). (RMH 5/9–13).
9. Hutchins to Groves, September 17, 1945. The Smyth report's Chap. XIII, paragraph 8, reads: "These questions are not technical questions; they are political and social questions, and the answers given to them may affect mankind for generations. . . . In a free country like ours, such questions should be debated by the people and decisions must be made by the people through their representatives. This is one reason for the release of this report" (RMH 5/9–13, LSP).

10. See Glenn T. Seaborg notes and article in the *Bulletin of the Atomic Scientists,* September 1985, pp. 31–33; and AKS/MIT, p. 94.
11. Benton to Szilard, September 20 and October 12, 1945 (LSP).
12. Remarks at Atomic Energy Control Conference, September 1945 (LSP 66/16).
13. AKS/MIT, p. 93. FBI report, December 23, 1946 (LSP).
14. Hewlett and Anderson, pp. 413ff. The Senate adopted the resolution on September 28, 1945.
15. "Round Table," September 30, 1945. This was Szilard's first opportunity to state his views on the bomb to a national audience.
16. In a 1955 television interview, Szilard dates this Princeton meeting as "shortly after the bomb was dropped at Hiroshima and Nagasaki," but his itinerary (as plotted by the FBI) shows him in Buffalo in late August and in New York in late September. Unless Szilard made a separate, unrecorded visit to Princeton in August, it is likely the meeting occurred as he was traveling to or from the New York "Round Table" broadcast during the last week of September. See interview transcript in AEP and in Sayen, p. 151.
17. "Reminiscences," p. 135. Szilard dictation, May 22, 1956 (LSP 107).
18. Congressional Quarterly Service, p. 242.
19. The bill numbers assigned were H.R. 4280 and S. 1463. Hewlett and Anderson, pp. 428ff. See also "Jurisdiction Row Again Blocks Atomic Energy Measure in Senate," *The New York Times,* October 5, 1945.
20. AKS/MIT, p. 137. *The Nation,* December 22, 1945, pp. 718–19. See also Szilard's May 22, 1956, dictation (LSP 107).
21. "Reminiscences," p. 135.
22. December 23, 1946, FBI report, p. 14 (LSP). AKS/MIT, p. 138.
23. "Reminiscences," pp. 136–37. See also Szilard's May 22, 1956, dictation, p. 30 (LSP 107).
24. AKS/MIT, p. 139. *Washington Post,* October 11, 1945, p. 6.
25. Anderson to Higinbotham, October 11, 1945 (copy in EWP).
26. FBI report, pp. 35–36 (LSP). Final outline in *Bulletin of the Atomic Scientists,* March 15, 1946, last page.
27. John A. Simpson interview, March 17, 1987. Hewlett and Anderson, p. 487.
28. Herbert Anderson interview, October 6, 1986.
29. John A. Simpson interview, March 17, 1987.
30. Hewlett and Anderson, p. 446.
31. Elizabeth Donahue, *PM,* October 17, 1945. "400 Experts . . ." *The New York Times,* October 14, 1945.
32. Bernard T. Feld interview, January 13, 1987.
33. Charles D. Coryell, oral-history interview, pp. 314–18, Columbia Oral History Research Office.
34. Howard J. Lewis to the author, December 30, 1985.

35. Hewlett and Anderson, pp. 440–41.
36. "Reminiscences," pp. 138–39. AKS/MIT, p. 138.
37. AKS/MIT, pp. 157–58. *The New York Times,* (U.P.) October 18, 1945.
38. Herbert Anderson interview, October 6, 1986.
39. US Congress, House, *Atomic Energy, Hearings* before the Committee on Military Affairs, House of Representatives, 79th Congress, 1st Session, on H.R. 4280, October 9 and 18, 1945 pp. 71–96. AKS/MIT, pp. 158–59. Telegram (LSP 71/7).
40. Photograph in Hewlett and Anderson, p. 449.
41. Stephen White, "Federal Control of Atomic Energy," *New York Herald Tribune,* October 23, 1945. Although most derogatory remarks were edited from the hearing transcript, they were reported at the time by journalists.
42. Hearings, pp. 898–901.
43. Ibid., pp. 913 and 920.
44. Ibid., p. 926.
45. Ibid. Stephen White, *New York Herald Tribune,* October 23, 1945.
46. AKS/MIT, p. 164.
47. Szilard dictation, May 22, 1956, p. 32 (LSP 107).
48. William S. White, *The New York Times,* October 19, 1945. *Newsweek,* October 29, 1945, p. 9. Szilard dictation, May 22, 1956 (LSP 107).
49. *The New York Times,* October 21, 1945.
50. *The New York Times,* October 23, 1945. Robert E. Nichols, *New York Herald Tribune,* October 31, 1945.
51. Raymond J. Blair, *New York Herald Tribune,* November 6, 1945.
52. (LSP 71/7). *PM,* November 9, 1945. Raymond J. Blair, *The New York Herald Tribune,* November 6, 1945. PM, *New York Herald Tribune,* and *New York Post,* November 9, 1945. *Congressional Record,* November 14, 1945, pp. A4877–78. W. H. *[sic]* Laurence, *The New York Times,* November 9, 1945.
53. *Newsweek,* December 3, 1945, p. 45.
54. William Higinbotham interview, June 5, 1984.
55. AKS/MIT, pp. 214–15.
56. *The New York Times,* November 16, 1945. Hewlett and Anderson, pp. 461–66.
57. Szilard to Farrington Daniels, November 17, 1945 (LSP 7/5).
58. Miller interview with Alice K. Smith, February 10, 1960.
59. *The New York Times,* November 20 and 21, 1945.
60. *New York Post,* November 24, 1945.
61. Anthony Leviero, *The New York Times,* November 28, 1945.
62. *The New York Times,* November 29 and 30, 1945.
63. Hewlett and Anderson, p. 451. Program in GSS #4, p. 73. In the Waldorf ballroom that night Szilard attended as a guest of honor, along with Manhattan Project chemist Harrison Brown, Condon, Fermi, von

Neumann, Oppenheimer, Rabi, Urey, and physicists John Dunning of Columbia and John A. Wheeler of Princeton. Among the officials and guests were several people who shared Szilard's political views and would work with him in the future, including Leo M. Cherne, Norman Cousins, Marshall Field, Jr., James G. Patton, and Michael Straight.

64. *The New York Times*, December 3, 1945. Text released by *The Nation* as a press release and reprinted as "We Turned the Switch," December 22, 1945, pp. 718–19 (LSP 34/9).

65. Ibid.

66. William Higinbotham interview with Alice K. Smith, in AKS/MIT, p. 248.

67. See Szilard's December 8, 1945, drafts (LSP).

68. *Atomic Energy*. Hearings, Part 2, on S. Res. 179, A Resolution Creating a Special Committee to Investigate Problems Relating to the Development, Use, and Control of Atomic Energy. Szilard's testimony appears on pp. 267–300 (LSP 71/7).

69. Anthony Leviero, *The New York Times*, December 11, 1945.

70. Anthony Leviero, "Scientists and Senators Puzzle over Atom Control," *The New York Times*, December 16, 1945.

71. AKS/MIT, p. 269.

72. Introduced as S. 1717.

73. Hewlett and Anderson, p. 476.

74. Press release from *The Nation*, December 3, 1945. "Address by Dr. Leo Szilard . . ." (LSP 34/9).

CHAPTER 20

1. Bronowski wrote: "As you know, Szilard failed [to prevent the atomic bombing of Japan], and with him the community of scientists failed. He did what a man of integrity could do. He gave up physics and turned to biology—that is how he came to the Salk Institute—and persuaded others too." *The Ascent of Man* (Boston: Little, Brown and Company, 1973), p. 370. See also "Jacob Bronowski: Life and Legacy," KPBS Television, San Diego, October 2, 1984, transcript, p. 13.

2. Max Lerner interview, January 23, 1986.

3. Szilard file, "Biographical Information for University Records," a questionnaire dated August 18, 1959, quote in "recreational or diversional interests." News and Information Office, University of Chicago.

4. US Atomic Energy Commission, *In the Matter of J. Robert Oppenheimer*, p. 172. According to Groves's biographer, Stanley Goldberg, the only other individual the general distrusted and disliked so intensely was the uranium merchant Boris Pregel, a Russian immigrant. Szilard and Pregel worked together secretly in the early 1940s to secure uranium for the US A-bomb

program, and during and after the war Szilard invested in Pregel's Eldorado Radium Corporation.

5. Nichols to Groves, January 12, 1946 (LSP).

6. *PM*, March 8, 1946, p. 3. *The New York Times,* March 9, 1946. *The New York Herald Tribune,* March 9, 1946, pp. 1 and 6.

7. Frances Henderson handwritten notes of March 8, 1946, interview. My thanks to Alice K. Smith for a copy of these notes. Typed notes in LSP.

8. Wigner is Jewish. Fermi was not Jewish but his wife was, and they left Italy in December 1938 because of anti-Semitism there.

9. Interviews with George Weil, April 7, 1983; Robert Wilson, November 21, 1985; and Herbert Anderson, October 6, 1986. Wigner, Fermi, and Anderson were quick to recognize that the buildup of boron in the Hanford reactors was squelching the chain reaction. But it was the Du Pont engineers' conservative overdesign of the reactors that allowed the problem to be corrected easily.

10. Ernest O. Lawrence, in Groves's view, was "not a genius" but "just a hard worker."

11. Henderson handwritten notes of the Szilard interview, beginning on p. 4.

12. The ceremony occurred on March 20, 1946. *The New York Times,* March 21, 1946.

13. *New York Post,* May 7, 1946 (LSP). At the time, Szilard had been a naturalized US citizen for more than three years and was qualified to continue nuclear research.

14. Daniels to A. H. Frye, Jr., May 3, 1946, and copy of Recommendation (MED).

15. Groves to district engineer, Oak Ridge. EIDM-WL-26, July 8, 1946. "Subject: Recommendation for Military Decoration (Dr. Leo Szilard)" (MED).

16. Szilard to Daniels, May 9, 1946 (LSP 7/5).

17. Daniels to Szilard, May 10, 1946 (LSP 7/5).

18. Szilard to Trude, May 12, 1946 (EWP). Civil Service Commission Investigation of July 6–9, 1959, p. 2 (LSP). See also Daniels to Szilard, May 10 and 15, 1946 (LSP 7/5).

19. US Atomic Energy Commission, *In the Matter of J. Robert Oppenheimer,* p. 172. Szilard's annotated copy of the Oppenheimer hearing transcript is now with MIT molecular biologist Maurice Fox.

20. L. Groves. Entry 10, Comments, interviews, and reviews. Box 6 RG 200 (MED). See Groves's copy of Margaret Gowing's *Britain and Atomic Energy 1939-1945,* US Military Academy Library.

21. p. 33, Groves's note a.

22. Ibid., p. 34, Groves's note a.

23. Ibid., p. 130, Groves's note b.

24. Ibid., p. 376, Groves's note c.
25. Giovannitti and Freed, pp. 23 and 48. Groves's copy is in the library of the US Military Academy.

CHAPTER 21

1. FBI director J. Edgar Hoover to Byron R. White, sent May 12, 1961 (LSP).
2. Conversation with Harold Agnew and Alvin Weinberg, April 26, 1989; see also "Remembering Herb Anderson" by Harold Agnew. "A Memorial Colloquium Honoring Herbert L. Anderson," August 31, 1988, Los Alamos National Labratory. Los Alamos, New Mexico (LALP-89-14), pp. 23–24.
3. Herbert Anderson interview, October 6, 1986.
4. Civil Service Commission Investigation of July 6–9, 1959, pp. 2–3 (LSP). *Chicago Sunday Tribune,* October 13, 1946. FBI report, December 23, 1946, p. 3 (LSP). Appointment letter from assistant comptroller to Szilard, October 25, 1946 (LSP). Complaint in Szilard to L. T. Coggshall, January 28, 1950 (LSP).
5. Szilard to Niels Bohr, November 7, 1950. MIT Vol. I, p. xix.
6. Hans Zeisel interview with Alice K. Smith, February 9, 1960.
7. Norene Mann interview, July 11, 1986.
8. After his death in 1964, Szilard's papers were assembled by his widow in La Jolla, at first cataloged by the container in which they were found. The first "find list" of the Szilard papers is organized by such categories as "Plaid Zipper Bag," "Blue Suitcase from Greens," and "Red Plaid Zipper Bag from Denver." Szilard kept his most important personal documents in the "Big Bomb Suitcase," so named because he wanted it with him in case of a nuclear war.
9. "The Facts About the Hydrogen Bomb," University of Chicago Round Table Radio Discussion, February 26, 1950. Interview with Harrison Brown, October 6, 1986.
10. Alvin Weinberg interview, August 9, 1988.
11. *Fortune* magazine's article about "The Great Science Debate" quoted Szilard in June 1946, p. 245.
12. AKS/MIT, pp. 337–38. It is unclear if Szilard arrived in New York in time to watch Baruch's UNAEC presentation. He bought a ticket in Los Angeles on June 13 and according to tax records was in New York by the fifteenth (LSP).
13. MIT Vol. I, pp. 178–89. See also University of Chicago press release, July 31, 1946, Szilard file, University of Chicago News and Information Office.
14. University of Chicago Archives, Quadrangle Club Records, Guest Register 1937–1955, 7/3 (JRL). Hans Zeisel interview with Alice K. Smith, February 9, 1960. Zeisel interview, February 25, 1987.
15. Foreword to MIT Vol. I, pp. xvi–xvii.

16. Jacob Marschak essay, prepared for MIT Vol. III ms but not used. I am grateful to the late Helen Hawkins for a copy of this manuscript. Szilard had first devised monetary reforms in 1919, and just after completing his doctorate in physics in 1922, he asked to start a second degree in economics. Discouraged at the time by a university official, Szilard nonetheless continued to think and talk about economics, first to save the world from monetary and fiscal crises in the 1920s and 1930s, later to multiply his own savings. See also Szilard's lecture on education at Brandeis, October 23, 1953 (LSP 42/20).
17. Milton Friedman's introduction to *The Midnight Economist* by William R. Allen (Chicago: Playboy Press, 1981), p. xiv. Interview with Milton Friedman, August 10, 1986. Friedman to the author, January 9, 1987.
18. Marschak essay for MIT Vol. III ms.
19. *The Voice of the Dolphins*, p. 78.
20. Aaron Novick interview, August 11, 1986.
21. Walter Volbach to the author, June 26, 1983.
22. *The Voice of the Dolphins*, pp. 115–22.
23. Ibid., pp. 105–14.
24. "Nicolai Machiavellnikow," July 1949 (LSP). This line is no doubt a reference to Szilard's own firing from the project in 1946.
25. Bernard Weissbourd interview with Alice K. Smith, February 10, 1960.
26. Szilard to Trude, August 16 and September 5, 1949 (EWP).
27. Szilard to Trude, September 9, 1949 (EWP).
28. Marschak to Edward Shils, October 4, 1964 (LSP).
29. Szilard to Trude, August 14 and 18, 1949 (EWP).
30. Szilard to Trude, September 5, 1949 (EWP). Gertrude Hausmann interview, October 1, 1986. "Kathy and the Bear" manuscript (LSP 27/2).
31. "Notes to Thucydides' History of the Peloponnesian War" (September 20, 1949), in MIT Vol. III, pp. 41–42.
32. Hans Zeisel interview, February 25, 1987.
33. Szilard to Peter Odegard, January 10, 1949. "Draft Memorandum of Proposals to Ford Foundation," January 26, 1949. Szilard to Ford Foundation, February 9, 1951. (LSP 7/14).
34. James Arnold interview, February 11, 1987.
35. Draft about "Atoms for Peace," MIT Vol. III, p. 144.
36. James Arnold interview, February 11, 1987. Lynn A. Williams, University of Chicago vice-president, to Szilard, June 6, 1951 (LSP). Rough draft of "Where will your children spend the war?" (LSP 34/16).
37. Joseph L. Rauh, Jr., interview, March 20, 1986.
38. Lerner to Szilard, January 9, 1952 (LSP 12/1).
39. See Szilard to Walter Blum, September 27, 1951 (LSP 4/33).

40. See Civil Service Commission Investigation of July 6–9, 1959, p. 3; "Atomic Scientist Named Professor at Brandeis U.," *The New York Times*, October 4, 1953; and Brandeis news release, October 4, 1953.
41. "On Education," October 23, 1953 (LSP 42/20).
42. "Talk on Education, June 15, 1954" (LSP 42/23).
43. Max Lerner interview, January 23, 1986.
44. Itinerary based on Szilard's October 1953 tally of earnings (LSP). Built in 1832 by New Jersey governor Charles S. Olden, the neocolonial white clapboard mansion called Drumthwacket has six Ionic columns at its front entry. The mansion was expanded in 1905 and remodeled after Spanel bought it in 1940. In 1967, Drumthwacket was sold to the state of New Jersey and designated again as the governor's mansion, although the first executive to return, James Florio, did not move in until 1990.
45. Egon and Renée Weiss interviews, March 4, 1986.
46. Frances Racker interview, November 22, 1985.
47. Nella Fermi Weiner interviews, March 17, 1987, and October 4, 1989.
48. At the University of Chicago, Szilard was reassigned to the Institute of Radiobiology and Biophysics at a salary of $10,500 a year. Civil Service Commission Investigation of July 6–9, 1959, p. 3. Grodzins to Szilard, July 12, 1954 (LSP 9/8). Comptroller John J. Kirkpatrick to Szilard, August 12, 1954 (LSP).
49. Theodore Puck to Szilard, August 23, 1954, and October 12, 1955 (LSP 15/25).
50. Szilard memo, October 12, 1955. His six sponsors were the University of Chicago, Rockefeller University, the National Institutes of Health, California Institute of Technology, New York University, and the University of Colorado.
51. Szilard to Lewis L. Strauss, March 19, 1956 (LSP 18/21).
52. "Memorandum" on power plants, June 14, 1956 (LSP). Draft letter to the *New York Times* on fusion, Summer 1956 (LSP).
53. Szilard to von Laue, February 8, 1957 (LSP). My thanks to Helmut Rechenberg of the Max Planck Institute in Munich for a copy of this letter. On October 7, 1957, Szilard spoke on molecular biology to the German Chemical Society (LSP 42/27).
54. Creutz to Szilard, January 10 and March 11, 1957; Szilard to Creutz, February 1, 1957 (LSP 6/37).
55. Szilard to Menke, February 28, 1957 (LSP 13/12). Creutz to Szilard, March 11, 1957; Szilard to Creutz, June 27, 1957; Creutz to Szilard, July 10, 1957 (LSP 6/37).
56. Edward Creutz interview, August 8, 1987.

57. February 11 and March 18 and 21, 1959. "Thermo Electric Generator" patent plans are based on thermionic emission in an alkali vapor atmosphere (LSP 39/10 and 11). Edward Creutz interview, August 8, 1987.
58. Von Neumann to Teller, 1947. Von Neumann Papers (LCM). My thanks to Barton J. Bernstein for this source.
59. Brillouin reprint (LSP 52/23). Darrow to Szilard, February 21, 1952; Szilard to Darrow, February 27, 1952 (LSP 4/10). See also Anatol Rapoport, "The Promise and Pitfalls of Information Theory," *Behavioral Science,* Vol. 1, 1956, pp. 13–17.

Szilard and Brillouin met in 1950 after an introduction by Warren Weaver of the Rockefeller Institute. Brillouin was then writing a paper on information and entropy, "Maxwell's Demon Cannot Operate: Information and Entropy I," *Journal of Applied Physics,* 22, pp. 334–37 (1951). See Weaver to Szilard, September 15, 1950. Warren Weaver Diary, RA 1.2 228/2205/200D University of Chicago Genetics Leo Szilard (RAC). See also Brillouin's *Science and Information Theory,* pp. xi and 176–83 (New York: Academic Press, 1956).
60. Norman Macrae interview, May 18, 1990. "Maxwell's Demon" by Werner Ehrenberg and "Demons, Engines and the Second Law" by Charles H. Bennett, *Scientific American,* vol. 317 [offprint] and vol. 257, no. 5, pp. 108–116 (1987), respectively.
61. MIT Vol. I, p. 726.
62. Szilard to Trude, March 17, 1952 (EWP). Maurice Fox interview, January 13, 1987. Hans Zeisel interview, February 25, 1987. Zeisel to Szilard, April 18, 1952; Szilard to Zeisel, May 21, 1952 (LSP 21/24).

CHAPTER 22

1. Maurice Fox interviews, November 12, 1985, and January 13, 1987. Stan Zahler interview, November 22, 1985. Herbert Anderson interview, October 6, 1986.
2. Harrison Brown and Herbert Anderson interviews, October 6, 1986.
3. Walter Blum interview with Alice K. Smith, February 18, 1960, and with the author, February 26, 1987.
4. Edward Teller interview (for *NOVA,* Tape 1 transcript, p. 1).
5. Szilard to Richard B. Gehman, April 8, 1952 (LSP 8/27).
6. Frances Racker interview, November 22, 1985.
7. See, for example, Szilard to Trude, July 8 and August 8, 1942, and July 1 and 27, 1943 (EWP).
8. Szilard to Trude, July 1, August 16, and September 22, 1943 (EWP).
9. Szilard to Trude, September 11, 1948 (EWP).
10. Szilard to Trude, September 15, 1942, March 19, 1944, and December 9, 1946 (EWP).

11. Szilard to Trude, August 9, 1942; May 2, June 23, and September 4 and 9, 1946; March 18, 20, and 30, 1947 (EWP).
12. Interviews with Frances Racker, November 22, 1985, and with Efraim and Frances Racker, March 4, 1986.
13. Szilard to Trude, August 29 and September 10, 1948 (EWP).
14. Victor Weisskopf interview, December 16, 1985.
15. Roland Detre interviews, October 1 and 2, 1986.
16. Gertrude Hausmann interview, October 1, 1986. Roland Detre interview, October 2, 1986.
17. Leigh Fenly, "The Agony of the Bomb and Ecstasy of Life with Leo Szilard," *San Diego Union,* November 19, 1978, pp. D-l and D-8. Gertrude Hausmann interview, October 1, 1986. Frances Racker interview, November 22, 1985.
18. Szilard to Trude, February 23, 1951 (EWP).
19. Szilard to Trude, February 20, 1951 (EWP).
20. Szilard to Trude, October 22, 1951 (EWP).
21. See, for example, Szilard to Trude, February 20 and April 24, 1951 (EWP). At Trude's urging, Szilard asked psychiatrist Franz Alexander in New York for the name of a Chicago psychiatrist and was given the name Dr. Block, but did not bother to learn his first name or to make an appointment. Szilard to Trude, October 21 and 22, 1951 (EWP).
22. See, for example, Szilard's later complaint to Trude, May 24, 1952 (EWP).
23. They were married by city clerk Murray W. Stand, witnessed by Heidi Thwaites; license (LSP 1/4).
24. Gertrude Hausmann interview, October 1, 1986. Helen Weiss interview, March 3, 1986.
25. Szilard to Trude, October 15, 1951 (EWP).
26. Szilard to Trude, October 16, 1951 (EWP).
27. Szilard to Trude, October 21, 1951 (EWP).
28. See, for example, Szilard to Trude, October 24 and 27, 1951 (EWP).
29. Szilard to Trude, October 22, 1951 (EWP).
30. Interview with Theodore Puck, David Hawkins, and Walter Orr Roberts, November 18, 1986. Szilard wrote: "I believe you should hurry to grow up if that is at all possible." Szilard to Trude, October 30, 1954 (EWP).
31. Szilard to Trude, June 1, 1952 (EWP).
32. Szilard to Trude, February 20, 1951 (EWP). Max Lerner interview, January 23, 1986.
33. Szilard to Trude, May 24, 1952 (EWP).
34. Szilard to Trude, November 9, 1953 (EWP). See also Margaret Spanel, "The Spanels of Drumthwacket," *Princeton History,* No. 3 (1982), p. 52.
35. Margaret Spanel to the author, April 22, 1987.
36. Szilard to Trude, September 2, 1952 (EWP).

37. John Marshall, Jr., interview, February 2, 1988. Szilard was in Bolivia from August 26 to September 2, 1952, and in Peru on September 2 and 3, according to passport stamps and letters to Trude.

38. Szilard to Trude, July 3, 1952 (EWP).

39. Szilard to Trude, November 5, 1953 (EWP).

40. Margaret Spanel interview, December 2, 1986.

41. Miki Pavelic to Szilard, Easter 1960 (LSP).

42. Margaret Spanel interview, December 2, 1986.

43. Szilard to Trude, May 24, 1952 (EWP).

44. Szilard to Trude, September 25, 1954 (EWP).

45. Szilard's will, signed in New York City on January 25, 1955, listed Trude as sole beneficiary, with his sister, Rose, and brother, Bela, as alternate recipients; executors and trustees included Walter Blum, James P. Hume, Edward Levi, Aaron Novick, and Herbert Anderson (LSP 1/7).

46. Szilard to Trude, October 7, 1956 (EWP).

47. US Patent No. 2,708,656. *The New York Times*, May 19, 1955; reproduced in MIT Vol. I, p. 386.

48. "Draft for Letter" by Szilard for Einstein, October 18, 1954; Nehru letter drafts by Einstein (LSP).

49. Einstein to Nehru, April 6, 1955 (AEP 32–751); Clark, 1972, p. 764.

50. Sayen, p. 151. Transcript in AEP. On a flight from Rome to Paris, Russell heard the pilot announce that Einstein had died and became dejected, thinking he had not signed their manifesto. Only later did Russell learn that Einstein had signed. The manifesto, released on July 9, 1955, laid the groundwork for the Pugwash Conferences on Science and World Affairs. Joseph Rotblat interview, September 21, 1985.

51. Roland Detre interviews, October 1 and 2, 1986.

52. Eva Zeisel interview, August 19, 1985.

53. Inge Feltrinelli to the author, Fall 1987.

54. Puck to Szilard, December 10, 1955 (LSP 15/25).

55. Maurice Fox interview, November 12, 1985. Victor Weisskopf interview for *NOVA* (NV1625 SR206).

56. Bernard Davis interview, December 17, 1985.

57. Szilard to Trude, November 9, 1957 (EWP).

58. On March 7, 1958, Dr. A. Ravin wrote to Szilard explaining the findings of a February 22 examination. The EKG revealed "evidence of a recent posterior-lateral myocardial infarction," but Szilard's condition improved between February 22 and 26. "My feeling is that Dr. Szilard has coronary artery disease." (LSP)

CHAPTER 23

1. Szilard wrote "Security and Equality" on October 28, 1950 (LSP 31/18). See Roberts correspondence (LSP 16/15).
2. "The AEC Fellowship: Should We Yield or Fight?" was written in June 1949 and published in Grodzins and Rabinowitch, pp. 410–13.
3. See September 13, 1950, CIA memo about Szilard and October 9 FBI memo by J. Edgar Hoover (LSP).
4. Walter Orr Roberts interview, October 7, 1986.
5. Szilard to Bernard Davis, April 2, 1952 (LSP 7/8).
6. John (Benjamin) to Szilard, December 29, 1950 (LSP 4/27). David and Frances Hawkins interview, November 19, 1986.
7. Warren Weaver Diaries, 1955, p. 35. RG 12 (RAC).
8. Executive Order 10450, issued April 27, 1953, required that federal employees and consultants be suspended if information exists that their employment may not be clearly consistent with the interests of national security.
9. See Philip M. Stern's foreword to US Atomic Energy Commission, *In the Matter of J. Robert Oppenheimer*, p. vi.
10. The piece was written after December 29, 1953, and published in the *Bulletin of the Atomic Scientists*, December 1954. See Szilard telegram to Trude, December 29, 1953 (EWP).
11. "Security Risk," *Bulletin of the Atomic Scientists*, December 1954, pp. 384–86 and 398.
12. Perhaps this was intended as an introduction to "Security Risk," since that story is also in the form of a letter.
13. "Cyclotron," January 2, 1954 (LSP 23/17).
14. Draft of a Statement, 1954. MIT Vol. Ill, p. 129.
15. FBI report, November 30, 1947, No. WFO 105-9968, p. 2 (LSP).
16. Szilard to Oppenheimer, May 19, 1948 (LSP 14/27).
17. Szilard correspondence about Oppenheimer (LSP).
18. "Urey, Other Scientists Hit Oppenheimer's Suspension," *Daily Worker*, April 16, 1954. See also Szilard's FBI file 62-59520-26 of August 5, 1954, p. 15 (LSP).
19. Ruth Adams to Szilard, April 30, 1954 (LSP).
20. Hewlett and Holl, p. 86; Rhodes review of Blumberg and Panos, *The New York Times Book Review*, February 11, 1990, p. 3, with Teller letter, March 21, and Rhodes reply, April 7.
21. See MIT Vol. III, p. xliii.
22. Teller to the author, September 16, 1988.
23. James Arnold interview, February 11, 1987.
24. Also in the group were Edward Levi of the University of Chicago Law School and Louis Wirth of the sociology department (LSP).

25. Teller to Strauss, April 19, 1954 (LLS AEC Teller 1954, January–June Box 111).

26. For example, on September 10, 1953, Teller sent a telegram to Strauss pleading that his name not be used in a forthcoming *Time* magazine article about the AEC chairman, perhaps fearing that publicity about their collaboration would harm his relatives' chances for release (LLS AEC Series, Teller, Edward).

27. "Introduction" to MIT Vol. III, p. xliii. Barton Bernstein interview, July 30, 1986.

Although this story could be true as told, documents in the Szilard Papers raise doubts about the timing and may refute it. Teller testified before the AEC board on the afternoon of April 28, 1954; according to the transcript, he began after 2:00 P.M., concluding at 5:50 P.M. Airline tickets and a hotel receipt in the Szilard Papers show that he flew from New York to Washington at 1:25 P.M. on that day, stayed the night, and returned to New York by plane at 4:25 P.M. on the twenty-ninth (LSP 32/2 and 79/7).

It is possible that Szilard tried to telephone Teller from New York on the night of April 27. It is also possible that Szilard thought Teller would testify again on the twenty-ninth, as he had ended the session late in the afternoon of the twenty-eighth, and in that case, Szilard could have been searching for Teller that night. A third possibility is that Szilard did not know when Teller would testify but impulsively boarded a plane for Washington to find him. This explanation would account for Trude's memory that Szilard pursued Teller and is also in accord with Szilard's mode of behavior.

28. US Atomic Energy Commission, *In the Matter of J. Robert Oppenheimer,* p. 710.

29. Ibid., p 1019.

30. Szilard to Oppenheimer, December 23, 1954. Oppenheimer Papers (LCM 70).

31. Teller interview with Gabor Pallo, December 6, 1983. My thanks to Dr. Pallo for his translation of this interview.

32. Strauss to Dulles, July 27, 1954 (LLS AEC Series. Teller, Edward).

33. "I believe you owe it to the scientific community to defend your views before your peers. Moreover I think it would be a healthy thing for you to do so. . . ." Szilard to Teller, November 29, 1957, and July 16, 1956 (LSP 18/35).

34. Explained in a copy of Szilard to Warren Johnson, 1957 (LSP 10/22). Teller to Szilard at the Hotel Kempinski in Berlin, December 15, 1957 (LSP 18/35).

35. Teller to Compton, June 26, 1956; Teller to Atoms For Peace nominees, January 27, 1957; Teller to Killian, September 25, 1957 (MIA, MC10, 5/245).

36. Blumberg and Owens, p. 182.

37. Szilard to Rem S. Krassilnikov, June 11 and 23, 1958 (LSP).
38. "Report on the Work of the Academy's Operating Committee on World Security Problems Raised by Nuclear Weapons for the Year 1958–1959 by Leo Szilard, Chairman," May 11, 1959, National Academy of Arts and Sciences (LSP 66/12).
39. Gertrud Weiss Szilard interview with Alice K. Smith, May 11, 1962. Harrison Brown to Smith, April 19, 1960. Brown interview, October 6, 1986.
40. Ann Moir, assistant editor, *Bulletin of the Atomic Scientists,* to Szilard, October 16, 1958 (ERP 12/30).
41. Lajos Janossy Papers, Budapest (526/958/Sz). My thanks to Gábor Pallo for a copy of this letter.
42. Gertrud Weiss Szilard interview with Alice K. Smith, May 11, 1962.

CHAPTER 24

1. *Life,* January 31, 1946. Shown is the fourth-floor walk-up office of the Federation of Atomic Scientists at 1621 K Street NW. See Hewlett and Anderson, p. 485.
2. Privately, however, Szilard's idea may have spread among the Russians. On January 4, 1946, Prof. Jacob Frenkel of the Leningrad Physico-Technical Institute wrote to Szilard's colleague Edward U. Condon, then president of the American Physical Society as well as director of the National Bureau of Standards. Frenkel sent New Year's greetings and suggested an exchange of Soviet and American scientists working in the nuclear area. Condon replied on February 14 that the society desires "to extend the basis of complete scientific co-operation between your physicists and ours" and added that "secrecy practices which were established during the war represent a perversion of the true spirit of science." Stephen White, *New York Herald Tribune,* February 15, 1946.
3. Hans Bethe interview, November 21, 1985.
4. AKS/MIT, p. 312.
5. Robert R. Wilson interviews, November 21, 1985, and May 1, 1987.
6. *The New York Times,* March 9, 1946.
7. The amendment passed on March 12, 1946, with McMahon as the sole dissenter. AKS/MIT, p. 312.
8. Philip Morrison interview, November 11, 1986.
9. Drew Pearson, *New York Daily Mirror,* March 18, 1946; item in *PM,* March 18, 1946. Frederick Seitz interview, May 20, 1987. Ralph Lapp interview, April 6, 1987.
10. *One World or None* (New York: Whittlesey House, McGraw-Hill Book Co., 1946), pp. 61–65.
11. *New York Herald Tribune Weekly Book Review,* March 17, 1946, pp. 1–2.

12. *The New York Times Book Review,* March 17, 1946, pp. 1 and 16.
13. Ibid. The next day, the *Herald Tribune*'s "Books and Things" by Lewis Gannett reported on *One World or None:* "Treaties, the scientists and Mr. Lippmann agree, are risky foundations for any hope; but they might give a temporary respite from fear and allow a little more time for mutual understanding to grow. 'We shall have to take risks,' says Leo Szilard, of the Chicago Metallurgical Laboratory."
14. *The New York Times,* March 19, 1946, pp. 1 +.
15. McMahon to Szilard, April 16, 1946 (LSP 13/8).
16. *Chicago Sun,* May 3, 1946, pp. 11 and 16.
17. *New York Herald Tribune,* May 24, 1946. Organized in May, with Einstein as its chairman, the Emergency Committee of Atomic Scientists had Szilard, Bethe, Urey, Hogness, and Weisskopf as board members by its formal incorporation in August.
18. Harold Oram interview, November 5, 1987. See also "Einstein Memorabilia" by Harold Oram, June 10, 1985.
19. AKS/MIT, pp. 337–38.
20. Szilard had already annoyed the film's producer, Samuel Marx, by insisting that he and Einstein could only be impersonated if M-G-M made a $100,000 donation to the FAS. Then Szilard trimmed the price to $5,000, before finally, on "principle," refusing to take any money from the studio. AKS/UC, pp. 316–17.

    Marx was prepared for dealing with Szilard, having just soothed two of the biggest egos in the Manhattan Project, Oppenheimer and Groves. By making the Oppenheimer character, played by matinee idol Hume Cronyn, "an extremely pleasant one with a love of mankind, humility, and a fair knack of cooking" and by making it clear in the film that Oppenheimer and not Groves ran the first test at Alamogordo, Marx was able to persuade Oppenheimer to sign his release in May. Groves, who was played by handsome tough guy Brian Donlevy, cared little about character development, although he did appreciate the actor's rugged good looks. For him the main concern was national security, and Marx had to let him and his aides pore over the script to prevent any secrets from flashing onto the silver screen. Reingold, "MGM Meets the Atomic Bomb."

21. AKS/UC, p. 316.
22. Marx also showed Szilard some of the scenes already shot, apparently to Szilard's satisfaction, because Marx then wrote Einstein that Szilard "was, to some extent, impressed that we are doing a sincere job and one that will reflect only credit on science." Marx to Einstein, July 1, 1946 (AEP 57 155–1 & 2).

    By mid-July, M-G-M's flamboyant president, Louis B. Mayer, assured Einstein that the film would portray the scientists favorably. "It must be realized," a memo Mayer sent as proof stated, "that dramatic truth is just as compelling a requirement on us as veritable truth is on a scientist." Nevertheless, tensions persisted all summer between the filmmakers and the scientists. One "consultant" invited by M-G-M at Oppenheimer's request was Los Alamos administrator David Hawkins, who, when shown some scripts, discovered absurd technical errors. In one scene the scientists' "radiation monitor" was a hutch of rabbits—the screenwriters had thought that if canaries could detect gas in mines, then rabbits could detect radiation in laboratories. Hawkins insisted that Geiger counters be substituted, one of his few recommendations that the studio followed in the scenes still to be shot. However, Hawkins never convinced the moviemakers that the sound of the distant A-bomb's explosion should be heard several seconds *after* the light from the blast and not at the same moment. David Hawkins interview, May 13, 1987.

23. "Regional Conference . . ." program and notes (LSP 70/4). Walter Orr Roberts interview, October 7, 1986.
24. *Life,* August 11, 1946.
25. Fielding, p. 292.
26. Lippmann to Aydelotte, October 28, 1946 (AEP 57 167–1 & 2).
27. *The New York Times,* November 18, 1946. ECAS "Princeton Speech," November 17, 1946 (LSP 42/10).
28. Szilard to Hutchins, January 2, 1947 (RMH 5/12). FBI report, December 23, 1946 (LSP 95/9). Also in report of April 14, 1947 (LSP 95/7).
29. FBI report, December 23, 1946, p. 18 (LSP 95/9).
30. MIT Vol. III, p. 20 n 1. This talk, which Szilard revised and gave many times thereafter, appeared as "Calling for a Crusade" in the April/May 1947 *Bulletin.*
31. AKS/UC, p. 318. Harrison Brown interview, October 6, 1986. In Washington, where the film officially premiered in February, M-G-M billed its creation as "the story of the most HUSH-HUSH secret of all time." But to reviewers it was merely ho-hum. *Time* found "the picture seldom rises above cheery imbecility" by treating "cinemagoers as if they were spoiled or not-quite-bright children." The public, which did not make the film a box-office hit, soon forgot whatever point M-G-M had tried to make, although the film did win an Oscar for its special effects. See Reingold, "MGM Meets the Atomic Bomb."
32. April 10, 1946, Bloomington speech (LSP 42/13).
33. April 21, 1947, Portland speech (LSP 42/14). April 23, 1947, Spokane speech (LSP 42/15).
34. *Saturday Review,* May 3, 1947, pp. 31–34. This article prompted a radio talk on "Atomic Energy Control" on New York City's WNBC, at 6:15 P.M., May 15, 1947.

35. Einstein statement (AEP 40 563–3). Hans Bethe oral-history interview, May 8–9, 1972, p. 44 (AIP).

36. Einstein statement (AEP 40 563–2).

37. Szilard and Weisskopf met with Einstein later, and the three—acting at a formal meeting of the ECAS trustees—agreed to increase the board from eight to ten members, voting Mayer and Frederick Seitz as members. They also agreed to hire Brown full-time to work for ECAS (AEP 40 563–61 to 5).

38. Meeting held June 18–21, 1947. AKS/UC, pp. 505–508; *The New York Times,* June 30, 1947.

39. Hans Bethe interview, March 3, 1986. The ECAS discord was publicized by the *New Republic,* which wrote a week later that, unlike Szilard, Urey believed the United Nations was inadequate and that only a world government could assure peace. Szilard condemned a proposed nuclear union with Russia, fearing it would lead to an arms race and war, and called for the atom's international control as part of overall reconstruction plans involving economic, social, and political elements—Secretary of State George C. Marshall had announced a US-financed economic-recovery scheme for Europe (the Marshall Plan) a month earlier. Other ECAS members quoted by the magazine thought they should simply stay out of politics. *New Republic,* July 7, 1947.

40. "Tentative draft to be submitted for approval," Szilard to Edward Levi, September 29, 1947. My thanks to Edward Levi for a copy of this memo. Later drafts in (LSP 27/5 and 6) and in "Letter to Stalin" (LSP 73/9).

41. MacDuffie Collection, notes of October 9, 1947, pp. 1 and 2–3, and October 10, 1947, p. 1–4. My thanks to historian Carol Gruber for sharing her research in the MacDuffie Collection at the Columbia University Library.

42. Einstein and Philip Morse to Marshall, October 11, 1947 (AEP, copy in LSP 7/27). The two wrote as chairman and acting executive director, respectively, of the Emergency Committee of Atomic Scientists.

43. George T. Washington to Hoover, November 10, 1947, and subsequent correspondence (LSP 95/1). Washington was then assistant solicitor general. Nothing derogatory was disclosed, but Clark later had copies of Szilard's FBI files sent to the State Department and the Atomic Energy Commission. Clark to Hoover, December 12, 1947 (LSP).

44. "Letter to Stalin," *Washington Post and Times Herald,* November 26, 1947.

45. The manuscript for "My Trial as a War Criminal" was typed before October 24, 1947 (see letter of that date to Rabinowitch in LSP) and was retyped on November 7. The story was first published in the *University of Chicago Law Review,* Autumn 1949, and was reprinted in *The Voice of the Dolphins* (pp. 75–86) in 1961.

46. Szilard to Hutchins, April 26, 1948, MIT Vol. III, pp. 35–37. Szilard to Hutchins, June 8 and 16, 1948 (LSP 10/9).

47. 'The Atom: What'll We Do?" *Newsweek,* October 31, 1949.

48. 'The Facts about the Hydrogen Bomb, University of Chicago Round Table," No. 623, February 26, 1950 (LSP 70/24).

49. Hans Bethe interviews, November 21, 1985, and March 3, 1986.

50. "Science: Hydrogen Hysteria," *Time,* March 6, 1950, p. 88.

51. "Cost of Suicide," *Newsweek,* October 30, 1950.

52. "Memorandum on 'Citizens' Committee.'" March 27, 1950. MIT Vol. III, pp. 95–102.

53. The Rosenbergs were convicted on March 29, 1951, and executed on June 9, 1953. US troops first fought North Korean forces on July 5, 1950, and the Chinese army joined the war on November 26 as US troops under UN authority approached the Yalu River dividing North Korea and China. The war ended in a cease-fire in July 1953 that divided North and South Korea along the 38th parallel.

54. See rough draft of "A Letter in the Open," August 31, 1950 (LSP 27/5).

55. Joseph Rotblat interview, September 21, 1985. See Rotblat's Pugwash History, MIT Press. Rotblat remembers the meeting as taking place on September 21, 1951. Rotblat to the author, November 29, 1986.

56. "You Do Not Want War with Russia?," pp. 1–3. Gertrud Szilard marked this manuscript "March ? 1952," but with references on p. 8 to political action in 1952, it may have been drafted later that year (LSP 34/19).

57. Ibid., p. 8. This proposal called for a 1 percent tithe, whereas his council in 1962 sought "two per cent for peace."

58. On August 2, 1953, the Soviet Union detonated a small lithium-deuteride device with a yield of less than one-half megaton (LLS AEC Disarmament Chronology 56–58, October 22, 1956, memo, p. 6). Drafts of "Meeting of the Minds" in the Szilard Papers are dated between June 1953 and January 1954. Dulles proposed "massive retaliation" on January 12, 1954, and Szilard's last draft of "Meeting" is dated January 26.

59. "Forty Cents a Head," in Matter of Fact by Stewart Alsop, *Washington Post,* May 6, 1955.

60. Szilard to Hubert H. Humphrey, August 2, 1955 (LSP 10/8).

61. The Einstein-Russell Manifesto was issued April 11, 1955. See "Proceedings of the First Pugwash Conference on Science and World Affairs," ed. Joseph Rotblat, published by the Pugwash Council in 1982, pp. 167–70.

62. Joseph Rotblat interview, September 21, 1985.

63. At the time of the Russell-Einstein Manifesto, in July 1955, Szilard had just finished proposing yet another of his working groups to study US-Soviet problems and was then drafting a sweeping essay that would eventually appear in the October 1955 *Bulletin* as "Disarmament and the Problem of

Peace," his ambitious link of economics and arms control as a way to gain a peace dividend.

About this time, Szilard also suggested that Sen. Hubert Humphrey, the Minnesota Democrat who chaired the Foreign Relations Subcommittee on Disarmament, hold conferences for his staff on arms-control policies with leading diplomats and journalists. In December 1956, Szilard wrote Soviet physicist Lev Landau, whom he had met in Berlin in the 1920s, proposing a scientists' discussion on arms control.

In January 1957, Szilard proposed creating a foreign policy club in Washington to discuss world problems informally. That March, Szilard drafted a memo on the H-bomb, arguing for a "foolproof inspection system" as the key to progress in arms control.

64. Szilard to Russell, May 23, 1957. MIT Vol. III, pp. 156–57.
65. Harrison Brown interview, October 6, 1986. "Proceedings of the First Pugwash Conference . . ." p. 46.
66. Ruth Adams interview, October 21, 1986.
67. Rotblat, pp. 27–32. On July 21, 1955, President Eisenhower had proposed a US-Soviet aerial inspection system to prevent a surprise nuclear attack.
68. Ibid., pp. 39 and 27.
69. "Afterdinner Speech at Pugwash," dictated August 7, 1957, pp. 3–4 (LSP 69/10).
70. Szilard to Topchiev, July 31, 1957 (LSP 19/4). Szilard to Morton Grodzins, August 9, 1957 (LSP 9/8). Szilard to Skobeltzyn, October 15, 1957 (LSP 18/5). Topchiev to Szilard, December 14, 1957 (LSP 19/4).
71. Hans Bethe interview, March 3, 1986.
72. Joseph Rotblat interview, September 21, 1985.
73. The Second Pugwash Conference met from March 30 to April 12, 1958.
74. In February 1958, *Life* magazine had published "The Compelling Need for Nuclear Tests," excerpts from Edward Teller's book *Our Nuclear Future* (New York: Criterion Books, 1958), which advocated developing "clean" nuclear weapons. "If we stop testing now and if we should fail to develop to the fullest possible extent these clean weapons, we would unnecessarily kill a great number of noncombatants in any future war," wrote Teller and his coauthor, Albert Latter. See Katherine Magraw, "Teller and the 'Clean Bomb' Episode," *Bulletin of the Atomic Scientists,* May 1988, pp. 32–37.
75. "Documents of the Second Pugwash Conference of Nuclear Scientists," transcript, pp. 430ff.
76. Ibid., pp. 450–51.
77. Ibid., p. 453.
78. Ibid., pp. 475–76.
79. Ibid., p. 493.
80. Ibid., p. 483.

81. "Memorandum" from Leghorn, Szilard, and Wiesner to Topchiev, April 6, 1958 (LSP 68/13).

82. For background, see "Report on the Work of the Academy's Operating Committee on World Security Problems Raised by Nuclear Weapons for the Year 1958–1959 by Leo Szilard, Chairman," May 11, 1959 (LSP 66/12).

83. "Pax Russo-Americana," first draft dated April 24, 1958; July 16, 1958, draft in (LSP 29/14).

84. Szilard's lecture and discussion were held on May 7, 1958 (LSP 42/28). Arthur D. Little Lecture transcript, p. 12 (MIA MC 167, 18/172).

85. Ibid., pp. 17-b and 21-b.

86. Ibid., pp. 40-b and 41-b.

87. See Szilard's May 11, 1959, "Report" on the committee's work (LSP 66/12). Philip Morrison interview, November 11, 1986.

88. John P. Holdren, "The Pugwash Conferences on Science and World Affairs: Background for Funders," n.d., pp. 4 and 5.

89. "Proceedings of Fourth Pugwash Conference of Nuclear Scientists, June 25—July 4, 1959, Baden, Austria," transcript, pp. 100–123. "Toasts and Notes from the Baden Conference," paragraph 2 (both LSP 69/13).

90. Szilard to Khrushchev, September 6, 1959. MIT Vol. III, p. 263. Russell to Szilard, December 21, 1959 (LSP 16/27).

CHAPTER 25

1. Schrödinger's essay has been mentioned by several physical scientists as a stimulant for their shift to biology. For different accounts of the work's influence, see Yoxen, pp. 17–52; Abir-Am, pp. 73–117; Perutz, pp. 242ff.

2. During the Manhattan Project, Novick had studied how carbon became displaced in the graphite structure within the plutonium-production reactors at Hanford, Washington, and discovered that annealing the graphite at high temperatures reversed this. Novick and Szilard became acquainted as they discussed this problem frequently. Novick to the author, March 1991.

3. Interview with Harold Agnew, August 8, 1988. Interview with Herbert Anderson, October 6, 1986. On some evenings, Szilard's conversations ran so late that Agnew, who slept on the couch, had to stay awake until they ended.

4. Aaron Novick interview, July 2, 1986. Novick to the author, August 25, 1986. See also Novick's "Phenotypic Mixing," in Cairns et al., eds., pp. 133–34.

5. The authors were J. Huxley, G. P. Wells, and H. G. Wells. Szilard had met H. G. Wells in London in 1929 and had tried to visit Huxley and J. B. S. Haldane at the same time. See, for example, Szilard to Michael Polanyi, April 1, 1929 (MPP 2/5).

6. *Nature* pp. 131, 421, 457 (1933). Stent and Calendar, pp. 24–25. See also Victor Weisskopf's review of *Redirecting Science* by Finn Aaserud in *Science,* Vol. 251, pp. 684–85 (February 8, 1991).

7. MIT Vol. II, p. 16.

8. Timofeeff-Ressovsky observed that the rate of mutation in fruit flies doubled with a 10°C temperature rise (a change typical of chemical reactions). Delbrück inferred from this result that mutations must involve a chemical reaction, making genes chemical structures that, in principle, are understandable and not unfathomable mysteries.

9. Salvador Luria and Max Delbrück, "Mutations of Bacteria from Virus Sensitivity to Virus Resistance," *Genetics* 28, p. 491 (1943).

10. A second advance occurred in 1944 when three researchers at the Rockefeller Institute in New York City—Oswald Avery, Colin MacLeod, and Maclyn McCarty—concluded that deoxyribonucleic acid (DNA), a large, stringlike molecule found in all living cells, was the carrier of genetic information. From then on, biologists wanted to learn more about DNA—especially, how it could express and transmit genetic information. Avery, MacLeod, and McCarty, "Studies on the Chemical Nature of the Substance Inducing Transformation of Pneumococcal Types. Induction of Transformation by a Deoxyribonucleic Acid Fraction Isolated from Pneumococcus Type III," *Journal of Experimental Medicine* 79, 137 (1944).

11. Judson, p. 605. See also "Max Delbrück," a review *of Thinking about Science.* "Max Delbrück and the Origins of Molecular Biology," in *Science,* December 23, 1988, pp. 1711–12.

12. Frances Racker interview, November 22, 1985.

13. MIT Vol. I, p. xv.

14. Novick to the author, March 1991.

15. Philip Morrison interview, November 11, 1986. Aaron Novick interview, July 2, 1986.

16. Novick, "Phenotypic Mixing," p. 135.

17. Judson, p. 50.

18. François Jacob interview, July 1, 1987. See also Jacob, p. 293.

19. Novick to the author, March 1991.

20. Notes for "Talk Comparing Research in Biology with Research in Physics" to the University of Colorado Medical School, November 11, 1950 (LSP).

21. Szilard and Novick returned in 1951 and 1953, and Szilard came alone in 1956, 1961, and 1963.

22. *Newsweek,* June 18, 1951.

23. John Platt to Alice K. Smith, February 24, 1960.

24. John Platt interview, January 12, 1987.

25. Aaron Novick interview, July 2, 1986. Edward Levi interview, March 19, 1987. "Introduction" in MIT Vol. I by Aaron Novick, pp. 389–92.

26. See Aaron Novick and Leo Szilard, "Virus Strains of Identical Phenotype but Different Genotype," *Science,* 113: 34–35 (1951), reprinted in MIT Vol. I, pp. 417ff. Novick to the author, March 1991. My thanks to Maurice Fox for clarifying this and other biological explanations.
27. John Platt to Alice K. Smith, February 24, 1960.
28. Walter R. Volbach to the author, June 26, 1983.
29. Szilard to Trude, August 16 and 26, 1948 (EWP).
30. Manuscript in LSP 23/23. See also *The Voice of the Dolphins,* pp. 87–102.
31. Szilard to Trude, September 7, 1948 (EWP).
32. Szilard to Trude, September 11, 15, and 18, 1948 (EWP).
33. If the washout rate exceeded the maximum growth rate of the bacteria, they would eventually be flushed from the growth vessel; but if the washout rate were set below the maximum growth rate, the bacteria population would increase in size until it began to exhaust whatever component of the nutrient supply was in least relative amount. Szilard wanted to provide a nutrient with an excess of all necessary elements save one, which would become the limiting growth factor. He anticipated that as the growing population tended to deplete the concentration of the limiting growth factor, the bacteria would respond by reducing their growth rate and their utilization of the factor.

    In this case, one could expect a balance, or "steady state," when the growth rate of the bacteria equaled the washout rate, leading to the maintenance of a population of constant size and a constant low concentration of the limiting growth factor. The concentration of the limiting growth substance would be determined only by the washout rate. Aaron Novick to the author, April 24, 1991.
34. MIT Vol. I, p. 390. Novick to the author, March 1991.
35. Stan Zahler interview, November 22, 1985.
36. Aaron Novick interview, July 2, 1986. Novick, "Phenotypic Mixing," p. 136. At times, Szilard numbered from the back of the lab book, beginning with 1 on the second-to-last left-hand page, 2 on the last right-hand page, etc.
37. Platt, pp. 152–53. Platt interview, January 12, 1987. Platt to Alice K. Smith, February 24, 1960.
38. Maurice Fox to the author, December 9, 1991.
39. *Newsweek,* June 18, 1951.
40. Maurice Fox interview, May 24, 1988.
41. John Platt to Alice K. Smith, February 24, 1960.
42. MIT Vol. I, p. xvii.
43. Earl Ubell, "Germ Changes Found Rising as Life Lengthens. Scientists' Discovery Backs Report of Shorter Lives for Older Women's Babies," *New York Herald Tribune,* September 7, 1951, p. 21.
44. Manuscript dated July 29, 1955 (LSP).

45. October 30 and 31 and November 6, 1958, drafts. See "The Broad Biological Theories" in *The University of Chicago Reports,* Vol. 12, No. 2, 1961, pp. 9a–12a.
46. "A Theory of How We Age," *New Scientist,* February 12, 1959, p. 346. "Genetics. Chilling Forecast," *Newsweek,* March 31, 1959.
47. See Tabin, MIT Vol. I, p. 531.
48. According to Novick, this patent was used for some commercial enterprises but yielded scant financial return. Melvin Cohn interview, August 11, 1987.
49. Szilard to Novick, February 24, 1954 (LSP 14/21). Maurice Fox interview, May 24, 1988.
50. Werner K. Maas interview, May 19, 1987; see also "Perspectives" by Maas, *Genetics* 128:489–494 (July 1991). And MIT Vol. I, pp. xvi–xvii.
51. Melvin Cohn interview, August 11, 1987.
52. William Doering interview, November 12, 1986.
53. Tristram Coffin, "Leo Szilard: The Conscience of a Scientist," *Holiday,* February 1964, p. 67.
54. John Platt, "Leo Szilard and His Ideas," from notes at the 1958 Conference on Biophysics, Boulder, Colorado, July 20–August 16, 1958, pp. 3 and 5. My thanks to John Platt for the use of these notes.
55. Szilard later explained that one reason he sought, and enjoyed, the company of younger scientists was their willingness to perform "really imaginative experiments" that do not "virtually guarantee publishable results." See Szilard to Livingston, October 26, 1958 (LSP). See also John Platt to Alice K. Smith, February 24, 1960.
56. Novick to the author, April 1991.
57. Hans Zeisel interview with Alice K. Smith, February 9, 1960.
58. Hans Zeisel interview, February 25, 1987. Lorenz to the author, May 25, 1987.
59. Bernard Davis interview, December 17, 1985. Maurice Fox interview, May 24, 1988.
60. Maurice Fox interview, May 24, 1988.
61. Harrison Brown interview, October 6, 1986.
62. Harrison Brown to Alice K. Smith, April 19, 1960.
63. The fertility necklace was made by University of Chicago mechanic John Hanacek, who charged Szilard $10.80 "For services rendered in making: one set of Beads and Clasp. 4 ½ hours" (LSP).
64. MIT Vol. I, p. 14.
65. Quoted in Gertrud Weiss Szilard to Puck, January 17, 1956. This draft letter was apparently never sent and was found in Trude Szilard's papers when she died (EWP).
66. Philip Marcus interview, December 3, 1986. Marcus to the author, October 27 and November 19, 1991, and March 13, 1992.

67. Theodore T. Puck and Philip I. Marcus, "A Rapid Method for Viable Cell Titration and Clone Production with HeLa Cells in Tissue Culture: The Use of X-irradiated Cells to Supply Conditioning Factors," *Proceedings of the National Academy of Sciences,* Vol. 41, No. 7, pp. 432–37 (July 1955). Marcus later told Puck and the author that this paper was a compromise that did not fully state Szilard's role in the process. Up to the time of this book's first publication, Marcus and Puck disagreed over the significance of Szilard's contributions to their work. Marcus has since documented Szilard's role in Genesis of the "Feeder Cell": Concept and Practice. Philip I. Marcus. In Vitro Cell.Dev.Biol.—Animal 42:235–237 (2006). See especially "Postcript" 237.

68. John Benjamin to Szilard, October 17, 1958 (EWP).

69. Puck to Szilard, December 10, 1955 (LSP 15/25) Contant Webb correspondence.

70. Irwin Weil interview, October 18, 1991.

71. "The Great Science Debate," *Fortune,* June 1946. Speculating about the future of his own profession, Szilard had written "Observations on the Provision of the Progress of Sciences," comparing scientific training in different countries and endorsing the proposed National Science Foundation (NSF).

    He urged that scientists try to study unrecognized problems, not those already identified, and he was pictured and quoted as one of four "Scientific Irregulars" (with Urey, Condon, and Harlow Shapley) in a comprehensive article in *Fortune* about the future of American science.

72. According to his expense records, Szilard met Salk on October 25, 1956. They met again on January 7–9, 1957, at the conference on "Cellular Biology, Nucleic Acids, and Viruses." In addition to Salk, speakers included Renato Dulbecco and Francis Crick.

73. MIT Vol. I, pp. 505–24.

74. January 11, 1957, "Memorandum" and its "Appendix" (LSP).

75. Among Szilard's candidates who did not join were Min Chueh Chang, working in reproduction studies at the Worcester Institute; Joshua Lederberg; Howard Green; Norton Zinder; Maurice Fox; Milton Weiner; and Theodore Puck.

76. Szilard to Salk, January 14, 1957 (LSP 17/6).

77. Szilard met Salk on January 26, 1957. See Szilard to Salk, January 25, 1957 (LSP 17/6), and Salk to Canfield [cc: Szilard], February 8, 1957 (LSP 70/9).

78. Szilard to Salk, February 12, 1957 (LSP 17/6).

79. Canfield to Szilard, February 14, 1957 (LSP 6/4).

80. Oak Ridge 10th Annual Research Conference, April 8–10, 1957.

81. Szilard to Salk, April 17, 1957 (JSP 23, General Correspondence, 1957, Sk–Sz).

82. Sydney Brenner interview, June 25, 1987.
83. "Memorandum on conversations with Jim Watson on May 4, and with Roger Revelle on May 5, regarding the hypothetical possibility of convincing Jonas Salk to set up a research institute for basic and applied biology at La Jolla in loose affiliation with the University of California at La Jolla," May 7, 1959 (LSP).
84. Szilard to Salk, May 8, 1959 (LSP 17/6).
85. Szilard to Salk, June 19 and August 4, 1959 (LSP 17/6).
86. Peter Krohn, ed., *Topics in the Biology of Aging.* A Symposium held at the Salk Institute for Biological Studies, San Diego, California, 1965 (New York: Interscience Publishers, John Wiley & Sons, 1966).
87. John Newhouse conversation, February 17, 1987. Newhouse visited Szilard at the club to discuss ghostwriting an autobiography, a project that never developed.
88. Herman Kalckar interview, January 13, 1987.
89. Szilard to Livingston, October 26, 1958 (LSP 12/9).
90. Frances Racker interview, November 22, 1985.
91. Livingston to Szilard, June 13, 1958 (LSP 12/9).
92. Szilard to Detlev Bronk, June 19 and 20, 1958 (LSP 5/10).
93. Herbert Anderson to Szilard, July 11, 1958 (LSP 4/12). Anderson interview, October 6, 1986.
94. Szilard to Livingston, July 21, 1958 (LSP 12/9).
95. John C. Lilly interview, February 8, 1987.
96. Szilard to Livingston, September 28 and 29, 1958 (LSP 12/9). Szilard to Detlev Bronk, September 29, 1958 (LSP 5/10).
97. John Benjamin to Szilard, October 17, 1958 (EWP).
98. Szilard to Livingston, September 28, 1958 (LSP 12/9). It is not clear that this letter was sent, as the copy in the Szilard Papers contains many handwritten corrections.
99. Szilard to Livingston, October 26, 1958 (LSP 12/9).
100. *Proceedings of the National Academy of Sciences,* January 1959, Vol. 45, No. 1, pp. 30–45. Szilard to Norris Bradbury, January 20, 1959 (LSP 5/2).
101. Szilard to Khrushchev, September 6, 1959. Szilard enclosed a copy of his essay "How to Live with the Bomb—and Survive," noting that he had also sent it to the White House. MIT Vol. III, p. 262.
102. *The Way of the Scientist: Interviews from the World of Science and Technology.* Selected and Annotated by the Editors of *Science and Technology* (New York: Simon & Schuster, 1966), pp. 29–31. See also *International Science and Technology,* May 1962, pp. 36–37.
103. Francis Reitel to Szilard, November 4, 1959 (LSP).
104. People and places from Szilard's "Application for Research Grant," June 23, 1959 (RG6876), pp. 15–16.

105. "The Control of the Formation of Specific Proteins in Bacteria and in Animal Cells," pp. 277–92, and "The Molecular Basis of Antibody Formation," pp. 293–302. *Proceedings of the National Academy of Sciences,* Vol. 46, No. 3 (March 1960).

CHAPTER 26

1. March 30, 1960, interview for *Small World,* CBS Television (LSP 108).
2. Ibid.
3. James J. Nickson note in Szilard's medical records. Nickson to Feinstein, June 29, 1962 (EWP). Szilard underwent a superpubic cystostomy and fulguration for two tumors.
4. Doctors boosted the maximum dose to 5,300 rem and never gave Szilard less than 4,500 rem. James J. Nickson note with June 29, 1962, Feinstein letter to Szilard (EWP). January 14, 1960, Memorial Center Follow-up and Progress Notes. Szilard medical records (EWP).
5. Nickson note and test result notes (EWP). No sign of neoplasm detected for urine samples on February 19, March 21 and 30, April 6, May 3, June 9, and August 15 and 29. See also Szilard to Joseph Barnes, August 17, 1960 (LSP).
6. "Remembrance of a Genius," *Life,* June 12, 1964.
7. Szilard to Urey, January 6, 1960 (Harold Urey Papers, Box 91, University of California, San Diego, Library).
8. See Szilard's November 21, 1959, note (EWP).
9. Conant Webb interview, January 13, 1987. William Doering interviews, November 12, 1986, and March 9, 1988. Naomi Liebowitz Wood and Ramsay Wood interviews, April 22, 1986.
10. Szilard to Joseph Barnes, August 21, 1960 (LSP). Szilard to Topchiev, August 29, 1960 (LSP 19/4).
11. Various details about Szilard's year at Memorial Hospital come from interviews with Baruj Benacerraf, January 15, 1987; Bernard Davis, December 17, 1985; Maurice Fox, January 13, 1987; Mariana Heller, January 19, 1988; Rollin Hotchkiss, January 16, 1987; and Werner Maas, May 19, 1987.
12. In January he signed an agreement with Marc Wood International and the National Center for Scientific Research for a venture with Jacob, Monod, Novick, and Meselson to manufacture and market genetic strains. January 12, 1960, contract (LSP).

    In May, Szilard's letter to *Nature* on "Dependence of the Sex Ratio at Birth on the Age of the Father" was published. This technical explanation of his general theory of aging noted statistics that suggested the ratio of boys to girls at birth decreased with the father's age. MIT Vol. I, pp. 495–96.

At the end of May, Szilard drafted and revised a "Memorandum on Enzyme Repression" (LSP). But Szilard wrote nothing else in microbiology during his busy hospital year, and except for conversations with visitors such as Jacob, Maas, Howard Green, Novick, Webb, Livingston, Fox, and Doering, he appears to have given the subject little thought.

13. "Memoirs" (LSP).
14. February 21, 1960, "Leo Szilard—Biographical Notes" (LSP).
15. Iseman to Cousins, April 27, 1960 (LSP). NAL, July 3, 1960, agreement and advance, and August 16, 1960, correspondence (LSP).
16. Robert Pick to Szilard, June 2, 1960 (LSP).
17. Szilard to Robert Pick, June 6, 1960 (LSP).
18. "Outline for Book," June 14 draft (LSP 30/9).
19. The little Szilard dictated about himself appeared after his death as his "Reminiscences" in a Harvard University collection on *Perspectives in American History*, Vol. II, 1968, *The Intellectual Migration: Europe and America, 1930–1960* (Fleming and Bailyn, eds.). Identical and related material appeared in MIT Vol. II in 1978.
20. "Leo Szilard: The Man Behind the Bomb, A Postscript with Gertrud Weiss Szilard," KPBS-TV, San Diego (LSP 108).
21. For an earlier version, see "Proceedings of Fourth Pugwash Conference of Nuclear Scientists, June 25–July 4, 1959, Baden, Austria," transcript, pp. 100–123.
22. *Bulletin of the Atomic Scientists*, February 1960. Reprinted in Vol. III, pp. 207–37.
23. *Edward P. Morgan and the News*, American Broadcasting Network, March 7, 1960 (LSP).
24. "THE H-BOMB: 'Rules' for Nuclear War," *Newsweek*, March 7, 1960, p. 88.
25. *Krokodil*, April 20, 1960, No. 11 (1589).
26. *Bulletin of the Atomic Scientists*, March 1960. Historically, Szilard's skepticism about governments' commitment to test bans is supported by Robert A. Divine, who concludes, in *Blowing on the Wind: The Nuclear Test Ban Debate 1954–1960* (New York: Oxford University Press, 1978), p. 27, that the prevailing pattern became one of "sudden interest in nuclear tests, intense debate and public discussion, and then the equally abrupt dropping of the issue."
27. "Szilard Bars Operation to Work for Peace. Sees 80% Chance of Nuclear War," Marguerite Higgins, *New York Herald Tribune*, April 6, 1960.
28. "Small World," May 22, 1960 (LSP 108).
29. Transcripts from April 2, 1960, interview with Teller, Watson, and Szilard (LSP).
30. MIT Vol. III, pp. 238–50. Transcript in (LSP 108). *New York Post* magazine, "Page Four," November 20, 1960.

31. "Somber Proposal from Dr. Szilard," Marquis Childs, *Washington Post,* February 17, 1960. "Life of a Man," Max Lerner, *New York Post* magazine, "Page Four," March 4, 1960, p. 34. Marguerite Higgins two-part interview, *New York Herald Tribune,* April 5 and 6, 1960.

32. *New York Post* magazine, "Page Four," November 20, 1960. *Newsweek,* March 7, 1960, p. 88. *Life,* May 23, 1960, 48:54. *U.S. News & World Report,* August 15, 1960, pp. 68–71.

33. The Einstein award was sponsored by a memorial foundation honoring Strauss's parents and was given that year for Szilard's A-bomb work that had contributed "effectively to the defense of the free world." *Science,* March 25, 1960. The "getting better" quote is from "I'm looking for a market for wisdom," *Life,* September 1, 1961, p. 76.

34. Killian to Szilard, March 29, 1960 (MIA, MC 10, 3/111). Four recipients, two for 1959 and two for 1960, shared $150,000. *The New York Times,* May 19, 1960. Nurse's Notes, March 28, 1960 (EWP). The Research Institute of America gave Szilard its Living History Award on April 27, 1960. See also Nurse's Notes for April 28, 1960 (EWP).

35. "The Presentation of the Atoms for Peace Award . . . at the National Academy of Sciences, Washington, D.C. May 18, 1960," program, pp. 34–37. Szilard's remarks at a press conference afterward were widely reported the next day. See "Two Scientists Wary on Policing of Atomic-Test Ban," *The New York Times,* May 19, 1960.

36. Szilard to Kennedy, April 15, 1960 (LSP 11/5). Kennedy to Szilard, April 27, 1960 (EWP). For details of the postelection study program, see Szilard to Bowles, April 14, 1960, and Bowles to Szilard, April 18, 1960 (LSP 4/36).

37. Conference to Plan a Strategy for Peace. June 2–5, 1960, Arden House, Harriman, New York. "Final Report," p. 20. First sponsored by philanthropists Tom Slick and Mrs. Albert Lasker, the session eventually became the Stanley Foundation's annual "Strategy for Peace" conference.

38. Whitmore notes, June 3, 1960, "Memorial Center Follow-up and Progress Notes, 5-8-60 Dr. Nickson" (EWP).

39. Follow-up and Progress Notes and Nurse's Notes, June 3, 1960 (EWP).

40. June 6, 1960, Health Report (LSP 1/26).

41. Bohlen to Szilard, June 14, 1960, and Szilard to Bohlen, June 16, 1960 (LSP).

42. Szilard to Khrushchev, June 27, 1960. MIT Vol. III, pp. 264–67.

43. Land to Szilard, June 14, 1960 (LSP).

44. MIT Vol. III, p. 204.

45. The 1960 *Bulletin* articles that formed the basis of Szilard's narrative were "How to Live with the Bomb—and Survive . . ." (February), "To Stop or Not to Stop" (March), and "The Berlin Crisis" (May).

46. *La Jolla Journal,* April 2, 1964, p. 3.

47. John C. Lilly interview, February 8, 1987.
48. *The Voice of the Dolphins*, p. 21.
49. Ibid., pp. 22–24.
50. Szilard also wrote an outline for a play, a television show, or a movie based on "Dolphins." See August 7, 1960, manuscript "Adaptation of The Voice of the Dolphins'—leaving out the Dolphins" (LSP) and *"Die stimme der delphine"* (LSP 34/6).

On August 9, 1960, Szilard signed a contract with Simon and Schuster for *The Voice of the Dolphins*. It was published in April 1961; printed in British, French, Italian, Argentinian, German, Dutch, Japanese, and Russian editions; and reprinted by Stanford University Press in 1992 (LSP 32/12–15, 33/1–9, and 34/1–8).

CHAPTER 27

1. *New York Herald Tribune,* May 10, 1960. *Washington Post,* May 13, 1960. *The* New *York Times,* June 12, 1960.
2. Szilard to Khrushchev, June 27, 1960. MIT Vol. III, p. 264.
3. Szilard to Khrushchev, August 16, 1960. MIT Vol. III, p. 268.
4. Khrushchev to Szilard, August 30, 1960. MIT Vol. III, p. 269.
5. Details of this meeting come from notes Szilard wrote on October 5 and dictated four days later and from Szilard's account to Maurice Fox (interviewed November 12, 1985) and Bela Silard. The order of topics follows the agenda that Szilard had prepared for the meeting, although not knowing how much time he had for their conversation, Szilard actually skipped around in order to raise the most important subjects first. See "Conversation with K on October 5, 1960," MIT Vol. III, pp. 279–87 (LSP 23/14).
6. According to historian and policy analyst William L. Ury, the idea of the Moscow-Washington "hot line" was developed by Harvard economist Thomas Schelling, who thought of it after reading Peter Bryant's *Red Alert* (New York: Ace, 1958), the novel on which the film *Dr. Strangelove* was based. Schelling mentioned the idea to Henry Owen, a State Department planner in the Eisenhower administration. Owen advanced the idea to Gerard C. Smith, then director of the policy planning staff, who argued for a telephone link but made little headway because of his colleagues' mistrust of the Russians.

Independently, in the fall of 1959, *Parade* magazine editor Jess Gorkin and a friend hit on the idea of a phone link, and in 1960, in an open letter in the magazine, Gorkin proposed an emergency communication link to Eisenhower and Khrushchev. His proposal was reprinted in the Soviet newspapers *Pravda* and *Izvestia*. Gorkin solicited statements of support for the hot-line idea from presidential candidates Nixon and Kennedy during the 1960 campaign, and at a reception in New York that fall asked Khrushchev about the proposal. Khrushchev called it "an excellent idea," Gorkin recalled.

Owen pressed for the hot line in the Kennedy administration, and the proposal was made to the Soviets at an Eighteen-Nation Conference on Disarmament in Geneva in April 1962. The Cuban Missile Crisis that fall gave new impetus to the idea, and on August 31, 1963, the hot line went into operation. See Ury, pp. 144–45.

7. Szilard to Khrushchev, "Memorandum (September 30, 1960)" in MIT Vol. III, pp. 273–78; and "Conversations with K . . ." (LSP 23/14).

8. Typescript of the speech "Can We Get Off the Road to War?," delivered May 3, 1962 (LSP 23/4).

9. Szilard to Khrushchev, October 8, 1960 (LSP 11/7).

10. *Pravda*, October 6, 1960, p. 1. Translation in MIT Vol. III ms, p. 1828.

11. William Kincade interviews, May 28, 1985, and March 22, 1988.

12. Szilard to Bohlen, October 6, 1960 (LSP). Herter to Szilard, November 10, 1960. MIT Vol. III, p. 290.

13. Notes, MIT Vol. III ms, p. 1840.

14. *"Je donne"* is French for "I deal" and is used in card games.

15. Gerard Piel interview, June 9, 1988.

16. MIT Vol. III, pp. 291–92.

17. Alexander Rich interview, October 29, 1987.

18. Szilard's "Diary" of the Moscow trip, p. 8 (LSP 30/14). Herbert Anderson interview, October 6, 1986. Alexander Rich interview, October 29, 1987.

19. Ruth Adams interview, October 21, 1986.

20. William Higinbotham interview, March 11, 1988. David Frisch interview, January 12, 1987.

21. Jerome Wiesner interview, November 10, 1987. William Higinbotham interview, March 11, 1988. Alexander Rich interview, October 29, 1987.

22. In the "Proceedings of the Sixth Pugwash Conference, Moscow USSR November 27 to December 5, 1960," Szilard's remarks are recorded in only four of the thirteen sessions for which transcripts were made.

23. The dolphin ashtray came to Semenov's son-in-law, Vitalii I. Goldanskii, who had first met Coryell at MIT in 1957. Goldanskii interviews, September 2 and 3, 1986.

24. Szilard to Khrushchev, December 2, 1960. MIT Vol. III, p. 293.

25. "Proceedings," Minutes of the Sixth Session, November 30, 1960, pp. 329–33.

26. "Proceedings," Minutes of the Eighth Session, December 2, 1960, pp. 472–75 and 521–22.

27. *The Voice of the Dolphins*, p. 46.

28. "On Flexibility, Communication, and an Arms Control Proposal," "Proceedings," pp. 608–29.
29. William Higinbotham interview, March 11, 1988.
30. While waiting in Moscow, Szilard wrote "The Postwar Events and the Russian Disarmament Proposals of 1960," a nine-page manuscript with a "Diary" for 1945 to 1960, and he drafted another account that carried world events to 1984, labeled "Appendix 1" (LSP 30/14).
31. Szilard to Khrushchev, December 20, 1960. MIT Vol. III, p. 294.
32. Szilard's 1960 "Diary," pp. 8 and 12 (LSP 30/14). Rose Scheiber and Alice Danos interviews, July 16, 1987. Josef Litván's son also reports that his father and Szilard met in Vienna at this time. George Litván to the author, August 18, 1987.
33. Inge Feltrinelli to the author, 1987.

CHAPTER 28

1. See Lyle to Gertrud Weiss Szilard, April 30, 1960 (Aaron Novick Papers).
2. Eva Zeisel interview, August 19, 1985.
3. Mike Wallace interview, February 27, 1961, WNTA-TV New York (LSP 40/22).
4. Jerome Wiesner interview, November 10, 1987.
5. See, for example, Szilard to McCloy, April 4, 1961; Szilard to McCloy et al. on the Berlin crisis, July 7, 1961; Szilard to McCloy et al. with twenty-one-page "Disarmament" memo; McCloy to Szilard, September 1, 1961; Szilard to McCloy, October 6, 1961 (LSP).
6. George Rathjens interview, November 10, 1986.
7. Szilard to McCloy, October 6, 1961 (LSP).
8. Glenn Seaborg, *Kennedy, Khrushchev and the Test Ban* (Berkeley: University of California Press, 1981), p. 61. Szilard to Seaborg, March 19, 1961 (LSP 17/16).
9. Lisbeth Bamberger Schorr interview, April 12, 1988.
10. "I'm Looking for a Market for Wisdom," *Life*, September 1, 1961, p. 76.
11. Louis Lerman interview, January 14, 1987. Thomas I. Emerson to the author, January 15, 1988.
12. Maurice Dolbier, "The Voice of Leo Szilard," *New York Herald Tribune*, May 14, 1961.
13. "Letter to the Editor," dated April 21, 1961 (LSP 27/7). *Washington Post*, April 20, 1961. *New York Herald Tribune* and *St. Louis Post-Dispatch*, April 27, 1961.
14. See Szilard to Kennedy, June 6, 1961 (JFK White House Names, Szilard Box 2745).

15. "Petition to the President of the United States" and May 10, 1961, Memorandum and letter to the President (JFK White House Names, Szilard Box 2745).
16. Marcus G. Raskin interview, April 18, 1988.
17. Szilard to Murrow, June 6, 1961 (LSP 13/26).
18. L. H. Thomas to Kennedy, May 15, 1961 (JFK White House Names, Szilard Box 2745).
19. Franck to Szilard, May 21, 1961 (JRL Franck Papers 9/4, Leo Szilard).
20. FBI director to Byron R. White, May 12, 1961 (LSP). "Request called by White's office 5-9-61" is written on Hoover's note. White's call may have been prompted by Szilard's interview by Dan Dixon on CBS Radio that day.
21. Szilard to O'Donnell, May 19, 1961 (JFK, Rostow Box 3, General Correspondence No–Z, Sto–Sz).
22. Wofford to Kenneth O'Donnell, May 22, 1961 (JFK President's Office Files, General Correspondence 1961 ST–SZ).
23. Bowles to Szilard, May 26, 1961 (LSP 4/36).
24. Isaacson and Thomas, pp. 608–609.
25. "Memo on Disarmament" by Szilard, July 18, 1961 (LSP).
26. "A Proposal by Professor Leo Szilard" (LSP 30/22). Szilard to Eaton, July 18, 1961 (LSP 7/24). Szilard to Kennedy, September 1, 1961. Kennedy replied with a noncommittal note on October 18, 1961 (LSP 11/5).
27. Memorandum, September 25, 1961. MIT Vol. III, p. 375. "Tentative Draft," September 21, 1961 (LSP).
28. "Draft Proposal," July 7, 1961, "Insert-Page 10-2-" (LSP).
29. Kissinger to Szilard, August 3, 1961 (EWP).
30. Kissinger to Bundy, with attached memos, undated from Kissinger and dated September 11, 1961, from Bundy (JFK Countries Box 56 C092/LG/Berlin).
31. Jerome Wiesner interview, November 10, 1987.
32. Life, September 1, 1961, pp. 75–77.
33. "A Message to You from the President," September 7, 1961. Life, September 15, 1961.
34. Ibid.
35. Martin Kaplan interview, September 4, 1986.
36. Joseph Rotblat interview, September 1, 1986. Bernard T. Feld interviews, September 1–3, 1986.
37. John Schrecker interview, November 11, 1987. Schrecker to the author, January 20, 1988.
38. Szilard to Khrushchev, September 20, 1961. MIT Vol. III, p. 296.
39. Draft letters to Bundy and Kaysen, September 25, 1961 (LSP).
40. Rauh to Szilard, October 18, 1961 (LSP 16/7). Yarmolinsky to Szilard, September 29, 1961 (LSP).

41. Barnet essay for Gertrud Weiss Szilard, undated (EWP).
42. Herbert York interview, January 24, 1986.
43. Szilard to Khrushchev, October 4, 1961. MIT Vol. III, pp. 297–99.
44. *Bulletin of the Atomic Scientists,* December 1961, pp. 407–12. *Newsweek,* December 25, 1961, p. 70.
45. Marcus Raskin interview, May 5, 1988.

CHAPTER 29

1. Draft dated October 30, 1961 (LSP).
2. Ibid, and "Are We On the Road to War?" draft, November 14, 1961, p. 14 (LSP).
3. Roger Fisher interview, November 13, 1986.
4. Szilard dictated an eight-page statement on October 30, 1961 (LSP).
5. Descriptions of the speech are from Barton J. Bernstein, in his introduction to MIT Vol. III and in personal interviews. "Are We On the Road to War?," November 14, 1961, draft (LSP).
6. *Holiday,* February 1964, p. 94.
7. Audiotape of questions and answers after "Are We On the Road to War?" speech, November 17, 1961. My thanks to Helen Weiss for a copy of this tape.
8. MIT Vol. III, p. lxii. Szilard was scheduled to meet students at Associated Harvard Clubrooms, Holmes Hall, between 10:30 A.M. and 4:30 P.M. Gar Alperovitz to Huberman, November 4, 1961 (LSP).
9. Bud Weidenthal, "First A-Bomb Builder Here for Peace Talk," Cleveland *Press and News,* November 29, 1961; "Atomic Pioneer Asks Campaign to End War," Cleveland *Plain Dealer,* November 30, 1961. "Responses to Date (February 24, 1962)," MIT Vol. III, p. 449.
10. "Responses to Date (February 24, 1962)," MIT Vol. III, p. 450.
11. Richard Lewis, "Szilard Urges People's Drive, Lobby to Work for Peace," *Chicago Sun-Times,* December 2, 1961; "Lobby Against War, A-Bomb Pioneer Urges," *Chicago Daily Tribune,* December 2, 1961; "Developer of Atom Bomb Proposes Scientist-Scholar Lobby for Peace," *Washington Post,* December 2, 1961; "Nineteen Years Later," an editorial on Szilard's council proposal, *Chicago Sun-Times,* December 3, 1961; Austin C. Wehrwein, "Scientist Would Form Council to Lobby for Abolishing War," *The New York Times,* December 3, 1961; "Lobby for Peace," *Commonweal,* December 15, 1961; "Human Nature: The Whys of War," Science section, *Newsweek,* December 11, 1961. Gertrud Weiss Szilard FOIA request 62-59520 (LSP).
12. Allan Forbes to Szilard, November 19, 1961 (LSP 8/13). Forbes interview, November 11, 1986.
13. Maurice Fox interview, May 24, 1988.

14. Salk to Szilard, December 20, 1962 (LSP 17/6). Ralph Pomerance to Szilard, December 24, 1962 (LSP 15/19).
15. Szilard to Marcus Kogel, December 29, 1962, declining an invitation to be commencement speaker at the Albert Einstein College of Medicine (LSP).
16. Theodore Irwin, 'The Legend That Is Dr. Szilard," *Pageant*, December 1961, pp. 53–59. My thanks to Irwin for a helpful reply to my author's query.
17. "Atom Age's 'Father' Lobbies Against War," San Francisco *Call Bulletin*, January 10, 1962. David Perlman, " 'A People's Plan' to End Arms Race," San Francisco *Chronicle*, January 10, 1962. Arthur Caylor column "Just 10 Years Left," San Francisco *Call Bulletin*, January 11, 1962. William B. Wood, "Szilard Suggests a New Plan of Political Action," *Stanford Daily*, January 16, 1962. Szilard pocket diary and correspondence with Joshua Lederberg (LSP 11/29). Clifford Grobstein interview, August 27, 1987.
18. Barbara Snader Public Service Announcement (LSP). *Christian Science Monitor*, January 24, 1962. "Special Note to the Speech 'Are We on the Road to War?' for Los Angeles Area Readers (January 18, 1962)," MIT Vol. III, p. 447.
19. Jack Roberts, "A-Scientist Offers Formula to Avoid Hot War," undated *Reporter* clip from Portland, Oregon. "Atomic Pioneer Bares Peace Plan," *Oregon Journal*, January 13, 1962. "Nuclear Expert Says World Rolling Along on Road to War," *Oregonian*, January 13, 1962. "Expert Says Use of Nuclear Weapons 'Inevitable' if Victory US Goal," undated Oregon clip. "Responses to Date (February 24, 1962)," Vol. III, p. 449.
20. WBAI talk in LSP. John C. Waugh, "Szilard Stumps for Strong Peace Movement," *Christian Science Monitor*, January 24, 1962.
21. Fermi, Allison, Lashof, and Meier to Szilard, January 16, 1962 (LSP).
22. Gibson to Szilard, January 11, 1962 (LSP 8/30).
23. Penn to Szilard, February 12, 1962 (LSP 15/9).
24. Bancroft to Szilard, February 18, 1962 (LSP).
25. MIT Vol. III, pp. 448ff Just in case he might have missed any clever arrangements, Szilard also borrowed the bylaws of the National Committee for an Effective Congress, a group organized in 1948 that designated worthy candidates and pooled contributions to support them. Maurice Rosenblatt interview, February 3, 1986.
26. "Council Mailing with a Letter to Prospective Members (June 11, 1962)," MIT Vol. III., pp. 456–72. Ruth Adams interview, May 1, 1986.
27. Jennifer Robbins interview, December 6, 1985. Forbes to Feld, August 1962 (Council for a Livable World Archives, JGC). Feld Papers (MIA, CLW file).
28. Ruth Pinkson interview, September 15, 1991.
29. Forbes to Feld, "*M. le Président du Conseil pour l'Abolissement de la Guerre*," August 1962, Feld Papers (MIA).
30. Michael Straight interview, April 14, 1988.

31. *Time,* March 23, 1962. *Newsweek,* March 26, 1962.
32. George McGovern interview, May 18, 1988.
33. "Camera Three" was broadcast on June 3 and 10, 1962, by CBS Television; transcript in LSP and excerpts in MIT Vol. III, pp. 381ff.
34. Kaplan, p. 315.
35. Allan Forbes interview, November 11, 1986.
36. Szilard quote from MIT Vol. III, p. 300. Bethe to Szilard, September 18, 1962 (LSP 4/29).
37. Szilard to Khrushchev, October 9, 1962. MIT Vol. III, pp. 300–330. "Confidential Memorandum," January 8, 1963. MIT Vol. III, pp. 314–17.
38. Szilard to Khrushchev, October 9, 1962. MIT Vol. III, pp. 300–330. Khrushchev replied November 4. Szilard to Livingston, October 12, 1962 (LSP 12/9). Copy in (JFK NSF Box 369, Disarmament Szilard, Leo #2a).
39. Allan Forbes interview, November 11, 1986.
40. Congressional Quarterly Service, p. 133.
41. Allan Forbes interview, November 11, 1986. See Szilard's undated rough draft on the Cuban Missile Crisis (LSP), excerpts in MIT Vol. III, p. 478.
42. *Holiday,* February 1964, p. 96ff.
43. Lisbeth Bamberger Schorr interview, April 12, 1988.
44. Marcus Raskin interview, April 18, 1988.
45. Graham T. Allison, *Essence of Decision: Explaining the Cuban Missile Crisis* (Boston: Little, Brown and Company, 1971), p. 131. Elie Abel, *The Missile Crisis* (Philadelphia: J. B. Lippincott Company, 1966), p. 155.
46. Joshua Lederberg interview, May 20, 1987. Richard Garwin interview, September 4, 1986.
47. Paul Marks interview, January 21, 1988. Szilard's October 25, 1962, notes (LSP).
48. Victor Weisskopf interview, December 16, 1985. Weisskopf NOVA interview (NV1625 SR206, p. 11).
49. Three weeks later, after exchanging several letters with Khrushchev, Kennedy announced that all Soviet missile installations had been dismantled and the US quarantine lifted. The United States quietly withdrew its Jupiters from Turkey. Only years later was it revealed, by former Secretary of State Dean Rusk, that Kennedy had a second fallback position; an offer through diplomat Andrew Cordier to have UN Secretary-General U Thant propose a mutual withdrawal of the Cuban and Turkish missiles. Eric Pace, "Rusk Tells a Kennedy Secret: Fallback Plan in Cuba Crisis," *The New York Times,* August 28, 1987, pp. Al and A9. Walter Pincus, "Transcript Confirms Kennedy Linked Removal of Missiles in Cuba, Turkey," *Washington Post,* October 22, 1987, p. A18.
50. Elias Blatter interview, July 6, 1987.
51. Livingston to Szilard, November 9, 1962 (LSP 12/9).

52. Khrushchev to Szilard, November 4, 1962. MIT Vol. III, p. 305.
53. Szilard to Khrushchev, November 15, 1962. MIT Vol. III, pp. 307–308.
54. Szilard's November 19, 1962, Memorandum. MIT Vol. III, p. 309.
55. Szilard's "Confidential Memorandum," January 8, 1963 (LSP).
56. Marcus Raskin interview, April 18, 1988. Trude's "Trip to Europe 1962," Gertrud Weiss Szilard Papers (LSP), recapitulated in Szilard to Livingston, November 30, 1962 (LSP 12/9).
57. Szilard to Khrushchev, November 25, 1962. MIT Vol. III, pp. 310–11.
58. Sergei Khrushchev, *Khrushchev on Khrushchev* (Boston: Little, Brown and Co., 1990), p. 128.
59. Allan Forbes interview, November 11, 1986.
60. Herken, 1987, p. 183. Thomas Powers, "Seeing the Light of Armageddon," *Rolling Stone,* April 29, 1982, pp. 13–17.
61. Szilard to Bundy, December 26, 1962 (JFK White House Names Box 2745 Szil). See also note on this letter, which was only that the two had met. McGeorge Bundy interview, December 8, 1987.
62. Szilard to Kaysen, January 7, 1963 (JFK NSF Box 369 #1, 1/63). Szilard's January 11, 1963, Memorandum (JFK NSF Box 369 Kaysen Disarmament Szilard, Leo 1/63 #3).
63. Szilard's January 15, 1963, Memorandum (LSP). York to Szilard, February 11, 1963 (LSP 21/19). Szilard to Kaysen, February 14, 1963 (JFK NSF Box 369 C Kaysen Disarmament Szilard, Leo 2/63 to 4/63 items # 3 and 3a).
64. Szilard's February 20, 1963, meeting with Bundy (JFK/CO55 White House Names Szil) and follow-up letter that day (JFK NSF Box 369 C Kaysen Disarmament Szilard, Leo 2/63 to 4/63 item 6a).
65. Szilard's March 20, 1963, "Modus operandi" (LSP). Szilard "Letter (Progress Report) to Council Members," March 25, 1963, MIT Vol. III, pp. 480–82.
66. Szilard remarks to the Women Strike for Peace, September 21, 1962. My thanks to Leo Orso for a tape recording of this event.
67. Bernard Gwertzman, "The World Has Six Years to Live," Sunday magazine, *Washington Star,* March 24, 1963, p. 7.

    In the 1962 elections, the council raised $79,318.65: $20,938.84 in general contributions and $36,379.81 in checks sent for transmittal to candidates. McGovern was the council's principal candidate, receiving $20,091.55 in direct contributions and $2,000 from the general fund—financing one-fifth of his campaign.

    The council also earmarked some $10,400 to Joseph S. Clark of Pennsylvania, $4,675 to Wayne L. Morse of Oregon, $2,500 to J. William Fulbright of Arkansas, and $1,500 to Frank Church of Idaho (JGC Council for a Livable World Files). George McGovern interview, May 18, 1988.
68. Szilard to Spitz, March 29, 1963 (LSP).
69. Harold Orlans interview, September 8, 1986.

70. The International Affairs Seminars of Washington, meeting of April 26, 1963; Szilard's speaking notes, pp. 3 and 5 (LSP).

71. Szilard to Wiesner, May 9, 1963 (LSP 21/3).

72. Szilard's "Why America May Come to Grief" said on March 27, 1963, that the United States and the Soviet Union "might conceivably live with saturation parity" (LSP 34/17). Szilard to Salk, March 29, 1963 (JSP Salk Institute Papers, Szilard, Leo).

73. Szilard to Salk, April 30, 1963 (LSP 17/6). In 1962, Szilard dabbled in biology by writing two papers: about how "spitting-image" children develop, "Homologous Suppression in Man" (LSP 26/1), and "On the Occasional Dominance of the 'Perceptible Phenotype' in Man" (LSP 29/13). He also edited "The Aging Process and the 'Competitive Strength' of Spermatozoa," a study related to his own vague theories about aging (LSP 22/6).

74. Szilard-Salk correspondence for May 21 and 23, 1963 (LSP 17/6).

75. See, for example, Szilard to C. F. von Weizsäcker, January 14, 1963 (CERN #1 cc. to Weisskopf, CERN Archive file 20683). See also CERN Report dated August 20, 1963 (No. 6808, CERN No. 1). Dakin to Weisskopf, July 19, 1963 (CERN book #1). Szilard managed to raise support for EMBO through his approaches to the French government and the Volkswagen Foundation.

76. See Kennedy-Hoagland correspondence, June 12 and 15, 1963 (JFK NSF Box 369 C Kaysen Disarmament Szilard, Leo 5/63 to 7/63 items 6–6a and numbers a–c). Kaysen to Fisher, ibid.

77. Details of the confusion appear in Szilard to Roger Fisher, July 27, 1963 (LSP 8/11). See also Szilard-Kaysen telegrams (JFK NSF Box 369 C Kaysen Disarmament Szilard, Leo 5/63 to 7/63 item 9a).

78. Szilard to Fisher, August 4, 1963 (LSP 8/11 and 71/4). MIT Vol. III, p. 329 n 1.

79. See September 5, 1963, draft (LSP).

80. Szilard to Edward Levi, July 6, 1963, Appendix (LSP 12/2). Pugwash Proceedings (LSP 60/17).

81. Szilard to Edward Levi, July 6, 1963, with Appendix (source of Washington speculation) and a later August 2 Appendix on his movements and activities in Europe (LSP 12/2).

82. Szilard to Edward Levi, August 1, 1963, Memorandum (LSP 12/2).

83. Szilard's September 26, 1963, Memorandum for the Vatican (LSP).

84. Szilard to Eldon Griffiths, May 19, 1964 (LSP). In London, Szilard met Dennis Healy, had lunch with Richard Crossman, and spoke with physicist Patrick M. S. Blackett. After Szilard's visit, maverick MP Anthony Wedgwood-Benn (later just 'Tony Benn") praised his ideas on minimal deterrence. Szilard to Muller, Healy, etc., December 6, 1963 (LSP).

85. Szilard to Kennedy, November 14, 1963; Bundy to Szilard, November 21, 1963 (JFK White House Names Box 2745 Szil).
86. December 16, 1963, draft of "Farewell to Arms Control" then changed to "Let Us Put Up, or Shut Up" (LSP 27/21).

CHAPTER 30

1. Patricia Murphy, "Szilard Papers in Her Trust," *Los Angeles Times*, April 18, 1973, Part IV, p. 12.
2. Szilard to Cecil Green, January 27, 1964 (LSP).
3. "Summary," p. 4. MIT Vol. III, p. 421.
4. The Del Charro was located northeast of where Torrey Pines Road and La Jolla Shores Drive meet.
5. Murphy, "Szilard Papers in Her Trust," p. 12.
6. Szilard to Bundy, February 7, 1964 (LSP).
7. Szilard pocket diary, February 11, 1964 (LSP).
8. Szilard to Robert McNamara, February 10, 1964 (LSP).
9. Attached to "Memorandum" (LSP).
10. Tristram Coffin, "Leo Szilard: The Conscience of a Scientist," *Holiday*, February 1964, pp. 64ff.
11. Szilard to Carl von Weizsäcker, March 2, 1964 (LSP 20/27).
12. Szilard to Kogel, March 16, 1964 (LSP).
13. Szilard began dictating this paper on March 3. See Transcription from Shorthand Notebooks of Dr. Leo Szilard's Paper "On Memory and Recall," p. 1 (SIA), hereafter called "Transcript." My thanks to Sylvia Bailey and June Gittings for a copy of this transcription.
14. "Memorandum on the Induction of B Galactosidase" (LSP) and "Memorandum on Antibody Formation" (Transcript, p. 4).
15. (Transcript, p. 5).
16. Szilard to Wigner, March 13, 1964 (LSP 21/4).
17. Jonas Salk interview, March 23, 1985.
18. Melvin Cohn interview, August 11, 1987.
19. Jonas Salk interview, March 23, 1985.
20. Melvin Cohn interview, August 11, 1987.
21. Ibid.
22. (Transcript, p. 99).
23. Rita Bronowski interview, January 22, 1986. Jacob Bronowski diary 1964.
24. (Transcript, p. 100). Szilard to Bernard L. Horecker, May 5, 1964 (LSP).
25. All quotations are from "On Memory and Recall," *Proceedings of the National Academy of Sciences*, Vol. 51, No. 6, pp. 1092–99 (June 1964). Reprinted in MIT Vol. I, pp. 497–504.
26. MIT Vol. I, p. 392.
27. Nicholas Kurti interview, May 26, 1986.

28. Bronowski to Salk, June 10, 1964 (Salk Institute Files, Dr. Szilard Resident Fellow, JSP).

29. Freeman Dyson interview, December 1, 1986. Melvin Cohn interview, August 11, 1987.

30. Melvin Cohn interview, August 11, 1987.

31. Salk to Szilard, March 2, 1964 (EWP).

32. Szilard to Spanel, May 6, 1964 (LSP).

33. "Proposed Conference on Research Prospecti in Biological Ageing," undated (LSP). Leslie Orgel interview, February 12, 1987.

34. Ralph O'Teary, "Dr. Teller's One Regret," *Houston Post*, May 19, 1963. Edward Teller, whose "one regret" in this interview was that he had received the prize before Oppenheimer did, had tried to arrange for Szilard to receive it in 1960. See Aaron Novick to Cyril Smith, January 14, 1960 (Aaron Novick Papers).

35. Tabin to Gertrud Weiss Szilard, July 31, 1964 (LSP). Julius Tabin interview, December 19, 1985.

36. Jonas Salk interview, March 23, 1985.

37. Invitation declined by Szilard, January 22, 1964; April 6, 1964, correspondence; Szilard to Bronowski, May 27, 1964 (LSP).

38. Gertrud Weiss Szilard to Herbert Anker, January 15, 1965 (LSP 86/7).

39. Szilard to F. C. Stewart, May 13, 1964 (LSP).

40. Szilard to Betty Goetz Lall, May 13, 1964 (LSP).

41. *Boston Herald*, March 3, 1964, p. 32. *Boston Herald*, March 9, 1964, p. 24.

42. H. A. Crosby to Szilard, March 10, 1964 (LSP).

43. Correspondence in LSP 6/35.

44. Szilard to Burdick, May 28, 1964 (LSP 5/18).

45. Anna Beck to Szilard, May 30, 1964. Beck to the author, 1983.

46. Whitmore follow-up note, June 1, 1964 (EWP).

47. San Diego County Certificate of Death, No. 8009, Dist. 3449; CC 940-64; signed June 1, 1964.

48. Associated Press dispatch A22LA, May 30, 1964.

49. Associated Press dispatch A30LA, May 30, 1964.

50. James M. Whisenand to Willet F. Whitmore, June 4, 1964 (EWP).

51. Whitmore to Whisenand, June 24, 1964. James J. Nickson to Trude Szilard, June 27, 1964 (LSP 85/4).

52. "Leo Szilard Dies, A-Bomb Physicist," *The New York Times*, May 31, 1964, pp. 1 and 77. London *Times*, June 1, 1964.

53. Alice Danos interviews, June 10 and September 4, 1986.

54. Wigner to Gertrud Weiss Szilard, May 31, 1964; John Polanyi to Gertrud Weiss Szilard, June 14, 1964 (LSP 85/5). "Special Report. This was Leo Szilard. Remembrance of a Genius," *Life*, June 12, 1964, p. 31.

55. *Disarmament and Arms Control*, Autumn 1964, p. 453.

56. Jonas Salk's June 1, 1964, statement (Salk Institute Files, JSP).

57. Gertrud Weiss Szilard release, June 1, 1964. J. C. Stoddart (La Jolla Mortuary) to Edwin Lennox, June 3, 1964 (Gertrud Weiss Szilard, May 30, 1964, file, EWP).

58. "Leo Szilard," June 13, 1964 (Salk Institute Files, Dr. Szilard Resident Fellow, JSP).

59. James Arnold interview, February 11, 1987. Leslie Orgel interview, February 12, 1987. Szilard to Trude, August 4, 1960 (EWP).

EPILOGUE

1. James D. Watson to the author, May 7, 1987. Szilard to Trude, August 4, 1960 (EWP).

2. Nobel Lecture by Evgeni Chazov, December 11, 1985. He was referring to Szilard's 1949 story "Calling All Stars," which appeared in *The Voice of the Dolphins and Other Stories.*

3. "A Message from Gorky" by Andrei Sakharov, *Bulletin of the Atomic Scientists,* June/July 1983, p. 2.

4. Nicholas Wade, *Science,* November 1972.

5. Jonas Salk interview, March 23, 1985.

6. Rita Bronowski interview, January 22, 1986.

7. "Ever See a Paper Grade a Scholar?" San Diego *Tribune* ad, *Editor & Publisher,* March 25, 1989, p. 2.; "The Salk Institute at a Crossroads," *Science,* June 27, 1990, p. 360.

8. Jonas Salk interview, March 23, 1985.

9. "Strong Inference" by John R. Platt, *Science,* October 16, 1964, pp. 347–53.

Bibliography

This list includes the published sources that I found most useful when researching and writing Szilard's biography and is not a compilation of all references cited.

ARTICLES

Abir-Am, Pnina G. "Themes, Genres and Orders of Legitimation in the Consolidation of Biology." *History of Science* 23 (1985): 73–117.

Ackland, Len. "Dawn of the Atomic Age." *Chicago Tribune Magazine,* November 28, 1982, pp. 10ff.

Bennett, Charles H. "Demons, Engines and the Second Law." *Scientific American,* November 1987, pp. 108–16 and 150–51.

Bernstein, Barton. "Roosevelt, Truman, and the Atomic Bomb, 1941–1945: A Reinterpretation." *Political Science Quarterly* 90, no. 1 (Spring 1975): 23ff.

Brown, Harrison. "The Beginning or the End: A Review." *Bulletin of the Atomic Scientists* 3 (March 1947): 99.

Ehrenberg, Werner. "Maxwell's Demon." *Scientific American,* November 1967, pp. 103–10.

Feld, Bernard T. "Nuclear Proliferation—Thirty Years After Hiroshima." *Physics Today,* July 1974, pp. 24–25.

Gruber, Carol S. "Manhattan Project Maverick: The Case of Leo Szilard." *Prologue* (Journal of the National Archives) 15, no. 2 (Summer 1983): 73–87.

McClaughry, John. "The Voice of the Dolphins." *The Progressive,* April 1965, pp. 26–29.

Reingold, Nathan. "MGM Meets the Atomic Bomb." *The Wilson Quarterly,* Autumn 1984, pp. 154–63.

Rider, Robin E. "Alarm and Opportunity: Emigration of Mathematicians and Physicists to Britain and the United States, 1933–1945." History of Science and Technology Program, The Bancroft Library, University of California at Berkeley. *Historical Studies in the Physical Sciences* 15:1 (1984).

Schaffner, Kenneth F. "Logic of Discovery and Justification in Regulatory Genetics." Salk Institute, February 17, 1971.

Shils, Edward. "Science and Scientists in the Public Arena." *American Scholar,* Spring 1987, pp. 185–202.

Szabadváry, Ferenc. "Leo Szilard's Studies at the Palatine Joseph Technical University of Budapest." *Periodica Polytechnica,* 31, no. 187 (1987): 187–90.

Teller, Edward. "Seven Hours of Reminiscences." *Los Alamos Science,* Winter/Spring 1983, pp. 190–95.

Weart, Spencer R. "The Road to Los Alamos." *Journal de Physique,* Colloque C8, supplement au n° 12, Tome 43, décembre 1982, C8-301–21.

Yavenditti, Michael J. "Atomic Scientists and Hollywood: The Beginning or the End?" *Film and History,* December 1978, Vol. 8, No. 4.

Yoxen, Edward. "Where Does Schrödinger's *What Is Life?* Belong in the History of Molecular Biology?" *History of Science* 17 (1979): 17–52.

PAMPHLETS, THESES, AND REPORTS

Azimov, Isaac. *Worlds Within Worlds: The Story of Nuclear Energy* (three volumes). US Atomic Energy Commission, 1972.

Badash, Lawrence, Elizabeth Hodes, and Adolph Tiddens. "Nuclear Fission: Reaction to the Discovery in 1939." Institute on Global Conflict and Cooperation, University of California, San Diego. IGCC Research Paper No. 1, 1985.

Forman, Paul. "The Environment and Practice of Atomic Physics in Weimar Germany: A Study in the History of Science." Ph.D. thesis, University of California, Berkeley, 1967. Ann Arbor: University Microfilms, Inc., 1970.

Krivatsy, Christina Maria. "Goethe's *Faust* and *The Tragedy of Man* by Imre Madach: A Comparative Analysis." Duke University, Department of Germanic Languages and Literature, 1973.

A Memorial Colloquium Honoring Herbert L. Anderson, August 31, 1988. Los Alamos National Laboratory. LALP-89-14, August 1989.

Titus, Alice K. "Collective Conscience: A Short Study of the Beginnings of the Scientists Movement." *Network,* pp. 6–7 and 10–11.

US Atomic Energy Commission. "Nuclear Terms, A Glossary," 1967.

US House of Representatives, Committee on Military Affairs, *Atomic Energy,* Hearings on H.R.4280, 79th Congress, 1st Session, October 9, 18, 1945.

US Senate, Special Committee on Atomic Energy, Hearings on S. Res. 179, Part 2, December 5, 6, 10, and 12, 1945.

Walker, Mark. "Uranium Machines, Nuclear Explosives, and National Socialism: The German Quest for Nuclear Power, 1939–1949." Department of History, Princeton University, October 1987.

BOOKS

Alperovitz, Gar. *Atomic Diplomacy.* New York: Simon & Schuster, 1965. New York: Viking Penguin Inc., 1985.

Amrine, Michael. *The Great Decision.* New York: G. P. Putnam's Sons, 1959.

Ashmore, Harry S. *Unseasonable Truths: The Life of Robert Maynard Hutchins.* Boston: Little, Brown and Company, 1989.

Badash, Lawrence. *Kapitza, Rutherford, and the Kremlin.* New Haven: Yale University Press, 1985.

Bernstein, Barton, ed. *Politics and Policies of the Truman Administration.* Chicago: Quadrangle, 1972.

_____. *The Atomic Bomb: The Critical Issues.* Boston: Little, Brown and Co., 1976.

Beveridge, William A. A *Defence of Free Learning.* London: Oxford University Press, 1959.

_____. *The London School of Economics and Its Problems 1919–1937.* London: George Allen and Unwin, 1960.

Blackett, P. M. S. *Fear, War and the Bomb: The Military and Political Consequences of Atomic Energy.* New York: McGraw-Hill, 1949.

Blumberg, Stanley A., and Gwinn Owens. *Energy and Conflict: The Life and Times of Edward Teller.* New York: G. P. Putnam's Sons, 1976.

Boorse, Henry A., and Lloyd Motz. *The World of the Atom.* 2 vols. New York: Basic Books, 1966.

Boyer, Paul. *By the Bomb's Early Light.* New York: Pantheon Books, 1985.

Brians, Paul. *Nuclear Holocausts: Atomic War in Fiction 1914–1984* (Kent, Ohio: Kent State University Press, 1987). [BAS March 86, p. 53].

Brillouin, Leon. *Science and Information Theory.* New York: Academic Press, 1965.

Brodie, Bernard. *The Atomic Bomb and American Security* (Yale University, Memorandum No. 18), 1945.

_____. *The Absolute Weapon.* New York: Harcourt Brace, 1946.

_____. *Escalation and the Nuclear Option.* Princeton, NJ: Princeton University Press, 1966.

Bronowski, Jacob. *The Ascent of Man.* Boston: Little, Brown and Co., 1973

Brown, Harrison. *The Challenge of Man's Future.* New York: Viking Press, 1954.

Budapest Anno. Gyorgy Klosz Archive. Budapest: Corvina, 1984.

Byrnes, James. *All in One Lifetime.* New York: Harper, 1958.

Cairns, J., G. S. Stent, and J. D. Watson, eds. *Phage and the Origins of Molecular Biology.* Cold Spring Harbor: Cold Spring Harbor Laboratory of Quantitative Biology, 1966.

Clark, R. W. *The Birth of the Bomb.* London: Phoenix House, 1961.

_____. *Einstein: The Life and Times.* New York: Avon Books, 1972.

_____. *The Scientific Breakthrough.* New York: G. P. Putnam's Sons, 1974.

Cline, Barbara L. *The Questioners: Physicists and the Quantum Theory.* New York: Thomas Y. Crowell Co., Inc., 1965.

Compton, Arthur H. *Atomic Quest.* New York: Oxford University Press, 1956.

Conant, James B. *Modern Science and Modern War.* New York: Columbia University Press, 1952.

Congressional Quarterly Service. *Congress and the Nation, 1945–1964.* Washington, DC, 1965.

Crick, Francis. *Life Itself, Its Origin and Nature.* New York: Simon and Schuster, 1981.

Davis, Nuel P. *Lawrence and Oppenheimer.* New York: Simon and Schuster, 1968.

Dictionary of Scientific Biography, vol. 13. New York: Scribners, 1971.

Doblin, Alfred. *Alexanderplatz Berlin.* New York: Frederick Ungar Publishing Co., 1931.

Dyke, Richard Wayne. *Mr. Atomic Energy: Congressman Chet Holifield and Atomic Energy Affairs, 1945–1974.* New York: Greenwood Press, 1989.

Esterer, A. K. and L. A. *Prophet of the Atomic Age: Leo Szilard.* New York: Julian Messner, 1972.

Evans, Medford. *The Secret War for the A-Bomb.* Chicago: Regency, 1953.

Everett, Susanne. *Lost Berlin.* New York: Gallery Books, 1979.

Feis, Herbert. *The Atomic Bomb and the End of World War II.* Princeton, NJ: Princeton University Press, 1966.

Fermi, Enrico. *Collected Papers of Enrico Fermi.* Chicago: University of Chicago Press, 1965.

Fermi, Laura. Atoms *in the Family.* Chicago: University of Chicago Press, 1950.

———. *Illustrious Immigrants.* Chicago: University of Chicago Press, 1971.

Fielding, Raymond. *The March of Time, 1935–1951.* New York: Oxford University Press, 1978.

Fisher, Ernst Peter, and Carol Lipson. *Thinking about Science: Max Delbrück and the Origins of Molecular Biology.* New York: Norton, 1988.

Fleming, Donald, and Bernard Bailyn, eds. *Perspectives in American History,* vol. 2. Lunenburg, VT: The Stinehour Press, 1968. In citations, this is referred to as Szilard's "Reminiscences."

Freedman, Lawrence. *The Evolution of Nuclear Strategy.* New York: St. Martin's Press, 1983.

Friedrich, Otto. *Before the Deluge.* New York: Harper & Row, 1972.

Frisch, O. R. *Working with Atoms.* New York: Basic Books, 1965.

Gelwick, Richard. *The Way of Discovery: An Introduction to the Thought of Michael Polanyi.* New York: Oxford University Press, 1977.

Giovannitti, Len, and Fred Freed. *The Decision to Drop the Bomb.* New York: Coward-McCann, Inc., 1965.

Glasstone, Samuel. *Sourcebook on Atomic Energy,* 3rd ed. New York: Van Nostrand Reinhold, 1967.

Goldberg, Stanley. *Understanding Relativity.* Boston: Birkhauser, 1984.

Goldschmidt, Bertrand. *The Atomic Complex.* LaGrange Park, IL: American Nuclear Society, 1982.

Goldsmith, Maurice. *Frédéric Joliot-Curie: A Biography*. London: Lawrence and Wishart, 1976.

Gowing, Margaret. *Britain and Atomic Energy, 1939–1945*. London: Macmillan, 1964.

Graetzer, Hans, and D. L., Anderson, eds. *The Discovery of Nuclear Fission: A Documentary History*. New York: Van Nostrand Reinhold, 1971.

Grodzins, Morton, and Eugene Rabinowitch, eds. *The Atomic Age*. New York: Simon and Schuster, 1965.

Groueff, Stephane. *Manhattan Project: The Untold Story of the Making of the Atomic Bomb*. New York: Bantam Books, 1968.

Groves, Leslie R. *Now It Can Be Told*. New York: Harper & Row, 1962.

Heisenberg, Werner. *Physics and Beyond*. New York: Harper & Row, 1971.

Herken, Gregg. *The Winning Weapon*. New York: Alfred A. Knopf, 1981.

_____. *Counsels of War*. New York: Oxford University Press, 1987.

Hewlett, Richard G., and Oscar E. Anderson. *The New World, 1939/1946*. Washington: US Atomic Energy Commission, 1972.

Hewlett, Richard G., and Francis Duncan. *Atomic Shield, 1947/1952*. Washington: US Atomic Energy Commission, 1972.

Hewlett, Richard G., and Jack M. Holl. *Atoms for Peace and War, 1953–1961*. Berkeley: University of California Press, 1989.

Hoagland, Mahlon. *Discovery: The Search for DNA's Secrets*. Boston: Houghton Mifflin Company, 1981.

Hogerton, John F. *The Atomic Energy Deskbook*. New York: Reinhold Publishing Co., 1963.

Inglis, David R. *To End the Arms Race*. Ann Arbor: University of Michigan Press, 1985.

International Biographical Dictionary of Central European Emigres 1933–1945, vol. 25, part 2. Munich: K. G. Saur, 1983.

Irving, David. *The German Atomic Bomb*. New York: Simon and Schuster, 1967.

Isaacson, Walter, and Evan Thomas. *The Wise Men*. New York: Simon and Schuster (Touchstone), 1986.

Jacob, François. *The Statue Within*. New York: Basic Books, Inc., 1988.

Jacobson, Harold, and Eric Stein. *Diplomats, Scientists, and Politicians: The United States and the Nuclear Test Ban Negotiations*. Ann Arbor: University of Michigan Press, 1962.

Janos, Andrew C, and William B. Slottman, eds. *Revolution in Perspective: Essays on the Hungarian Soviet Republic*. Los Angeles: University of California Press, 1971.

Jones, Vincent C. *MANHATTAN: The Army and the Atomic Bomb*. Washington, DC: Center of Military History, United States Army, 1985.

Judson, Horace Freeland. *The Eighth Day of Creation*. New York: Simon and Schuster (Touchstone), 1979.

Jungk, Robert. *Brighter Than a Thousand Suns*. New York: Harcourt, Brace and Co., 1958.

Kaplan, Fred. *The Wizards of Armageddon*. New York: Simon and Schuster (Touchstone), 1984.

Kelman, Peter, and A. Harris Stone. *Ernest Rutherford: Architect of the Atom*. Englewood Cliffs, NJ: Prentice-Hall, Inc.

Kevles, Daniel. *The Physicists: The History of a Scientific Community in Modern America*. New York: Alfred A. Knopf, 1978.

Kuhn, Thomas S. *The Structure of Scientific Revolutions*. Chicago: University of Chicago Press, 1970.

Kurzman, Dan. *Day of the Bomb: Countdown to Hiroshima*. New York: McGraw-Hill, 1986.

Kuznick, Peter J. *Beyond the Laboratory: Scientists as Political Activists in 1930s America*. Chicago: University of Chicago Press, 1987.

Lamont, Lansing. *Day of Trinity*. New York: Atheneum, 1965.

Lapp, Ralph E. *The Weapons Culture*. Baltimore: Penguin Books, 1969.

———. *Atoms and People*. New York: Harper & Row, 1956.

Libby, Leona Marshall. *The Uranium People*. New York: Crane Russak and Charles Scribner's Sons, 1979.

Lifton, Robert Jay, and Richard Falk. *Indefensible Weapons*. New York: Basic Books, 1982.

MacPherson, Malcolm. *Time Bomb: Fermi, Heisenberg, and the Race for the Atomic Bomb*. New York: Dutton, 1986.

Madách, Imre. *The Tragedy of Man*. Translated by J. C. W. Horne. Budapest: Corvina Press, 1963.

McCagg, William O., Jr. *Jewish Nobles and Geniuses in Modern Hungary*. New York: Columbia University Press. 1986.

Meyer, Henry Cord, ed. *The Long Generation: Germany from Empire to Ruin, 1913–1945*. New York: Walker and Company, 1973.

Moore, Ruth. *Niels Bohr*. New York: Alfred A. Knopf, 1966.

National Academy of Sciences. *Biographical Memoirs: Leo Szilard*. Vol. 40. New York: Columbia University Press, 1969.

Nelkin, Dorothy. *Technological Decisions and Democracy*. London: Sage, 1977.

Nicolson, Nigel. *Diaries and Letters 1930–1939*. New York: Atheneum, 1966.

Perutz, Max. *Is Science Necessary? Essays on Science and Scientists*. New York: Oxford University Press, 1991.

Pfau, Richard. *No Sacrifice Too Great: The Life of Lewis L. Strauss*. Charlottesville: University of Virginia Press, 1984.

Platt, John Rader. *The Step to Man*. New York: John Wiley & Sons, Inc., 1966.

Powers, Thomas. *Heisenberg's War*. New York: Alfred A. Knopf, 1993.

Rhodes, Richard. *The Making of the Atomic Bomb*. New York: Simon and Schuster, 1986.

Rotblat, Joseph. *Scientists in the Quest for Peace—A History of the Pugwash Conference.* Cambridge, MA: MIT Press, 1972.

_____, ed. *Proceedings of the First Pugwash Conference on Science and World Affairs.* London: The Pugwash Council, 1982.

Sayen, Jamie. *Einstein in America.* New York: Crown Publishers, 1985.

Schmitt, Francis O., ed. *Biophysics—a New Discipline.* MIT Press, 1959.

Schonland, Basil. *The Atomists (1805–1933).* New York: Oxford University Press, 1956.

Seaborg, Glenn T. *The Transuranium Elements.* New Haven: Yale University Press, 1958.

Segré, Emilio. *Enrico Fermi: Physicist.* Chicago: University of Chicago Press, 1970.

Sherwin, Martin J. *A World Destroyed: The Atomic Bomb and the Grand Alliance.* New York: Alfred A. Knopf, 1975.

Shirer, William L. *The Rise and Fall of the Third Reich.* New York: Simon and Schuster, 1960.

Singh, Jagjit. *Great Ideas in Information Theory, Language and Cybernetics.* New York: Dover Publications, 1966.

Smith, Alice K. *A Peril and a Hope.* Chicago: University of Chicago Press, 1965.

_____. *A Peril and a Hope.* Cambridge, MA: MIT Press, 1970.

Smyth, Henry DeWolf. *Atomic Energy for Military Purposes: The Official Report on the Development of the Atomic Bomb under the Auspices of the United States Government, 1940–1945.* Princeton, NJ: Princeton University Press, 1945.

Stent, Gunther S., and Richard Calendar. *Molecular Genetics.,* 2nd ed. San Francisco: W. H. Freeman and Company, 1978.

Strauss, Lewis L. *Men and Decisions.* Garden City, NY: Doubleday, 1962.

Stuewer, Roger, ed. *Nuclear Physics in Retrospect: Proceedings of a Symposium on the 1930s.* Minneapolis: University of Minnesota Press, 1979.

Szilard, Leo. *The Voice of the Dolphins and Other Stories.* New York: Simon and Schuster, 1961; reissued by Stanford University Press, 1992.

The Collected Works of Leo Szilard (Cambridge, MIT Press):

Vol. I: Feld, Bernard T., and Gertrud Weiss Szilard, eds. *The Collected Works of Leo Szilard: Scientific Papers.* 1972.

Vol. II: Weart, Spencer, and Gertrud Weiss Szilard, eds. *Leo Szilard: His Version of the Facts.* 1978.

Vol. III: Hawkins, Helen, G. Allen Greb, and Gertrud Weiss Szilard, eds. *Toward a Livable World.* 1987.

Teller, Edward, and A. Brown. *Legacy of Hiroshima.* Garden City, NY: Doubleday, 1962.

_____. *Energy from Heaven and Earth.* San Francisco: W. H. Freeman, 1979.

Thomas, Lewis. *The Youngest Science.* New York: Viking Press, 1983. 1954.

Ury, William L. *Beyond the Hotline: How Crisis Control Can Prevent Nuclear War.* Boston: Houghton Mifflin Company, 1958.

US Atomic Energy Commission. *In the Matter of J. Robert Oppenheimer.* Hearing transcript. Cambridge, MA: MIT Press, 1971.

Wagner, Francis S. *Hungarian Contributions to World Civilization.* Center Square, PA: Alpha Publications, Inc., 1977.

_____. *Eugene P. Wigner: An Architect of the Atomic Age.* Epilogue by Edward Teller. Toronto, 1981.

Watson, James D. *The Double Helix.* New York: New American Library, 1968.

Weart, Spencer. *Scientists in Power.* Cambridge, MA: Harvard University Press, 1979.

_____. *Nuclear Fear: A History of Images.* Cambridge, MA: Harvard University Press, 1988.

Weeks, Mary E. *Discovery of the Elements.* Easton, PA: Chemical Education Publishing Co., 1968.

Wells, H. G. *The World Set Free.* London: Macmillan, 1914.

Wigner, Eugene, as told to Andrew Szanton. *The Recollections of Eugene P. Wigner.* New York: Plenum Publishing Corp. 1992.

Williams, Robert C, and Philip L. Cantelon. *The American Atom: A Documentary History of Nuclear Policies from the Discovery of Fission to the Present, 1939–1984.* Philadelphia: University of Pennsylvania Press, 1984.

Wilson, Jane, ed. *All in Our Time: The Reminiscences of Twelve Nuclear Pioneers.* Chicago: Bulletin of the Atomic Scientists, 1975.

Wyden, Peter. *Day One.* New York: Simon and Schuster, 1984.

York, Herbert F. *The Advisors.* San Francisco: W. H. Freeman, 1976.

RECENT PUBLICATIONS

ARTICLES AND TALKS

Dannen, Gene. "The Einstein-Szilard Refrigerators." *Scientific American,* January 1997, 90-95.

Frank, Tibor. "Ever Ready to Go: The Multiple Exiles of Leo Szilard," *Physics in Perspective* 7 (2005): 204-252.

Lanouette, William. "Leo Szilard: A Comic and Cosmic Wit." A talk at the International Szilard Seminar, Leo Szilard Centenary Celebration, Eötvös University, Budapest, 9 February 1998. Published in *Leo Szilard Centenary Volume.* Budapest: Eotvos Physical Society, 1998, 103-108.

_____, "Why We Dropped the Bomb." *Civilization* (The Magazine of the Library of Congress) Vol. 2, No. 1 January/February 1995, 28-39. Reprinted in *Annual Editions: World History, Volume II (Fourth Edition).* Guilford, Connecticut: Brown & Benchmark 1996, 159-165.

_____, "The science and politics of Leo Szilard, 1898-1964: evolution, revolution, or subversion?" *Science and Public Policy*, Vol. 33, No. 8, 613-617. October 2006.

_____, "A Narrow Margin of Hope: Leo Szilard in the Founding Days of CARA." [Council for Assisting Refugee Academics, formerly the Academic Assistance Council] Chapter 2 of *IN DEFENCE OF LEARNING: The Plight, Persecution, and Placement of Academic Refugees, 1933-1980s*. Proceedings of the British Academy 169. Oxford: Oxford University Press, 2011, pp. 45-57.

_____, "Civilian Control of Nuclear Weapons" *Arms Control Today*, May 2009 (Vol. 39, No. 4), pp. 45-48.

_____, "The Manhattan Project and its Cold War Legacy" Woodrow Wilson International Center for Scholars, 20 February 2008. Podcast.

_____, "The Odd Couple and the Bomb" *Scientific American* November 2000 (Vol. 283 No. 5), 86-91. Reprinted as "Fermi, Szilard und der erste Atomreaktor"*Spektrum der Wissenschaft* Januar 1 / 2001, 78-83; "La extrana pareja y la bomba" *Investigacion y Ciencia* Enero 2001, 18-23; "Bombowa Wspolpraca" *Swiat Nauki* Maj 2001, 70-75;. Japanese Edition of *Scientific American* February 2001 (Vol. 31, No. 2), 114-121.

_____, "Reason and Circumstances of the Hiroshima Bomb" A talk about the US scientists' struggle to control the A-bomb they had created, presented at a UNESCO Conference on World Cultural Heritage of the 20th Century "Modernity and Barbarism." The Bauhaus, Dessau, Germany. 4 June 1999. Published as "Grunde und Hintergrunde des Bombenabwurfs auf Hiroshima" in *Bauhaus und Brasilia, Auschwitz und Hiroshima. Weltkulturerbe des 20.Jahrhunderts: Modernitat und Barbarei*. (Edited by Walter Prigge, translated by Marie Neumullers) Berlin: Edition Bauhaus, Jovis Verlag, 2003, 187-195.

Telegdi, Valentine. "Szilard as Inventor: Accelerators and More." Physics Today 53 (10):25 (2000). (See also Gene Dannen letter, Physics Today 54, 3, 102 (2001).

PAMPHLETS, THESES, AND REPORTS

Boyer, Patrick. "Pugwash: Confronting the Peril of Nuclear Weapons." Bracebridge, ON: Canadian Shield Communications Corporation, 2006.

Kelly, Cynthia C. and Robert S. Norris. "A Guide to Manhattan Projects Sites in Manhattan." Washington, DC: The Atomic Heritage Foundation, 2008.

BOOKS

Alperovitz, Gar, *The Decision to Use the Atomic Bomb, and the architecture of an American myth*. New York: Alfred A. Knopf, 1995.

Balogh, Brian. *Chain Reaction: Expert debate and public participation in American commercial nuclear power, 1945-1975.* Cambridge and New York: Cambridge University Press, 1991.

Bergman, Jay. *Meeting the Demands of Reason: The Life and Thought of Andrei Sakharov.* Ithaca and London: Cornell University Press, 2009.

Bird, Kai and Lawrence Lifschultz (eds.) *Hiroshima's Shadow: Writings on the Denial of History and the Smithsonian Controversy.* Stony Creek, CT: The Pamphleteer's Press, 1998.

_____, and Martin J. Sherwin. *American Prometheus: The Triumph and Tragedy of J. Robert Oppenheimer.* New York: Alfred A. Knopf, 2005.

Brown, Andrew. *The Neutron and the Bomb: A Biography of Sir James Chadwick.* Oxford, New York, and Tokyo: Oxford University Press, 1997.

_____. *Keeper of the Nuclear Conscience: The Life and Work of Joseph Rotblat.* Oxford: Oxford University Press, 2012.

Conant, Jennet. *Tuxedo Park: A Wall Street Tycoon and the Secret Palace of Science that Changed the Course of World War II.* New York: Simon and Schuster, 2002.

Cooper, Ray M. (ed.) *Refugee Scholars: Conversations with Tess Simpson* .Leeds: Moorland Books, 1992.

Fermi, Rachel and Esther Samra. *Picturing the Bomb: Photographs from the Secret World of the Manhattan Project.* New York: Harry N. Abrams, Inc., 1995.

Francis, Sybil, Burton C. Hacker, and Gabor Pallo. George Marx (ed.) *The Martians: Hungarian Émigré Scientists and the Technologies of Peace and War 1919-1989.* Budapest: Eötvös University, 1997.

Grandy, David A. *Leo Szilard: Science as a Mode of Being* (New York: University Press of America, Inc., 1996)

Hargittai, Istvan. *The Road to Stockholm: Nobel Prizes, Science, and Scientists.* Oxford: Oxford University Press, 2003.

_____, *The Martians of Science: Five Physicists Who Changed the Twentieth Century.* Oxford: Oxford University Press, 2008.

_____, *Judging Edward Teller: A Closer Look at One of the Most Influential Scientists of the Twentieth Century.* Amherst, NY: Prometheus Books, 2010. (With an Afterword by Richard Garwin.)

Herken, Gregg. *Brotherhood of the Bomb: The Tangled Lives and Loyalties of Robert Oppenheimer, Ernest Lawrence, and Edward Teller.* New York: Henry Holt and Company, 2002.

Hershberg, James. *James B. Conant: Harvard to Hiroshima and the Making of the Nuclear Age.* New York: Alfred A. Knopf, 1993.

Holloway, David. *Stalin and the Bomb.* New Haven & London: Yale University Press, 1994.

Kelly, Cynthia C. (ed.) *Remembering the Manhattan Project: Perspectives on the Making of the Atomic Bomb and its Legacy.* Hackensack, NJ and Singapore: World Scientific Publishing Co. Pte. Ltd., 2004.

_____, (ed.) *The Manhattan Project: The Birth of the Atomic Bomb in the Words of its Creators, Eyewitnesses, and Historians.* New York: Black Dog & Leventhal Publishers, Inc., 2009.

Lapp, Ralph E. *My Life With Radiation: The Truth About Hiroshima.* Madison, WI: Cogito Books, 1995.

Lopez, George A. and Nancy J. Myers (eds.) *Peace and Security: The Next Generation.* Lanham, MD: Rowman & Littlefield Publishers, Inc., 1997.

Marks, Shula, Paul Weindling and Laura Wintour. *In Defence of Learning: The Plight, Persecution, and Placement of Academic Refugees 1933-1980s.* London: The British Academy (by Oxford University Press), 2011.

Marton, Kati. *The Great Escape: Nine Jews Who Fled Hitler and Changed the World.* New York: Simon & Schuster, 2006.

Marx, Gyorgy. *Szilard Leo.* Budapest, Akademiai Kiado, 1997. [In Hungarian]

_____, *The Voice of the Martians.* Budapest: Akademiai Kiado, 1997.

_____, (ed.) *Leo Szilard Centenary Volume.* Budapest: Eötvös Physical Society, 1998.

McMillan, Priscilla J. *The Ruin of J. Robert Oppenheimer, and the Birth of the Modern Arms Race.* New York: Viking, 2005.

Newman, Robert P. *Truman and the Hiroshima Cult.* East Lansing, MI: Michigan State University Press, 1995.

Norris, Robert S. *Racing for the Bomb: General Leslie R. Groves, the Manhattan Project's Indispensable Man.* South Royalton, VT: Steerforth Press, 2002.

Ottaviani, Jim, Janine Johnston, Steve Lieber, Vince Locke, Bernie Mireault, and Jeff Parker. *Fallout: J. Robert Oppenheimer, Leo Szilard, and the Political Science of the Atomic Bomb.* Ann Arbor, MI, 2001.

Perutz, Max. *I Wish I'd Made You Angry Earlier.* Cold Spring Harbor, NY: Cold Spring Harbor Laboratory Press, Expanded Edition 2003.

Rhodes, Richard. *Dark Sun: The Making of the Hydrogen Bomb.* New York: Simon & Schuster, 1995.

Rowe, David E. and Robert Schulmann. *Einstein on Politics: His Private Thoughts and Public Stands on Nationalism, Zionism, War, Peace, and the Bomb.* Princeton: Princeton University Press, 2007.

Seabrook, Jeremy. *The Refuge and the Fortress: Britain and the flight from tyranny.* Basingstoke: Palgrave Macmillan, 2009.

Teller, Edward with Judith Shoolery. *Memoirs: A Twentieth-Century Journey in Science and Politics.* Cambridge, MA: Perseus Publishing, 2001.

Walker, J. Samuel. *Prompt and Utter Destruction: Truman and the Use of Atomic Bombs Against Japan.* Chapel Hill: University of North Carolina Press, Revised Edition 2004.

Watson, James D. *Genes, Girls, and Gamow: After the Double Helix.* New York: Vintage, 2003.

Index

About the Author

WILLIAM LANOUETTE has covered atomic energy and arms-control topics for more than four decades and was a senior analyst for energy and science policy at the US Government Accountability Office. He has written for *the Atlantic, the Economist, New York* magazine, *Scientific American,* and *Smithsonian* magazines and was a staff member at *Newsweek, the National Observer, National Journal,* and the *Bulletin of the Atomic Scientists.* He holds a BA in English from Fordham College, and MSc and PhD degrees in political science from the London School of Economics. He lives with his wife in San Diego, California, where he enjoys (in no special order) writing, rowing, exploring the West Coast, and caring for grandkids.

BELA SILARD, Leo Szilard's brother, simplified his name as an immigrant to the United States in 1938. He translated his parents' memoirs, compiled family records and photographs, and drafted recollections for this book. A retired electrical engineer and inventor, he lived with his wife in Pleasantville, New York, until his death in 1994.